PROTECTED CULTIVATION

Structural Design, Crop Management Modeling, and Automation

PROTECTED CULTIVATION

Structural Design, Crop Management Modeling, and Automation

Edited by

Mahesh Chand Singh, PhD

K. K. Sharma, PhD

APPLE ACADEMIC PRESS

First edition published 2024

Apple Academic Press Inc.
1265 Goldenrod Circle, NE,
Palm Bay, FL 32905 USA

760 Laurentian Drive, Unit 19,
Burlington, ON L7N 0A4, CANADA

CRC Press
2385 NW Executive Center Drive,
Suite 320, Boca Raton FL 33431

4 Park Square, Milton Park,
Abingdon, Oxon, OX14 4RN UK

Library and Archives Canada Cataloguing in Publication

Title: Protected cultivation : structural design, crop management modeling, and automation / edited by Mahesh Chand Singh, PhD, K.K. Sharma, PhD.
Names: Singh, Mahesh Chand, editor. | Sharma, K. K. (Plant pathologist), editor.
Description: First edition. | Includes bibliographical references and index.
Identifiers: Canadiana (print) 20230513948 | Canadiana (ebook) 20230514022 | ISBN 9781774914168 (hardcover) | ISBN 9781774914175 (softcover) | ISBN 9781003402596 (ebook)
Subjects: LCSH: Greenhouse management.
Classification: LCC SB415 .P76 2024 | DDC 631.5/83—dc23

Library of Congress Cataloging-in-Publication Data

CIP data on file with US Library of Congress

ISBN: 978-1-77491-416-8 (hbk)
ISBN: 978-1-77491-417-5 (pbk)
ISBN: 978-1-00340-259-6 (ebk)

About the Editors

Mahesh Chand Singh, PhD
Scientist (Soil & Water Engineering), Department of Soil and Water Engineering, COAE&T, Punjab Agricultural University, Ludhiana, Punjab, India

Dr. Mahesh Chand Singh holds a PhD degree and has a background in Soil and Water Engineering. Early in his career, he worked as an Assistant Professor at Mata Gujri College, Fatehgarh Sahib, Punjab, India. He has held positions at various institutions, including Dr. Y. S. Parmar University of Horticulture and Forestry in Solan, Himachal Pradesh, where he was appointed as a scientist in the Department of Soil Science and Water Management. He has also worked as a research associate at NASF, ARSB, New Delhi. He gained international experience by working as a Research Associate with IRRI, Philippines, under an IFAD-funded research project. Further, Dr. Singh has worked on water conservation techniques, micro-irrigation and irrigation scheduling. He has developed several mathematical models for predicting greenhouse microclimate under soilless cropped conditions. As of the most recent information available, Dr. Singh is a Scientist in the Department of Soil and Water Engineering at the College of Agricultural Engineering and Technology of Punjab Agricultural University (PAU) in Ludhiana, Punjab, India. He is actively involved in teaching at both undergraduate and postgraduate levels. Dr. Singh's research interests span various areas including irrigation and drainage systems, remote sensing, GIS applications, and protected cultivation. He has been involved in several research projects related to these topics. He has made significant contributions to the field with over seventy research papers published in internationally reputed journals, four book chapters, four popular articles, and one practical manual. He has received awards such as the Best Scientist Award and PAU Gold Medal for Doctoral Research in recognition of his research achievements. His work on "Soilless Media Cultivation Practices for Seedless (Parthenocarpic) Cucumber" has been included in the package of practices of PAU. He has also participated as a reviewer for various scientific journals, including

Archives of Agronomy and Soil Science, Scientia Horticulturae, and Journal of the Science of Food and Agriculture (Wiley). He is a life member of the Indian Society of Agricultural Engineers (ISAE), which indicates his active involvement in the professional community.

Dr. Mahesh Chand Singh's career showcases a commitment to advancing agricultural engineering and soil science through research, teaching, and collaboration with various national and international institutions. His contributions have been recognized through awards, publications, and active participation in academic and professional activities.

K. K. Sharma, PhD
Plant Pathologist, Dr. D.R. Bhumbla Regional Research Station (Punjab Agricultural University), Ballowal Saunkhri, Balachaur, SBS Nagar, Punjab, India

K. K. Sharma, PhD, is a Plant Pathologist at Dr D.R. Bhumbla Regional Research Station of Punjab Agricultural University, Ballowal Saunkhri, Balachaur, Saheed Bhagat Singh Nagar, Punjab, India. Previously, he was an assistant professor (Plant pathology) at Doon P.G. College of Agricultural Science and Technology, Dehradun, Uttarakhand, India. He has been engaged for more than 10 years in research (both academic and corporate sectors) and four years in teaching. He is also involved in transfer of technology through different extension activities. He has published more than 20 research papers in national and international journals of repute, seven practical e-manuals, several book chapters and popular articles. Dr. Sharma is associated in development and testing of technology and has three recommendations to his credit. In addition to teaching and research, he is also engaged in various agriculture extension activities, such as agro advisory services, transfer of technology through farmers' trainings, and field days. He is a recipient of young scientist awards, best poster awards, etc., for his research contributions by reputed societies. He also served as a member of the editorial board of *American Journal of Agriculture and Forestry (AJAF)*, *Archives of Agriculture and Environmental Science,* and also served as reviewer for reputed journals. He is also a member of agricultural and phytopathological societies in India.

Contents

Contributors

Poonam Batra
Regional Plant Quarantine Station, Ministry of Agriculture and Farmers Welfare, Rangpuri, New Delhi, India

Sunaullah Bhat
Department of Zoology, Kumaun University-SSJ Campus, Almora, Uttarakhand, India

Aanchal Chauhan
Department of Vegetable Science, Dr. Yashwant Singh Parmar University of Horticulture and Forestry, CoHF, Thunag, Mandi, Himachal Pradesh, India

Shilpi Chawla
Agricultural Research Station, College of Agriculture, Virginia State University, Virginia, USA

Subrata K. Das
School of Biological Engineering and Life Sciences, Shobhit Institute of Engineering and Technology, (Deemed-to-be-University), Meerut, India

A. R. Desai
ICAR - Central Coastal Agricultural Research Institute, Old Goa, Goa, India

H. Fatnassi
The International Center for Biosaline Agriculture (ICBA), Academic City, Dubai, United Arab Emirates

Jagreeti Gupta
Department of Floriculture and Landscape Architecture, Dr. Yashwant Singh Parmar University of Horticulture and Forestry, CoHF, Neri, Hamirpur, HP, India

Mathala Juliet Gupta
ICAR-Central Coastal Agricultural Research Institute, Old Goa, Goa, India

Navnath Indore
ICAR-Central Institute of Postharvest Engineering and Technology, Ludhiana, India

Alpana Joshi
School of Biological Engineering and Life Sciences, Shobhit Institute of Engineering and Technology, (Deemed-to-be-University), Meerut, India

Dilip Singh Kachwaya
Department of Agriculture, Mata Gujri College, Fatehgarh Sahib, Punjab, India

Sakharam Kale
ICAR-Central Institute of Postharvest Engineering and Technology, Abohar, Punjab, India

Pankaj Kannaujia
ICAR-Central Institute of Postharvest Engineering and Technology, Abohar, Punjab, India

Pratap Singh Suresh Khapte
ICAR-National Institute of Abiotic Stress Management, Baramati, Maharashtra, India

Anjani Khulbe
Indian Council of Agricultural Research, Krishi Bhawan, New Delhi, India

Aekesh Kumar
Soil and Water Engineering, College of Agriculture and Research Station, Chhuikhadan, Indira Gandhi Krishi Vishwavidyalaya, Raipur, Chhattisgarh, India

Sandeep Kumar
Department of Zoology, Kumaun University-SSJ Campus, Almora, Uttarakhand, India

Poonam Kumari
CSIR-IHBT, Palampur, Himachal Pradesh, India

Gopal Ramdas Mahajan
ICAR - Central Coastal Agricultural Research Institute, Old Goa, Goa, India

A. López-Martínez
CIAMBITAL (Centro de Investigación en Agrosistemas Intensivos Mediterráneos y Biotecnología Agroalimentaria), University of Almería, Almería, Spain

Hasan Mirzakhaninafchi
Department of Farm Machinery and Power Engineering, Punjab Agricultural University, Ludhiana, Punjab, India

Shideh Mojerlou
Department of Horticulture and Plant Protection, Faculty of Agriculture, Shahrood University of Technology, Shahrood, Iran

F. D. Molina-Aiz
CIAMBITAL (Centro de Investigación en Agrosistemas Intensivos Mediterráneos y Biotecnología Agroalimentaria), University of Almería, Almería, Spain

Prerna Nath
ICAR-Central Institute of Postharvest Engineering and Technology, Abohar, Punjab, India

Vishnu Prasad
Water Technology Center, Indian Agricultural Research Institute, New Delhi, India

R. Ramesh
ICAR - Central Coastal Agricultural Research Institute, Old Goa, Goa, India

Harish Chandra Raturi
Department of Agriculture, Mata Gujri College, Fatehgarh Sahib, Punjab, India

Homraj Sahare
Department of Horticulture, School of Agriculture, Lovely Professional University, Phagwara, Punjab, India

Akansha Saxena
School of Biological Engineering and Life Sciences, Shobhit Institute of Engineering and Technology, (Deemed-to-be-University), Meerut, India

Anurag Sathpati
College of Agriculture, G. B. Pant University of Agriculture and Technology, Pantnagar, Uttarakhand, India

Ravi S. C.
ICAR-Central Institute for Sub-Tropical Horticulture (CISH), Lucknow, Uttar Pradesh, India

Deepa Sharma
Department of Vegetable Science, Dr. YSPUHF, CoHF, Neri, Hamirpur, Himachal Pradesh, India

K. K. Sharma
Dr. D. R. Bhumbla Regional Research Station (Punjab Agricultural University), Ballowal Saunkhri, Balachaur, SBS Nagar, Punjab, India

Rishu Sharma
Department of Floriculture and Landscaping, Punjab Agricultural University, Ludhiana, Punjab, India

Mahesh Chand Singh
Department of Soil and Water Engineering, Punjab Agricultural University, Ludhiana, India

R. K. Singh
ICAR-Central Institute of Postharvest Engineering and Technology, Ludhiana

Anupreet Singh Tiwana
Department of Agriculture, Mata Gujri College, Fatehgarh Sahib, Punjab, India

Mohd Tufail
Department of Geography, Govt. P. G. College, Rajouri, Jammu & Kashmir, India

D. L. Valera
CIAMBITAL (Centro de Investigación en Agrosistemas Intensivos Mediterráneos y Biotecnología Agroalimentaria), University of Almería, Almería, Spain

S. M. Vanitha
ICAR-Indian Institute of Soil and Water Conservation (IISWC), Research Centre, Udhagamandalam, Tamil Nadu, India

Chaudhari Ganesh Vasudeo
ICAR - Central Coastal Agricultural Research Institute, Old Goa, Goa, India

Arti Verma
Krishi Vigyan Kendra, Langroya, Punjab Agricultural University, Ludhiana, Punjab, India

Abrar Yousuf
Regional Research Station of Punjab Agricultural University, Ballowal Saunkhri, Balachaur, SBS Nagar, Punjab, India

Abbreviations

AICRP	All India Coordinated Research Project
ANN	artificial neural network
BCAs	biocontrol agents
BCR	benefit–cost ratio
BER	blossom end rot
C/N	carbon-to-nitrogen
CAD	computer-aided design
CCPs	critical control points
CDB	coconut development board
CEA	controlled environment agriculture
CFD	computational fluid dynamics
CGMMV	cucumber green mottle mosaic virus
CH_4	methane
CIH	Central Institute for Horticulture
CMV	cucumber mosaic virus
CNL	compensated no leakage
CNN	convolutional neural network
CO_2	carbon dioxide
CPU	central processing unit
CSKHPKV	Chaudhary Sarwan Kumar Himachal Pradesh Krishi Vishvavidyalaya
DBNs	deep belief networks
DFT	deep flow technique
DL	dead load
DLI	daily light integral
DWC	deep water culture
EC	electrical conductivity
EIL	economic injury level
ENs	electronic noses
ETL	economic threshold levels
FAO	Food and Agriculture Organization
FDI	foreign direct investment
FRP	fiber glass-reinforced plastic
GA3	gibberellic acid 3

GERES	Groupe Energies Renouvelables, Environment et Solidarités
GI	galvanized iron
GPRS	general packet radio service
GPS	global positioning system
GPUs	graphic processing units
GSM	system for mobile communications
GT	greenhouse technology
H_2	hydrogen
HACCP	hazard analysis of critical control points
HID	high-intensity discharge
HMNEH	Horticulture Mission for North East & Himalayan States
HPS	high pressure sodium
HR	hypersensitive reaction
IARI	Indian Agricultural Research Institute
ICAR	Indian Council of Agricultural Research
ICT	information and communication technologies
IDM	integrated disease management
IGES	initial graphics exchange specification
IIHR	Indian Institute of Horticultural Research
INM	integrated nutrient management
IoT	internet of things
IPCC	intergovernmental panel on climate change
IPM	integrated pest management
IR	infrared
IRGA	infrared gas analyzer
IRR	internal rate of return
ISM	industrial scientific and medical
LAN	local area network
LDPE	low-density polyethylene
LDR	light-dependent resistor
LED	light-emitting diode
LEHO	Ladakh Environmental Health Organization
LL	live load
LREDA	Ladakh Renewable Energy Development Agency
LVQ	learning vector quantization
MEMS	micro electromechanical systems
MGNREGA	Mahatma Gandhi National Rural Employment Guarantee Act
MH	maleic hydrazide
MIDH	Mission for Integrated Development of Horticulture

MIMO	multiple-input multiple-output
MLP	multilayer perceptron
MSP	minimum support price
MTD	maximum temperature differences
N_2O	nitrous oxide
NAA	naphthalene acetic acid
NABARD	National Bank for Agriculture and Rural Development
NBM	National Bamboo Mission
NCIPMC	North Central Integrated Pest Management Center
NCPA	National Committee on use of Plastics in Agriculture
NFT	nutrient film technique
NHB	national horticulture board
NHM	national horticulture mission
NIR	near-infrared
nm	nano meter
N:P:K	nitrogen:phosphorous:potassium
NPV	net present value
O_2	oxygen
PAMS	prevention avoidance monitoring and suppression
PAR	photosynthetically active radiations
PCT	protected cultivation technology
PDDKBSY	Pandit Dean Dyal Upadhaya Krishak Bagwan Samridhi Yojna
PE	polyethene
PET	plasticulture engineering and technology
PGR	plant growth regulators
PMFBY	Pradhan Mantri Fasal Bhima Yojana
PMMoV	pepper middle mottle virus
PMOs	plant maintenance operations
ppm	parts per million
PPN	plant parasitic nematode
PVC	polyvinyl chloride
PVY	potato virus Y
RBF	radial basis function
RF	radio frequency
RFID	radio-frequency identification
RH	relative humidity
RKN	root knot nematode
RKVY	Rashtriya Krishi Vikas Yojana
Rn	net radiation

SAS	safety access system
SF_6	sulphur hexafluoride
SG	solar greenhouses
SMP	soil matric potential
SSG	sunken solar greenhouse
STAAD	structural engineering, analysis, and design software
T6P	trehalose-6-phosphate
TIBA	triiodobenzoic acid
TIR	thermal infrared
TMV	tobacco mosaic virus
TSS	total soluble solids
TSSM	two-spotted spider mite
TSWV	tomato spotted wilt virus
TWSNs	terrestrial wireless sensor networks
TYLCV	tomato yellow leaf curl virus
UAV	unmanned aerial vehicle
UV	ultraviolet
UWB	ultra-wide band
UWSNs	underground wireless sensor networks
VPD	vapor pressure deficit
VWC	volumetric water content
WDG	water dispersible granule
Wi-Fi	wireless fidelity
WLAN	wireless local area network
WSN	wireless sensor network
WUE	water use efficiency

Preface

Globally, climate change has become one of the key determinants of agricultural production due to its negative impact on crop growth pattern, productivity and quality of produce, particularly in relation to rising temperatures and higher carbon dioxide (CO_2) levels. Moreover, the use of agricultural chemicals (i.e., pesticides/fungicides) affects the quality of high-value crops such as cucumber, capsicum, tomato, and strawberries, and also contributes to environmental pollution. Further, the occurrence of insects, diseases, and nematodes under traditional methods (soil as a growing media) of growing crops negatively affects the plant's performance, both quantitatively and qualitatively. To safeguard crop plants from the above-mentioned issues, crop production in soilless-based systems inside greenhouses/polyhouses/glasshouses has become the need of the hour as it offers improved crop growth conditions.

The soilless-based growing system offers improved control on nutrient supply to plants and disease incidence, in relation to having better control on availability of water, pH, EC, and nutrient status within the plant's root zone. Thus, under the current scenario of increasing demands for fresh produce (vegetables and fruits) and significantly decreasing land sizes, the adoption of soilless-based growing systems in protected cultivation has become essential for obtaining improved crop yields and quality with no environmental pollution. A protected structure helps to offer year-round (off-season) cultivation of different crops when encouraging micro-climatic conditions are maintained, particularly in soilless cultivation systems through automation.

Protected cultivation, particularly in soilless-based systems results in higher crop productivity, improved quality, and encouraged prices of the produce. Further, the production of high-value crops cultivated inside protected structures can be increased to many times as compared to open field cultivation. The automation of irrigation and nutrient management also becomes easy in soilless-based protected cultivation. The economic analysis of protected cultivation under differently cropped conditions should also be done at different levels (e.g., farmer and local level). To understand the different components of protected cultivation for improving the crop growth performance and production (yield and quality), a complete guide is needed in one place in the form of a book.

Keeping the above facts in view, we designed this comprehensive book *Protected Cultivation: Structural Design, Crop Management Modeling, and Automation,* particularly for undergraduate and postgraduate students, course instructors, and researchers. This book contains chapter contributions from several eminent experts of the world. A total of 15 chapters are included beginning from worldwide status of protected cultivation, comparison of different types of structures, construction and design, soilless growing systems, irrigation and fertigation systems of protected cultivation, cropping sequences, integrated management of insects and pests followed by biological control of insects and pests, climate control, modeling, automation, and economics of protected cultivation. We are confident that this book will prove very useful for the students, teachers, and researchers in the field of protected cultivation.

The authors of the different chapters in this book are engaged in either teaching or research or extensive services. The book includes well-known scientists as authors from around the globe. All the authors have incorporated their experience and knowledge along with the latest information in their respective chapters.

We trust that this book will serve as a useful resource and guide for students, teaching professionals, and researchers connected directly or indirectly with protected agriculture/protected cultivation.

We are thankful to the various authors for their valuable time and for contributing valued materials for this book. We are also thankful to AAP for publishing this book so that it can be placed in the hands of esteemed readers all around the globe.

—Editors

Introduction

Climate change has significant impact on agricultural production through its negative effects on crop growth patterns. The crop production and quality are undermined under augmented exposure to high levels of carbon dioxide (CO_2) and temperature. Any variation in the meteorological parameters such as temperature (maximum and minimum) and rainfall may impose negative effects on the growth of crops and production. Hence, cultivation of high-value fruits, vegetables, and ornamental plants/flowers such as strawberries, cucumber, capsicum, tomato, and cut-flowers under protected structures becomes desirable to obtain their highest yield potential and quality under artificially modified environment. Protected cultivation using a variety of structures is practiced and is now continuously expanding in subtropical and temperate regions of the world, including India.

When supported with optimal microclimatic conditions, a structure allows year-round cultivation of diverse high-value crops in the subtropical and temperate regions across the globe including the northern plains of India. It also improves the crop performance in terms of enhanced yield, quality, and profit. Yields of high-value crops cultivated inside protected structures can be increased many times due to better control over climatic conditions. Moreover, protected cultivation helps minimize the use of fertilizers and pesticides. There is continuing demand of fruits and vegetables from dwindling holdings of arable lands in different countries, especially India. Protected cultivation offers a great scope for cultivating high-value crops using low-cost naturally ventilated polyhouses, particularly in the hills where mild climatic conditions exist.

A growing media has significant effect on the germination of seed, emergence, development, and ultimately the quality of seedlings. Thus, selecting an appropriate substrate for growing a crop is highly significant as it affects the performance of the plants/seedlings to a great extent. A growing media helps to support the plants physically and stores nutrient solution in it to be used by the plants through their rooting system. Apart from the effects of microclimate on the nursery, where plants are raised inside protected structures, the substrate/media also affects the quality of the plants in the nursery and even after transplanting to a great extent. Moreover, prevalence of soil-borne diseases and insects has restricted cultivation of vegetable crops in soil under both open field conditions and protected constructions. Soilless crop-growing techniques are highly efficient in regulating the nutritional requirements of plants and insect-pest/diseases occurrence due to their competence in controlling the available water, pH, EC, and nutrients in the root zone of the plants. Greenhouse cultivated plants indicate faster growth rates (vegetative and reproductive) with a high rate of water and nutrient uptake, particularly in soilless-based growing systems. Drip/trickle irrigation is considered the most efficient technique for watering fruits and vegetables in both soil- and soilless-based cultivation methods. It helps reduce the percolation of chemical/nutrient solution through root zone of plants. The yield and water use efficiency of a crop under drip irrigation is significantly higher than that obtained under surface irrigation methods viz. furrow irrigation, check basin, and border irrigation. Thus, advanced know-how of interaction between growing media, irrigation or fertigation method, water and fertilizer application rate, plant root distribution, water nutrient uptake rates, and microclimate is crucial for better management of water and other input resources.

Microclimate is an association of several climatic factors developing nearby the plants. It affects the metabolic activities of the plants and hence the yield/production. Solar radiation and outside air temperature are two key climatic factors required to assess the climate appropriateness for protected cultivation in an area. Microclimate in the cropped area inside a greenhouse structure must be as desired by the crops grown. Air temperature and humidity considerably affect the growth and development of the crops, consequently quality and the yield. Thus, a better understanding of the relationships amid microclimate and plants is highly desirable to maintain an optimal environment for improved performance of the crop grown inside protected structures. During summer season, when temperature keeps on rising and humidity goes down, ventilation, either forced or natural via wind or buoyancy or both can play an important role in maintaining desirable temperature inside protected structures whereas it can expel the unnecessary vapor and offer an appropriate climate during winter. In addition, fogging (evaporative cooling) also becomes essential particularly under summer climate to enhance the relative humidity to the desired range with a reduction in vapor pressure deficit. Predicting greenhouse microclimate may help the farmers/growers to manage the crop production and engineers to improve the cooling (ventilation) and heating systems. Mathematical modeling is one such tool for the prediction of greenhouse microclimate. Thus, it becomes imperative to develop models for predicting microclimate of a greenhouse (natural or forced ventilated), particularly when a crop is cultivated in soilless system and accordingly suggest the best climate settings for growing different crops based on microclimatic parameters. The economic analysis also becomes essential to compare the total investment and the profits made for working out economic viability of protected cultivation.

Keeping the above information in view, this book was planned on "Protected Cultivation: Structural Design, Crop Management Modeling and Automation." The compiled book deals with several components of protected cultivation, including climate control, modeling, automation, and economics. The chapters of the book are focused on selecting site-specific appropriate protected structures, design and construction of structure, adoption of improved crop cultivation techniques (shifting from conventional to modern soilless cultivation approaches), cropping sequence of vegetable crops, fertigation and automation, climate monitoring and management, insect-pests and diseases of common occurrence and their management including biological control under protected cultivation, selection, and optimization of insect-proof screens for greenhouses, modeling greenhouse microclimate/environment (such as ventilation rate), automation, and economic analysis.

The compiled chapters are substantiated with novel figures of good quality. The chapters provide suitable and recent references. Information provided on different components of protected cultivation will be helpful for academicians and researchers. Further, this book will also provide adequate information for undergraduate and postgraduate studies.

—Editors

CHAPTER 1

Protected Cultivation: Technology, Constraints, and Its Global Status

K. K. SHARMA[1], SHILPI CHAWLA[2], and MAHESH CHAND SINGH[3]

[1]*Dr. DR Bhumbla Regional Research Station (Punjab Agricultural University), Ballowal Saunkhri, Balachaur, SBS Nagar, Punjab, India*

[2]*Agricultural Research Station, College of Agriculture, Virginia State University, Virginia, USA*

[3]*Department of Soil and Water Engineering, Punjab Agricultural University, Ludhiana, India*

ABSTRACT

The current scenarios of climate change, dwindling land holdings, and the prevalence of soil-borne diseases have begun to negatively affect agricultural production globally, both quantitatively and qualitatively. However, the demand for quality produce of fresh vegetables and fruits is increasing tremendously with time. Moreover, the fresh water resources are depleting with time due to their over exploitation for meeting the demand of agriculture, rapid urbanization, and growing population. Thus, it becomes imperative to adopt the improved production technologies like protected cultivation (both conventional and soilless), which could constantly enhance the overall profitability of the agricultural sector to make a country self-sufficient and provide food security. The global area under protected agriculture has been estimated to be about 5,630,000 ha in 2018, out of which about 497,815 ha has been reported under greenhouse vegetables alone. The total area of horticultural crops under protected cultivation in India has been estimated to be about 40,000 ha, covering the highest area in Northern India. The

Protected Cultivation: Structural Design, Crop Management, Modeling and Automation.
Mahesh Chand Singh and K. K. Sharma (Eds.)

polyhouse cultivation of vegetables such as cucumber, capsicum, tomato, cauliflower, spinach, lettuce, broccoli, cabbage, etc., fruits like grapes, strawberry, cherries, apricot, peach, plum, citrus, etc., and cut-flowers such as chrysanthemum, carnation, gerbera, lilium, etc. has set up a practical example throughout the world including India. It plays the crucial role for attaining objectives of protected cultivation where the yield of vegetables, fruits, and cut-flowers can be easily achieved by 3–5 folds as compared to open-field cultivation under diverse agro-climatic regions. Farmers are now able to produce vegetables, fruits, and cut-flowers round the year and can make higher profits from their produce. In India, the technology of vegetable production in naturally ventilated low-cost polyhouses and walk-in poly tunnels is the best gift for farmers having small land holdings in the hills as well as plain areas. The technology of protected cultivation needs a careful planning such as selection of varieties for off-season cultivation to increase economics of quality production (Singh et al., 2020b), which also entails precise management of production and harvest time according to that period when there is shortfall in availability with high market prices of vegetables, fruit, and cut-flowers. Moreover, the youth, particularly in developing countries, may also get employment opportunities through the technology of protected soilless cultivation. Furthermore, technology of microclimate control, modeling, and automation using artificial intelligence to modify temperature, humidity, light, etc., inside the greenhouse as per crop requirements assure maximum yield of quality produce.

1.1 INTRODUCTION

Protected cultivation technology refers to grow the plants in a partially to fully controlled environment by modifying the requirements for light, temperature, humidity, CO_2, and air circulation inside the protected structures to protect the crop from abnormal weather, achieve optimal plant growth, and thus obtain improved yield (both quantitatively and qualitatively). In other words, technology of protected agriculture relies on an idea of modifying the natural environment that surrounds a crop for obtaining quality production with an extended growing season (Jensen and Malter, 1995). The structures, such as polyhouse, glasshouse, poly-tunnels, walk-in-tunnels, poly net-house, are widely used in protected cultivation. All these structures work on the principle of "greenhouse effect" (Ghosh, 2009). Controlled environment agriculture (CEA) is another term for modern greenhouse production. It has made possible to cultivate food-producing plants in places and at times when climate conditions would be detrimental or even prevent them from growing. The greenhouse technology, a subset of controlled environment agriculture, has recognized a rapid expansion during past couple of decades, due to climate change impacts and secondly, the growing demand for high-quality fresh fruits, vegetables, herbs, and cut-flowers throughout the globe. This technology can be used singly or in integration with others to offer a suitable environment for plants to withstand harsh weather and to extend the growing period or developing off-season crops.

The initial thought to grow crop plants under controlled environment goes to Roman gardeners in early part of Roman Empire (14–37 AD) who used methods like protected system of movable beds for growing vegetables for their daily needs (Janick et al., 2007). Those beds were placed outside during favorable days and inside when weather conditions became harsh. The beds were covered with transparent sheets of mica or alabaster (Wittwer and Castilla, 1995). In the 17th century, Netherlands and England evolved with the concept of protected cultivation, but however failed to maintain adequate temperature inside the structures (Muijzenberg, 1980). Nowadays, the scenario has changed, as numerous biggest polyhouses in the world are owned by Netherlands, having a huge capacity to produce vegetables annually. Later, a translucent greased paper and glass were utilized in protected technology to cultivate and heat plants during severe cold environment (Jensen and Malter, 1994). It was first time in 1948 when Professor E. M. Emmert from University of Kentucky replaced the expensive glass panels with polythene sheets as covering material in greenhouse which reduced the cost of construction (Espi et al., 2006). During the 13th century in Italy, true greenhouses for the first time were constructed known as "botanical gardens" in such a way so that they may capture sufficient sunlight to produce heat for plant growth and store exotic tropical plants for medical research. In India, technology of protected cultivation was introduced during the 1980s for research only and first polyhouse was installed in 1985 at Leh (Jammu and Kashmir). The start of growing ornamental plants in protected structures was credited to Indo-American Hybrid Seed Companies who made greenhouses before 1970 for the cultivation of such plants. Greenhouse cultivation of carnations was initiated by M/s Feroz Masani and Sons, Nasik during 1980, while in the late 1980s, M/s Pune Flowers initiated to grow roses in a greenhouse (1000 m^2) on an artificial growing media (rockwool). Off-season indigenous and exotic vegetables, export-quality cut-flowers, and quality seedlings are now being cultivated at commercial level in greenhouses. Over the last decade or so, protected cultivation of high-value vegetables and cut-flowers has shown considerable promise which opens up new avenues in agriculture with an add-on of advanced technological tools. These innovations helped not only for big-commercial growers, but also for those having smaller landholdings by increasing the productivity of quality produce, income, and a significant reduction in pesticide usage also for healthier produce, which keeps agriculture economically viable.

The greenhouse cultivation of vegetables such as cucumber, tomato, and capsicum is a practical example of achieving the objectives of protected cultivation which are widely grown giving higher returns (Chandra et al., 2000; Nagarajan et al., 2002). The off-season cultivation of cucumbers in Himachal Pradesh in low-cost polyhouses makes the availability of its fresh produce round the year, which offers an alternative during the period of paucity (Sharma et al., 2009). Protected cultivation of high-value cash crops, vegetables, and cut-flowers is managed under controlled environment with desired conditions to crop growth, resulting in higher productivity of quality produce and profitability (Choudhary, 2016). The protected structures are also used for raising of seedlings, runner production in strawberry, preparing quality planting material through budding or grafting in case of mango, aonla, guava, etc., and protecting plants from frost, heat, rain, hailstorm, and insect-pest attacks. Protected cultivation is one of the most productive and safe technologies, with an ability to increase yields of vegetable and flower crops by several times. Growers can get benefit of higher prices of vegetables and cut flowers by producing them round the year irrespective of seasons. The sensor-based modern tools of climate control and automation provide favorable environment to crop growth inside the greenhouse which assures higher return by regular supply of quality produce during off-season (Wani et al., 2011; Hemming et al., 2019).

1.2 RESOURCE USE EFFICIENCY OF PROTECTED STRUCTURES

In protected cultivation, land, irrigation, agrochemical, and energy resources are intensively utilized to obtain improved crop yields and quality of produce. Provision of subsidies from the government for the establishment of a high-tech protected structure and employment generation round the year are the key aspects observed by farmers adopting polyhouses (Ghanghas et al., 2018). The innovative greenhouses of the future with modified design and management will have almost zero environmental impact which may fulfill the necessity for the safe and cost-effective use of energy, water, and pesticides (Stanghellin and Montero, 2012).

Greenhouse cultivation reduces grower's dependence on rainfall with the optimal use of resources. Moreover, less water is required under protected cultivation as compared to open field as it is supplied through micro-irrigation methods (e.g., drip or micro-sprinklers). Approximately one-third of the crop loss is observed in open-field conditions due to insect-pests and diseases, whereas 2–3 times improved quality produce can be harvested under poly-house cultivation without considerable crop loss. The farmers adopting protected cultivation grow multiple crops and generate their income round the year including premium price for off-season cultivation of vegetables (Kumar et al., 2017).

1.3 CRITERIA FOR THE CONSTRUCTION AND DESIGN OF A POLYHOUSE/GREENHOUSE

1.3.1 SITE SELECTION

Choosing a suitable site including soil type and its profile is very crucial for the construction of a greenhouse/polyhouse, as doing so may create differences in their performance and operational activities. Moreover, the soil should be fertile having the pH from 5.5 to 6.5 and free from any infestation. The leveled ground is always preferred as a construction site and should be located far away from industrial and densely populated areas. The location should not be adjacent to any kind of buildings and trees to ensure maximum light intensity (Sabir et al., 2011). An adequate source of regular supply of quality water (pH 5.5–7.0) and electricity should be available closer to the greenhouse construction site. A ground slope should also be considered during site selection which is very significant to divert excess of surface water from the greenhouse including proper drainage to a lower area around the site. A facility of good road transport entails connectivity with nearby markets and availability of easy and cheap laborers is also desirable before constructing the protected structure.

1.3.2 DESIGN AND CONSTRUCTION

The design of a greenhouse differs according to locality, whether it is in the desert or tropics or temperate zone (Jensen, 2002). In freezing temperatures, growing of vegetables was rather impossible (Albright, 2002), so highly controlled greenhouses jumped up first in temperate regions, however comparatively simple greenhouses equipped with nominal system of climatic control which aided in the production of an economic yield of vegetable crops (Enoch, 1986). Glasshouses are preferred in temperate regions of the world, whereas greenhouses provide "shading effect" and "windbreak effect" in subtropics and tropics. Rain shelters are the most commonly protected structures in rainy tropical regions to avoid flooding (Garnaud, 1987), while in dry and arid areas, the temperature and moisture inside the greenhouse provide an "oasis effect" in comparison with the dry heat outside. Glasshouses, polyhouses, insect-proof net houses, low tunnel polyhouses, and zero energy polyhouses are all examples of greenhouses used for the production of vegetables, fruits, and cut-flowers. There are various types of protected structures such as low cost or naturally ventilated polyhouse/greenhouse with a low initial investment by using locally available materials like bamboo, timber, etc., without control devices to regulate environment inside the structure, appropriate during cold weather, particularly in hilly areas. The medium cost or partially climate controlled polyhouse/greenhouse having structural frame of GI pipes, exhaust fans for ventilation, thermostatically controlled, cooling pad to humidify the air coming into the greenhouse, which is appropriate for growing vegetables throughout mild winter and summer at lower hills of North Western Himalayan region. The high-tech or costly polyhouse/greenhouse with systems of full climate control which is five to six times costier, consists of sensors, frame of iron or aluminum may be dome shaped or cone shaped and highly durable. The soilless growing media (peat, perlite, vermiculite, rockwool, etc.), fertigation, pesticide sprays by fogging machine etc. are also used here. In addition to above, structures such as plastic low tunnels are small temporary structures which keep the plants safe from external agencies such as rains, winds, low temperature, frost, etc. and best way for off-season vegetable nursery production and hardening of seedlings which are easy to transport. The net houses of simple framed with a net of mesh size 40 and higher restricts entry of flying insect to save crop from viral disease. The mulches for covering the soil around plants with a mulch of organic substance (foliage, straw, grasses, etc.) or artificial/synthetic material (white plastic of 30–40 μ used in summers and black plastic in winters and yellow plastic mulch to decrease the incidence of viruses in numerous vegetable crops) are other examples of protected cultivation. Form or shape (lean-to-style, even-span, ridge, or furrow type), climate control mechanisms, design (wooden, pipe, or truss framed), and covering materials are all the factors to be considered when choosing a greenhouse structure (glass, fiberglass, plastic-film, etc.). Acrylic, polycarbonate, fiberglass reinforced polyester, polyethylene film, and polyvinyl chloride film are examples of plastic film covering materials. The plastic glazed greenhouses have a number of advantages, and the cost is most important. Plastic can be adapted to different greenhouse designs, as it is normally break resistant, lightweight, and relatively simple to use. The majority of greenhouse crops thrive well in 400–700 nm wavelength of light; hence, the glazing materials used should be highly transparent. Positioning of the single-span greenhouse may be kept in any direction, while it should be in the north-south direction for multispan greenhouses, to avoid the chances of continuous shading of its certain portions.

1.4 PRODUCTION TECHNOLOGY

1.4.1 GROWING MEDIA

For raising a disease and pest-free nursery and subsequent crop, it is crucial to ensure the health of nursery plants through proper fertigation, its development through hormonal care, accurate sterilization of substrate media, and finally greenhouse management. Protected cultivation in both soil and soilless systems assures profitable crop yield with quality produce under regulated microclimatic conditions (Singh et al., 2019a, 2019b), including off-season cultivation with an efficient fertigation management and a precalculated controlled application of water and nutrient (Singh et al., 2018b). In a conventional system, crops are raised in natural soil. The major drawbacks of this system are the occurrence of soil-borne diseases, insect-pests, and weeds. The excess of nutrients in solution form may leach into the ground during crop cultivation. Moreover, flood irrigation may decrease aeration causing reduced root growth. Hence, it needs a replacement of greenhouse soils, paving the way for soilless cultivation.

Soilless cultivation refers to growing of vegetables, fruits, or ornamental plants in the media other than the soil such as coco-peat, perlite, vermiculite, saw dust, rock wool, peanut hulls, rice hulls, and other mixtures. One of the most common methods of raising the nursery is plug-tray technology, which is now being used to grow disease-free and vigorous off-season seedlings. Coco-peat is the most common substrate for growing nurseries via soilless method; however, it has to be replaced with other substrates that have the same effect due to its extraction (Vaughn et al., 2011). Soilless growing media is prepared by mixing perlite, vermiculite, and coco-peat in a fraction of 1:1:3 in which bio-agents, hydrogels, and organic amendments are also added. This not only helps in the production of high-density seedlings, but also protects against soil-borne diseases (Sabir et al., 2011) and insect-pests (Jovicich et al., 2009). The types of soilless media used for raising the nursery have been standardized with their ingredients. Seedlings are watered when required, while liquid fertilizers are applied on the basis of leaf color and growth rate. According to Litvinov (2006) and Mamatha and Baghyaraj (2002), correct pH of the substrate and application of optimum nutritional supplements ensure the seedling vigor and uniform growth. This technology promotes the rapid root growth of healthy nursery plants and reduces mortality during transplantation, also protecting them from biotic and abiotic stresses (Figure 1.1).

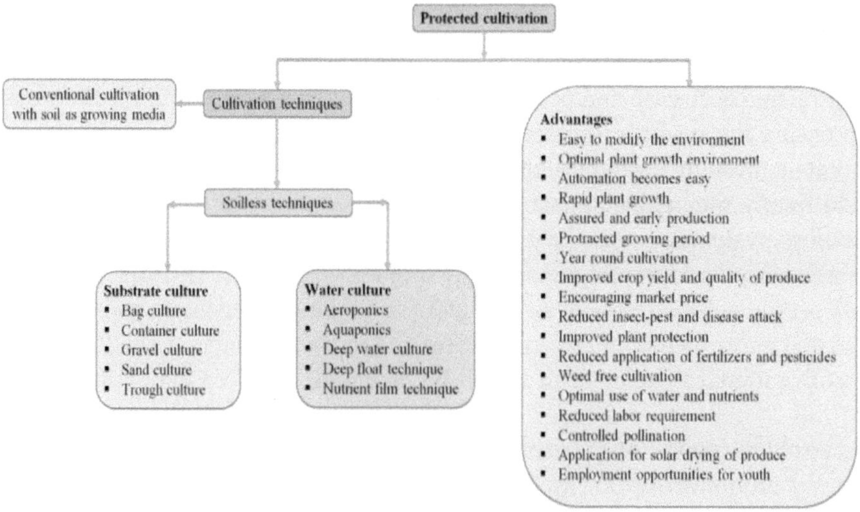

FIGURE 1.1 Protected cultivation technology and its advantages.

Furthermore, the term "soilless cultivation" also refers to "hydroponics" and "aeroponics" that relies on an artificially made soil of inert ingredients like sand, sawdust, peat moss, perlite, vermiculite, coir, rock wool, etc. or an aerated solution of nutrients. Vegetables can be grown in greenhouses through a most recent cultivation technology of hydroponics, comprising planting of seedlings in a system where nutrient solution is continuously supplied through the roots of the plants (see detailed description in Chapter 4). A mineral solution or an inert medium such as perlite or gravel is essentially needed to grow terrestrial plants with their roots. In aeroponic cultivation technology, rooted plants are placed and grown in a special type of box or container with computer-controlled humid atmosphere and sprayed with a mist of nutrient solution. It is a comparatively new production system for protected cultivation, used particularly for research purposes.

Controlled environment agriculture production of high-value crops, providing human health benefits, can be grown through hydroponics by changing the nutritional composition and controlling the temperature that provides useful information on agricultural technology management. Managing the chemical composition of hydroponic culture solutions and changing the physical environment in which the plants grow will help the crops to perform better. This greenhouse cultivation system is mainly being used in developed countries, as it is very costly. Only nitrogen can be recovered completely from traditional soil fertilizers, but all of the nutrients given to plants can be retained in hydroponics. To avoid environmental contamination, it is important to reuse nutrient solution drained out from greenhouse crops grown under hydroponics. As a result, disinfecting the drainage nutrient solution before reusing is critical for maintaining the optimum nutrient concentration and a high root density in the moisturized media (Bar Yosef and Sagiv, 1982). An advantage of increase in yield with quality produce is usually guaranteed in protected cultivation with soilless systems; however, in India it has not gained much attention due to some limitations from farmer's point of view.

1.4.2 IRRIGATION AND FERTIGATION

Fertigation refers to the application of precalculated and precise amount of fertilizers including additional water-soluble nutrients or soil amendments via irrigation systems to minimize adverse effects through reduction in leaching of agrochemicals beneath the root-zone. Drip irrigation is highly productive, water efficient, and environment friendly, and growers may expect an additional remuneration for high-quality produce by adopting this technology (Jensen, 2002). It provides uniform application with lesser amount of fertilizers and hence reduces the cost. But, there is a possibility of contamination threat from equipment malfunctions and careful handling is also required for liquid fertilizers. Fertilizers are provided from various sources such as urea, ammonium nitrate, ammonium sulfate for nitrogen with a concentration of 150–200 ppm; phosphoric acid and mono-ammonium phosphate for phosphorus at 50 ppm; potassium chloride, potassium nitrate, potassium sulfate, and mono-potassium for potassium with a concentration between 200 and 400 ppm through fertigation to fulfill the nutritive need of the particular crop.

According to Choudhary and Verma (2018), for obtaining high yield of capsicum under greenhouses, it is essential to supply nutrients at optimal level with N:P:K in the ratio of 5:3:6. For this purpose, ammonium nitrate, potassium nitrate, and phosphoric acid are applied through drip fertigation as the source of nitrogen, potash, and phosphorus, respectively. For the growing of cucumbers in greenhouses during winter, irrigation requirement is 2.0–2.5 and 2.5–3.0 $m^3/1000\ m^3/day$ at the vegetative stage and after the initiation of picking, respectively, whereas in summer it goes up to 3.0–4.0 and 3.5–4.0 $m^3/1000\ m^3/day$ at the vegetative stage and after the initiation of picking, respectively. The NPK fertilizers namely nitrogen, phosphorus, and potash are applied through fertigation at 80–100, 60–70, and 100–120 ppm/m^3 of water, respectively, during the vegetative stage and 100–150, 80–100, and 120–150 ppm/m^3 of water, respectively, after the start of picking (Choudhary and Verma, 2018).

1.4.3 DISEASE AND PEST MANAGEMENT

In the last 2–3 decades, area under protected cultivation has increased tremendously including cultivation of high-value crops with a sharp growth. Any unfavorable variation in temperature inside the greenhouse causing stress to the plants and higher humidity (Shishido, 2011) may encourage the incidence of diseases and pests. Despite this, a very little concern has been paid to prevent disease and insect infestations by manipulating greenhouse environment. Moreover, there might be soil degradation due to continuous and mono-cropping over a period of time causing high incidence of soil-borne diseases and pests in the greenhouse, which becomes very difficult to control after getting established. In greenhouses, the prevailing warm and humid environment, and the year-round availability of food, encourages the occurrence of pathogens, insects and pests. The examples of diseases in protected cultivation are powdery mildew, downy mildews, gray mold (Shishido, 2011; Goldammer, 2019), root rot, Sclerotinia rot (Malathrakiset and Goumas, 1999; Shishido, 2011; Goldammer, 2019), fungal, and Bacterial wilt (Malathrakiset and Goumas, 1999; Shishido, 2011). Among viral diseases, tobacco mosaic virus, cucumber mosaic virus, tomato yellow leaf curl virus, and tomato spotted wilt virus are common in greenhouses (Pundt, 2010), *Botrytis cinerea* and Fusarium fruit rots in case of papaya. Root knot nematode (RKN) is a severe problem, which is faced not only in open-field cultivation of numerous crops, but also has become a serious threat in polyhouses. The four species of root knot nematode namely *Meloidogyne javanica*, *M. incognita*, *M. hapla*, and *M. arenaria* cause at least 90% damage among RKNs in field and horticultural crops including grassland and grasses (Kepenekcia et al., 2018). Unfortunately, the available strategies for the management of RKN have not been found very effective. Major insect pests in greenhouses include aphids, thrips, whiteflies, caterpillars, leaf miner, and mites, causing damage on various crops such as chilli, capsicum, cucumber, tomato, okra, and gerbera, ranging from 15 to 37% (Dhandapani et al., 2003; Vashisth, 2009; Ibrahim et al., 2011; Sood et al., 2012; Sabir and Singh, 2013; Gavkare et al., 2014; Chinniah et al., 2016; Rathee et al., 2018; Javed et al., 2019; Ghongade, 2020; Singh and Joshi, 2020).

The chemical control is still a reliable and widely adopted method; however, conventional and some organic methods are also common for the control of insect pests and diseases under greenhouses. Moreover, considering the hazardous effects of chemicals and increasing demand for safe food worldwide, in some European countries like America and Japan, showed great interest in the management of pests and diseases with bio-control (Sharma, 2018; Sharma et al., 2019) in protected cultivation, which is now also being followed by other countries (Bueno et al., 2010). In European countries, use of bio-agents under protected cultivation for the control of pests in cucumber, tomato, and sweet pepper accounts for more than 80%, whereas it is around 90% for all commercialized natural enemies that are used in all horticultural crops. The bio-agents help not only in the management of several soil-borne pathogens including plant parasitic nematodes (PPN) in an ecofriendly way, but also induce plant growth (Cakmakci et al., 2006). The use of biological control as an integrated approach to cultural practices would provide further benefits in the future. Sesveren et al. (2011) and Fenoll et al. (2010) advocated sanitation, soil-solarization, fumigation, or mulching of greenhouse soil to control incidence of soil-borne pathogens, weeds, and insect pests. Some eco-friendly techniques prevention measures including spraying of *Bacillus thuringiensis* and synthetic insecticides for banana moth are used (Galan Sauco et al., 2007). Ge et al. (2011) and Gelsomino et al. (2010) found that improving soil structure and fertility with manuring, composting, and other organic amendments also increases soil microbial activity, nitrogen absorption by plants, and reduces some soil-borne diseases, all of which lead to more yield. In countries like Korea and Japan, the grafting technique is being widely employed in vegetable crops of family solanaceae and cucurbitaceae to manage many soil-borne diseases and nematodes as these are constantly and intensively cultivated in greenhouses. In India, this technique has enough scope in greenhouses where wilt and nematode infestations are faced. The CSKHPKV, Palampur and IIHR,

Hessarghatta, Bangalore have started their research entailing manual and robotic grafting against the above-mentioned problems (Pathania, 2018).

1.4.4 CLIMATE CONTROL AND MODELING

The several factors affecting the microclimate inside a protected structure (Table 1.1) include, solar radiation, light interception above plant canopy, air temperature, relative humidity, CO_2, etc. Among these parameters, temperature and relative humidity are considered crucial (when not in optimal range), which negatively affect the water and nutrient requirements of crops and hence the microclimate around the plants (Singh et al., 2021a, 2021b). The microclimate (assemblage of temperature, humidity, light, etc., forming around a plant) inside a greenhouse should be managed as per the requirement of the crops grown (Singh et al., 2016, 2017b). To control temperature and light, shade-nets of agricultural grade are used. Adequate air circulation inside the greenhouse around the plants is desirable and the carbon-dioxide (CO_2) concentration more than 300 ppm should be ensured. The atmospheric CO_2 concentration is 0.03% (300 ppm), which is utilized by the plants for photosynthesis. During the night hours, photosynthesis does not occur and CO_2 is given out from the plants through respiration. The accumulation of this CO_2 around plants continues and hence the CO_2 concentration becomes more inside as compared to outside which is then utilized by the plants for rapid photosynthesis. The CO_2 requirement inside a greenhouse is always greater than open-field conditions (Table 1.1). The CO_2 usage by plants is variable with environmental conditions, crop types, and nutrition and crop types. The CO_2 level for greenhouse crop cultivation should range from 700.0 to 900.0 ppm (De Pascale and Maggio, 2008). Ventilation in greenhouses may help in improving CO_2 level through entry of fresh outside air.

Temperature and moisture are generally higher inside the polyhouse than the outside, which help to improve photosynthesis and result in uniform growth of the plants (Palni and Rawat, 2000). The temperature ranging from 26°C to 30°C is required during daytime for optimal and healthy plant growth, whereas it should range from 15°C to 18°C during night hours. Usually, humidity and temperature are maintained from 60 to 80% and 18 to 24°C, respectively, avoiding the temperature limits of more than 35°C and less than 12°C. Furthermore, temperature inside the polyhouse/greenhouse can increase up to 40°C as it is covered by polythene sheet. There are several methods and climate control devices used to regulate

TABLE 1.1 Factors Affecting the Performance of Protected Cultivation.

Factor/parameter	Effect on plant response
Sunlight (solar radiation)	
• Sunlight as photo-synthetically active radiation (PAR) is essentially required for photosynthesis. • Mainly visible spectrum of light (400–700 nm) is used in photosynthesis process. • Light, CO_2, and water participate in the photosynthesis process to form carbohydrates which in turn regulate the plant growth and reproduction. • Light intensity can be controlled artificially using shading. • Artificial lightening can also be done as per crop's requirement.	• The plant growth gets reduced with diminishing light intensity due to decelerating of photosynthesis process. $\text{Light} + CO_2 + \text{Water} \xrightarrow[\text{Plant nutrients}]{\text{Chlorophyll}} \text{Carbohydrates} + \text{Oxygen}$ or $\text{Sunlight} + 3CO_2 + 6H_2O \xrightarrow[\text{Plant nutrients}]{\text{Chlorophyll}} C_6H_{12}O_6 + 3O_2$ • The light intensity beyond optimal range also decelerates plant growth due to injury to the chloroplasts.
Temperature	
• Temperature is another key factor which affects the crop performance by affecting the biochemical reactions in the plant. • Diverse ranges of temperature are required for different crops. Enzymes are responsible for controlling all the biochemical reactions in the plant. • Temperature can be controlled using shading, ventilation, and other methods such as forced evaporative cooling methods. • White washing can also be done at the top of the greenhouses for reducing the inside air temperature. • Under hot climatic conditions, the inside air temperature can be controlled using shade net, ventilation, and evaporative cooling.	• Generally, an air temperature >35°C is not suitable for greenhouse crops. • A temperature below optimal range stops the life process of plants through ice formation within plant tissue and cells. • At a temperature beyond optimal range, enzymes become inactive and plant life process stops.

TABLE 1.1 *(Continued)*

Factor/parameter	Effect on plant response
Relative humidity (RH)	
• Generally, process of evapo-transpiration increases the moisture inside a greenhouse and therefore, inside RH is greater than the outside. • Low RH is easier to control by keeping greenhouse closed (minimum ventilation) and using evaporative cooling techniques. • If outside RH reaches at lower level than inside RH, ventilation may be initiated to decrease the inside RH. • Alternatively, inside air can also be De-humidified chemically, however it is costly. • The processes such as humidification or de-humidification can be employed to control RH inside greenhouses. • In general, RH ranges from 60 to 90% was best suited for most of the crops. • The desired RH around 90% is required to propagate the plants.	• Relative humidity (RH), when not in optimal range negatively affects the plant development through encouraging the prevalence of diseases and insect-pest attack. • RH value above optimal range may result in disease incidence (Botrytis) in fruits and flowers, and the infamous Downy Mildew in roses and other crops.
Ventilation (natural/forced)	
• Ventilation is performed in greenhouses to reduce air temperature, to replenish CO_2 supply and to control RH of air. • Adequate ventilation can help to bring the greenhouse air temperature within range. • Natural ventilation is economical (cost-effective) and easy in operation. • The maximum area under natural ventilation should be about 60% of floor area with minimum top and side ventilation areas of 10% and 20% of the floor area. • Forced ventilation (use of fan) is essentially practiced to have a precise control over air temperature, RH, and CO_2 levels.	• Inappropriate ventilation may negatively affect: • Air temperature and RH. • Cooling efficiency. • CO_2 concentration (may decrease and go below the outside concentration).

TABLE 1.1 *(Continued)*

Factor/parameter	Effect on plant response
CO_2 concentration	
• The CO_2 used by plants varies with microclimatic conditions, nutrition, and crop types. • The optimum range of CO_2 concentration varies from 700 to 900 ppm for the cultivation of greenhouse crops. • However, numerous crops respond positively to higher concentration of CO_2 ranging from 1000 to 1200 ppm. • The ventilation or infiltration of outside air may help to elevate CO_2 concentration inside the greenhouse.	• The CO_2 concentration below ambient levels (300–500 ppm) will certainly retard the plant growth. CO_2 enrichment may be required CO_2 enrichment $= CO_{2(PU)} + CO_{2(IL)}$ $CO_{2(PU)} = CO_2$ used by plants $CO_{2(IL)} = CO_2$ loss through infiltration $I_L = V_g \times N \times (CO_{2(D)} - 300)$ V_g = Volume of greenhouse (m^3) N = Air exchanges/h $I_L = CO_2$ loss through infiltration (m^3/h) $CO_{2(D)}$ = Designed CO_2 level

the temperature inside the polyhouses. Climate control involves either cooling or heating (passive or active) of greenhouses as per the outside climatic conditions or season (Singh et al., 2018b). The most common climate control techniques are ventilation (natural or mechanical), shading (internal above plant canopy or external above the covering material), and evaporative cooling such as fogging and fan-and-pad system (Singh et al., 2021a). For heating, there are heating systems installed in greenhouses. One of the most critical factors in polyhouse is the light (Singh et al., 2017a, 2018a). The minimum quantity of light between 5 and 6 thousand lux is essential for healthy crop growth in a polyhouse. The several tools or appliances used for monitoring the climate inside a protected structure are listed in Table 1.3.

Modeling is one of the effective techniques for quantifying the greenhouse microclimate through energy and mass transport processes (Yang et al., 1990; Zhang et al., 1997; Gijzen et al., 1998; Singh et al., 2018e, 2018f) and maintaining desirable microclimatic conditions as per the growth stages of the crops cultivated. Modeling plant growth (such as leaf area, leaf area index (LAI), dry matter, fruit yield, etc.) and microclimate (temperature, vapor pressure deficit (VPD), crop transpiration, etc.) help in predicting the plant performance and microclimate inside the structure, thereby helping to modify the climatic conditions as per crop requirement for achieving the optimal yield potential of the crops (Singh et al., 2018c, 2019c).

1.5 ADVANTAGES OF PROTECTED CULTIVATION

There are numerous advantages (Figure 1.1) of protected cultivation technology where the vital inputs such as water and nutrients are efficiently utilized and better quality of produce is obtained. Other benefits of protected farming include growing of early nurseries and crops, and off-season cultivation of vegetables and ornamentals through improved monitoring of growth conditions and crop management techniques. In addition to the efficient disease and pest control, crop is also protected from adverse weather conditions like heavy rains, snow loads, hails, and birds. Moreover, the expenditure on crop management for weed, disease, and pest control is reduced. About 15-fold per acre yield of produce can be achieved in greenhouses with more than 90% of it as marketable, whereas it is only 40–60% when cultivated in open-field conditions.

1.6 ECONOMICS OF PROTECTED CULTIVATION

There arises a need of crop diversification to attain food security by adopting modern methods of crop cultivation such as protected farming and several countries have already started promoting it. In India, the state and central governments are continuously encouraging the farmers to grow the vegetables and cut-flowers under protected structures for extended growing period.

Several studies pertaining to evaluating the impact of protected cultivation have been conducted so far (Mahesh, 1996; Tiwari, 2014; Spehia, 2015; Choudhuri, 2016; Singh et al., 2019a, 2019b). According to one of the studies, the vegetables and flowers including high-value cash crops grown under controlled environment have resulted in higher productivity and profitability, and hence protected cultivation has turned into an emerging agri-entrepreneurship in Himachal Pradesh under the support of state and central governments (Choudhuri, 2016). A study conducted to assess the economics of protected cultivation for ornamental flower crop including its marketing in Madhya Pradesh showed that the total cost of gerbera cultivation on an area of 1200 m^2 was Rs. 757,672, and it was worked out in the ratio of 35:65 for operational and fixed cost (Tiwari, 2014). Furthermore, total expenditure for the protected cultivation of rose (including its more than two-third part on fixed cost) was calculated to be Rs. 735,431, which was at par with gerbera. The gerbera and rose cultivation under natural ventilated simple polyhouse gave annual net profits of Rs. 529,868 and Rs. 345,288 with benefit-cost ratios of 1:70 and 1:46, respectively. Thus, the existing production technology yields sufficient profit to the cut flowers' growers.

1.7 CONSTRAINTS IN THE ADOPTION OF PROTECTED CULTIVATION TECHNOLOGY

Protected cultivation is a developing technology to grow high-value cash crops, vegetables, and cut- flowers; however, some constraints are also faced by the farmers while adopting this technology (Table 1.2). Despite the assured crop return and enhanced productivity, the high initial investment for constructing polyhouse is one of the utmost concerns to the farmers for adopting this technology (Jadhav and Rosentrater, 2017). The high cost and short life of polyethylene-covering sheets, infestation of insect-pests, diseases, and nematodes after a couple of years are the major production constraints. Moreover, some of the major marketing constraints namely absence of MSP (minimum support price) and highly fluctuating prices with lack of information about market (Kumar et al., 2016), non-payment for produce by commission

TABLE 1.2 Constraints of Protected Cultivation.

Problem classification		
1. High initial cost	Fixed cost	Variable cost
Fixed cost	Construction of structure	Cost of field preparation
Variable cost	Drip-irrigation and fertigation systems	Cost of fertilizers
Maintenance cost	Shading net	Cost of irrigation (if any)
	Fogging system	Electricity cost (if any)
	Water-pumping unit	Cost of soilless system (soilless substrate, hydroponic system, aeroponic, etc.)
	Filtration system	
	Cost of atomization of irrigation or fertigation (if any)	Cost of plant support systems (roller hooks, over-head wires, troughs, supporting trays, etc.)
	Cost of artificial cooling (e.g., fan and pad system) or heating system (if any)	Cost of plant protection (Chemicals: pesticides, insecticides, weedicides, etc.)
	Installation costs.	Cost of cooling or heating (if any)
		Maintenance and miscellaneous cost
		Cost involved in a regular repair of structure, its components, and others.
2. Affordability by small farmers.		
3. Nonavailability of different materials of polyhouse construction in the local market.		
4. Short life-span of the covering or cladding materials/polyethylene sheets.		
5. Lack of technical education or trainings.		
6. Poor handling or technical failure	Poor control on microclimatic parameters, that is, sunlight, temperature, humidity, CO_2, and other parameters in inappropriate range.	
7. Operational failure	Improper working of different system components such as cooling or heating system, fogging system, shading mechanism, and others.	

TABLE 1.2 *(Continued)*

Problem classification			
8.	Climate control and other problems	There are numerous constraints which affect the production of horticultural crops under protected structures. Some factors are listed below:	
		Inappropriate sunlight	Atmospheric carbon-dioxide (CO_2)
		Temperature (T_{max}/T_{min})	Moisture deficits or excesses
		Relative humidity (RH_{max}/RH_{min})	Weed growth
		Wind velocity (particularly under naturally ventilated conditions)	Soil nutrients deficiency
			Soil-borne diseases (e.g., insect-pest and nematode infestation).
9.	Delayed subsidies.		
10.	Marketing constraints: No well-defined market (or market information) for selling of produce, lack of minimum support price (MSP), and uneven selling prices or high price fluctuations.		
11.	Storage problems.		

TABLE 1.3 Commonly Used Climate Monitoring Tools or Appliances.

Temperature	Mercury thermometer: Common device to measure temperature (0–110°C) with least count of 0.1°C.
	Thermistor: It is useful for dynamic temperature measurement.
	Thermocouple: Most common method of electrical temperature measurement.
	Thermo-hygrometer.
	Infrared thermometer: To measure temperature of distant objects such as plant surface.
Solar radiation	Pyrheliometer (or Actinometer): To measure direct solar radiation from sun and small fraction of the sky around sun (beam radiation).
	Pyranometer: To measure total hemispherical solar radiation (beam+diffuse) or measurement of global solar radiation received from entire hemisphere on a horizontal surface.
Sunshine	Sunshine recorder (to measure sunshine duration in hours).
Light intensity	Lux meter.
RH	Hygrometer (hair hygrometer), RH sensors.
Wins speed	Anemometer.
Wind direction	Wind vane.
Photosynthesis rate	Photosynthesis analyzer (LI-COR quantum sensor), infrared gas analyzer, and electrochemical gas sensor: To measure the photosynthesis rate.
LAI	SunScan of Delta-Tdevices, LAI meter (LAI-2000): To measure LAI and PAR.
pH	Digital pH meter (laboratory use or portable): To measure pH (basicity or acidity) of nutrient solution and growing media.
EC	Digital EC meter (laboratory use or portable): To measure EC of nutrient solution and growing media.
CO_2	CO_2 analyzer, nondispersive infrared CO_2 sensors, and potable CO_2 meter/CO_2 sensor: To measure CO_2 concentration.
Data storage	Data logger: To store the data on different parameters connected to the logger.

agents due to grading conflict or de payments, and high transportation cost (Wadkar et al., 2006) are also faced by the farmers. According to Ghanghas et al. (2018), the greenhouse growers also face weather irregularities like frequent windstorms and hailstorms, lacking of cold storage facilities at village level, and costly transporting vehicle with refrigerated facilities. Furthermore, exponential increase in population of mites and whiteflies under protected environment are other major constraints realized by the polyhouse-owned farmers. In addition to the higher initial cost of adopting the technology, lack of technical know-how, unavailability of quality pesticides and fertilizers in market at cheaper prices are also observed as major production constraints to the farmers (Malik, 2017). As far as commercial cultivation is concerned, the high cost and unavailability of several parts are yet again the limitation to adopt the polyhouse technology (Sanwal et al., 2004). Different structures of protected cultivation for diverse locations and agro-climatic regions of the country are not standardized precisely.

1.8 STATUS OF PROTECTED CULTIVATION IN INDIA AND OVERSEAS

1.8.1 INDIAN SCENARIO

Agriculture is the backbone of India's economy and its growth trend over the last 50 years has experienced a number of significant achievements on one hand and missed opportunities on the other hand. The land holding constraints due to rapid urbanization and ever-increasing population has created several challenges including decline in crop production and increased demand (Singh et al., 2017). Furthermore, the demand for quality agricultural produce, particularly vegetables, is increasing tremendously. If India has to remain self-sufficient to provide food security and also be able to export high-quality fruits and vegetables, new and efficient production technologies are prerequisites, so that the productivity and profitability of the agricultural sector can be improved regularly. Here, protected cultivation has provided better opportunities for the Indian farmers to produce more on a limited area. Protected cultivation is widely practiced in the developed countries, but its use in India is limited. The commercial-level technology of protected cultivation in India is comparatively new and just three decades old, while it is an old technique in developed countries.

The total area of horticultural crops under protected cultivation has been estimated to be about 30,000–40,000 ha in India (Singh, 2014; Shweta et al., 2014; Beniwal, 2015; Datta et al., 2017). Protected cultivation of fruits in India is in the nascent stage as strawberry is the only fruit crop which is being grown commercially under greenhouses. Mainly Maharashtra, Karnataka, Uttarakhand, Jammu and Kashmir, Himachal Pradesh, Haryana, and Kerala are known for the greenhouse cultivation in India, whereas a variety of vegetable crops such as tomato, capsicum, cucumber, leafy vegetables, onion, cabbage, cauliflower, knoll-khol, radish, and coriander are successfully grown in diverse types of protected structures throughout the country (Figure 1.2). At present, more than 250 private companies are also known for greenhouse production of ornamental flower crops in India and a total area of more than 700 ha has been estimated under greenhouse cultivation of ornamentals. Integration of plastic mulching with fertigation has proved to be a way of futuristic protected vegetable cultivation.

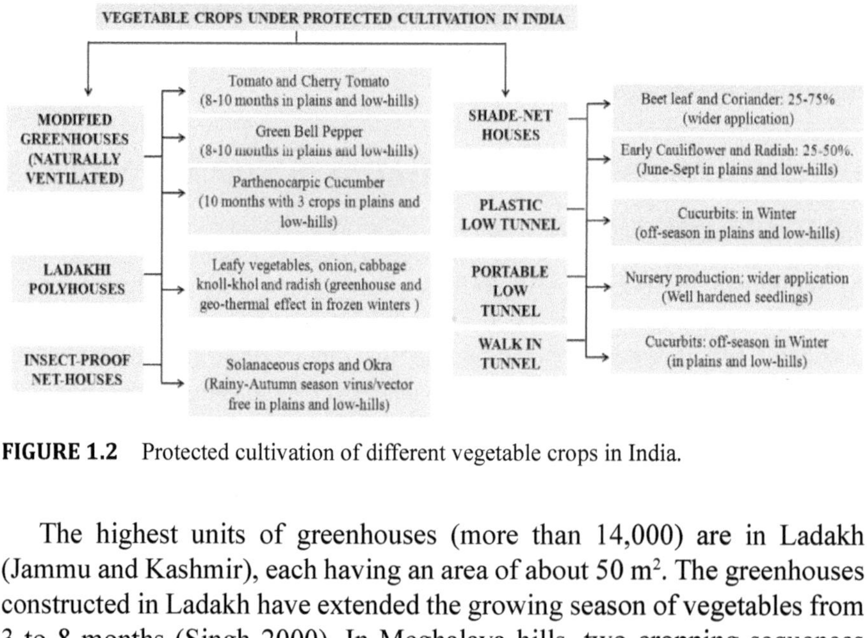

FIGURE 1.2 Protected cultivation of different vegetable crops in India.

The highest units of greenhouses (more than 14,000) are in Ladakh (Jammu and Kashmir), each having an area of about 50 m^2. The greenhouses constructed in Ladakh have extended the growing season of vegetables from 3 to 8 months (Singh 2000). In Meghalaya hills, two cropping sequences viz. capsicum–tomato–lettuce and tomato–frenchbean–cabbage in a bamboo greenhouse have resulted in the net annual return of rupees 109.10 and 76.0/m^2, respectively (Bhatnagr, 2015). The cropping sequence of five vegetable crops in mid-hills of Uttarakhand cultivated in a naturally ventilated polyhouse has resulted in net return of rupees 66.70–113.58/m^2 with a benefit cost ratio of 2.57–3.94 (Figure 1.3).

In Himachal Pradesh, the National Horticulture Mission (NHM) achieved an area of about 1.5 lakh ha under protected cultivation in 2014–2015, out of which maximum area (about 70%) was occupied by plastic mulching

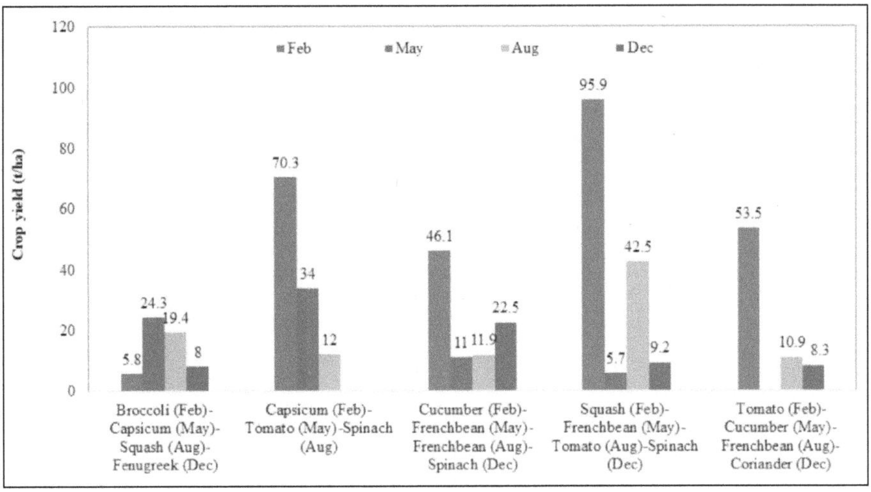

FIGURE 1.3 Greenhouse vegetable yields of five cropping sequences in Uttarakhand.
Source: Adapted from Bhatnagar (2015)

and approximately 20% was under the greenhouses furnished with natural ventilation and fan-pad system (Punera et al., 2017). The Himachal Pradesh has more than 300-hectare area under protected cultivation with productivity of about 18.9 t/ha, having maximum area in Bilaspur followed by Kangra, Una, and others. The total number of polyhouses in Himachal Pradesh built till March, 2019 by Department of Agriculture was 17,350, covering an area of about 221.53 ha under two schemes, Pandit Dean Dyal Upadhaya Krishak Bagwan Samridhi Yojna (PDDKBSY) and Dr. YSP Krishak Samridhi Yojna (YSPKSY). The construction of maximum number of polyhouses was recorded in Mandi (3180), whereas the maximum area was observed in Kangra (38 ha) district. The major vegetables namely tomato, cucumber, capsicum, cauliflower, frenchbean, spinach, coriander, potato, pea, and broccoli were grown under polyhouse cultivation in the state. The maximum area under polyhouse was covered by tomato crop (29.94%), followed by cucumber (18.81%) with productivity of 422 q/ha and 522 q/ha (Figure 1.4a, 1.4b), respectively (Sharma, 2020). In Himachal Pradesh, cropping sequences of capsicum-cabbage-green onion and capsicum-cucumber-frenchbean produced the maximum yield of 13.28 and 10.42 quintal/100 m^2/year in a greenhouse, but reverse trend of the benefit cost ratio was recorded due to variation in selling prices. In a case study conducted (during 2016–2017) in Sirmor, Himachal Pradesh on "Classic Floritech Project" established in 2008–2009 and assisted by NABARD for polyhouse cultivation of roses in an area of 3 acres, owner Mohmand Mustaq shared the outcomes of this project. He told that a gross income of rupees 16.29 lakh and net income of rupees 10.29 lakh per acre (expenditure of rupees 6.0 lakh) was received over 3 years of the project from selling of 81,467 number of flowers at an average rate of rupees 20/flower.

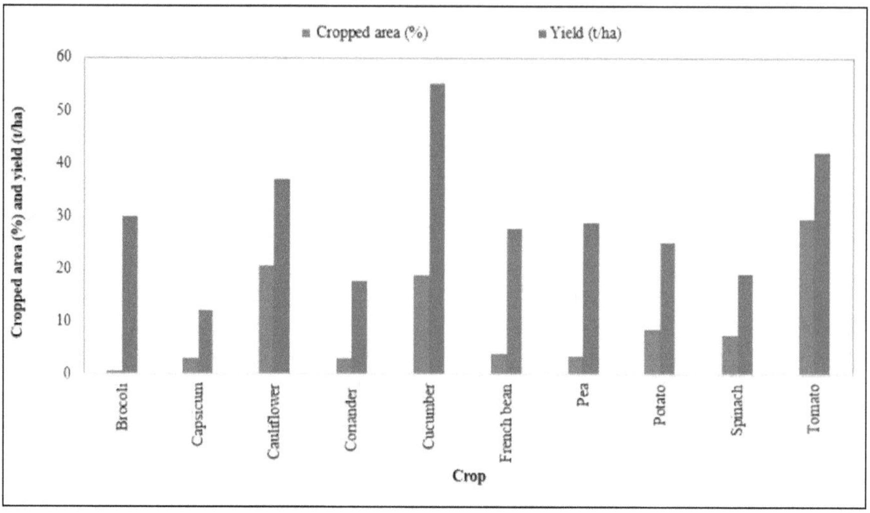

FIGURE 1.4 Area under different crops under polyhouse cultivation and their yields in Kangra district of Himachal Pradesh.

Source: Adapted from Sharma, 2020

In the Haryana state, the total number of polyhouses has also increased from 1356 to 1589 (Kumar et al., 2018). Furthermore, Sonepat is the leading district to adopt the technology with the maximum number of polyhouses, followed by Karnal and Bhiwani, whereas Mahendergarh is at the bottom with the minimum number of polyhouses. In another case study, Sardar Surjeet Singh Virk of Pakka Kheda in Karnal district started growing cucumber and capsicum in net-houses in an area of 5 acre in 2015, which was increased to 7 acre in 2017 with the assistance from NABARD. The structures were prepared with 65% subsidy on initial construction cost of rupees 21.0 lakh/acre. The first crop (September–December) resulted in an average yield of 275 quintal/acre with gross income of rupees 5.50 lakh/acre at a selling price of rupees 20/kg, while second crop (February–June) yielded 500 quintal/acre with gross income of rupees 7.50 lakh at a selling price of rupees 15/kg. The farmer received the total gross income of rupees 13.0 lakh with 40% of expenditure, ultimately earning annual net profit of rupees 8.20 lakh/

acre from one crop in a year. Similarly, when the farmer cultivated green capsicum, he obtained a yield of 400 quintal/acre from August-September planted crop, earning a gross income of rupees 8.0 lakh/acre (selling at an average rate of rupees 20/kg) and the involved profit was rupees 3.50 lakh. In this way, the farmer earned an annual net profit of rupees 4.50 lakh/acre from green capsicum.

In Punjab, the high-value vegetable crops namely tomato, capsicum, cucumber, and lettuce are grown with inside the greenhouses to increase economic viability of protected cultivation. In a case study conducted during 2016–2017 at Narayan Garh farm, Ambala (established in 2013) with the assistance from State Government for growing medicinal plants (Medicinal Plants PC Herbs) under net-houses over an area of 6 acres with an average of 1.0 lakh plants of Stevia/acre, an annual net income of rupees 44.0 lakh/year has been obtained.

The Indian farmers with small landholdings in hills as well as plain areas realized the benefits through adoption of low and walk-in poly tunnels for vegetable production. In Arunachal Pradesh, poly-tunnels are utilized for raising world's hottest chili, the Naga/King, Chili/Bhut, Jhulukia, which is used in Chili Bomb and to disperse crowd. Insect-proof net-house, an inexpensive, viable, and closed production system of 40 mesh size with 20% shade, provides a physical barrier for the entry of insect pests to growing crops. The shade-net structures providing 20, 40, and 60% shading are also economically viable. The Orchids in Sikkim and Arunachal Pradesh are cultivated in shade-net structures, which has become a worthy business for small farmers. The Orchid variety, Lady's Slipper (Sheetal 1 Sikkim), gives flowering during October–February/March with durability of more than 4 months. Greenhouses are mostly used as rain shelters, particularly in high rainfall areas of India such as North-Eastern states and coastal regions. In Kerala, vegetable farming in rain-shelters is also done. The various agencies/departments/universities promoting protected cultivation in India are mentioned in Table 1.4.

1.8.2 GLOBAL SCENARIO

Most of the countries, with the exception of a few, require managed environment agriculture to protect crops from climatic and environmental extremes. While the greenhouse industry has built a great deal of technology for Earth's temperate climate regions, there is an enormous possibility of enhancing safe agriculture in more severe climates. High solar radiation is available in tropical and arid climates, but temperature and relative humidity are extreme. Most of the protected structures are open in nature to allow for ventilation and heat dissipation, but they are vulnerable to pests and diseases. The structures in polar climates and urban areas are relatively enclosed to trap heat and allow effective use of space. Protected agriculture technologies provide solutions for growers in extreme climates who want to produce high yields of high-value crops (Singh et al., 2018d, 2020a). Europe has been transitioning from a "sustained agriculture" to a "sustainable agriculture" structure for last few decades. Recently, protected cultivation systems in North-Western Europe and the Mediterranean have been described in terms of resource use, production efficiency, environmental impact, and related political and social-economic issues.

TABLE 1.4 Agencies or Departments and Universities Leading or Promoting Protected Cultivation in India.

Abbreviated name	Full name
AICRP-PET	All India Coordinated Research Project on Plasticulture Engineering and Technology.
DRDO	Defence Research Development Organization.
DAC&FW	Department of Agriculture, Cooperation and Farmer's Welfare.
MA&FW	Ministry of Agriculture and Farmer's Welfare.
NCPAH	National Committee on Plasticulture Applications in Horticulture under MAFW.
NHB	National Horticulture Board.
NHM	National Horticulture Mission.
PFDC	Precision Farming Development Centre under NCPAH.
ATMA	Agricultural Technology Management Agency under DAC&FW.
CAU and SAU	Central and State Agricultural Universities (more than 70).

The global area under protected agriculture was estimated to be about 5,630,000 ha in 2018 and it was about 497,815 ha under greenhouse vegetables across the world (https://www.hortidaily.com/article/9057219/world-greenhouse-vegetable-statistics-updated-for-2019/). Greenhouse production of vegetables is being done at commercial level in more than 55 countries (Figure 1.5) across the world and the number is steadily increasing (Kottiswaran, 2018). Soilless systems including hydroponics contribute 95,000 ha of the total global greenhouse area under vegetables (Hickman, 2011). As per year 2015, the total global area under polyhouses and glass greenhouses has been estimated as more than 307,000 ha (Beniwal, 2015).

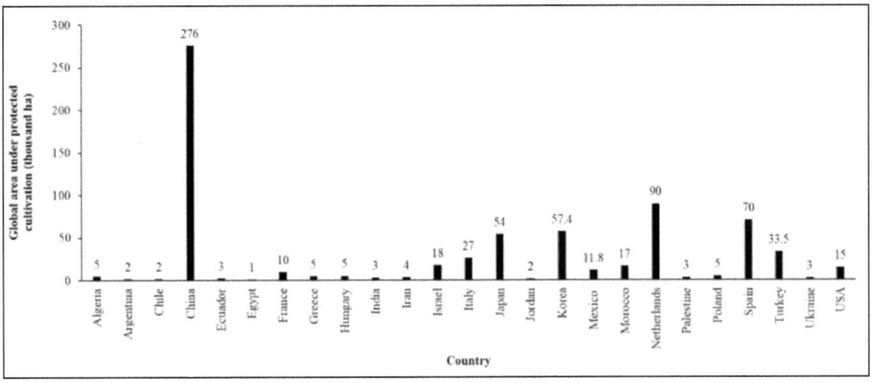

FIGURE 1.5 Countries with at least 2000 ha area under protected cultivation.

Source: Adapted from Choudhary and Verma, 2018

A variety of fruits such as grapes, strawberry, cherries, apricot, peach, plum, and citrus have been fruitfully grown under protected cultivation in the world. In subtropics, three single-stemmed crops namely banana, pineapple, and papaya have been cultivated under greenhouse. Since light is one of the utmost vital factors for the production of vegetables under greenhouse, the South-West has emerged as an ideal location for future growth of this industry, particularly during the winters when prices of cucumber and tomato are at an all-time high. According to the Agricultural Census, 2017 of United States Department of Agriculture, the three most common crops viz. tomatoes, herbs, and cucumbers are produced under greenhouses and high tunnels, contributing annual sale of 401, 71, and 78 million US dollar, respectively (Figure 1.6a, 1.6b).

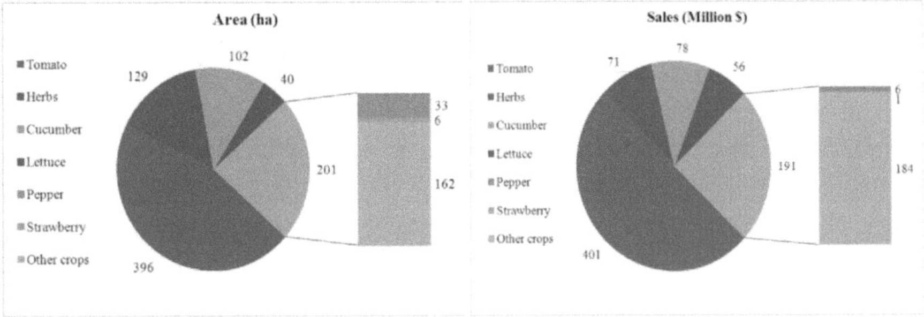

FIGURE 1.6 (a) Protected cultivation area under various crops and (b) sales in the United States of America.

Source: Adapted from USDA, Agricultural Census, 2017

In China, protected cultivation came in existence with full boom during the 1990s. In China, four million ha of vegetable production including melon and watermelon was recorded under protected farming in 2010 that was more than 95% of the total area under protected cultivation of the country, and it was more than 80% of global production of all vegetables under protected environment. Protected cultivation technologies in Asia represent diverse structures varying from solar heating dependent, reasonably priced poly-tunnels to highly expensive and complex hi-tech structures, which rely exclusively on artificial energy, sophisticated climate control system, and artificial intelligence. A number of crops including exotic one, totally dissimilar stages of plants such as transplants, plants to reap fruits, and vegetables can be produced successfully under various types of protected structures. In Asia, constant progress in the exploration of crops, their production, and overall development within related industries is being noticed due to high level of acceptance for the implementation of protected farming in the continent (Kang et al., 2013). Israel has set an example of potentiality of this technology by taking a great advantage where a variety of quality fruits, vegetables, and flowers are being produced under water-scarce desert area. Japan has the area of more than 42,000 ha in the world under protected cultivation excluding mulching, floating mulching, and row covering. The PVC polyhouses are popular in Japan due to their superior thermal insulation against long-wave radiations and more than 60% of the total area is not heated. Floating mulching using plastic nets or films permeable to water vapor is gaining more popularity as it protects from wind, rain, and birds, while also improving the thermal environment. Mango crop is cultivated in greenhouses chiefly in Japan and its commercial cultivation was also started by Israel. Hydroponics has regained prominence with "nutrient film technology" (NFT) and rockwool cultivation. Factory-style vegetable production has been adopted for the processing of radish sprouts and lettuce. Seedling production is not yet a separate industry, but it is gaining traction as a result of advances in robotics and fine environmental control. Greenhouse computer systems are powered by expert systems and local network systems. At present, Dutch-protected cultivation is one of the best examples in the world of intensive farming with highest level of output achieved through the use of cutting-edge technology (Goncharova, 2004). Almeria city in Spain has 40,000 ha of greenhouses at one place and exporting the produce to the entire Europe.

1.9 CONCLUSIONS

The global scenario of aberrant weather events due to climate change and continuously increasing population has created an everlasting demand of vegetables and fruits from the smaller land holdings. Under these situations, economic returns from the high-value agricultural produce can be increased substantially when grown in greenhouses under optimal environmental conditions. As we know, protected farming itself has a huge assembly comprising various aspects related to agriculture. So, an effort has been made to prepare this book covering the current scenario of protected cultivation in Indian and global context including design and construction, production technologies, irrigation and fertigation, cropping sequences, diseases and pests, climate control and modeling, automation, and economics of protected cultivation.

KEYWORDS

- **protected cultivation**
- **hydroponics**
- **climate control**
- **fertigation**
- **economics**

REFERENCES

Albright, L. D. Controlling Greenhouse Environments. *Acta Hortic.* **2002,** *578*, 47–54.

Bakkes, J. A.; Grosskurth, J.; Idenburg, A. M.; Rothman, D. S.; van-Vuuren, D. P. *Descriptions of Selected Global Models for Scenario Studies on Environmentally Sustainable Development*; RIVM Report No. 402001018, National Institute of Public Health and the Environment: Bilthoven, Netherlands, 2000; p 55.

Bar Yosef, B.; Sagiv, B. Response of Tomatoes to N and Water Applied via a Trickle Irrigation System. *I. Nitrogen. Agron. J.* **1982,** *74*, 633–637.

Beniwal, R. S. Disease Management of Crops in Protected Cultivations/Greenhouses. In *Protected Cultivation of Horticultural Crops*; Singh, et al., Eds.; CCS HAU: Hisar, 2015; pp 61–72.

Bhatnagar, P. R. Strategies for Protected Cultivation for Small and Marginal Farmers in India. In *Agriculture: Towards a New Paradigm of Sustainability*; Excellent Publishing House: New Delhi, India, 2015; pp 158–166. ISBN 978-93-83083-64-0. http://www.krishisanskriti.org/vol_image/07Sep201505091128.pdf (accessed on Aug 24, 2021)

Bueno, V. H. P.; van Lenteren, J. C. In *Biological Control of Pests in Protected Cultivation: Implementation in Latin America and Successes in Europe*, Memorias, XXXVII Congreso Sociedad Colombiana de Entomologia, Bogota, D. C., 2010; pp 261–269.

Çakmakçi, R.; Dönmez, F.; Aydın, A.; Şahin, F. Growth Promotion of Plants by Plant Growth-Promoting Rhizobacteria under Greenhouse and Two Different Field Soil Conditions. *Soil Biol. Biochem.* **2006,** *38*, 1482–1487.

Chandra, P.; Sirohi, P. S., Behera T. K.; Singh, A. K. Cultivating Vegetables in Polyhouse. *Indian Hort.* **2000,** *45*, 17–25.

Chinniah, C.; Ravikumar, A.; Kalyanasundaram, M.; Parthiban, P. Management of Sucking Pests, by Integration of Organic Sources of Amendments and Foliar Application of Entomopathogenic Fungi on Chili. *J. Biopest.* **2016,** *9* (1), 34–40.

Choudhary, A. K. Scaling-up of Protected Cultivation in Himachal Pradesh, India. *Curr. Sci.* **2016,** *111* (2), 272–277.

Choudhary, B. R.; Verma, A. K. Greenhouse Cultivation of Capsicum. In *Prospects of Protected Cultivation in Hot Arid Region*, Technical Bulletin No. 69, 2018; p 41.

Datta, V.; Chattopadhyay, K. S.; Roy, D.; Majumder, D. An Economic Analysis of Protected Cultivation under MIDH in Sikkim, Study No.-186, Agro-Economic Research Centre (For the States of West Bengal, Sikkim and Andaman & Nicobar Islands), Visva-Bharati, Santiniketan, West Bengal, 2017; pp x+66.

De Pascale, S.; Maggio, A. Plant Stress Management in Semiarid Greenhouse. *Acta Hortic.* **2008,** *797*, 205–215.

Dhandapani, N.; Shelkar, U. R.; Murugan, M. Bio-intensive Pest Management (BIPM) in Major Vegetable Crops. *J. Food Agric. Environ.* **2003,** *1* (2), 333–339.

Espi, E.; Salmeron, A.; Fontecha, A.; Garcia Y.; Real A. I. Plastic Films for Agricultural Applications. *J. Plast. Film Sheeting* **2006,** *22*, 85–102.

Fenoll, J.; Ruiz, E.; Hellín, P.; Navarro, S.; Flores, P. Solarization and Biosolarization Enhance Fungicide Dissipation in the Soil. *Chemosphere* **2010,** *79* (2), 216–220.

Garnaud, J. C. A Survey of the Development of Plasticulture. Questions Still to be Answered. *Plasticulture* **1987,** *74*, 5–14.

Gavkare, O.; Surjeet, K.; Yuvraj, S.; Bhopale, S. P. Abundance of the *Myzus persicae* (Sulzer) under Protected Environment in Himachal Pradesh, India. *J. Ind. Pollut. Control* **2014,** *30* (2), 285–287.

Ge, T.; Nie, S.; Wu, J.; Shen, J.; Xiao, H.; Tong, C.; Huang, D.; Hong, Y.; Iwasaki, K. Chemical Properties, Microbial Biomass, and Activity Differ Between Soils of Organic and Conventional Horticultural Systems under Greenhouse and Open Field Management: A Case Study. *J. Soils Sediments* **2011,** *11*, 25–36.

Gelsomino, A.; Petrovičová, B.; Zaffina, F.; Peruzzi, A. Chemical and Microbial Properties in a Greenhouse Loamy Soil after Steam Disinfestation Alone or Combined with CaO Addition. *Soil Biol. Biochem.* **2010,** *42*, 1091–1100.

Ghanghas, B. S.; Malik, J. S.; Yadav, V. P. S. Sustainable Vegetables and Flowers Production Technology (Poly House): Problems and Prospects in Haryana. *Indian Res. J. Ext. Educ.* **2018,** *18* (2), 12–16.

Ghongade, D. S. Population Abundance and Management of Whitefly and Red Spider Mite on Parthenocarpic Cucumber Grown under Protected Conditions. Ph.D Dissertation, Punjab Agricultural University, 2020, p 154.

Ghosh, A. General Idea of Greenhouse Technology. In *Greenhouse Technology (The Future Concept of Horticulture)*, Kalyani Publishers: New Delhi, 2009; pp 18–28.

Goldammer, T. Greenhouse Management: A Guide to Operations and Technology. Apex Publishers, USA, 2019; p 406.

Goncharova, N.; van-der Vlist, A.; Verstegen, J. A. A. M. Changes in the Horticulture Sector in the Netherlands. *Acta Hortic.* **2004,** *655*, 319–331.

Hall, A. J.; Bockett, G.; Taylor, S.; Sivamohan, M. V. K.; Clark, N. Why Research Partnerships Really Matter: Innovation Theory, Institutional Arrangements and Implications for Developing New Technology for the Poor. *World Dev.* **2001,** *29* (5), 783–797.

Hemming, S.; de Zwart, F.; Elings, A.; Righini, I.; Petropoulou, A. Remote Control of Greenhouse Vegetable Production with Artificial Intelligence-Greenhouse Climate, Irrigation, and Crop Production. *Sensors* **2019,** *19*, 1807. https://doi.org/10.3390/s19081807

Hickman, G. W. Greenhouse Vegetable Production Statistics: A Review of Current Data on the International Production of Vegetables in Greenhouses. Cuesta Roble Greenhouse Consulting: Mariposa, CA, 2011; p 72. wwwcuestaroble.com

Ibrahim, L.; Hamieh, A.; Ghanem, H.; Ibrahim, S. K. Pathogenicity of Entomopathogenic Fungi from Lebanese Soils Against Aphid, Whitefly and Non-target Beneficial Insects. *Int. J. Agr. Sci.* **2011,** *3* (3), 156–164.

Jadhav, H. T.; Rosentrater, K. A. In *Economic and Environmental Analysis of Greenhouse Crop Production with Special Reference to Low Cost Greenhouses: A Review*, An ASABE Meeting Presentation, 2017; vol 701178, pp 1–6.

Janick, J.; Paris, H. S.; Parrish, D. C. The Cucurbits of Mediterranean Antiquity: Identification of Taxa from Ancient Images and Descriptions. *Ann. Bot.* **2007,** *100*, 1441–1457.

Javed, K.; Javed, H.; Mukhtar, T.; Qiu, D. Pathogenicity of Some Entomopathogenic Fungal Strains to Green Peach Aphid, *Myzus persicae* Sulzer (Homoptera: Aphididae). *Egypt J. Biol. Pest Control* **2019,** *29* (92), 1–7.

Jensen, M. H. In *Deep Flow Hydroponics: Past, Present and Future*, Proceedings of the National Agricultural Plastics Congress, National Agricultural Plastics Association: Madison, WI, 2002; vol 30, pp 40–46.

Jensen, M. H.; Malter, A. J. Protected Agriculture: Global Review, World Bank Technical Paper number 253, 1994; p 160.

Jovicich, E. *Leaf Miner Pest-Generic Incursion Management Plan for the Australian Vegetable Industry;* Project Report, Horticulture Australia Limited, 2009; p 126.

Kang, Y.; Chang, Y. A.; Choi, H.; Gu, M. Current and Future Status of Protected Cultivation Techniques in Asia. *Acta Hortic.* 2013, *987*, 33–40.

Kepenekcia, I.; Hazirb, S.; Oksalc, E.; Lewisd, E. E. Application Methods of *Steinernema feltiae, Xenorhabdus bovienii* and *Purpureocillium lilacinum* to Control Root-Knot Nematodes in Greenhouse Tomato Systems. *Crop Prot.* **2018,** *108*, 31–38.

Kottiswaran, S. V. Technological and Ecological Approaches for Designing Protected Structures in India. In *Technologies and Sustainability of Protected Cultivation for Hi-Valued Vegetable Crops*; Kumar, S., Patel, N. B., Saravaiya, S. N., Patel, B. N., Eds.; 2018; pp 18–29.

Kumar, P.; Chauhan, R. S.; Mehta, V. P. In *Current Scenario of Poly Houses in Haryana*, International Conference on Sustainable Agriculture, Energy, Environment and Technology, MDU, Rohtak, 2018; A-623.

Kumar, P.; Chauhan, R. S.; Rohila, A. K. Grover, R. K. Poly House Technology in Vegetable Production for Nutritional Security in Haryana: Status, Production and Constraints. *Agric. Econ. Res. Rev.* **2016,** *29*, 207.

Kumar, P.; Chauhan, R. S.; Tanwar, N.; Grover, R. K.; Kumar, R. Doubling of Farmer's Income by Vegetable Crop Production under Poly House. *Indian J. Econ.* **2017,** *13* (2a), 295–300.

Mahesh, K. Variability Studies in Carnation (Dianthus caryophyllus L.). M. Sc. (Agril.) Thesis, University of Agricultural Sciences, Bangalore, Karnataka (India), 1996.

Malathrakis, N. E.; Goumas, D. E. Fungal and Bacterial Diseases. In *Integrated Pest And Disease Management In Greenhouse Crops;* Albajes, et al., Eds.; Kluwer Academic Publishers: Netherland, 1999; pp 34–47.

Malik, K. Economic Viability of Cucumber Cultivation in Greenhouses. *Int. J. Innov. Res. Multidiscip. Field.* **2017,** *3* (6), 366–368.

Mamatha, G.; Bagyaraj, D. J. In *Different Levels of VAM Nursery Bioinoculants for Sustainable Agriculture and Forestry*, Proceedings of National Symposium, Feb 16–18, 2002; pp 43–48.

Muijzenberg, E. W. B. van den. In *A History of Greenhouses*; Institute for Agricultural Engineering: Wageningen, Netherlands; 1980; p 435.

Nagarajan, M.; Senthilvel, S.; Planysamy, D. Material Substitution in Green House Construction. *Kisan World* **2002,** *11*, 57–58.

Palni, L. M. S.; Rawat, D. S. Simple Technologies for Capacity Building and Economic Upliftment of Women in Mountains, Paper Presented at the Indian Science Congress 3–7 January, 2007, Pune.

Pathania, N. K. Strategies and Techniques for Sustainable Greenhouse Vegetable Production in India. In *Technologies and Sustainability of Protected Cultivation for Hi-Valued Vegetable Crops*; Kumar, S., Patel, N. B., Saravaiya, S. N., Patel, B. N., Eds.; 2018; pp 31–43.

Pundt, L. Some Virus Diseases of Greenhouse Crops, Integrated Pest Management, University of Connecticut, 2010. http://www.hort.uconn.edu/ipm/greenhs/htms/gcrpvirus.htm

Punera, B.; Pal, S.; Jha, G. K.; Kumar, P. Economics and Institutional Aspects of Protected Cultivation of Carnation in Himachal Pradesh. *Agric. Econ. Res. Rev.* **2017,** *30* (1), 73–80.

Rathee, M.; Singh, N. V.; Dalal, P. K.; Mehra, S. Integrated Pest Management under Protected Cultivation: A Review. *J. Entomol. Zool. Stud.* **2018,** *6* (2), 1201–1208.

Sabir, N.; Singh, B. Protected Cultivation of Vegetables in Global Arena: A Review. *Indian J. Agric. Sci.* **2013,** *83* (2), 123–135.

Sanwal, S. K.; Patel, K. K.; Yadav, D. S. Vegetable Production under Protected Conditions in NEH Region. *Envis Bull. Himal. Ecol.* **2004,** *12* (2), 1–7.

Sesveren, S.; Kırda, C.; Kaman, H. Effects of CO2 Enriched Air and Soil Tillage on Greenhouse Solarization. *J. Food Agric. Environ.* **2011,** *9* (1), 550–556.

Sharma, K. K. Trichoderma in Agriculture: An Overview of Global Scenario on Research and its Application. *Int. J. Curr. Microbiol. Appl. Sci.* **2018,** *7* (8), 1922–1933.

Sharma, K. D. *Upscaling Technological Support for Addressing Emerging Problems of Vegetable Crops under Protected Environment: Economics and Marketing (Kangra District)*; Research Report No. 89; Department of Agricultural Economics, Extension Education and Rural Sociology, College of Agriculture: CSK HPKV, Palampur, 2020; p 34.

Sharma, L.; Shukla, S. K.; Jaiswal, V. P.; Sharma, K. K. Characterization of Cultural Beneficial Bacterial Diversity for Plant Growth Promoting Attributes Associated with Rice Rhizosphere. *J. Environ. Biol.* **2019,** *40*, 1180–1187.

Shishido, M. Plant Disease Management in Protected Horticulture. *Hort. Res.* **2011,** *65*, 7–18.

Singh, B. In *Protected Cultivation of Horticultural Crops in India: Challenges and Opportunities*, 3rd International Conference on Agriculture & Horticulture, Oct 27–29, 2014; Hyderabad International Convention Centre, India, http://dx.doi.org/10.4172/2168-9881.S1.010

Singh, B. Vegetable Production in Cold Desert of India: A Success Story on Solar Greenhouses. *Acta Hortic.* **2000,** *534*, 1.

Singh, H.; Joshi, N. Management of the Aphid, *Myzus persicae* (Sulzer) and the Whitefly, *Bemisia tabaci* (Gennadius), using Biorational on Capsicum under Protected Cultivation in India. *Egypt J. Biol. Pest Control* **2020,** *30*, 1–9.

Singh, M. C.; Kachwaya, D. C., Kalsi, K. Soilless Cucumber Cultivation under Protective Structures in Relation to Irrigation Coupled Fertigation Management, Economic Viability and Potential Benefits-A Review. *Int. J. Curr. Microbiol. App. Sci.* **2018b,** *7* (03), 2451–2468.

Singh, M. C.; Singh, J. P.; Pandey, S. K.; Cutting, N. G.; Sharma, P.; Shrivastav, V.; Sharma, P. A Review of Three Commonly used Techniques of Controlling Greenhouse Microclimate. *Int. J. Curr. Microbiol. App. Sci.* **2018a,** *7* (1), 3491–3505.

Singh, M. C.; Singh, J. P.; Pandey, S. K.; Mahay, D.; Srivastava, V. Factors Affecting the Performance of Greenhouse Cucumber Cultivation-a review. *Int. J. Curr. Microbiol. App. Sci.* **2017a,** *6* (10), 2304–2323.

Singh, V. P.; Singh, P. K.; Singh, C. P. In *Problems and Prospects of Greenhouse Cultivation in Fruit Crops*, Proceedings of 30th National Convention of Agricultural Engineers & National Seminar on 'Technological Innovations for Enhancing Profitability of Small and Marginal Farms' Organized by Pantnagar Local Centre, The Institutions of Engineers (India), Feb 27–28, 2017; pp 188–193.

Singh, M. C.; Singh, J. P.; Singh, K. G. Optimal Operating Microclimatic Conditions for Drip Fertigated Cucumbers in Soilless Media under a Naturally Ventilated Greenhouse. *Ind. J. Ecol.* **2017b,** *44*, 821–826.

Singh, M. C.; Singh, K. G.; Singh, J. P. Indirect Method for Measurement of Leaf Area and Leaf Area Index of Soilless Cucumber Crop. *Adv. Plants Agric. Res.* **2018c,** *8* (2), 188–191.

Singh, M. C.; Singh, K. G.; Singh, J. P. Yield of Soilless Cucumbers Planted under Partially Controlled Greenhouse Environment in Relation to Deficit Fertigation. *Ind. J. Hort.* **2018d,** *75* (2), 259–264.

Singh, M. C.; Singh, J. P.; Singh, K. G. Development of Mathematical Models for Predicting Vapour Pressure (Deficit) Inside a Greenhouse Independently from Internal and External Climate. *J. Agrometeorol.* **2018e,** *20* (3), 238–241.

Singh, M. C.; Singh, J. P.; Singh, K. G. Development of a Microclimate Model for Prediction of Temperatures Inside a Naturally Ventilated Greenhouse under Cucumber Crop in Soilless Media. *Comput. Electron. Agric.* **2018f,** *154*, 227–238.

Singh, M. C.; Singh, K. G.; Singh, J. P. Nutrient and Water use Efficiency of Cucumbers Grown in Soilless Media under a Naturally Ventilated Greenhouse. *J. Agric. Sci. Technol.* **2019b,** *21* (3), 193–207.

Singh, M. C.; Singh, J. P.; Singh, K. G. Modeling Plant Growth and Microclimate of a Protected Structure-A Review. *Ann. Hort.* **2019c,** *12* (1), 47–54.

Singh, M. C.; Singh, K. G.; Singh, J. P. Interactive Effects of Fertigation and Varieties on Pant Growth Attributes and Yield of Soilless Cucumbers. *Int. J. Sci. Technol. Soc.* **2020a,** *8* (3), 60–65.

Singh, M. C.; Singh, J. P.; Singh, K. G. Mathematical Modeling of Greenhouse Microclimate under Vertically Trained Soilless Cropped Conditions. *Agric. Res.* **2021a,** *10*. https://doi.org/10.1007/s40003-021-00609-0

Singh, M. C.; Singh, J. P.; Singh, K. G.; Gupta, O. P.; Kumar, G. Computation of Vapour Pressure Deficit and Crop Transpiration via Development of a Web Based Computer Module. *J. Agrometeorol.* **2021b,** *23* (4), 243–248.

Singh, M. C.; Singh, K. G.; Singh, J. P.; Mahal, A. K. Performance of Soilless Cucumbers in Relation to Differential Fertigation under Naturally Ventilated Greenhouse Conditions. *J. Plant. Nutr.* **2019a,** *42* (11–12), 1316–1332.

Singh, M. C.; Singh, A.; Singh, J. P.; Singh, K. G. Economic Viability of Soilless Cucumber Cultivation under Naturally Ventilated Greenhouse Conditions. *Ind. J. Hort.* **2020b,** *77* (2), 315–321.

Singh, M. C.; Yousuf, A.; Singh, J. P. Greenhouse Microclimate Modeling under Cropped Conditions-A Review. *Res. Environ. Life* **2016,** *9* (12), 1552–1557.

Sood, A. K.; David, B. V. The Greenhouse Whitefly, *Trialeurodes vaporarriorum (*Westwood*).* In *The Whitefly or Mealywing Bugs*; David, B. V., Ed.; LAP Lambert Academic Publishing Gmbh & Co. KG: Germany, 2012; pp 147–169.

Sphehia, R. S. Status and Impact of Protected Cultivation in Himachal Pradesh, India. *Curr. Sci.* **2015,** *108* (12), 2254–2257.

Stanghellin, C.; Montero, J. I. Resource Use Efficiency in Protected Cultivation: Towards the Greenhouse with Zero Emissions. *Acta Hortic.* **2012,** *927*, 91–100.

Tiwari, Y. Economic of Production of Marketing of Flower under Protected Condition in Jabalpur District of MP, 2014; p 79. URL: HTTP//Krishikosh.egranth.ac.in/handle/1/68788

Vashisth, S. Insect and Nematode Complex Associated with Some Polyhouse Crops. CSK Himachal Pradesh Krishi Vishvavidyalaya, Palampur (Unpublished), M.Sc. Thesis, 2009.

Vaughn, S. F.; Deppe, N. A.; Palmquist, D. E.; Berhow, M. A. Extracted Sweet Corn Tassels as a Renewable Alternative to Peat in Greenhouse Substrates. *Ind. Crops Prod.* **2011,** *33*, 514–517.

Wadkar, S. S.; Bagade, S. R.; Veerkar, P. D. A Study on the Economics of Cutflower Production under Polyhouse. *Agric. Econ. Res. Rev.* **2006,** *19*, 214.

Walker, S.; Joukhadar, I. Greenhouse Vegetable Production, College of Agricultural, Consumer and Environmental Sciences, New Mexico State University, Circular 556, 2019.

Wani, K. P.; Singh, P. K.; Amin, A.; Mushtaq, F.; Dar, Z. A. Protected Cultivation of Tomato, Capsicum and Cucumber under Kashmir Valley Conditions. *Asian J. Sci. Technol.* **2001,** *1* (4), 056–061.

Weil, R. Defining and Using the Concept of Sustainable Agriculture. *J. Agronomic Educ.* **1990,** *19*, 126–130.

Wittwer, S. H.; Castilla, N. Protected Cultivation of Horticultural Crops Worldwide. *Hort. Technol.* **1995,** *5* (1), 6–23.

Web Sources:

https://www.hortidaily.com/article/9057219/world-greenhouse-vegetable-statistics-updated-for-2019 (accessed on Aug 24, 2021)

CHAPTER 2

Classification and Suitability of Protected Structures

JAGREETI GUPTA[1], DEEPA SHARMA[2], and AANCHAL CHAUHAN[3]

[1]*Department of Floriculture and Landscape Architecture, Dr. YSPUHF, CoHF, Neri, Hamirpur, Himachal Pradesh, India*

[2]*Department of Vegetable Science, Dr. YSPUHF, CoHF, Neri, Hamirpur, Himachal Pradesh, India*

[3]*Department of Vegetable Science, College of Horticulture, Dr, YSPUHF, Nauni, Solan, HP*

ABSTRACT

Cultivating crops under protected conditions for obtaining optimal plant growth and yield is termed protected cultivation. There are various types of protected structures viz., polyhouses, glasshouses, shade nets, high or low tunnels, hot beds, cold frames, lath house, passive solar greenhouse, and retractable roof greenhouse. These structures protect the plant from adverse climatic conditions and therefore assist in improving the production both quantitatively and qualitatively. Commercially greenhouses are used to produce off season vegetables, good quality cut flowers, medicinal plants, and for raising nursery. The major commercial flower crops grown in greenhouses are rose, carnation, gerbera, lilium, anthurium, chrysanthemum, orchids, etc., and the vegetables are capsicum, tomato, cucumber, leafy, and exotic vegetables. Year-round crop production is possible under protected structures even under harsh conditions. Greenhouses are classified into many types on the basis of their shape, size, cladding material, environment control, cost involved, and so on. Shade-net/screen-houses are efficient for raising healthy

Protected Cultivation: Structural Design, Crop Management, Modeling and Automation.
Mahesh Chand Singh and K. K. Sharma (Eds.)

nursery/grafts and hardening of various horticultural crops. Plastic tunnels and walk in tunnels are low-cost structures which can be efficiently utilized for healthy nursery and off-season crop production. Plastic mulching prevents the weed growth, regulates the soil temperature, and is available in different colors, such as black, silver-black, red, yellow, and white-black. Hydroponics and aeroponics systems can also be used to grow the plants under protected environment. In hydroponics and aeroponics, plants are grown in soilless system in nutrient-rich solution or nutrients are directly sprayed to the root system of plants. These systems are helpful in urban areas having insufficient land or problematic soils that are not suitable for cultivation due to various soil-borne diseases and abiotic stresses like soil salinity, alkalinity, drought, and floods. These are the various methods of protected cultivation, used to provide favorable climate for plant growth, for improving yield, quality of produce, and grow crops in off-season to fetch higher returns.

2.1 INTRODUCTION

Due to growing population, rising pressure on natural resources (i.e., depleting land and water resources), climate change, and the demand for high-value produce (such as tomato, capsicum, cucumber, strawberry, etc.), there is a need to shift toward modern crop production technologies. It is a technique of providing favorable climate for the plant growth and protects the crop from adverse climatic conditions of cold, precipitation, wind, hail, excessive heat, and other factors with an aim to enhance crop production, efficiency, and value of end product (Critten and Bailey, 2002; Mishra et al., 2010). Under enclosed structures, the climate nearby plants is regulated either completely or partially (Singh et al., 2017a, 2017b; 2018a; 2021). The initiative of raising crops (vegetables) under protected/controlled conditions was taken by the Roman farmers/gardeners, who used to cultivate vegetables in protected structures for daily consumption. (Janick et al., 2007). Protected cultivation of high value crops is spread all over the globe, especially in Europe, Asia, and USA. In USA, greenhouses are mainly used for growing flowers, whereas, in Italy and Spain, these are mostly used for vegetable production. In Spain, tunnel-type greenhouses are commonly used for producing flowers as well as vegetables. In the Netherlands, the greenhouse industry is most advanced and is a leading producer of greenhouse-grown vegetables and flowers in the world. Asia, China, and Japan have considerable area under protected cultivation. Chinese people mostly use local materials for frames and polythene films as cladding material.

In 1980, greenhouse technology was started in India and was used for research purposes initially. Keeping in view the different agro-climatic conditions of the country different types of protected structures were suggested. Glasshouses were constructed in Ladakh and other cold regions to extend vegetable production during winter months. Greenhouses have been constructed mainly as rain shelters in North East, whereas in the coastal region, naturally ventilated polyhouses are more suitable for efficient and year-round crop production. In the northern region of the country, cooling and heating systems are required to be used in protected structures for climate regulation. Net houses of at least 40 mesh have been successfully utilized in the summer season to grow capsicum, tomato, beans, etc. Therefore, the local climatic conditions and topography play an important role in the designing/ construction of greenhouses and selection of covering materials.

Protected cultivation is mainly used for the exotic and off-season production of vegetables, quality flowers, medicinal plants, and also for raising quality nursery. This technology not only generates avenues at a higher level but also improve economic status of small farmers (Sabir and Singh, 2013). It also helps in year-round cultivation of high-value vegetables and flower crops especially (Murthy et al., 2009; Choudhary, 2016; Kumar et al., 2017; Singh et al., 2018b) and thus results in more yield and income (Cheema et al. 2004; Navnath et al., 2019) as compared with the open-field cultivation. In the open-field conditions, insect-pests attack and adverse climatic conditions may lead to the crop loss which is nearly one-third, whereas in polyhouse farming, we can achieve more yields (2–3 times higher). Cheema et al. (2004) analyzed that the yield of different vegetable crops is higher inside the polyhouse and this was mainly due to the better microclimate around the plant. The highest yield in relation to the increased number of fruits was obtained in protected structures as compared to the open fields. Different protected technologies exhibit greater net return and benefit-cost ratio (B: C) in a polyhouse compared to the open fields (Singh et al., 2011; Singh et al., 2020). Sharma et al. (2014) investigated the economic feasibility of commercial flower crops viz. carnation, rose, lilium, gerbera, and chrysanthemum in Himachal Pradesh, India. All the crops were cultivated in the low-cost protected structure provided with natural ventilation. The values of B:C for carnation, roses, chrysanthemum, lilium, and gerbera were observed to be 2.37, 2.89, 2.39, 1.89, and 2.01, respectively. Sanjeev et al. (2015) investigated the economic feasibility of cucumber under low-cost naturally ventilated polyhouses and observed B:C values of 1.4 and 0.6 for the years 2013 and 2014, respectively.

2.2 TYPES OF PROTECTED STRUCTURES

Protected structures are utilized for nursery production, crop protection, growing quality flowers, and vegetables. Keeping in view the importance and suitability of protected structures, these are discussed in detail.

2.2.1 GREENHOUSE

The structure enclosed semitransparent material like glass, polyethylene, and poly carbonate, which permits ample sunlight for crop production and has provisions for partial control of plant environment. Protected cultivation has a potential to generate food security which has become one of the greatest challenges faced by mankind (Sanchez et al., 2020). A greenhouse, depending upon the transparency of the glazing material, admits sunlight which is absorbed by the crop, equipment, structure, and the floor. These objects in turn emit thermal radiations that are partially transmitted out of the greenhouse. As a result of this, a part of the solar energy is retained in the greenhouse, leading to increase in temperature. This rise in temperature is utilized during winter months to grow crops with or without supplementary heating system. Similarly, during summers, the greenhouses are cooled and relative humidity can also be increased or decreased as per the crop requirement. In general, crops in the greenhouses will receive favorable climatic conditions and protection against insects and diseases. Cultivating off-season vegetables, ornamentals, and nursery raising are remunerative practices in enclosed structures. Among vegetables, capsicum, tomato, lettuce, cucumber, and cabbage, whereas among flower crops, carnation, gerbera, rose, chrysanthemum, poinsettia, and orchids are grown in greenhouses.

2.2.1.1 ADVANTAGES

- Round the year production of crop.
- Crop production is possible under harsh environmental conditions.
- High-quality production of export.
- Early production of quality seedlings and planting materials is possible.
- More production per unit area as compared to the open-field cultivation.
- Pest and weed management is easier in the greenhouses as compared to the open fields.
- Quality protection of plant germplasm.

- Stem hardening of plants in tissue culture.
- Opportunities for employment.
- Greenhouses can provide substantial income for cultivators having small land holdings.

2.2.1.2 CLASSIFICATION OF GREENHOUSES

There are various types of greenhouse structures used successfully for crop production (Navnath et al., 2019; Bhattacharjee, 2006). Each type has their own advantages for a particular application. Categorization of greenhouses can be done on the basis of climate control mechanism, shapes, utilities, construction materials, covering materials, and cost (technology).

2.2.1.2.1 Greenhouses Based on Climate Control Mechanisms

2.2.1.2.1.1 Naturally Ventilated Greenhouse/Open-Circuit Ventilation System

The parameters such as temperature, light, humidity, and carbon dioxide are maintained and/or regulated either naturally (e.g., ventilation/shading) or mechanically (e.g., using fans). This structure is suitable for areas having mild climate and for coastal areas having high humidity (Gupta et al., 2007).

2.2.1.2.1.2 Forced Ventilated Greenhouse/Closed-Circuit Ventilation System

In such types of greenhouses, the climatic parameters such as light, temperature, humidity, and carbon dioxide are maintained and/or controlled through forced air circulation by using fan and pad systems (for hot regions)/heaters (for temperate regions), foggers, and curtain actuators. These structures require continuous power supply and backup to function automatically. Maitra et al. (2020) observed that the gerbera cultivar Dana Ellene, Dune, and Artist perform better under forced ventilated polyhouse.

2.2.1.2.2 Greenhouses Based on the Shape/Structure

The following are the types of greenhouse structures based on their shapes:

2.2.1.2.2.1 Lean-to Type

A lean-to type design is constructed where the roof of the greenhouse is faced toward south and is placed against the south side of an existing building wall. This type of structure (width of 7–12 feet) is restricted to one- or two-row benches of plants. It is easy to construct and relatively cheap also, as it does not additionally require support to the roof. The sunlight is best utilized in this type of structure. This type of greenhouse has some disadvantages also like limited space, inadequate ventilation, and light. It is difficult to control the temperature inside the greenhouse because the building wall may collect the sun's heat, whereas the transparent cover including walls possibly loses the heat quickly.

2.2.1.2.2.2 Even-Span Type

In this the two roof slopes are equal on both sides of the greenhouse and are a standard type, full-size structure. This design is constructed on a leveled ground. This structure involves higher cost as compared to lean-to type. The design of this structure is flexible and can accommodate more number of plants. In this type of structures, higher cost may be involved in heating as they lose more heat during winter due to larger exposed area. This structure includes two walks, two side-benches, and a wide bench in the center. The width, length, and height of this type of structure are kept as 5–9 m, 24 m, and 2.5–4.3 m, respectively.

2.2.1.2.2.3 Uneven-Span Type

On hilly or sloppy terrain, this type of greenhouse is constructed. Due to unequal width of the roofs, the structure is easy going with hillside slopes. However, it is not adaptable for automation so seldom used nowadays. These are used to increase natural cooling especially in hilly or in some windy areas.

2.2.1.2.2.4 Ridge- and Furrow-Type

These are A-designed greenhouses linked with each other lengthwise at the eaves. To carry rain water and melted snow away, the eaves serve as gutter or furrow. Due to the elimination of side wall between the spans, it forms a single building of big interior. Merging of the interior space reduces labor, fuel cost, lowers the automation cost, and improves personal management. This structure can hold excessive quantity of snow. These greenhouses are efficiently used in Canadian, European, and Indian conditions.

2.2.1.2.2.5 Saw-Tooth Type

These structures are like ridge and furrow-type greenhouses but provision for natural ventilation is kept which favors specific natural ventilation flow path.

2.2.1.2.2.6 Quonset Type

The structure in built by means of pipe arcs or trusses which are receiving support from purlin running through the length of the greenhouse. The shape of this type is semiround extended to ground. In this type of greenhouses, generally polyethylene sheet is used as cladding. Such greenhouses are more economic than the gutter connected greenhouses. This type of structures can be made either in free, standing style or can be interlocked in ridge and furrow style. This interlocking-type greenhouse area is efficiently utilized as truss members overlapping sufficiently over each other. This allows growing of plants in a bed amid overlapped portions.

2.2.1.2.3 Greenhouses Based on Utility

This classification is based on functions involved such as artificial cooling and heating. However, these types of greenhouses are very expensive. A detailed description of these types is given as follows.

2.2.1.2.3.1 Greenhouses for Active Heating

In temperate region or areas where outside temperature falls below 10°C frequently, a heating system is required. In this area to provide optimum growth conditions to the plants and to avoid cold injury, some amount of heat needs to be supplied through artificial means. The rate of heat supply depends upon the rate with which the heat-loss to the outside environment takes place. Heat losses can be minimized in different ways, that is, by using a double-layer polyethylene sheet, thermo-pane glasses and by using different heating systems, such as electric heater, coal or gas heater, hot water, and solar heating system. Solar systems of heating with seasonal thermal energy storage have better performance as compared to the common heat pump heating systems (Ozturk, 2005; Xu et al. 2014).

2.2.1.2.3.2 Greenhouses for Active Cooling

In tropical regions temperature inside greenhouses goes very high and sometimes it may result in heavy loss to the growing plants. To avoid heat injury

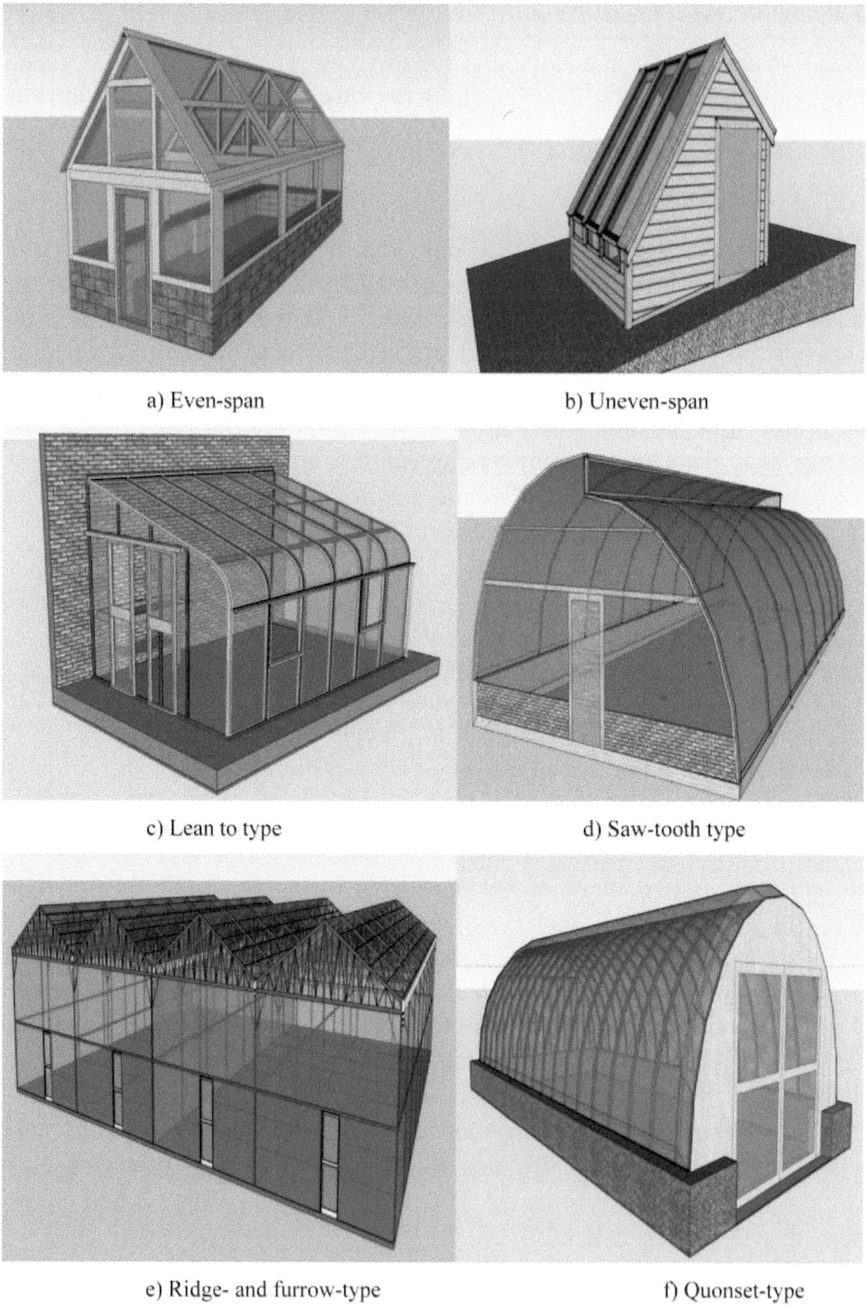

FIGURE 2.1 Greenhouse structures on the basis of their shapes.

to the plants excessive heat inside the greenhouse needs to be replaced with cool and humid environment. This can be achieved by selecting an appropriate type of structure, proper orientation, glazing and cladding materials, use of shade nets, etc. In case of small greenhouses (<6 m wide), natural ventilation can be used effectively during spring and autumn season. During summer season, fan and pad system, fog cooling, and roof shading are useful to reduce the greenhouse temperature.

2.2.1.2.4 Greenhouses Based on Construction

Type of construction and span of greenhouse highly influence the microclimate inside the greenhouse. Strength of constructing material should be kept in mind before deciding the span. Stronger constructing materials like sturdy truss-type frames are required for multi-span greenhouses as compared to single span. Simplified designs such as hoops may be utilized for greenhouse construction with smaller span. Such a greenhouse is categorized as pipe-framed, wooden-framed, and truss framed on the basis of construction material involved.

2.2.1.2.4.1 Wooden-Framed Structures

Wooden-framed structures are more suitable for greenhouse having span less than 6 m. These structures are constructed by using side posts and wood columns without using a truss. Most commonly pine wood and bamboo are utilized in these structures as they are easily available, inexpensive and possess good strength. For the construction of these types of structures, locally available timbers having enough toughness, strength, and machinability are used.

2.2.1.2.4.2 Pipe-Framed Greenhouse

In these structures, pipes made up of galvanized mild steel are utilized for manufacturing. Pipe-framed structures are generally suitable for greenhouses having a span of 12 m. The purlins, side-posts, cross ties, and columns are built by means of pipes and trusses are not used. This structure is cost effective, but it gets damaged by wind and snow (Moriyama et al., 2015).

2.2.1.2.4.3 Truss-Framed Greenhouse

Truss frames are used when we want to construct a greenhouse having large span or around 15 m span. These structures are made by using flat or tubular steel or angular iron (a truss is formed including chords, struts, and rafters). Strut and chords are used to provide support under compression and tension, respectively. Angle iron purlins are bolted on each truss to run along the length of the structure. Columns are suitable for truss-frame houses of very wide ($\geq$21.3 m). Most of the glass houses fall under the category of truss-framed type, as these are best suited for prefabrication.

FIGURE 2.2 Wooden-framed structure.

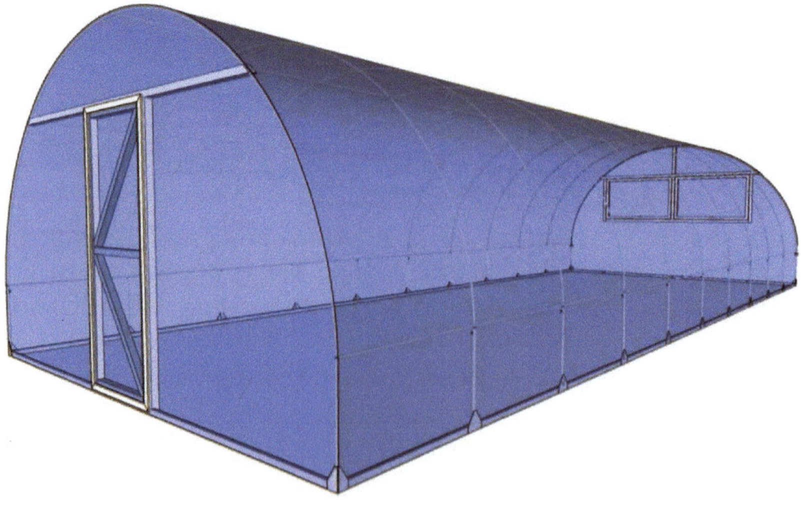

FIGURE 2.3 Pipe-framed structures.

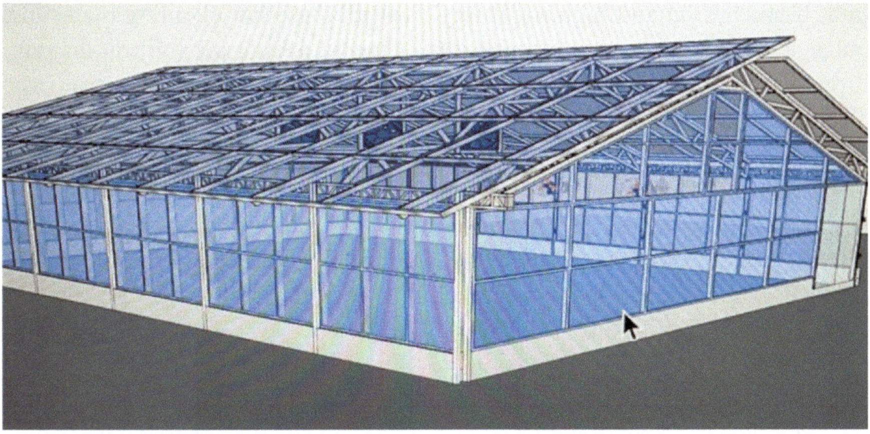

FIGURE 2.4 Truss-framed greenhouse.

2.2.1.2.5 Greenhouses Based on Use of Covering Material

Covering material directly influences the microclimate of the greenhouse by controlling the light and heat. Greenhouse simply operates on the greenhouse effect. Sunlight having short wave radiations passes through transparent or translucent covering materials like polyethylene and glass. As sunlight falls on an impervious surface like floor of greenhouse, plant surface/leaves, and other objects, it gets converted into heat and as a result temperature inside the greenhouse is increased. Covering material can transmit sunlight into the greenhouse but not heat and therefore it alters the temperature. Hence, selecting an appropriate cladding is significant while constructing a greenhouse as it influences the light transmission. Apart from light transmission, cost of construction, durability, and insulation are other important points to be considered while choosing a greenhouse cladding. Insulation is an important factor which is often overlooked and it can prove a damaging and costly mistake later. Improper insulation during winter months will increase the heating cost which will lower down the profits. On the basis of covering material used for construction greenhouses are classified as glass structure, plastic-film structure, and rigid panel structure. Different types of frames are used to fix different covering materials.

2.2.1.2.5.1 Glass Greenhouses

Until the development of plastic film in 1950, glass had been the most popular covering material for greenhouse construction. It gives up to 92%

light transmission and high durability over all the other cladding materials, but it is very expensive and breakable. Due to a high air infiltration rate, it results in lesser humidity inside the structure and improved control over disease occurrence. Due to heavy weight and massive framework, the initial cost is more, but due to its long life it is cheaper as compared to the frequently replaced polythene films and therefore it is still right choice for the places having lower sunshine levels. Even-span, lean-to type, and ridge-and furrow-type of structural designs are suitable for constructing glasshouses. In the North-European countries such as Holland, greenhouses are mostly glass-covered (Likoskoufis, 2011).

TABLE 2.1 Different Types of Cladding/Covering Materials (adapted from Slathia et al., 2018;

Material	Life (years)	Transmission ratio		Maintenance
		For light	For heat	
Fiber glass	7–10	0.90	0.05	Low
Tedlar-coated fiber-glass	15	0.90	0.05	Low
Double-strength glass	20	0.90	0.05	Low
Polycarbonate	≥10	0.80	0.05	Very low
UV-stabilized polyethylene	2–3	0.90	0.70	High
Polyethylene	≤1	0.90	0.70	Very high

Source: Reprinted from Slathia et al., 2018. Open Access

2.2.1.2.5.2 Plastic-Film Greenhouses

These greenhouses are made by using flexible plastic films like polyester, polyethylene, and polyvinyl chloride (PVC) as cladding materials. These plastic films are available in a wide range of thicknesses and transparencies. It is the least expensive cladding material. These days use of plastics films as cladding material has gained popularity, as they are more economical and heating cost is also less as compared to other covering materials. There is only one drawback in using plastic films which is its short life. The life span of UV-stabilized polyethylene film is maximum of 4 years. Quonset and gutter-connected designs are more suitable for using plastic films as claddings.

2.2.1.2.5.3 Rigid-Panel Greenhouses

In these types of greenhouses rigid panels of polyvinyl chloride (PVC), fiber glass-reinforced plastic (FRP), acrylic, and polycarbonate are used as the covering material. Quonset type or ridge and furrow-type structures are more suitable for constructing these structures. PVC has higher resistance to breaking and intensity of light remains even distributed inside the structure as compared to the glass or plastic. PVC films have lesser effect of oxidation and therefore have long life span. PVC films prevent the entry of long-wave (higher wavelength) infra-red radiation and thus minimize heat loss during night. These are high-grade panels having a long life span of about 20 years. Use of a fiberglass-reinforced plastic (FRP) rigid panel as greenhouse coverings has been very old practice, but now its popularity seems to decline day by day. Although these are considered rigid plastic, but they can be bent to form arch or quonset types of greenhouses. Poly-carbonate sheets are also one of the most widely used cladding material as it possesses good light transmission, more flexibility, rigidity, and heat retention characteristics. The initial costs of polycarbonate sheets are high, but it has a long life span of 10 to 15 years. The major drawback of these panels is that with time dust is deposited on these, which creates obstruction in the light transmission. These panels are very sensitive to fire hazard also.

2.2.1.2.6 Greenhouse Types Based on Cost Involved

Two main components that affect the cost of greenhouse are frame material and environmental control system used. The cost of a greenhouse is considerably affected by the choice of covering material, as in many cases it is fixed (i.e., poly-ethylene). Covering material required will change a bit depending upon the design dimensions and sites. Greenhouse construction cost generally varies from INR 450 to 2500. Its cost depends upon factors like the quality of material used, size, shape, transport, and structure.

2.2.1.2.6.1 Low-Cost Greenhouse

Low-cost greenhouses are the simplest structures constructed with low initial cost. It is fabricated by using low-cost wooden logs or bamboos. These structures lack heating and cooling facilities and provide basic protection to the crops plants. Such a structure is found unsuitable under harsh climatic conditions. The protection of wooden structures from insects especially from termites is a major challenge in wooden-framed structures. These structures do not last longer and they are smaller in dimensions. They are more suitable for cold climatic zones and sites of low wind velocity. The approximate expenditure involved in constructing such greenhouse units varies as Rs. 450–620 m^{-2}. In Northeast regions, these simple bamboo structures are being successfully utilized in the Northeastern states of India for commercial vegetables production. Rajender et al. (2017) had reported that a low-cost greenhouse, fabricated using casuarina wood (locally available) having coated with coal-tar as structural material and bamboos as framework costs around INR 467.6/- m^{-2}.

2.2.1.2.6.2 Medium-Cost Greenhouse

These types of structures are generally made using galvanized iron (GI) square or rectangular 0 pipes which increase the cost of construction as compared to low-cost greenhouses. These structures are suitable to low and mid hills of North India and are capable of tolerating high winds (up to 140 km/h). Height of these structures varies from 6.5 to 7 m. These structures are mostly naturally ventilated and the temperature is regulated by favoring air circulation by opening or closing of side vents. Foggers or misters can be used to maintain the humidity and shade nets can be used to check the light intensity. The cost involved in establishing such naturally ventilated polyhouse units ranges between INR 900 and 1000 m^{-2} as per the size of greenhouse.

2.2.1.2.6.3 High-Cost Greenhouse

Sometimes, medium-cost greenhouses cannot fulfill the required climatic conditions for the production of temperature sensitive, off-season, exotic crops. Therefore, greenhouses with precise control over climatic and nutritional needs of the plants are desired. In such types of greenhouses, temperature is regulated by fan and pad and other cooling/heating arrangements. A sensor, comparator, and an operator constitute such greenhouses. There is automatic control over temperature, humidity, and in these structures. The approximate cost of establishing such a greenhouse is higher than a low-cost greenhouse and it ranges between INR 1500 and 2500 m^{-2}.

2.2.2 TUNNELS

This falls under low-cost type of greenhouses. It is a temporary structure ventilated naturally (sides and ends) and involves less construction and maintenance cost. These are movable structures as they can be shifted to desired locations from the original place as per crop need. The tunnel structure raises the soil and air temperatures during winter climatic conditions and encourages the growth of plants. These tunnels vary from low (< 6 ft) to high (≥ 6 ft).

2.2.2.1 LOW TUNNEL

Also called "miniature greenhouses" and "row covers," generally cover the rows of plants in the field. These are made by stretching clean plastic films or nets over low-wire arcs or hoops up to 1.0 m high to provide protection to plants from adverse climatic conditions. The arcs/hoops are prepared of wire (steel) or cane. These hoops are covered by polythene sheets of about 50 microns' thickness and are provided with ventilation holes on the side opposite to the solar movement. The total ventilated area is kept as 4% approximately. Low tunnels provide a passive control of plant micro climate, that is, use of specific plastic material to control radiation and provision of natural ventilation. Plastic mulches and drip irrigation may also be used with low tunnels. There are several methods of low tunnel formation and operation. The low tunnels permit early and high yield for spring crops. A low tunnel is a polyethylene sheltered trench. The trench may be 20 to 40 cm deep. When the plants come in flowering to enable pollination by insects, the polyethylene sheet is removed from the trenches. Such trenches are showing excellent results under the cold desert conditions of Leh (Jammu and Kashmir, India) as moisture is also conserved to a great extent (Sanwal et al., 2004).

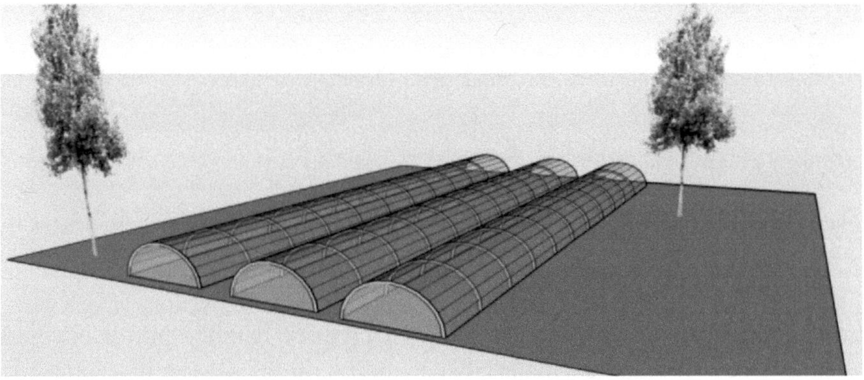

FIGURE 2.5 Low tunnel.

2.2.2.2 HIGH TUNNEL

These are similar to low tunnels in design and function, except that the height of these tunnels is more. Several rows of crop plants can be easily covered under single tunnel and this tunnel remains stabilized in one place for several growing seasons (Wells, 1996). These are passively ventilated structures constructed by using wood, metal (galvanized steel or iron), or plastic (polyvinyl chloride pipes), which is further covered with at least one layer of polyethylene film and insect-proof screens. The height is usually ranged as 10–16 feet, and it is recommended for indeterminate crops that do not grow well under the low tunnels. Its height also allows for the movement of workers and large agricultural machinery, in contrast to the low tunnels. The high tunnel has been used successfully in the production of fruits and vegetables worldwide (Carey et al., 2009). These are used for the production of melons, tomatoes, and other cash crops requiring high temperature in the Middle East and Europe (Wittwer and Castilla, 1995).

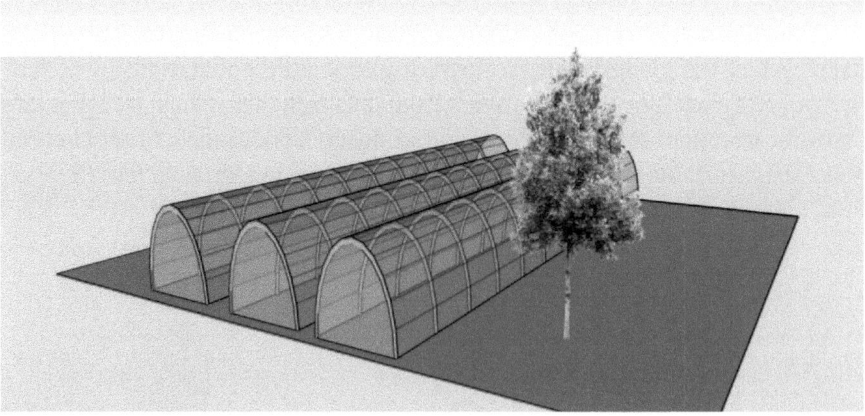

FIGURE 2.6 High tunnel.

2.2.3 WALK-IN TUNNEL

It is a temporary structure made using GI pipes or bamboo and is covered with a variety of cladding materials depending on the season in which it is planted. They are used to grow seasonal vegetables and flowering plants. They offer the advantage of better off-season product prices, offering more benefits in each area. The desirable size of the walkway is 60–75 sq. m, with a width of 2–2.5 m and a length of up to 30 m with a central height of 2–2.5 m. All in all, the height should be enough for the workers to move comfortably during field operations. Sanwal et al. (2004) noted that in cooler climates, farmers may upsurge their profits by planting early crops in this type of protected area. This type of tunnel is best suited for producing melons in the off season (Singh et al., 2005).

2.2.4 NET HOUSES

Depending on the material used for the cover, the net houses can be classified as insect proof, shade net, chromatic net, spider net, and hail netting houses.

2.2.4.1 INSECT-PROOF NET HOUSE

An insect-proof net house can be made in a variety of designs and that structure can be permanent or temporary. It can be in a walk in tunnel design and shape, with double door facility at one end of the structure. The smallest area of an insect-proof poly-net-house is kept as 100 m^2. A flat roof design with a height of 3.5–4 m and a dome-shaped design with a central height of 4.5–5.0 m are the two permanent designs. Pest control requires 25–60-mesh insect-proof screens. Nets with 40 or more mesh are sufficient to control most of the flying insects. These nets are also used for controlled pollen. Tomato planting under net houses was found to reduce the incidence of *Helicoverpa armigera* and *Aphis gossypii* which are critical pests of the tomato crop under open-field cultivation in the Punjab (Cheema et al., 2004). According to Kaur et al. (2004), brinjal cultivation under a net house increases the salable yields and reduces the pest infestations. In potato planting under a net house, virus-free seed production can be undertaken successfully (Karim et al., 2010).

2.2.4.2 SHADE-NET HOUSE

Shade nets are beneficial in providing crop protection from excessive sunlight, cold, frost, wind, and insects/birds (Takte et al., 2003; Maikhuri et al., 2007). It is designed primarily to protect plants from the sun's most powerful rays. The framework of this structure is made with stones, wooden poles (bamboo), and GI. Whenever wood or bamboo is used, then the poles should be treated with turpentine and tarred on one side before placing them on the ground/surface. The cover material used on the top and sides of the building is usually a shade net. Shade nets are available in different colors with different percentages of shading. Generally, shade nets are used for hardening of planting materials fruit orchard. All nets are UV stabilized, highly resistant to tear, and light in weight for easy and quick handling with a range of shades of 30–90% shading factor. The most widely grown vegetables under shade nets are ridge gourd, chilly, okra, broccoli, cucumber, and capsicum and among the flowers of anthurium, gerbera, rose, and ferns are grown mainly under shade nets.

2.2.4.3 TYPES OF SHADE NET

1. Tape net: These types of nets are physically weak and suitable for the sites receiving rich amount of light.
2. Monofilament Nets: These types of nets are physically strong, have small cords, and are suitable for animal protection.

2.2.5 MIST CHAMBER

The main purpose of this structure is to create high humidity and mist availability for propagating delicate soft wood cuttings of vegetable crops, and ornamentals. Cuttings are misted from time to time instead of continuous watering. A high-pressure pump, pipeline system, and a timer switch are used for intermittent water misting. Fog pipes are connected to the main pipelines for the proper misting of the plant material that grows inside the mist chambers. A fog room of area 15–25 m^2 is enough for a mist chamber. The temperature and the type of propagules affect the frequency of misting. By using these chambers, the planting materials can be produced throughout the year.

2.2.6 PLASTIC MULCHES

In agriculture, the use of plastic mulches has grown appreciably over the past few decades. This increase is due to the benefits of using them to increase the soil temperature; reducing weed growth, conserving moisture, reducing the incidence of certain pests, higher yields, and better utilization of soil nutrients. Natural mulches help to conserve soil organic matter and tilth and also provide shelter and food for soil microorganisms. However, natural resources are rarely available at reasonable prices, their quality is poor, and these are labor intensive to use. A natural mulch carries seeds of weeds and a lot delays the warming of the soil in the spring season. This condition can delay the growth and maturation of the warm-season vegetable crops. The use of straw mulches often contaminates the soil with seed of weeds and thus nitrogen in the soil depletes due to high carbon-to-nitrogen (C/N) ratio. Therefore, natural coverings could not be used effectively for year-round crop cultivation. To overcome these problems, paper and plastic have been developed and finding a widespread use in agriculture. Synthetic mulches consisting of small sheets of plastic, paper, and petroleum materials offer several benefits. Among these synthetic mulches, the polyethylene film mulch is found to be the most effective one (Kasirajan and Ngouajio, 2012).

Thermal properties of the plastic such as reflectivity, absorption, heat transfer, and exposure to sunlight significantly affect the soil temperature. In cold climate, the soil temperature increases by the use of clear plastic mulch and its use also promotes or improves the germination and emergence of many plants. Soil temperature can be adjusted to encourage faster plant growth. Different types and colors of plastic covers have ocular characteristics that alter the intensity of light rays reaching the soil, causing an increase or decrease in soil temperature. Among various colors of material, black and clear materials have a very strong soil warming potential. High soil temperatures improve nutrient uptake and availability, increase the activity and number of micro-organisms in the soil, and accelerate plant growth. The production of tomatoes, peppers, eggplant, watermelon, muskmelon, cucumber, squash, rose, carnation, gerbera, chrysanthemum, anthurium, and orchid has been increased by the use of plastic mulch with drip irrigation. Plastic covers offer a range of weed control, depending on the quantity of light being transmitted through them. By using plastic cover, it has been found that the weed growth is reduced by 64.0–98.0%. Black mulch is more effective at weed control, but generally provides less soil warmth than clear mulch. Nowadays, colored plastic sheets have been used to prevent or delay the onset of various insect-borne diseases as several insects utilize visual cues to identify the host plants. Large numbers of western flower thrips are attracted by the low bright colors of UV white, blue, and yellow. Small population of thrips have been reported on the green, balck, red, and UV-reflective substrates. The aphids also show an attraction to certain colors, such as green and yellow.

A major limitation on the commercial use of plastic mulch is their disposal after use, which leads to environmental pollution problem (Divya and Sarkar, 2019). Therefore, several photo and biodegradable plastic mulches have been developed that can eliminate the need for removal, with no negative impact on soil health and quality (Kasirajan et al., 2012). However, these perishable plastics often degrade too quickly or show incomplete degradation.

2.2.7 HOT BEDS AND COLD FRAME

These are used for the propagation of vegetables, flowers, and nursery production. Hot beds are used for growing nursery plants and a cold frame for hardening of plants in field conditions before transplanting. A hot bed is a bed where the heat is generated by the decomposition of fresh manure, hot water, smoke, and power lines. The heat generated is used for seed germination, which leads to the early production of nursery plants in addition to protecting them from the cold weather, especially during the winter. The size is 5–6 m wide and 6–12 m long. The final size will depend on planting needs, space availability, crop varieties, and number of plants. If a glass sash is to be used for covering, the length is usually multiplied by 3 feet; however, plastic coverings do not limit size. Many beds are made of wooden walls; however, permanent beds can be made of poured concrete or with building blocks.

A cold frame (of glass covering) is used to protect plants from freezing temperature and strong wind speed. These are used in winter for raising herbaceous annuals, biennials, cut flowers, and nursery of other perennial plants (Singh, 2012).

2.2.8 LATH HOUSE

Lath means a small piece of wood. It is an asset in the production of nursery before putting it in the field. It acts as an intermediate step between greenhouses, hotbeds or cold frames, and field planting. Accurate growth of other ornamental plants such as Azalea and hydrangea may require the use of lath shade during the summer months in many areas (Prasad and Kumar, 2007).

2.2.9 PASSIVE SOLAR GREENHOUSE

This type of heating is widely used in cold climates, where a heating system is required. Heating is done by solar energy without additional heating based on fuel. This type of greenhouse absorbs as much as heat from sunlight all through the daytime for heating purposes during winter climatic conditions (Dwivedi et al., 2016). Hot temperatures inside the greenhouse can be avoided by ventilation. A passive solar greenhouse is built along east-west direction with length of southern surface rising and angled project the major area in the sun.

FIGURE 2.7 Lath house.

FIGURE 2.8 Passive solar greenhouse.

2.2.10 LADAKHI GREENHOUSE

This is a small- to medium-sized polyhouses having mud walls on north, east, and west sides with a polyethylene cover facing the south side and is commonly known as the Ladakhi greenhouse. It is the most widely used passive solar heating system for crop production in the Ladakh region. It is popular because of its easy construction and high heat retention capacity at night especially in winter. Suitable alterations have been made to the Ladakhi greenhouse by various agencies and are named as GERES greenhouse, LEHO greenhouse, LREDA greenhouse, etc. However, the use of polyethylene sheet as cladding material, angled wooden roofs on the north, and mud walls on three sides are the most common features of all these greenhouses. The wooden roof is covered with a layer of straw and soil for protection.

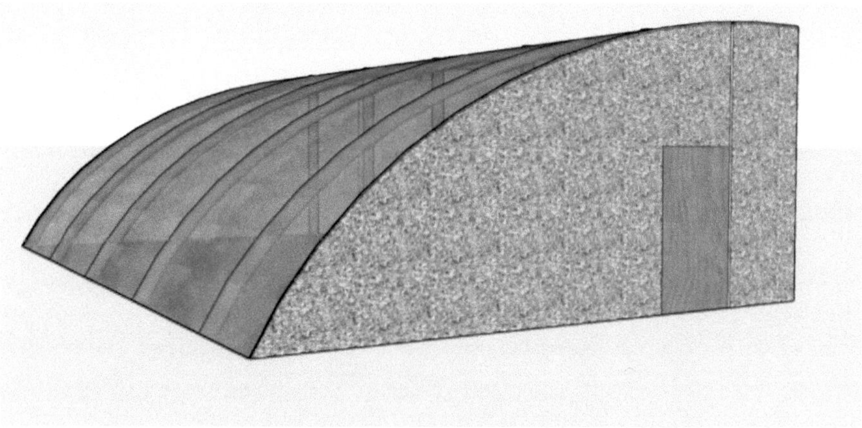

FIGURE 2.9 Ladakhi greenhouse.

2.2.11 TRENCH GREENHOUSE

A trench is a simple structure for growing vegetables during the harsh winter. Trench cultivation harnesses the soil and solar heat to create suitable climatic conditions for the growth of leafy vegetables like spinach, fenugreek, coriander, etc. This type of greenhouse is cheaper and is mostly used to grow vegetables during the extreme winter. It is a rectangular underground trench (30'× 10' × 3'; L × W × D) oriented in the north-south direction (Angmo et al., 2017). Five cylindrical GI pipes (5 cm wide, 13 feet long) are placed horizontally on the east and west at a distance of 6 m above the ground level to hold the covering material. In severe winters, a sheet of polyethylene is placed over a cladding material at night to retain heat inside the trench. The covering material is partially opened in May and October and fully removed during summer months (June–September) to control the excessive heat buildup inside the trench (Angchok and Srivastava, 2012).

2.2.11.1 POLYTRENCH GREENHOUSE

Polytrench is an improved greenhouse with measured dimensions of 30' × 10' × 3'; L × W × D in the North–South directions. It aims to overcome some of the limitations of trench greenhouse, especially its openness and the absence of a retaining wall above the ground making it vulnerable to animals. It has a thick metal tunnel framework over which the covering material is connected. A door with an entrance on the north side allows farm workers to enter the greenhouse. The cladding material is not removed even in the summer season. Natural air intrusion takes place whenever the entrance is opened and the plastic sheet is a little raised from the south end of the greenhouse. It is easier to build a greenhouse that retains a high temperature at night. The greenhouse is made of locally available materials, with the exception of the polyethylene coating. It is one of the most efficient solar storage plants for growing vegetables in winter (Singh, 2000).

2.2.12 RETRACTABLE ROOF GREENHOUSE

This type of greenhouse can create both outdoor and modified greenhouse environments by changing the position of the roof, curtains, and walls. The variety of possible roofs and side walls design allows them to be used either as polyhouse, shade houses, or wind barriers. Retractable greenhouse roof can also be used as a tool to adjust the climate and improve light availability, provided the outdoor and indoor conditions are considered. Compared to the open-field cultivation or cultivation in traditional protected structures, the retractable-roof structures can offer several benefits such as direct-solar radian and reduction in use of water, chemicals, and electricity (Berruto et al., 2008).

2.3 CONCLUSIONS

Under the current scenarios of shrinking land holdings, globalization of markets, and climate change, protected cultivation technology has developed as a significant alternative for achieving improved crop yield, quality, and returns. Protected cultivation refers to the growing of potential crops in the controlled atmosphere for ensuring optimum growth of plants without any or least stress. There are various types of protected structures that are constructed for growing high-value vegetables, flowers, preparing nurseries of different vegetable crops, ornamental and medicinal plants. Protected structures are of many types based on their shape, utility, span, construction material, climate control, and cost involved. Use of these structures varies with different locations. For example, even span greenhouses are preferred for plains, while quonset types for hills and underground trenches are used in cold deserts. While going to select the protected structures, one should have proper knowledge about different structures of greenhouses and their utility to obtain the higher economic returns.

FIGURE 2.10 Retractable roof greenhouse.

KEYWORDS

- **protected structure**
- **greenhouse**
- **tunnel**
- **net house**
- **mulching**

REFERENCES

Angchok, D.; Srivastava, R. B. Technology Intervention and Repercussion Among High Altitude Community of Ladakh: A Case Study of Trench Greenhouse. *Ind. Res. J. Ext. Edu.* **2012,** *1*, 268–271.

Angmo, S.; Angmo, P.; Dolkar, D.; Norbu, T.; Paljor, E.; Kumar, B.; Stobdan, T. All Year Round Vegetable Cultivation in Trenches in Cold Arid Trans-Himalayan Ladakh. *Def. Life Sci. J.* **2017,** *2*, 5–58.

Berruto, R.; Busato, P.; Debenedetti, A. Retractable Roof Greenhouse Systems for Sloped Areas: Comparison of Two Systems. *Acta Hort.* **2008,** *801*, 441–448.

Bhattacharjee, S. K. Greenhouse and Glasshouse for Ornamental Crops under Different Climatic Condition. In *Advances in Ornamental horticulture*; Chandra, P., Gupta, M. J., Pointer Publishers: Jaipur, Rajasthan, 2006; pp 82–94.

Carey, E. E.; Jett L.; Lamont Jr, W. J.; Nennici, T. T.; Orzolek, M. D.; Williams, K. A. Horticultural Crop Production in High Tunnels in the United States: A Snapshot. *Horttechnol.* **2009,** *19*, 37–43.

Cheema, D. S.; Kaur, P.; Kaur, S. In *Off-Season Cultivation of Tomato Under Net House Conditions*, VII International Symposium on Protected Cultivation in Mild Winter Climates: Production, Pest Management and Global Competition, 2004; vol 659, pp 177–181.

Cheema, D. S.; Kaur, P.; Kaur, S. Off-Season Cultivation of Tomato under Net House Conditions. *Acta Hort.* **2004,** *659*, 177–181.

Choudhary, A. K. Scaling-up of Protected Cultivation in Himachal Pradesh, India. *Curr. Sci.* **2016,** *111*, 272–277.

Critten, D. L.; Bailey, B. J. A Review of Greenhouse Engineering Developments During the 1990s. *Agric. For. Meteorol.* **2002,** *112*, 1–22.

Divya, V. U.; Sarkar, N. C. Plastic Mulch Pollution and Introduction of Biodegradable Plastic Mulches-A Review. *Agric. Rev.* **2019,** *40*, 314–318.

Dwivedi, S. K.; Paljor, E.; Ahmed, Z. Designing and Evaluation of a Low Cost Underground (Trench) Greenhouse Structure for Crop Production in Cold Arid Zones. *J. Rural Tech.* **2016,** *3*, 34–36.

Gupta, Y. C.; Verma, V. K.; Sharma, P. Manual on Greenhouse Cultivation of Flower Crops. Dept. Floriculture and Landscaping, Dr YS Parmar University of Horticulture and Forestry, Nauni, Solan, H.P., 2000; p 52.

Janick, J.; Paris, H. S.; Parrish, D. C. The Cucurbits of Mediterranean Antiquity: Identification of Taxa from Ancient Images and Descriptions. *Ann. Bot.* **2007,** *100*, 1441–1457.

Karim, M. R.; Hanafi, M. M.; Shahidullah, S. M.; Rahman, A. H. M. A.; Akanda, A. M.; Khair, A. Virus Free Seed Potato Production Through Sprout Cutting Technique under Net-House. *Afr. J. Biotechnol.* **2020,** *9*, 5852–5858.

Kaur, S.; Bal, S. S.; Singh, G.; Sidhu, A. S.; Dhillon, T. S. In *Management of Brinjal Shoot and Fruit Borer, Leucinodes orbonalis Guenee Through Net House Cultivation*, VII International Symposium on Protected Cultivation in Mild Winter Climates: Production, Pest Management and Global Competition, 2004; vol 659, pp 345–350.

Kasirajan, S.; Ngouajio, M. Polyethylene and Biodegradable Mulches for Agricultural Applications: A Review. *Agron. Sustain. Dev.* **2012,** *32*, 501–529.

Kumar, P.; Chauhan, R. S.; Tanwar, N.; Grover, R. K.; Kumar, R. Doubling of Farmer's Income by Vegetable Crop Production under Poly House. *Ind. J. Eco. Dev.* **2017,** *13*, 295–300.

Likoskoufis, J. H. Development of a System to Decrease Relative Humidity in Greenhouses Using GSHP and Hygroscopic Materials. PhD Thesis, Agricultural University of Athens, Greece, 2011.

Maitra, S.; Shankar, T.; Sairam, M.; Pine, S. Evaluation of Gerbera (*Gerbera jamesonii* L.) Cultivars for Growth, Yield and Flower Quality Under Protected Cultivation. *Ind. J. Natl. Sci.* **2020,** *10*, 20271–20276.

Mishra, G. P.; Singh, N.; Kumar, H.; Singh, S. B. Protected Cultivation for Food and Nutritional Security at Ladakh. *Defence Sci. J.* **2010,** *61*, 219–225.

Moriyama, H.; Sase, S.; Okushima, L.; Ishii, M. Which Design Constraints Apply to a Pipe-Framed Greenhouse?: From Perspectives of Structural Engineering, Meteorological Conditions and Wind Engineering. *Jpn. Agric. Res. Q.* **2015,** *49*, 1–10.

Murthy, D. S.; Prabhakar, B. S.; Hebbar, S. S.; Srinivas, V.; Prabhakar, M. Economic Feasibility of Vegetable Production Under Poly House: A Case Study of Capsicum and Tomato. *J. Hort. Sci.* **2009,** *4*, 148–152.

Navnath, I.; Kannujia, P.; Kale, S. J.; Panda, N.; Malathi, A. N. Offseason Vegetable Production Under Protected Cultivation. *Int. J. Eng. Res.* **2019,** *9*, 54–57.

Ozturk, H. H. Experimental Evaluation of Energy and Energy Efficiency of a Seasonal Latent Heat Storage System for Greenhouse Heating. *Energy Conv. Manag.* **2005,** *46*, 1523–1542.

Prasad, S.; Kumar, U. Greenhouse Construction: Related Aspects. In *Greenhouse Management for Horticultural Crop*, Agrobios: India, 2007; p 51.

Rajender, G.; Sushanth, K.; Mithun, K.; Devender, B; Raju, D.; Anoosha, K. Design and Development of Low Cost Greenhouse to Raise Different Cultivars. *Int. J. Agric. Sci. Res.* **2017,** *7*, 29–36.

Rawat, L. S.; Maikhuri, R. K.; Negi, V. S.; Bahuguna, A.; Rao, K. S.; Agarwal, S. K.; Sexena, K. G. Managing Natural Resources with Eco-friendly Technologies for Sustainable Rural Development: A Case of Garhwal Himalaya. *Inter. J. Sust. Dev. World Eco.* **2010,** *17*, 423–430.

Sabir, N.; Singh, B. Protected Cultivation of Vegetables in Global Arena: A Review. *Ind. J. Agric. Sci.* **2013,** *83*, 123–135.

Sanchez, J. A. A.; Munoz, J. F. V.; Felices, B. L.; Sanchez, I. M. R. An Analysis of Global Research Trends on Greenhouse Technology: Towards a Sustainable Agriculture. *Int. J. Environ. Res. Public Health* **2020,** *17*, 1–22.

Sanjeev, K.; Patel, N. B.; Saravaiya, S. N.; Desai, K. D. Economic Viability of Cucumber Cultivation under NVPH. *Afr. J. Agric. Res.* **2015,** *10*, 742–747.

Sanwal, S. K.; Patel, K. K.; Yadav, D. S. Vegetable Production under Protected Conditions in NEH Region. *Envis Bull. Himal. Ecol.* **2004,** *12*, 1–7.

Sharma, M.; Thakur, R.; Mehta, P. Economic Feasibility Analysis of Major Flower Crops in Himachal Pradesh State of India. *Int. J. Adv. Sci. Res. Manag.* **2014,** *3*, 24–40.

Singh, P. K. Hot Bed & Cold Frame Construction and Use. *Asian J. Agric. Rural Dev.* **2012,** *2*, 447–445.

Singh, B. Studies on Suitability of Various Structures for Winter Vegetable Production at Sub Zero Temperatures. *Acta Hort.* **2000,** *517*, 309–314.

Singh, M. C.; Kachwaya, D. C., Kalsi, K. Soilless Cucumber Cultivation under Protective Structures in Relation to Irrigation Coupled Fertigation Management, Economic Viability and Potential Benefits-A Review. *Int. J. Curr. Microbiol. App. Sci.* **2018b,** *7* (03), 2451–2468.

Singh, B.; Kumar, M.; Sirohi, N. P. S. *Protected Cultivation of Cucurbits under Low-Cost Protected Structures: A Sustainable Technology for Peri-urban Areas of Northern India*, III International Symposium on Cucurbits, 2005; pp 731.

Singh, A. K.; Singh, B.; Gupta, R. Performance of Sweet Pepper (*Capsicum annum*) Varieties and Economics Under Protected and Open Field Conditions in Uttarakhand. *Ind. J. Agric. Sci.* **2011,** *81*, 973–975.

Singh, M. C.; Singh, J. P.; Pandey, S. K.; Cutting, N. G.; Sharma, P.; Shrivastav, V.; Sharma, P. A Review of Three Commonly used Techniques of Controlling Greenhouse Microclimate. *Int. J. Curr. Microbiol. App. Sci.* **2018a,** *7* (1), 3491–3505.

Singh, M. C.; Singh, J. P.; Pandey, S. K.; Mahay, D.; Srivastava, V. Factors Affecting the Performance of Greenhouse Cucumber Cultivation-A Review. *Int. J. Curr. Microbiol. App. Sci.* **2017a,** *6* (10), 2304–2323.

Singh, M. C.; Singh, J. P.; Singh, K. G. Optimal Operating Microclimatic Conditions for Drip Fertigated Cucumbers in Soilless Media under a Naturally Ventilated Greenhouse. *Indian J. Ecol.* **2017b,** *44,* 821–826.

Singh, M. C.; Singh, J. P.; Singh, K. G. Mathematical Modeling of Greenhouse Microclimate under Vertically Trained Soilless Cropped Conditions. *Agric. Res.* **2021a**, *10*

Singh, M. C.; Singh, K. G.; Singh, J. P. Interactive Effects of Fertigation and Varieties on Pant Growth Attributes and Yield of Soilless Cucumbers. *Int. J. Sci. Technol. Soc.* **2020a,** *8* (3), 60–65.

Singh, M. C.; Singh, J. P.; Singh, K. G.; Gupta, O. P.; Kumar, G. Computation of Vapour Pressure Deficit and Crop Transpiration via Development of a Web Based Computer Module. *J. Agrometeorol.* **2021b,** *23* (4), 243–248.

Singh, M. C.; Singh, A.; Singh, J. P.; Singh, K. G. Economic Viability of Soilless Cucumber Cultivation under Naturally Ventilated Greenhouse Conditions. *Ind. J. Hort.* **2020b,** *77* (2), 315–321.

Slathia, D.; Nisa, M. U.; Reshi, M.; Dolkar, T.; Hussain, S. Protected Cultivation of Ornamentals. *Global J. Bio. Sci. Bitotechnol.* **2018,** *7,* 1–10.

Takte, R. L.; Ambad, S. N.; Kadam, U. S.; Dhawale, B. C. In *Greenhouse Cladding Material Shade Nets Ventilation etc*, Proceedings of All India on Seminar Potential and Prospects for Protective Cultivation, Organized by the Institute of Engineers; Ahmednagar, 2003; pp 117–119.

Wells, O. S. Row Cover and High Turnnel Growing Systems in the United States. *Horttechnol.* **1996,** *20,* 822–826.

Wittwer, S. H.; Castilla, N. Protected Cultivation of Horticultural Crops Worldwide. *HortTechnol.* **1995,** *5,* 6–23.

Xu J.; Li Y.; Wang R. Z.; Liu W. Performance Investigation of a Solar Heating System with Underground Seasonal Energy Storage for Greenhouse Application. *Energy* **2014,** *67,* 63–73.

CHAPTER 3

Design and Construction of Greenhouse Structures

SAKHARAM KALE[1], NAVNATH INDORE[2], PRERNA NATH[1], PANKAJ KANNAUJIA[1], and R. K. SINGH[2]

[1]ICAR-Central Institute of Postharvest Engineering and Technology, Abohar, Punjab

[2]ICAR-Central Institute of Postharvest Engineering and Technology, Ludhiana, Punjab

ABSTRACT

Greenhouse technology has several advantages in crop production systems due to which it is being widely adopted all over the world. Greenhouse structures are categorized according to shape, utility, cost, material, roof, etc. Selection of the appropriate type of greenhouse is critical to make greenhouse production a gainful business. Simultaneously, the scientific design of a greenhouse structure is very crucial to prevent the structure from damage, to provide desired microclimate, to reduce the cost of the structure, and to increase the profitability of the greenhouse production system. Greenhouse performance is primarily affected by its design, while greenhouse design is dependent on climate at the site, latitude, availability of construction materials, prevailing loading conditions, etc. Literature reports that several greenhouse location-specific designs have been developed for different climatic zones. Knowledge of the design procedure of a location-specific greenhouse is indispensable to determine precise specifications of framing materials under prevailing loading conditions. Generally, under-designed structures fail under loads, whereas over-designed structures require a higher

Protected Cultivation: Structural Design, Crop Management, Modeling and Automation.
Mahesh Chand Singh and K. K. Sharma (Eds.)

cost. On the contrary, optimal sizes and weights of the structural members reduce the risk of structural damage in addition to the cost of the structure. Consequently, an effort has been made in the present chapter to illustrate the methodology of greenhouse design with the help of examples. Methods for calculation of wind load, dead load, live load, and load combinations are included, and the determination of the precise sizes of the structural members has been illustrated through examples.

3.1 GREENHOUSE FOR TROPICAL REGIONS

Vegetable farming in tropical regions provides a major source of earnings for small landholders. It also provides fresh vegetables to a hefty segment of the population. However, open-field farming fetches more than a few problems in hot climates of tropical regisons, causing reduced yield and quality of the produce. Under this circumstance, greenhouse technology provides an opportunity to obviate the vagaries of nature and produce more crops from smaller landholdings. Greenhouse technology has the competence to boost the crop yield and bring more profits to the growers (Singh et al., 2019a, 2019b; 2020a, 2020b). In tropical areas, greenhouse technology is one of the eventual preferences due to the limited availability of cultivable land and water.

The cultivation of crops in greenhouse structures is fundamentally a temperate phenomenon (Herry and Yayu, 2007). The greenhouse has a natural tendency to raise the inside temperature, which rationalizes its applicability in temperate cold regions. Yet, a greenhouse is not a new technology in tropical regions and is being used for vegetable farming in several tropical countries, including India. Literature reveals that lots of research works have been conducted to make suitable greenhouse technology for tropical regions (Baudoin and Zabeltitz, 2002; Connellan, 2002; Kumar et al., 2009; Kumar et al., 2010; Emekli et al., 2010; McCartney and Lefsrud, 2018; Villagran et al., 2020). The key designs of temperate greenhouses include glasshouse/glass greenhouse, Quonset design, and tunnel-type structures. These designs are not suitable in tropical regions as they cause a rise in inside temperature beyond the crop requirement. Hence, tropical greenhouses are required to be designed with temperature control provisions such as shading nets (Mohammad et al., 2014), natural ventilation, force ventilation, fan-pad system, fogging system, and misting to maintain the desired temperature and relative humidity inside the greenhouse (Herry and Yayu, 2007).

On the whole, greenhouses are the frames of inflated structures made of galvanized iron (GI) or mild steel (MS) pipes enclosed with a transparent material inside which the crops are grown in a controlled or partially-controlled atmosphere. In tropical areas, greenhouses are designed such that they offer better control over temperature, relative humidity, solar radiation, and carbon dioxide inside the greenhouse. Although greenhouse technology has plentiful advantages, the major reasons for its less adoption in developing countries are a lack of awareness of the technology, lack of knowledge of the technology, and unavailability of suitable designs (Singh et al., 2018a, 2018b, 2018c). Presently, in India, diverse greenhouse designs are being supplied by different manufacturers. Materials, their specifications, shape, etc., may differ supplier-wise. Location-specific, region-wise, standard designs are yet to be made available. It may be stated, in the case of India, that a single greenhouse design cannot be embraced all over the country due to spatial climatic variability, topography, and location (Giacomelli et al., 2008; Indore et al., 2020). One design may be successful in one region but may fail in others. Hence, there is a need to generate knowledge on the design of location-specific greenhouses. Greenhouse technology itself is a wide domain, and it involves different subdomains under it. However, greenhouse design is the single most key aspect that decides its performance in maintaining the desired microclimate inside the structure (Indore et al., 2020). Therefore, an attempt has been made in the present chapter to discuss the design and construction of greenhouse structures for tropical regions.

3.2 TROPICAL REGIONS

Global climate is divided into six major groups, namely, tropical humid, dry, subtropical, temperate, boreal, and polar. Tropical humid zones have a temperature of the coolest month >18°C and 10–12 wet months, whereas dry zones include desert and arid regions having 8–10 months with average temperature >10°C. Therefore, it is indispensable to be acquainted with the climatic conditions of tropical regions before designing a greenhouse in such regions. Factors that influence the local climate in tropical regions include temperature, precipitation, solar radiation, and evapotranspiration.

Tropical regions are further subdivided into six regions on the basis of climatic conditions, primarily rainfall. These regions are (1) Tropical wet climates (Amazon River basin, equatorial West Africa, peninsular area of Southeast Asia); (2) Tropical humid and dry climates (North and South of the equator between 5° and 20° latitudes in all the continents); (3) Subtropical humid climates (coastal regions of East China, Uruguay, south Japan, Southeast Australia, and southeastern regions of USA); (4) Dry climates (arid and semi-arid regions); (5) Tropical highlands (Colombia, Kenya); and (6) Subtropical dry summer/Mediterranean (lands near Mediterranean sea, the southern part of South Africa, etc.

3.3 CLASSIFICATION OF GREENHOUSES

Greenhouses can be classified as per their shapes, cooling methods, span, utility, construction materials, etc. (Dalai et al., 2020). Information on the most common types of greenhouses has been summarized in Table 3.1.

3.3.1 GREENHOUSE DESIGNS SUITABLE FOR TROPICAL REGIONS

Tropical greenhouse structures may be grouped into the following two major groups on the basis of varying climatic conditions in different types of tropical climates:

1. Greenhouse with permanently opened ventilators and large volume for better ventilation efficiency. This design is suitable for (1) wet and (2) tropical wet and dry climates.
2. Greenhouse with closable ventilators, mainly suitable for subtropical humid climates in tropical highlands and dry areas if the temperature drops too low at night.

The basic greenhouse designs and shapes suitable for tropical regions have been illustrated by Zabeltitz and Baudoin (2005). By and large, the saddle-roof design is suitable for regions where ventilators are to be closed at night due to low temperatures. This design has side ventilators but no top ventilator. Other designs, such as single-span, multi-span, and sawtooth, have ventilators at the ridge besides two side ventilators. The sawtooth design has good ventilation efficiency due to the ridge ventilator. Two or more greenhouses can also be connected to obtain a larger interior space. The plastic film is stretched over the gaps and functions as a gutter. In another connected greenhouse design, the plastic film is stretched from column to column, and it is stretched in the middle by a plastic pipe filled with water. In all the designs, the roof is extended beyond the sidewall and ventilation at the ridge. This provision prevents the direct entry of rainwater into the greenhouse. All these designs come under the category of a naturally ventilated greenhouse.

TABLE 3.1 Major Types of Greenhouse Structures on the Basis of Different Criteria.

Criteria	Type	Characteristic
Shape	Lean to type	Constructed against one or more sides of an existing building.
	Even span type	It is a standard type and full-size structure. The slopes of the two roofs are equal.
	Uneven span type	Constructed on hilly/sloppy terrain. Roofs are of unequal width and pitch.
	Ridge and furrow types	It uses two or more A-frame structures connected to one another along the eave.
	Sawtooth type	Similar to ridge and furrow except that at natural ventilation provision at the top.
	Quonset	Made of pipe arches or trusses. Arches are supported by a pipe running along the length of greenhouse.
Utility	Greenhouses for active cooling	Used where the cooling arrangement is required. Modifications are made in the greenhouse so that cooled air can be drawn into the greenhouse.
	Greenhouses for active heating	Some amount of heat is supplied to avoid the cold bite on plants in cold climates. Artificial heating of greenhouse is achieved by means of heating systems such as heaters, radiant heat, solar heating systems, latent heat storage, etc.
Construction	Wooden framed structure	Wooden frames are used when a clear span < 6 m.
	Pipe framed structure	Pipe frames are used when the clear span is around 12 m.
	Truss framed structure	Truss frames are used when the clear span of the greenhouse is ≥ 15 m. Flat steel, steel pipe, or iron angles are welded together to form a truss.
Covering material	Glass	Glass offers greater interior light intensity inside the greenhouse. It offers a higher air infiltration rate, thereby lowering interior humidity and better disease prevention.
	Flexible plastic films	Polyethylene, polyester, and polyvinyl chloride (PVC) films are used as covering materials. They are cheap, and the cost of heating is less compared to glass structures.
	Rigid panel	PVC rigid panels, fiberglass-reinforced plastic (FRP), acrylic, and polycarbonate rigid panels are used as covering material in the ridge and furrow type or Quonset type structures. Rigid panels are more resistant to breakage, rupture, and tearing, and offer uniform light intensity.

TABLE 3.1 *(Continued)*

Criteria	Type	Characteristic
Cost of construction	High-cost greenhouse	The height of the sidewall is at least 4 m, and the height of the roof height at the center is up to 8 m from ground level. They offer excellent environmental performance and better crop production.
	Medium cost greenhouse	They are characterized by vertical walls. The height of the sidewall is more than 2 m but less than 4 m. The roof height at the center is less than 5.5 m.
	Low-cost greenhouse	The center height of these greenhouses is less than 3 m. The most common type low-cost greenhouses include tunnel houses.

3.4 SITE SELECTION CRITERIA

The selection of an appropriate site is very important before designing a greenhouse structure. Selecting an upright location for greenhouse construction is very critical as it decides the performance of the greenhouse. Several criteria should be considered prior to the design and construction of a greenhouse. Table 3.2 illustrates the desirable characteristics of a good site for greenhouse construction.

3.5 GREENHOUSE COMPONENTS

The greenhouse structure is made of assembling a variety of parts and components. Each component has a definite role in the structure. Loads on the component, number, section, and size of each component need to be designed during the process of greenhouse design. Figure 3.1 describes some of the important components of a free-standing greenhouse structure.

3.5.1 STRUCTURAL COMPONENTS

- **Foundation:** The main rationale behind the provision of a foundation is to prevent the uprooting and overturning of the structure. Lateral and vertical pressures generated attributable to wind and other loads are transferred to the foundation. The foundation is generally made of concrete. 2500 PSI or a greater mix of concrete is selected for the foundation. The concrete foundation can be 30 cm × 30 cm square cross-section or 30 cm diameter and about 100 cm deep. Foundation pipes are inserted in the concrete foundation to support the vertical columns.
- **Sidewall:** Poles (side posts) of the sidewall are set in the concrete foundation. The sidewall provides support to the trusses placed at specific intervals along the length of the structure. It takes the weight of the structure. Side posts are generally spaced 1–1.5 m apart.
- **Curtain wall:** It is the portion of the sidewall just above the ground level up to a certain height. It is made of UV-stabilized plastic sheets of higher strength, or some poured concrete, concrete blocks, bricks, or treated lumber, etc.
- **Sill:** The top of the curtain wall is termed a sill.
- **Eave:** It is the line that joins the top of the sidewalls with the lower edge of the roof.

TABLE 3.2 Desirable Characteristics of a Good Site for Greenhouse Construction.

S. No.	Criteria	Desirable/undesirable characteristic
1	Elevation	It influences summer maximum and winter minimum temperatures. Apposite elevation minimizes the cost of heating and cooling of the greenhouse.
2	Water	Good quality water is necessary for irrigation and cooling of the structure through the evaporative cooling system.
3	Solar radiation	A site with year-round high solar radiations and light intensity is desired.
4	Microclimate	
	1) Latitude	Latitude affects the heating and cooling requirements of the greenhouse.
	2) Large water bodies	Large water bodies influence ambient and greenhouse temperatures. Hence, the site should be away from large water bodies.
	3) Trees and mountains	The site should be away from trees and mountains. Threes may cast shadows on the greenhouse. Mountains may affect wind and/or storm patterns.
	4) Clouds and fog	Certain areas such as the leeward side of the mountain, have clouds or fog during certain times of the day, which may reduce light intensity and solar radiation.
	5) High wind areas	High winds affect greenhouse temperature adversely and can cause damage to the structure.
	6) Dust and sand	High winds can blow up dust or sand, especially in arid and desert regions, which can spoil greenhouse glazing.
5	Pest pressure	The site should be away from such existing agricultural areas, which may harbor insect pests in fields.
6	Level and stable ground	The sound on which the greenhouse is to be erected should be well graded to drain the surface water if required.
7	Utilities	Readily availability of utilities such as water, electricity, telephone services, fuel, etc., is desired.
8	Roads	Good access to roads is a desirable thing for speedy marketing.
9	Greenhouse orientation	1) Free-standing greenhouse:Eeast–West orientation.
		2) Free-standing greenhouse should have long axis perpendicular to the wind direction.
		3) Multi-span greenhouses: Gutter should be oriented North–South.
		4) Naturally ventilated greenhouse: Ventilator should open on the leeward side.
10	Capability of expansion	The site should be such that future expansion is possible with minimum cost.
11	Availability of labor	Availability of skilled labor at/near the site is desired.
12	Management residence	The manager/owner's house should be as close to the greenhouse as possible for proper monitoring and control.

- **Truss:** It is the combination of structural members, namely, struts, rafters, and chords. It takes the weight of the greenhouse roof and relocates it to the vertical poles/columns.
- **Purlins:** They run along the length of the greenhouse and seize the roof trusses aligned.
- **Ridge:** It is the line where two halves of the roof come together at the top of the greenhouse. Naturally ventilated greenhouses are provided with ventilation at the ridge.
- **Side posts and columns:** They are vertical supports that transfer the structural load to the foundation. The height of the vertical columns decides the overall height of the structure.
- **Sash bars:** They are perpendicular to the purlins and grip the glazing material in place.

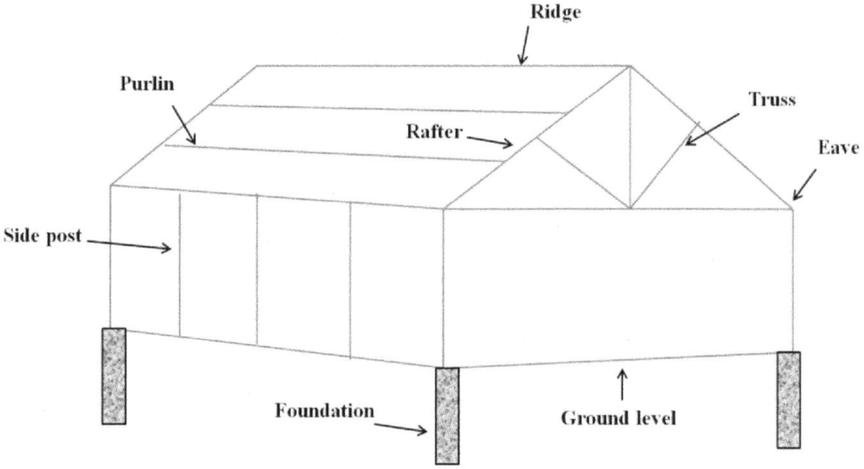

FIGURE 3.1 Components of free-standing greenhouse structure.

3.5.2 *FRAMING MATERIALS*

Criteria for choosing the framing materials for greenhouse construction are mentioned below (Zabeltitz and Baudoin, 2005):

- Should be a domestic material.
- Should be easily available in local markets.
- The cheapest material should be chosen as far as possible.

Structural components are made of framing materials. Information on structural materials to be used in the construction of a greenhouse is very imperative while designing a greenhouse structure. The number and size of the columns, purlins, and other members vary according to the materials to be used in the greenhouse. Material having higher strength (e.g., steel) is required in lesser quantity as it is able to bear heavy loads easily, whereas material having lesser strength (e.g., bamboo) is required in larger quantity to sustain under the loads. Aluminum is widely used in greenhouse construction as a framing material. It possesses good malleability and ductility and can be shaped as per the requirements. Aluminum frames allow light reflectance also. They are lighter and possess the longest life span. Besides aluminum, Galvanized Iron (GI) steel is commonly used in the greenhouse framing. In India, most greenhouse structures are made of GI steel. Some of the low-cost greenhouses are made of bamboo or other woods (Rajender et al., 2017), but bamboo and woods deteriorate with time, and hence their life span is very short as compared to GI or aluminum.

3.5.3 COVERINGS

Most common covering materials include glass, UV-stabilized flexible plastic films, and rigid panels. Glass is mainly used in temperate greenhouses. It allows excellent light intensity and solar radiation as compared to other materials. However, it breaks easily due to which it needs to be replaced more frequently. Moreover, the weight of the glass is higher, which requires heavy supporting structural members accounting for more cost. The most commonly used covering material in tropical greenhouses is UV-stabilized flexible films. In India, most of the greenhouse structures are covered with UV-stabilized poly sheets. These films are cheaper than glass and rigid panels. The overall cost of the structure is also lowered due to the lightweight of the films. However, the light transmission of such films is not as good as of glass, and they allow lesser light intensity in the greenhouse. Moreover, they get torn easily under storms. Rigid panels include polycarbonate sheets, fiberglass, and acrylic sheets. They are rigid, long-lasting, and obtainable in several light transmission levels. Rigid panels can be placed between glass and flexible films when compared based on light transmission, durability, self-weight, and cost. In general, glass and rigid panels are used when the structure is designed for a longer life span of nearly 20 years or more, whereas flexible films are used in temporary greenhouses.

3.6 CRITERIA FOR DESIGN OF COMMONLY ADOPTED GREENHOUSE STRUCTURES

The design of a particular greenhouse is affected by the climate at the site, latitude, availability of construction materials, etc. The following four criteria can be considered for the design of greenhouse structures in tropical regions:

i. Maximization of light intensity in the greenhouse.
ii. Minimization of structural members to avoid self-shadows.
iii. Ensuring better insulation to decrease heat gain/losses.
iv. Affordable and reasonable costs of the greenhouse.

Zabeltitz and Baudoin (2005) also enlisted the design criteria for plastic-film greenhouse as mentioned in the following:

i. Tearing of plastic film due to fluttering in the wind must be avoided. Films should be stretched tightly by using simple stretching devices.
ii. Simple methods for changing film.
iii. Avoidance of drop fall of water condensed at the covering material.
iv. Insulation of parts that are heated by solar radiation. Steel members and parts which are in direct contact with the film should be insulated. An insulating white plastic strip may be placed between steel parts and film.
v. Efficient ventilation must be provided to the structure. Under high-temperature conditions, ventilators should be as large as possible. If insect-proof netting/disease netting is necessary, ventilators should be larger to achieve the same ventilation efficiency.
vi. In climates where the temperature drops noticeably at night, ventilators must be closable.
vii. Reduction of too high inside relative humidity by preventing excessive evaporation from the soil.
viii. Greenhouse volume should be as large as possible.
ix. Provision of the gutters to remove rainwater safely, and its collection for irrigation purposes.
x. Structure should be framed without too many screws, nails, and bolts.
xi. Design should be such that growers should be able to construct structures themselves using domestic materials.
xii. The structure should be wind resistant.
xiii. Width of plastic film available in countries limits the width of individual spans in multi-span greenhouses.

3.7 DESIGN OF A GREENHOUSE STRUCTURE

Several studies have been carried out worldwide to design greenhouse structures for diverse regions of the world (Kendirli, 2006; Saribiyik and Saribiyik, 2011; Ubolsook and Thepa, 2011; Yun et al., 2014). Many such attempts have been made in India also (Nayak and Rao, 2014; Indore et al., 2020). Most of the Indian land falls in tropical regions. Hence, the design procedure deliberated in the present chapter has been embraced as prescribed by Indian standards. Examples of the design of greenhouse structures at selected sites in India have also been discussed here to increase the understanding of greenhouse design. These examples have been adopted from our previous studies by Kale (2014) and Indore et al. (2020). It must be noted that the greenhouse structure is the whole structure counting structural support and enclosure elements. The greenhouse structural system can be broadly categorized as follows:

1. Primary roof system: Rigid frame, truss, arch, or similar system.
2. Secondary structural system: Enclosure support components such as purlins, glazing bars, ridge beams, and gutters, and bracing elements.
3. Columns: End walls, columns, enclosing walls, or other elements.
4. Foundations: It may be a spread footing, continuous concrete footing, or pier-type foundation placed directly in the earth. In greenhouse design, pier-type footings are generally designed as unrestrained foundations.
5. Cladding: It is of many materials such as glass, polycarbonates, fiberglass, or polyethylene.

3.7.1 DESIGN LOADS

The greenhouse is designed in view of the prevailing loading circumstances on the structures. Major loads on the greenhouse include dead load, imposed loads, installations, wind load, snow, and seismic load. However, in tropical greenhouses, dead load, live load, and wind load are the major contributors to structural damage, whereas snow load and seismic load are negligible.

In India, the design loads are determined as per IS 875-1 (1987), IS 875-2 (1987), IS 875-3 (1987), IS 875-4 (1987 and 2003), and IS 1446 (1997).

1. Dead load: Self-weight of structure which includes all fixed services equipment such as ventilating, heating, electrical, air circulation, lighting, watering, and energy conservation covers. Long-duration

crops, for instance, tomatoes and cucumbers, sustained by the structure, may also be measured as dead loads. For the estimation of dead load, the actual mass of building materials and components should be taken into account.

2. Live loads: Live loads are temporary loads. They should include the mass of hanging plants and repair persons.
3. Snow load: Snow load may not be considered while designing a greenhouse for tropical regions. However, to be on the safer side, some manufacturers design a greenhouse to resist the snow load also. In areas where snowfall occurs, a minimum distance of 3 m should be provided between two adjacent greenhouses to allow snow accumulation and to avoid side wall crushing due to snow sliding off the roof. The snow load of greenhouse structures can be estimated as per IS 875 (Part 4).
4. Wind load: The wind load of greenhouse structures should be estimated as per IS 875 (Part 3). Since deadweight of the most greenhouses structures is minor, unusual consideration should be given to make sure that enough foundation support is provided to resist the upward lift force formed by the wind.

3.7.1.1 DEAD LOAD ON THE STRUCTURE

The dead load can be calculated by multiplying the unit weight of the structural material to be used with the volume of material. Another way is to use the standard codes. If the calculation of the actual dead load is not easy, the dead load may be considered as 0.35 kN/m^2 as specified in IS 875 (Part 1) (1987).

3.7.1.2 LIVE LOAD ON THE STRUCTURE

Live load on the structure, including crop load, can be considered as 0.45 kN/m^2 as prescribed in IS: 14462 (1997) (Kale, 2014; Indore et al., 2020).

3.7.1.3 WIND LOAD ON THE STRUCTURE

Wind load is the chief force accountable for the performance of greenhouse structures in tropical regions. Hence, the detailed procedure of wind load and stress caused due to collective loading is explained in the following section. Wind load on a greenhouse structure can be determined as per the procedure prescribed in IS 875 (Part 3) (1987).

Basic wind speed, V_b (m/s), for a specific location in India can be obtained from Table 3.3.

TABLE 3.3 Basic Wind Speed at Different Locations in India (IS: 875 (part 3) - 1987)

Place	Basic wind speed (m/s)	Place	Basic wind speed (m/s)	Place	Basic wind speed (m/s)
Agra	47	Dehradun	47	Nagpur	44
Ahmadabad	39	Delhi	47	Nainital	47
Ajmer	47	Durgapur	47	Nasik	39
Almora	47	Gangtok	47	Nellore	50
Amritsar	47	Guwahati	50	Panjim	39
Asansol	47	Gaya	39	Patiala	47
Aurangabad	39	Gorakhpur	47	Patna	47
Bahraich	47	Hyderabad	44	Pondicherry	50
Bengaluru	33	Imphal	47	Port Blair	44
Barauni	47	Jabalpur	47	Pune	39
Bareilly	47	Jaipur	47	Raipur	39
Bhatinda	47	Jamshedpur	47	Rajkot	39
Bhilai	39	Jhansi	47	Ranchi	39
Bhopal	39	Jodhpur	47	Roorkee	39
Bhubaneswar	50	Kanpur	47	Rourkela	39
Bhuj	50	Kohima	44	Shimla	39
Bikaner	47	Kurnool	39	Srinagar	39
Bokaro	47	Lakshadweep	39	Surat	44
Mumbai	44	Lucknow	47	Tiruchirappalli	47
Calcutta	50	Ludhiana	47	Trivandrum	39
Calicut	39	Chennai	50	Udaipur	47
Chandigarh	47	Madurai	39	Vadodara	44
Coimbatore	39	Mandi	39	Varanasi	47
Cuttack	50	Mangaluru	39	Vijayawada	50
Darbhanga	55	Moradabad	47	Vishakhapatnam	50
Darjeeling	47	Mysore	33		

Source: Adapted from IS: 875 (part 3) - 1987

Design wind speed, V_z (m/s) is determined using eq 3.1.

$$V_z = k_1 k_2 k_3 V_b \tag{3.1}$$

where k_1 is a probability factor or risk coefficient obtained from IS 875 (Part 3) (1987); k_2 is terrain, height, and structure size factor obtained from IS 875 (Part 3) (1987); k_3 is a topography factor obtained from IS 875 (Part 3) (1987).

Design wind pressure, P_z (N/m^2) is determined using eq 3.2.

$$P_z = 0.6 \times V_z^2 \tag{3.2}$$

where P_z is the design wind pressure in N/m^2 at height "z," and V_z is the design wind velocity in m/s at height "z."

In our study (Kale, 2014), six different Indian sites representing six different wind zones of the country were selected, and wind loads on the structures of different sizes were calculated. Table 3.4 indicates the design wind speed and design wind pressure calculated at selected sites. It is apparent from Table 3.4 that the increase in basic wind speed was accompanied by an increase in design wind pressure.

TABLE 3.4 Design Wind Pressure at Representative Sites of Six Wind Zones of India.

Site	Basic wind speed (m/s)	Design wind speed (m/s)		Design wind pressure (N/m²)	
		Quonset and sawtooth GH of 2000 m² area	Quonset and sawtooth GH of 4000 m² area	Quonset and sawtooth GH of 2000 m² area	Quonset and sawtooth GH of 4000 m² area
Bengaluru	33	30.40	28.85	554.48	499.35
Ahmadabad	39	35.16	33.37	741.84	668.07
Mumbai	44	39.24	37.24	923.83	831.97
Delhi	47	41.45	39.34	1031.06	928.53
Kolkata	50	44.1	41.85	1166.89	1050.85
Darbhanga	55	47.97	45.42	1380.73	1243.43

Indore et al. (2020) have also attempted to determine the design wind speed and design wind pressure at nine different sites representing different climatic conditions in India. Table 3.5 indicates the design wind speed and design wind pressure determined for selected sites by Indore et al. (2020).

3.7.1.3.1 Wind Load on Individual Component

When calculating the wind load on single structural components such as roofs and walls, and individual cladding units and their fittings, it is essential to take account of the pressure difference among opposite faces of such components or units. For clad structures, it is thus required to know the internal pressure in addition to the external pressure. The wind load "F" acting in a way normal to the single structural element or cladding unit is given by eq 3.3:

$$F = (C_{pe} - C_{pi}) \times A \times P_z \quad (3.3)$$

where C_{pe} is an external pressure coefficient [(IS 875 (Part 3) (1987)], C_{pi} is the internal pressure coefficient, A is the surface area of the structural or cladding unit, and P_z is the design wind pressure.

TABLE 3.5 Design Wind Pressure at Nine Different of India.

Region	Site	Climatic zone	Basic wind speed (m/s)	Design wind speed (m/s)	Design wind pressure (N/m²)	Polyhouse design
Kashmir	Srinagar	Mountain	39	35.88	772.42	Quonset
Uttarakhand	Almora		47	43.44	1121.81	Gothic
Sikkim	Gangtok		47	57.52	1985.69	Gable frame
Meghalaya	Barapani	Humid subtropical	50	61.20	2247.26	
Punjab	Ludhiana		47	42.30	1073.57	Double arch/ sawtooth
Rajasthan	Udaipur	Arid	47	43.56	1138.95	
Gujarat	Junagarh	Semi-arid	50	45.00	1215.00	Arc/ semi-arc
Karnataka	Raichur		39	35.88	772.42	
Jharkhand	Ranchi	Humid subtropical	39	36.96	819.46	

Internal air pressure in the structure depends on the degree of permeability of cladding to the airflow. The internal air pressure may be positive or negative conditional to the direction of flow of the air relative to openings in the structure. Internal air pressure coefficient (C_{pi}) differs according to the type of greenhouse structure, as shown in Table 3.6.

i. For low permeability (<5%), C_{pi} is ±0.2,
ii. For medium permeability (5–20%), C_{pi} is ±0.5, and
iii. For high permeability (>20%), C_{pi} is ±0.7.

Tables 3.7 and 3.8 depict the wind loads on the wall and roof of sawtooth shape and Quonset shape structures, respectively. It is apparent from Table 3.7 that sawtooth structure had two types of wind loads, one is windward and another is leeward. The windward load was found to be higher than the leeward load. The sawtooth shape roof had two slopes which credited the windward and leeward loads whereas the Quonset shape roof had only one load as it is an arch (Kale, 2014).

TABLE 3.6 Internal Air Pressure Co-efficient of Different Types of Greenhouses (Adapted from Indore et al., 2020).

Greenhouse type	Opening area (%)	Internal air pressure coefficient
Arc/Quonset	<5	±2
Gothic	<5	±2
Gable type freestanding	5–20	±5
Double arc	5–20	±5
Multi-span sawtooth	> 20	±7

Source: Adapted from Indore et al., 2020

TABLE 3.7 Wind Load on Wall and Roof of Sawtooth Shape Greenhouse (250 m^2).

Site	Maximum wind load on wall (kN/m^2)		Maximum wind load on roof (kN/m^2)	
	Windward	Leeward	Windward	Leeward
Bengaluru	0.78	−0.67	−0.79	−0.72
Ahmadabad	1.04	−0.89	−1.06	−0.96
Mumbai	1.29	−1.11	−1.32	−1.20
Delhi	1.44	−1.24	−1.47	−1.34
Kolkata	1.63	−1.40	−1.67	−1.52
Darbhanga	1.93	−1.66	−1.97	−1.79

TABLE 3.8 Wind Load on Wall and Roof of Quonset Shape Greenhouse (250 m^2).

Site	Maximum wind load on wall (kN/m^2)		Maximum wind load on roof (kN/m^2)
	Windward	Leeward	
Bengaluru	0.78	−0.67	−0.89
Ahmadabad	1.04	−0.89	−1.19
Mumbai	1.29	−1.11	−1.48
Delhi	1.44	−1.24	−1.65
Kolkata	1.63	−1.40	−1.87
Darbhanga	1.93	−1.66	−2.21

3.7.1.3.2 Effect of Number of Span on the Wind Load of Roof

It is easy to compute the wind load on the roof of single span greenhouse as there is an almost absent or very limited frictional drag involved (Figure 3.2). On the other hand, when the number of span upsurges beyond three, the frictional effect of wind on the roof surface needs to be taken into account. Figure 3.2 and Table 3.9 embody the effect of a number of spans on the wind load on a structure. IS 875 (Part 3) (1987) provides the code to overcome the frictional effect of wind on the roof. It is also found that the frictional drag effect is considered in the case of sawtooth shape structures only.

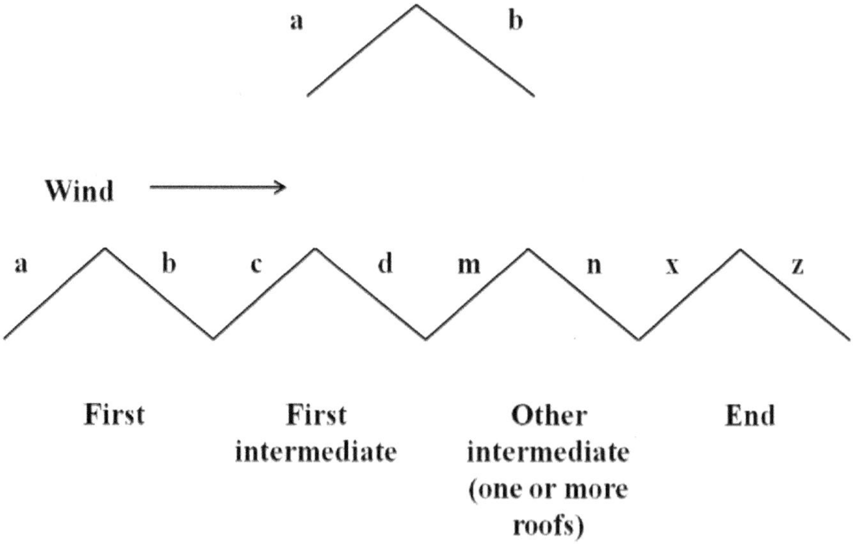

FIGURE 3.2 Effect of wind on multi-span sawtooth shape greenhouse.

3.7.1.4 LOAD COMBINATIONS

Combinations of various loads can be computed as prescribed in I: 875 (Part 5). These combinations would be,

- Dead load + Live load
- Dead load + Wind load
- Dead load + Live load + Wind load

TABLE 3.9 Wind Load (kN/m^2) on Sawtooth Shape Greenhouse of Area $\geq$ 1000 m^2 (4 Span Structure).

Site	First span		First intermediate span		Other intermediate span		End span	
	a	B	c	D	m	n	x	Z
Bengaluru	0.72	−0.78	−0.78	−0.61	−0.61	−0.72	−0.72	−0.55
Ahmadabad	0.96	−1.04	−1.04	−0.82	−0.82	−0.96	−0.96	−0.74
Mumbai	1.20	−1.29	−1.29	−1.02	−1.02	−1.20	−1.20	−0.92
Delhi	1.34	−1.44	−1.44	−1.13	−1.13	−1.34	−1.34	−1.03
Calcutta	1.52	−1.63	−1.63	−1.28	−1.28	−1.52	−1.52	−1.17
Darbhanga	1.79	−1.93	−1.93	−1.52	−1.52	−1.79	−1.79	−1.38

As already mentioned, wind load is the single most important factor that causes damage to the structure. Permissible stresses increase in the structure by almost 33% when wind load is considered. As per IS 875, just one load should be considered at a time when both live load and snow loads prevail at a site. Load combination gives the total load on the structure.

3.8 STRENGTH ANALYSIS AND SELECTION OF STRUCTURAL MEMBERS UNDER LOAD COMBINATIONS

It is one of the most critical steps in greenhouse design. On calculating the total load on the structure due to total load under prevailing conditions, stress, strain, deflection, drift, deformation, etc., generated in the members due to loading is determined. This exercise can be done through theoretical calculations by adopting a method of joint and section, which involves manual calculations or can be done using structural analysis software such as STAAD Pro and ANSYS. In the present chapter, the use of STAAD Pro and ANSYS has been explained seeing that the manual calculations are very tedious and often involve errors.

3.8.1 STRUCTURAL ANALYSIS OF THE MEMBERS USING STAAD PRO SOFTWARE

Steps involved in the structural analysis and determination of optimum size of the structural members are given in the following:

1. Calculation of the total load on the structure.
2. Determination of the reactions at the ends of purlin using STAAD Pro software. The reaction at the ends of the purlin is the load supported by the truss frame.
3. Drawing a truss in the STAAD Pro and assigning engineering properties to the material, specifications, and loads to each member of the truss.
4. Applying design command to check code.
5. Performing analysis and checking whether a member is FAIL or PASS under a set of loading conditions. If FAIL, go for a higher pipe size and perform analysis.

In our previous study (Kale, 2014), the aforementioned method was followed to analyze and design the structural members of sawtooth and Quonset shape greenhouses of 250 m^2 area. Structural members such as purlin, truss members, column, and foundation pipes were designed to resist the prevailing load conditions. Different loads working on the structures are presented in Table 3.10. A negative sign of the wind load depicts that the load acts away from the roof and tries to lift it. It is evident from the tables that both wall wind load and roof wind load increased with an increase in basic wind speed. The major portion of the total load was shared by wind load alone. Roof wind load (vertical load) was found higher than that on the wall (horizontal load) in all the cases. It is also evident that the roof wind load value of the sawtooth shape greenhouse structure was higher than that of the corresponding Quonset shape structure. Dead load (DL) and live load (LL) was kept constant for all the cases, as all other parameters except wind conditions were the same in all the cases. The nodal load on the truss is the load that is transferred to the truss by purlins.

The optimum size of the structural members can be determined by carrying out the structural analysis in STAAD Pro software. Members may be of steel, aluminum, or any other material. The material of the member and its engineering properties needs to be defined in the software prior to analysis. The optimum size of the steel members determined under nodal loads in our study is given in Table 3.11. It can be observed that an increase in wind speed from 33 to 55 m/s did not affect the member diameters significantly. Only the member thickness increased with an increase in wind speed. It can also be noted that purlin and truss members should be 33.7 mm in size. The column is a major load-bearing member which is acted by roof load as well as the wind on the load on the wall. Consequently, its outside diameter was 60.3 mm, the highest among all the structural members. The foundation pipe takes the compressive load of the whole superstructure but there is no tangential wind load on it. The outside diameter of the foundation pipe was found to be 48.3 mm.

TABLE 3.10 Loads on Greenhouse Structure at Representative Sites.

Site	Basic wind speed (m/s)	Wall wind load (kN/m^2)	Roof wind load (kN/m^2)	DL+LL (kN/m^2)	Total load (kN/m^2)	Nodal load on truss (kN)
Quonset shape						
Bengaluru	33	0.78	−0.79	0.8	1.59	3.98
Ahmadabad	39	1.04	−1.06	0.8	1.86	4.65
Mumbai	44	1.29	−1.32	0.8	2.12	5.30
Delhi	47	1.44	−1.47	0.8	2.27	5.68
Kolkata	50	1.63	−1.67	0.8	2.47	6.18
Darbhanga	55	1.93	−1.97	0.8	2.77	6.93
Sawtooth shape						
Bengaluru	33	0.78	−0.89	0.8	1.69	4.23
Ahmadabad	39	1.04	−1.19	0.8	1.99	4.98
Mumbai	44	1.29	−1.48	0.8	2.28	5.70
Delhi	47	1.44	−1.65	0.8	2.45	6.13
Kolkata	50	1.63	−1.87	0.8	2.67	6.68
Darbhanga	55	1.93	−2.21	0.8	3.01	7.53

3.8.2 STRUCTURAL ANALYSIS AND DESIGN OF THE MEMBERS USING ANSYS SOFTWARE

Steps involved in the structural analysis and determination of optimum size of the structural members using ANSYS are given in the following:

1. Design and drawing of the structural members in computer-aided design (CAD) software such as solid works. Conversion of CAD drawings into Initial Graphics Exchange Specification (IGES) platform to make them accessible in ANSYS software.
2. Simulation and modeling of the structure under prevailing loading conditions using imported drawings in ANSYS workbench. It consists of three steps, namely, preprocess, process/solution, and postprocess.
3. In preprocess step, the opposite structural model is identified, and engineering properties are assigned to the selected model components. In India, the majority of the structural members are made of GI

TABLE 3.11 Specifications of Structural Members of Greenhouse Structure (250 m^2).

Site		Purlin		Truss members		Column		Foundation	
	Basic wind speed (m/s)	OD (mm)	t (mm)	OD (mm)	t (mm)	OD (mm)	t (mm)	OD (mm)	t (mm)
Quonset shape									
Bengaluru	33	33.7	2.65	33.7	2.65	60.3	2.9	48.3	2.9
Ahmadabad	39	33.7	2.65	33.7	2.65	60.3	2.9	48.3	2.9
Mumbai	44	33.7	2.65	33.7	2.65	60.3	2.9	48.3	2.9
Delhi	47	33.7	3.25	33.7	3.25	60.3	3.25	48.3	2.9
Kolkata	50	33.7	3.25	33.7	3.25	60.3	3.25	48.3	2.9
Darbhanga	55	33.7	3.25	33.7	3.25	60.3	3.25	48.3	2.9
Sawtooth shape									
Bengaluru	33	33.7	2.65	33.7	2.65	60.3	2.9	48.3	2.9
Ahmadabad	39	33.7	2.65	33.7	2.65	60.3	2.9	48.3	2.9
Mumbai	44	33.7	2.65	33.7	2.65	60.3	3.25	48.3	2.9
Delhi	47	33.7	3.25	33.7	3.25	60.3	3.25	48.3	2.9
Kolkata	50	33.7	3.25	33.7	3.25	60.3	3.25	48.3	3.25
Darbhanga	55	33.7	3,25	42.4	2.65	60.3	3.65	48.3	3.25

OD, outside diameter.

steel pipes, the foundation is made of concrete, whereas ultraviolet (UV)-stabilized low-density polyethylene (LDPE) poly sheet is used as cladding material. The properties of these materials are already established. The next step is meshing which has three types, namely, fine, medium, and course. The meshing type depends on the number of members and a number of nodes. Meshing is the representation of the differential equations and boundary conditions. After deciding the meshing type, the estimated load/total load is assigned to the members at two wind angles, namely, 0° (vertical load) and 90° (horizontal load). Here, the vertical load may be called roof wind load, whereas the horizontal load may be termed wall wind load.

4. In the process step, simultaneous equations are solved under mathematical modeling in the software. The structure is analyzed to obtain the optimum sizes of the members under prevailing loads. Stress and deformation generated in the members can be observed through analysis.
5. Postprocess is the final step in the evaluation of stress and deformation. Analysis of all the members and nodes is produced in postprocessor, and final results are generated in the form of colored images. Usually, a color scheme is used to represent the stress and deformation, for example, dark red color represents the member under maximum stress/deformation, whereas dark blue color represents the minimum value of them.

In our study (Indore et al., 2020), different structures located in the different climatic zones of India were selected, prevailing loads were determined, and structural analysis of the members was performed using ANSYS software. An example of the double arc multi-span greenhouse has been discussed here. The structure was analyzed by considering two wind angles, namely, 0° and 90°. Members of two dissimilar sizes (column 76 mm, 42 mm; and another column 60 mm and 48 mm) were used. It was observed that total deformation was maximum at a wind angle of 0° (vertical load) as compared to 90° (horizontal load) (Figure 3.3). On the basis of the results of the simulation in ANSYS software, it was recommended by this study to use 60 mm member size for columns and 48 mm size for main purlin. The deep red color in Figure 3.3 shows that these members may FAIL due to deformation. Hence their specifications should be changed to subsequently available sizes. This study revealed that there is a requirement to revise the existing standards of greenhouse designs to prevent deformation and structural damage under particular agro–climatic conditions. The need for location-specific, region-wise greenhouse design standards has also been proven by Nayak and Rao (2014). The specifications of the structural members of the multi-span greenhouse gained after a design by means of ANSYS software are given in Table 3.12.

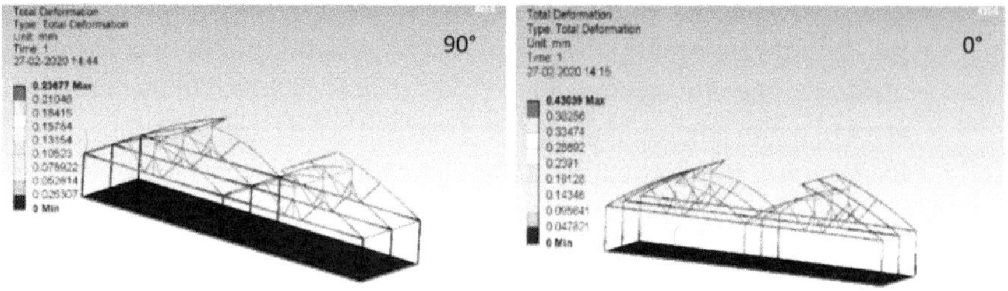

FIGURE 3.3 Deformation in multi-span structure and wind angles at 90° and 0° (adapted from Indore et al., 2020).

Source: Adapted from Indore et al., 2020

TABLE 3.12 Specifications of Structural Members of Multi-Span Greenhouse Obtained After Design Using ANSYS Software (Adapted from Indore et al., 2020).

Member	Outside diameter (mm)	Thickness (mm)	Weight per meter length (kg)
Column	60	2	3.75
Top purlin	48	2	2.00
Gutter purlin	42	2	2.00
Top arches of the truss	42	2	2.00
Bottom chord of the truss	60	2	3.00
Internal bracings of the truss	33	2	1.60
Corridors/balconies	60	2	3.00
Curtain runner	42	2	2.00
Flap control pipe	21	2	1.00
Curtain shaft	27	2	1.30
Cross bracing	33	2	1.60

Source: Adapted from Indore et al., 2020

3.9 CONCLUSIONS

Fundamentally greenhouse technology is suitable for crop production in temperate regions, but it can be adopted in tropical regions with modifications.

Several greenhouse designs have been developed for different climate zones of tropical regions. Literature reveals that the performance of the greenhouse structure is prominently influenced by its design. Likewise, the selection of an appropriate type, site, framing materials, covering, and cooling methods also affect the cost, performance, life span, and productivity of the greenhouse. Information on the design of a location-specific greenhouse is very imperative to find out the precise specifications of the framing materials under prevailing loading conditions. The under-designed structure is prone to failure under loads, whereas over-designed structures involve higher initial investment. Optimal sizes and weights of the structural members reduce the risk of structural damage and also reduce the cost of the structure.

KEYWORDS

- **greenhouse**
- **design**
- **location-specific**
- **specifications**
- **prevailing loads**
- **structural damage**

REFERENCES

Baudoin, W. O.; von Zabeltitz, C. H. R. Greenhouse Constructions for Small Scale Farmers in Tropical Regions. *Acta Hort.* **2002,** *578*, 171–179.

Castilla, N. In *Geenhouse Technology and Management*, CABI, Wallingford Publisher, Oxfordshire, UK: Cambridge, MA, 2013.

Connellann, G. J. Selection of Greenhouse Design and Technology Options for High Temperature Regions. *Acta Hort.* **2002,** *578*, 113–117.

Dalai, S.; Tripathy, B.; Mohanta, S.; Sahu, B.; Palai, J. B. Green-Houses: Types and Structural Components. In *Protected Cultivation and Smart Agriculture*; Maitra, S., Gaikwad, D. J., Shankar, T., Eds.; New Delhi Publishers: New Delhi, 2020; pp 09–17.

Emekli, N. Y.; Kendirli, B.; Kurunc, A. Structural Analysis and Functional Characteristics of Greenhouses in the Mediterranean Region of Turkey. *Afr. J. Biotechnol.* **2010,** *9* (21), 3131–3139.

Giacomelli, G.; Castilla, N.; Henten, E. V.; Mears, D.; Sase, S. Innovation in Greenhouse Engineering. *Acta Hort.* **2008,** *801*, 75–88.

Herry, S.; Yayu, R. Research on the Greenhouse Application in the Tropics. *J. Agric. Eng.* **2007,** *21* (4), 313–322.

Indore, N.; Kale, S. J.; Akhoon, A. B.; Singh, R. K.; Singh, H. Structural Analysis of Common Existing Greenhouses Designs in Different Agro Climatic Zones of India. *Int. J. Agric. Eng.* **2020,** *13* (1), 80–89.

IS:875. Indian Standard Code of Practice for Design Loads Other than Earthquake for Buildings and Structures, Part-2 Imposed Load, 1987.

IS: 875. Indian Standard Code of Practice for Design Loads Other than Earthquake for Buildings and Structures, Part-3 Wind Load, 1987.

IS: 875. Indian Standard Code of Practice for Design Loads Other than Earthquake for Buildings and Structures, Part-4 Snow Load, 1987.

IS: 875. Indian Standard Code of Practice for Design Loads Other than Earthquake for Buildings and Structures, Part-5 Special Loads and Combination, 1987.

IS:875. Code of Practice for Design Loads (Other than Earthquake) for Buildings and Structures. Part 4 Snow Loads, 2003.

IS:875. Indian Standard Code of Practice for Design Loads Other than Earthquake for Buildings and Structures, Part-1 Dead Load.

IS:1446. Indian Standard Code for Surface Covered Cultivation Structure, Glossary Terms, 1997.

IS:4462. Indian Standard recommendation for Layout, Design and Construction of Greenhouse Structure, 1997.

Kale, S. J. *Analysis and Design of Saw-Tooth and Quonset Shape Greenhouse Structures for Different Regions of India*; Report on Professional Attachment Training, Submitted to Agricultural and Food Engineering Department, Indian Institute of Technology, Kharagpur, West Bengal, India, 2014.

Kendirli, B. Structural Analysis of Greenhouses: A Case Study in Turkey. *Build. Environ.* **2006,** *41*, 864–871.

Kumar, K. S.; Jha, M. K.; Tiwari, K. N.; Singh, A. Modelling and Evaluation of Greenhouse for Floriculture in Subtropics. *Energy Build.* **2010,** *42*, 1075–1083.

Kumar, K. S.; Tiwari, K. N.; Jha, M. K. Design and Technology for Greenhouse Cooling in Tropical and Subtropical Regions: A Review. *Energy Build.* **2009,** *41*, 1269–1275.

McCartney, L.; Lefsrud, M. G. Protected Agriculture in Extreme Environments: A Review of Controlled Environment Agriculture in Tropical, Arid, Polar, and Urban Locations. *Appl. Eng. Agric.* **2018,** *34* (2), 455–473.

Mohammad, A. T.; Mat, S. B.; Alrubaih, M. S.; Al-abidi, A. A. Optimization of Cooling Load for Different Greenhouse Models in Malaysia. *Int. J. Renew. Energy Res.* **2014,** 42–48.

Nayak, A. K.; Rao, K. V. R. Estimation of Wind Load on a Greenhouse and Evaluation of its Structural Stability. *Int. J. Agric. Eng.* **2014,** *7* (2), 461–466.

Rajender, G.; Sushanth, K.; Mithun, K.; Devender, B.; Raju, D.; Anoosha, K. Design and Development of Low Cost Greenhouse to Raise Different Cultivars. *Int. J. Agric. Sci. Res.* **2017,** *7* (3), 29–36.

Saribiyik, A.; Saribiyi, M. Design of Permanent Greenhouse Structure by Pultruded Glass Reinforced Plastic. *Sci. Res. Essays* **2011,** *6* (17), 3629–3637.

Singh, M. C.; Kachwaya, D. C., Kalsi, K. Soilless Cucumber Cultivation Under Protective Structures in Relation to Irrigation Coupled Fertigation Management, Economic Viability and Potential Benefits-A Review. *Int. J. Curr. Microbiol. App. Sci.* **2018b,** *7* (03), 2451–2468.

Singh, M. C.; Singh, J. P.; Pandey, S. K. Cutting, N. G.; Sharma, P.; Shrivastav, V.; Sharma, P. A Review of Three Commonly used Techniques of Controlling Greenhouse Microclimate. *Int. J. Curr. Microbiol. App. Sci.* **2018a,** *7* (1), 3491–3505.

Singh, M. C.; Singh, K. G.; Singh, J. P. Yield of Soilless Cucumbers Planted under Partially Controlled Greenhouse Environment in Relation to Deficit Fertigation. *Ind. J. Hort.* **2018c,** *75* (2), 259–264.

Singh, M. C.; Singh, K. G.; Singh, J. P. Nutrient and Water use Efficiency of Cucumbers Grown in Soilless Media under a Naturally Ventilated Greenhouse. *J. Agric. Sci. Technol.* **2019b,** *21* (3), 193–207.

Singh, M. C.; Singh, K. G.; Singh, J. P. Interactive Effects of Fertigation and Varieties on Pant Growth Attributes and Yield of Soilless Cucumbers. *Int. J. Sci. Technol. Soc.* **2020a,** *8* (3), 60–65.

Singh, M. C.; Singh, K. G.; Singh, J. P.; Mahal, A. K. Performance of Soilless Cucumbers in Relation to Differential Fertigation under Naturally Ventilated Greenhouse Conditions. *J. Plant. Nutr.* **2019a,** *42* (11–12), 1316–1332.

Singh, M. C.; Singh, A.; Singh, J. P.; Singh, K. G. Economic Viability of Soilless Cucumber Cultivation Under Naturally Ventilated Greenhouse Conditions. *Ind. J. Hort.* **2020b,** *77* (2), 315–321.

Ubolsook, P.; Thepa, S. In *Structural Analysis of Bamboo Trusses Structure in Greenhouse,* 2nd International Conference on Environmental Science and Technology IPCBEE vol. 6, IACSIT Press: Singapore, 2011.

Villagran, E.; Leon, R.; Rodriguez, A.; Jaramillo, J. 3D Numerical Analysis of the Natural Ventilation Behavior in a Colombian Greenhouse Established in Warm Climate Conditions. *Sustainability* **2020,** *12* (8101), 1–26.

Yun, S. W.; Choi, M. K.; Chan, Yu.; Lee, S. Y.; Yoon, Y. C. Analysis of Working Load Based on Wind and Snow Load for Greenhouse Foundation, *J. Agric. Life Sci.* **2014,** *48* (1), 183–188.

Zabeltitz, C. V.; Baudoin, W. O. In *Greenhouse and Shelter Structures for Tropical Regions*; Published by Arrangement with FAO, Published by Daya Publishing House: New Delhi, 2005; pp 65–67.

CHAPTER 4

Soilless Cultivation Techniques in Protected Structures

ARTI VERMA

Krishi Vigyan Kendra, Langroya, Punjab Agricultural University, Ludhiana, Punjab, India

ABSTRACT

The term "soilless culture" refers to cultivating plants without using soil as a rooting media. In this system, inorganic nutrients are supplied along with irrigation water and absorbed by the roots. Soilless cultivation has become popular throughout the globe over the last three decades, particularly from the research point of view. The primary goal of soilless cultivation methods is to eradicate problems related to the greenhouse soils such as soil-borne diseases, salinity, and low soil fertility. The great potential provided by these cultivation methods is unquestionable, which includes productive, qualitative, and environmental benefits mainly due to efficient usage of water and nutrients. At present, the soilless culture systems are more and more being accepted as a main hi-tech component in modern greenhouses for achieving the highest productivity of the crops both quantitatively and qualitatively. This chapter involves the different soilless techniques (water and substrate cultures) used for cultivating high-value crops under protected structures.

4.1 INTRODUCTION

With the rapid growing population and industrialization, the biggest challenge in front of the agricultural sector is to increase the production that satisfies

Protected Cultivation: Structural Design, Crop Management, Modeling and Automation.
Mahesh Chand Singh and K. K. Sharma (Eds.)

the consumers demand. For agricultural production, the most commonly and widely used growing medium is soil, which meets all the basic requirements of a crop like nutrients, water, and air. Though, soil also hampers the crop growth in relation to poor fertility, poor drainage, nutrients imbalance, diseases, nematode infestation, etc., which makes it difficult to obtain the highest yield potential of the crops. In that situation, soilless cultivation has become a substitute to get better crop yield (quantitatively and qualitatively) in relation to improved control on the microclimate inside the protected structures (Singh et al., 2018a; 2021a,b) and to avoid uncertainties in nutrient and water status of the soilless growing media. A soilless growing system refers to any method of cultivating plants that does not include the usage of soil as a rooting medium, where inorganic nutrients are supplied along with irrigation water and absorbed by the roots (Savvas, 2003; Gruda et al., 2016a). The water-soluble fertilizers, which contain the proper amount of micro- and macro-nutrients are dissolved in irrigation water, and the resultant is called as nutrient solution (Singh et al., 2019a,b). In this technique, the crop can be raised either in porous media (substrates) in which nutrients are supplied along with irrigation water, or directly in a nutrient solution deprived of a solid phase. With the aim to optimize the plant nutrition, supply of nutrients in solution form has become a common cultural practice in recent years, in both soilless and soil-based crop cultivations.

Over the last three decades, soilless cultivation has become popular with its primary goal to eradicate problems related to the soils, like soil-borne diseases, salinity, low soil fertility, etc. The development of appropriate growing media with suitable physical, chemical, and hydraulic properties, as well as improvement in irrigation and fertigation systems and automation technology, have helped to establish soilless cultivation system as an important agricultural technology in modern greenhouses. Soilless cultivation is considered to be the safest and most effective technique to reduce the chance of disease and pest infestation maximally occurring due to infested soil. Thus, it becomes important for all types of protected cultivation structures to be designed to enhance favorable microclimatic conditions (Singh et al. 2018a,b).

Presently, the soilless culture technique is more and more being accepted as a main hi-tech component in modern greenhouses because of their benefits. A key benefit of soilless system is the eradication of soil-borne diseases such as nematode infestation. One more advantage of soilless cultivation systems is the reservation of cultivatable lands for major crops, saving about 90% of irrigation water, obtaining higher productivity compared to conventional agriculture, controlled use of nutrients and the protection against insect-pest. Moreover, this technology can be adopted in terraces, roof-tops, and uncultivated lands.

For cultivating vegetables and cut flowers, the most widely used media are rockwool, perlite, and coconut (coco-peat, coco-coir, coco-chips, etc.). For growing green leafy vegetables, most commonly used water culture systems are float hydroponics, nutrient film technique, and aeroponics. For nourishing and watering plants, completely automated fertigation heads are used which prepare and distribute nutrient solution judiciously. In soilless cultivation, it is easy to collect and recycle the nutrient solution extracted from the root zone of plants, which in turn increases water and nutrient use efficiency, and decreases the negative impact of fertilizer residues on environment. The use of recycled effluents may lead to the spread of pathogens, which make it important to use appropriate disinfection program before reusing the effluents, primarily based on ultra-violet radiation, slow sand filtration, or chemical treatment.

In soilless culture system, the nutrient solution composition must be adapted to the type of plant species, the stage of growth, cropping season, and the climatic conditions. Irrigation water can contain significant amounts of macronutrients, micronutrients, and other non-nutrient ions (Sonneveld and Voogt, 2009). These nutrients should be considered while calculating the amount of fertilizers for preparing a nutrient solution. Therefore, to prepare a nutrient solution of a particular composition the amount of fertilizer required should be calculated individually for each grower (Singh et al., 2019a,b). The soilless culture method permits more efficient control of soil-borne pathogens by eliminating the soil, as well as optimizing both physical and chemical characteristics in the root zone. Consequently, higher yields can be achieved at a lower cost of production, with less pesticide usage and better product quality (Gruda et al., 2018; Singh et al., 2018c, 2020).

4.2 BENEFITS OF USING SOILLESS SYSTEMS FOR CROP GROWING

1. **Increased productivity:** The improved control over crop growth factors such as nutrients (i.e., micro- and macro-nutrients), pH of both nutrient solution and soilless growing media and microclimatic parameters (e.g., light or solar radiation, temperature [air, plant, growing media, and covering material], relative humidity, carbon dioxide, and oxygen) in soilless growing systems greatly improves

the crop growth, yield and the quality of the produce (Singh et al., 2018c).

2. **Improved water management:** In soilless cultivation systems, irrigation water use is controlled more precisely as compared to the earlier method of irrigation. These systems save labor and time to look after (i.e., cleaning of dripper laterals and regular inspection of drippers that can be simply choked by calcium carbonate or other compounds, demanding extra cost, labor, and time).
3. **Improved monitoring of plant nutrition:** In soilless system, the elements that can be harmful to plants when available in excess which should be kept within safe range. The pH and electrical conductivity (EC) of a nutrient solution can be regulated according to the plant requirements and environmental conditions, which is being expensive and difficult in case of conventional soil cultivation.
4. **Minimum or no use of pesticide:** Soilless culture systems under controlled microclimatic conditions helps to prevent the spread of weeds, diseases, and pests; hence, use of pesticide that ultimately pollutes the environment is usually not required.
5. **Improved monitoring of root zone environment:** By using soilless cultivation systems, it becomes easy to regulate the nearby environment and root temperature.
6. **More crops per year:** In soilless culture, the interval between crops is almost negligible due to absence of cultivation operations; therefore, number of cultivated crops annually is more as compare to those cultivated in soil systems which results in higher income.
7. **Unsuitable and infested agriculture land:** The soilless culture provides an ideal alternative for cultivation in protected environment, where the soil in a particular locality is found not suitable either due to the presence of pathogens, or salinity or sodicity, etc.
8. **Reduced labor requirement:** In soilless culture system, all traditional cultural practices are not performed, which saves labor and time.

Apart from the several benefits of soilless growing systems, there are some disadvantages also:

1. **High initial cost:** The initial cost of the soilless culture system is high, but rapid crop growth, higher yield, and the improved quality of produce help to pay back within the first 3–4 years from beginning of the system.

2. **Technical skill requirement:** For growing crops in soilless culture systems, skilled manpower is required for monitoring, handling, and crop management.
3. **Risk of pathological injuries:** Injury in open system of soilless culture is low, while in closed systems, it is high and requires significant care and strict sanitation.

4.3 TYPES OF SOILLESS CULTURES

There are mainly two types of soilless cultivation systems viz. closed and open systems. Water culture systems are actually closed systems, as the outflow of a nutrient solution from the root zone is not limited by the water holding capacity (WHC) of a porous growing media. Therefore, using water culture as an open system could lead to significant wastage of nutrient solution. In substrate-grown crops in open systems, the amount of nutrient solution that runs off as drainage solution varies from 10 to 30%. Whereas in the closed systems, the nutrient solution is not wasted through free drainage (Savvas, 2002). Leaching of nutrients into groundwater gets eliminated if the drainage solution (leachate or waste nutrient solution) is reused (i.e., using the closed system) in hydroponic systems and additionally, about 40–50% of total feed (nutrient solution) can be saved (Savvas, 2002).

4.3.1 CLOSED SOILLESS CULTURE SYSTEM

In closed systems, the drained nutrient solution is recovered, replenished, filtered, and disinfected before re-using to prevent the spread of pathogens (Wohanka, 2002) (Figure 4.1). Compared to the open soilless culture system, it requires more detailed and regular control of a nutrient solution and technical knowledge, as it is more sensitive to operational errors. The drained nutrient solution should be treated to re-establish its original composition and eradicate the occurrence of disease spread via recycled leachate. In addition, the spread of root-borne diseases is possible; hence, disinfection of a nutrient solution should be done to kill the germs. In substrate-grown crops, different technologies of disinfecting such as UV-treatment, heat pasteurization, slow sand filtration, microfiltration, and chemical treatment with ozone or chlorination can be used (Postma et al., 2008). In the case of the water culture system, only the chemical method of disinfection is economically viable. The growing crops in closed hydroponic systems help to reduce fertilizers and water use, as well as water contamination which is caused by nitrates and phosphates from fertigation wastes (Savvas, 2002).

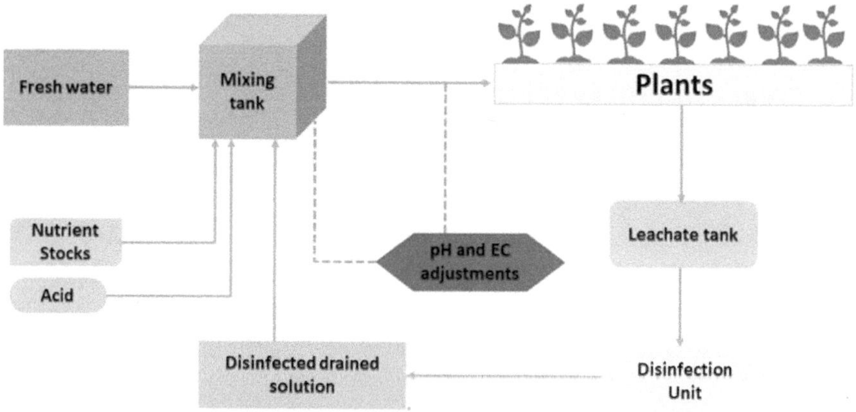

FIGURE 4.1 Closed soilless culture system.

Source: Adapted from Savvas, 2002

4.3.2 OPEN SOILLESS CULTURE SYSTEM

In the open soilless culture technique, new dissolved supplements are involved during the irrigation cycle (Figure 4.2). Water and nutrients are supplied to the crop according to conventional soil culture, while the leachate can be extracted and recycled, but it is often lost, posing risk to the atmosphere. Open-loop systems determine the nutrient solution to supply in conjunction with leaching, i.e., the volumetric ratio of the leachate to the applied nutrient solution. Every soilless culture uses only substrates and dribble frameworks have a place in open soilless culture.

4.4 SOILLESS CULTURE SYSTEMS

Substrate culture (artificial, mineral, or organic growing media, or a combination of these) and water or hydroponic culture system, in which the roots are immersed partially or completely in a nutrient solution are two common classifications for soilless cultures based on type of plant support (Figure 4.3).

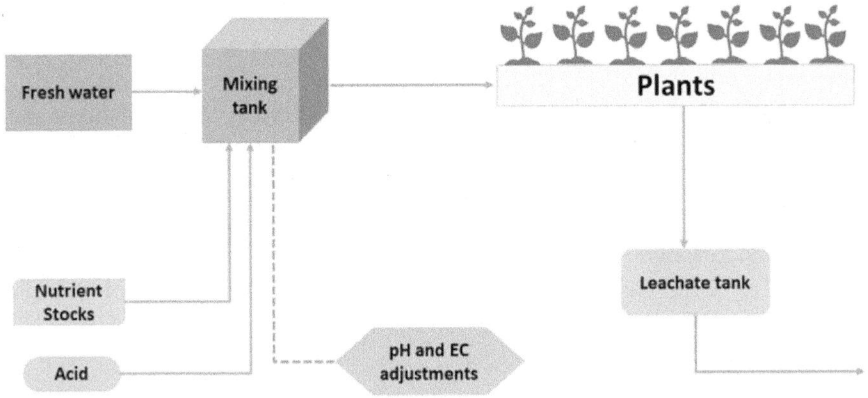

FIGURE 4.2 Open soilless culture system.

Source: Adapted from Savvas, 2002

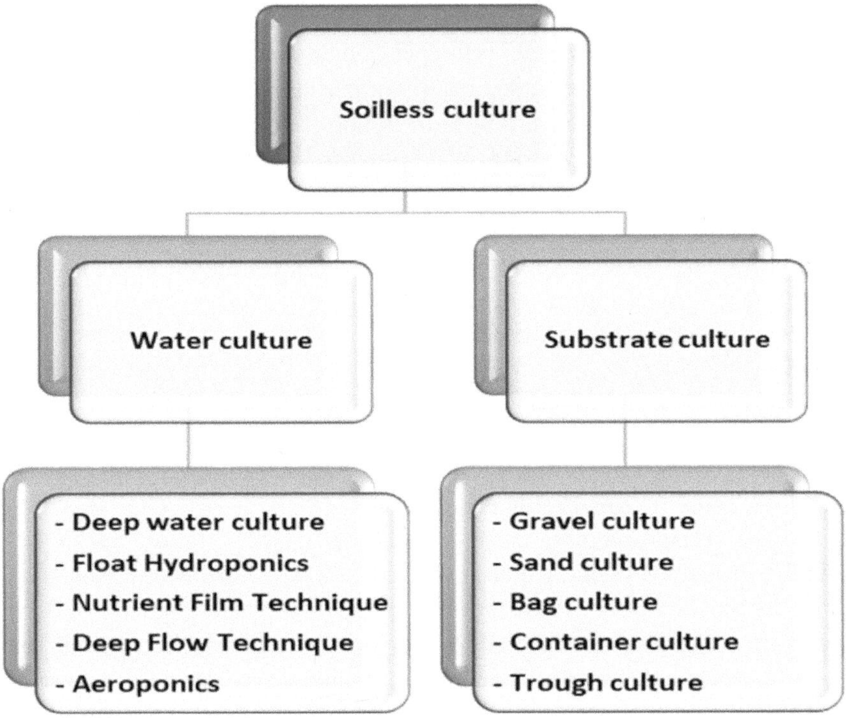

FIGURE 4.3 Soilless cultures in relation to the type of plant support.

Source: Adapted from Savvas, 2002

4.4.1 WATER CULTURE SYSTEMS

Deep water culture (DWC), float hydroponics, nutrient film technique (NFT), deep flow technique (DFT), and aeroponics are some of the available water culture systems. There is a worldwide rapid growth of water culture systems over the past couple of decades, may be due to their several advantages like the absence of soil-borne pathogens, acting as a safe substitute to soil disinfection, precise nutrient control, opportunity to cultivate crops even in saline or sodic soils, reduced fertilizer use, and prevention of nutrient leakage from greenhouses to the environment.

4.4.1.1 DEEP WATER CULTURE (DWC)

The first hydroponic system, i.e., DWC was developed by Professor W. F. Gericke of the University of California in 1929 for commercial crop production. It consists of a bucket covered with a net and cloth in which a sand layer of 1 cm is placed to support the plants and filled with nutrient solution. The roots of plants are immersed wholly or partially in a nutrient solution up to 15 cm depth. However, this system is not commercially viable because of hypoxic conditions occurring in the rhizosphere, owing to the restricted area of air–water exchange, in comparison to the volume of the solution, and the low diffusion coefficient of oxygen in the water. This limitation can be improved by using air pumps to oxygenate the solution of nutrients or by applying recirculating DWC systems. This system can be easily set up in glassware containers, ice boxes, plastic boxes, concrete containers, or in engraved containers covered with polypropylene sheets. The most convenient plants that grow well in this system are strawberry, lettuce, and herbs.

4.4.1.2 FLOAT HYDROPONICS

In this hydroponic system, plants are raised on trays suspended in tanks filled with nutrient solution. In 1980, the University of Arizona designed and developed a floating system, especially for growing seedlings. The system is of low-cost set-up and management, requiring minimal mechanization to monitor and regulate the nutrient solution. In a float hydroponic system, the sowing is done directly on the substrate containing polystyrene/styrofoam panels or trays/viols. Panels are allowed to float in tanks filled with nutrient solution after seed germination (Pardossi et al., 2005). The tanks can be constructed of low-cost materials (wooden planks, bricks, and concrete) or dug into the greenhouse up to a depth of 0.20–0.30 m. The tanks are sealed and filled at a rate of 150–250 l/m^2 with nutrient solution (Savvas et al., 2013). In floating systems, proper aeration, management of the nutrient solution, and pool depth should all be considered. During the whole growing season, the concentration of oxygen in a nutrient solution should be maintained between 5 and 6 mg/l. Pumps moves part of the nutrient solution through a pipe into which a Venturi tube is placed to insufflate air which is a simple mode to oxygenate the nutrient solution. However, in order to avoid root damage and recirculation of plant exudates, the airflow should never be very strong. The water culture systems permit more plants per unit area. It also diminishes the risk of sanitation and nitrate accumulation and enhances the development and automation of cultural practices. It is the best hydroponic system for growing vegetables such as lettuce, lamb's lettuce, spinach, watercress and aromatic plants.

4.4.1.3 NUTRIENT FILM TECHNIQUE (NFT)

The nutrient film technique (NFT) is a water culture technique in which a thin film of nutrient solution passes into channels such as gullies, troughs, or gutters, where the bare roots of plants lie. A storage tank containing a nutrient solution, a circulating pump, and a system of pipes that supply the solution to the heads of the growing troughs, from which solution falls down the storage tank through gravitational force are the main components of NFT (Savvas et al., 2013). The channels are provided with 1.2–3.0% slope and the solution descends into the channels that keep the roots moist. The nutrient solution is pumped by irrigators at the top of every sloping pipe, which returns back to the tank from the lower end of the channels. As a result, the nutrient solution is nonstop recycled. The nutrient film technique works best on plants with small roots such as lettuce, strawberry, and herbs and short-term crops, as these plants get ready for harvesting before their roots fill the channel. NFT offers several benefits such as simplification of irrigation, uniform nutrient application, better control of root temperature, reduced water and nutrient loss, and reduced adverse environmental impacts.

4.4.1.4 DEEP FLOW TECHNIQUE (DFT)

Deep flow technique (DFT) is the method in which the plant roots are exposed continuously to a running nutrient solution. In this technique, especially for aeration, a solution of nutrients is pumped around the troughs. The troughs are always filled with a nutrient solution, where the roots are kept suspended. The depth of the continuously flowing nutrient solution is kept as 5–15 cm which is more than the nutrient film technique (Savvas et al., 2013). In the deep flow method, the channel diameter is kept about 1 m. Plants are planted on polystyrene trays that float in the water or rest on walls of the channel. The high water volume allows for better nutrient solution management and temperature buffering, making this system ideal for areas where temperature fluctuations in the nutrient solution are a concern.

4.4.1.5 AEROPONIC SYSTEM

Aeroponics is undoubtedly the most advanced type of hydroponic system in which plants are planted in an opaque container or trough, where the plant roots are suspended and washed in a nutrient mist. Plants are planted in holes in expanded polystyrene panels placed parallel or on a slope and secured over a metal frame. Under the panel, the plant roots are held in mid-air and mounted in a spray box. In order to prevent algal growth from obstructing plant growth and polluting the system, the box must be closed such that the roots stay in the dark. Water and nutrients are provided by spraying on the hanging roots of the plant with nutrient solution by using sprayers and foggers or misters from time to time. The spray flow rate can vary from 35 to 70 l/h, while the spacing is determined by the design and size of the cropping modules. The system is usually turned on for 30–60 s for every 2–3 min, and its frequency varies depending on the species, growing season, plant growth stage, and the daytime. As the roots are exposed to air, the interruption in misting cycles results into quickly drying of the roots. Aeroponics permits a significant decrease in water and fertilizer use and assures sufficient root respiration. The disadvantage of this system is the failure to buffer interruptions in the flow of a nutrient solution and the plants can suffer from extreme thermal stresses, especially in summer.

4.4.2 SUBSTRATE CULTURE

In the substrate culture technique, a solid medium acts as a supportive mediator of the plants to prevent them from falling. The role of growth media is to provide oxygen to the roots, and deliver water and nutrients to the plant through an irrigation system (El-Kazzaz and El-Kazzaz, 2017). A variety of substrates can be utilized as growing media in soilless culture system and their selection primarily depends on water retention and water dynamics. The substrates maintain a reservoir of nutrient solution and protect the roots from temperature fluctuations. The horticultural industry typically employs a combination of growing media constituents and additives. Fertilizers, liming compounds, and bio-control or wetting agents are examples of additives while growing media constituents include combinations of different materials. These constituents can be of organic nature (bark, coconut coir, rice husk, peat, sawdust, and wood chips) or inorganic nature (sand, gravel, perlite, rockwool, volcanic stones, etc.) (Gruda et al., 2013).

Organic growing media may be synthetic, such as polyurethane or natural, such as peat or wood-based substrates. Peat (Schmilewski, 2009), composts (Raviv, 2013), bark (Maher and Thomson, 1991), and wood residues are the most widely available and applicable organic materials (Gruda and Schnitzler, 2004). Presently, there has been a growing interest in biochar (Nemati et al., 2014) and hydrochar. Inorganic growing media are all derived from natural sources, with only a small percentage undergoing industrial processing prior to their use. All over the world, rockwool has become the most common growing medium for fruit and vegetable production in greenhouses (Gruda et al., 2016b). Perlite is a well-known growing medium in Europe, and it is commonly used in the Mediterranean region because of its low cost (Grillas et al., 2001). A variety of rare growing media, such as pumice, zeolite, tuff, volcanic porous rock, and vermiculite have been used in addition to rockwool and perlite (Gruda et al., 2016b). Both organic and inorganic materials have their advantages and disadvantages when used as growing media constituents (Table 4.1).

Root aeration is essential for effective soilless cultivation. Understanding the aspects that contribute to air exposure in growing media is essential for effective substrate-grown crop management. The physical properties of commonly available growing media (Raviv, 2008) are given in Table 4.2. Growing media with a small pore space is prone to oxygen deficiency, particularly if the plants grow quickly along with intensive root respiration. In that case, keeping the substrate in layers of at least 20 cm in bags, containers or troughs can be a good technique to prevent aeration issues.

4.4.2.1 GRAVEL CULTURE

In gravel culture, growing beds are created with a slope of 0.2–0.3% either by soil digging or as aboveground troughs. These growing beds are lined with

TABLE 4.1 Advantages and Disadvantages of using Organic and Inorganic Materials as a Growing Media.

Growing media	Advantages	Disadvantages
Organic		
Bark (well-aged): By-product or waste of wood manufacture	Good aeration, good WHC, good total pore space (75–90% v/v), sub-acid to neutral pH (5.0–7.0), average volume weight (0.320–0750 g/cm^3) and long lasting.	Higher variability, requires time to minimize C:N ratio and terpenes concentrations, and higher cost as it is used as a fuel substitute and in landscaping.
Bio-char and hydro-char: Obtained from biomass hydrolysis or biomass pyrolysis.	Energy-neutral production, aids carbon sequestration, biologically very stable, lower bulk density, higher CEC and nutrient holding capacity. It can be used to make hydro-char from wet material.	Higher cost of production, bio-char often has high pH, and can be dusty.
Coconut coir: By-product of coconut fiber processing	Physically stable, lighter in weight (0.065–0.110 g/cm^3), sub-acid to neutral pH (5.0–6.8), It has good total pore space (94–96% v/v) and water retention ability.	It may contain high salt levels and energy consumption during transport.
Green compost: Obtained from composted plant residues	Higher potassium and micro-nutrient contents, suppression of diseases and good WHC.	Variable compositions, higher volume weight (0.600–0.950 g/sm^3), requires sufficient time to be composted and becomes easily waterlogged.
Peat: Natural anaerobically processed plant residues	Physically stable, good water retention capacity, high total pore space (85–97% v/v), low microbial activity, low volume weight (0.060–0.200 g/cm^3), low pH, and cost-effective.	Limited resource, CO_2 emissions, and environmental risks. It may be highly acidic and shrinking may result in substrate hydro-repellence.
Inorganic		
Perlite: Siliceous volcanic mineral that has been sieved and heated to 1000°C	Light volume weight (0.090–0.130 g/cm^3), sterile, neutral pH, no decay, and good total pore space (50–75% v/v).	Low nutrient capacity, energy consuming, and costly product.
Pumice: Light silicate mineral of volcanic material	Light volume weight (0.450–0.670 g/cm^3), good total pore space (55–80% v/v), cost-effective, extended life, and eco-friendly.	High transportation costs. The pH level could be high.

TABLE 4.1 *(Continued)*

Growing media	Advantages	Disadvantages
Rockwool: Melted silicates at 1500–2000°C	Light volume weight (0.080–0.090 g/cm^3), high total pore space (95–97% v/v), ease to handle, inert and nutrition can be controlled.	Disposal problems and higher energy consumption during production.
Sand: Naturally available with particle size of 0.05–2.0 mm.	Cost-effective and good drainage ability.	High volume weight (1.40–1.60 g/cm^3), low nutrient and WHC, and low porosity.
Vermiculite: Magnesium, aluminum and iron silicate sieved and heated to 1000°C.	Light volume weight (0.080–0.120 g/cm^3), high nutrient and moisture retention capacities, good pH buffering capacity and good total porosity (70–80% v/v).	Contraction takes place on heating. It is energy-inefficient and costly.

Source: Adapted from Gianquinto et al., 2006 and Gruda et al., 2016b thick plastic film having 30–40 m length. In many gravel culture systems, sub-irrigation is applied. Otherwise, as in the nutrient film technique, a nutrient solution is supplied at the upper end, descended into the trough, and drained at the lower end to retrieve and recirculate. This amended gravel culture system is called as the gravel film technique.

TABLE 4.2 Commonly Available Growing Media with their Physical Properties (Adapted from Raviv, 2008).

Substrate	Bulk density (g/cm3)	Total porosity (v/v)	Air-filled porosity (v/v)
Coco fibers	0.04–0.08	≈ 0.86	0.10–0.15
Coir	0.04–0.08	0.94–0.98	≈ 0.70
Peat	0.08–0.18	0.50–0.95	0.30–0.70
Perlite	0.065–0.13	0.91–0.97	0.35–0.70
Rockwool	0.05–0.10	0.92–0.97	≈ 0.82
Sand	1.50–1.80	0.30–0.45	≈ 0.32
Tree bark	0.10–0.30	0.80–0.82	0.40–0.65

Source: Adapted from Raviv, 2008

4.4.2.2 SAND CULTURE

In sand culture, sand is used primarily to support plants in the grow bed or a tray. Plastic-lined beds are used or sand is spread over the entire greenhouse floor. In this culture, the plants are grown in sand and nutrient solution is added time to time. Each plant is fed separately using a drip irrigation system and waste nutrient solution is rarely recycled.

4.4.2.3 BAG CULTURE

Plastic bags loaded with porous slabs or loose granules are utilized to cultivate plants in bag culture. Growers may make these bags themselves or buy them as ready-to-use bags. Rockwool and coir are the most widely used growing media (slab type). The rockwool slab of 90 cm long, 15–20 cm wide, and 8–10 cm high can be suitable for crop growing. The slabs of 15 cm width are most suitable for growing crops like pepper and tomato. Plants that require a solid base and a large root capacity need wider slabs. After rehydration or saturation, the coir slabs expand in size (length of 90–110 cm, width of 15–20 cm, and height of 6–12 cm). In Mediterranean countries, the most commonly used materials in the bag culture are peat, perlite, coir, pumice, or a mixture.

With the purpose of collecting the drainage solution, bags are placed in channels or panels. On the top of them, planting holes are cut and the number of planting holes varies based on the crop types. A dripper is placed in each hole to supply the nutrient solution and the substrate is saturated for 24–48 h with the same. Small cuts or holes are made along the bottom of the plastic envelope to allow the drainage of excess nutrient solution. Saturation is used to extract air and deliver the uniform moisture to the growing soilless media, providing optimum EC and pH conditions in the plants root zone. Whether the nutrient solution is recycled or not, but an environmentally appropriate method of disposing of the effluent from the substrate is needed in open-loop systems. Bags and slabs used over and over again should be discarded.

4.4.2.4 CONTAINER CULTURE

Various containers such as PE, PVC, or polystyrene buckets or pots of capacities of 12–18 l are used. Generally, a container more than 20 cm deep is required for proper root development and plant growth. The depth of the container depends on the crop, crop duration, and type of substrate. For deeper container, the air-to-water ratio is higher in the substrate. To provide nutrients to each plant, a drip irrigation system is used, and drainage is generally ensured by an opening at the bottom of the container. The most widely used growing media in this culture are peat, coir, and perlite (Savvas et al., 2013). The optimal physical properties required for container soilless media (Buck 2018) are given in Table 4.3.

TABLE 4.3 Ideal Physical Properties of Soilless Growing Media (Adapted from Buck, 2018).

Parameter	Optimal range
pH	5.5–6.0
Bulk density (g/cm^3)	0.19–0.70
Water holding capacity (v/v)	0.45–0.65
Total pore space (v/v)	0.60–0.85
Air-filled pore space (v/v)	0.10–0.20

Source: Adapted from Buck, 2018

4.4.2.5 TROUGH CULTURE

Trough culture refers to the cultivation of plants in plastic or plastic-lined troughs that are made above the ground. Trough depths range from 10 to 35 cm, depending on the particle size of the substrate. For an instance, particles of 0.3–0.5 mm size require troughs with at least 35 cm depth. The troughs should have a 0.5% uniform slope. From one side to the other, a discharge pipe is mounted at the bottom of the trough. The plants are placed at normal spaces and irrigated with a drip system to provide nutrients to each plant separately. Perlite, sawdust, coir, sand, and gravel are the most widely used growing media in trough culture (Savvas et al., 2013).

4.5 FERTILIZATION IN SOILLESS CULTURE SYSTEMS

In the fertilizer program of the soilless culture system, all essential plant nutrients are provided in the form of a nutrient solution, except carbon and hydrogen, which are obtained from air as carbon dioxide and water, respectively. Diverse plant species can be effectively cultured using a single nutrient solution composition, as proposed by Hoagland and Arnon (1950) due to the plant's ability to selectively absorb nutrients according to their nutritional needs. The composition of the nutrient solution mainly depends on the following factors:

1. **Type of plant species:** To meet the precise nutritional needs of each plant species, scientists have wisely analyzed the specific nutrient solution compositions (Sonneveld and Voogt, 2009; Savvas et al., 2013).
2. **Growth stage of plant:** The nutritional requirements of plant species may vary at different stages of growth (Sonneveld, 2002; Gianquinto et al., 2013). Therefore, in commercial soilless culture systems, the nutrient solution composition needs to be adjusted periodically during a cropping period.
3. **Type of culture system:** The composition of a nutrient solution works best when the mean uptake ratios between individual nutrients is equal to the relative proportions between the same nutrients in a given nutrient solution, which is strictly obeyed by closed hydroponic systems to evade nutrient accumulation and/or depletion (Sonneveld and Voogt, 2009; Savvas et al., 2017). However, a part of the nutrient solution is exhausted in open hydroponic systems. Therefore, the ratio between some nutrients present in the solutions is different from those taken in, which more or less resemble to their concentrations in the whole plant tissue. The ratios of K: Ca and K: Mg in the root region which should be considerably lesser than

the corresponding uptake ratios to ensure sufficient bivalent cation uptake (Sonneveld and Voogt, 2009).

4. **Climatic conditions:** Under various climatic conditions, the nutrient/water absorption ratio may change, so the optimal composition of a nutrient solution can also be affected by local climatic conditions. As a result, when a nutrient solution composition that has been proven effective in one region is implemented in another region with different climatic conditions with further adjustment if needed.

4.6 MODIFICATION OF CROP NUTRITION IN A SOILLESS CULTURE SYSTEM

The concentration of nutrients in the root zone is very important because plants interact and respond to the nutrients present in their roots. A typical nutrient solution formulation for plants as per Cooper (1988) is given in Table 4.4. The composition of a nutrient solution in the root zone is also essential because it is the primary tool for attaining and sustaining target nutrient concentrations in the same (Savvas, 2001; Sonneveld and Voogt, 2009). Therefore, the concentration of nutrients in a given nutrient solution can be altered several times during the cropping period to more closely match the target concentrations in the root zone, even though the latter remains similar to a particular stage of plant development. However, since different plant parts can have different relative nutrient compositions, crop requirements for a particular nutrient can change during different stages of its growth (Adams, 2002; Gianquinto et al., 2013). As a result, it is better to propose a variety of nutrient solution compositions at the vegetative and reproductive stages in plants with a long harvest time.

The EC of a nutrient solution is an estimate of the total ionic concentration in the solution, considered to be one of the most vital aspects to affect the suitability of the nutrient solution for fertigating the crops. A very low value indicates a lack of nutrient supply, whereas a very high value indicates salt stress (Savvas and Gruda, 2018). The EC in the root zone which is required for ideal plant growth and yield can decrease or exceed the range because plants uptake nutrients selectively. Using quality irrigation water to prepare nutrient solution is a very effective technique for preventing extremely high EC levels in the plant root zone. In some cases, a very high EC in the root zone can be adjusted by increasing the frequency of irrigation (Sonneveld, 2000).

TABLE 4.4 Macro- and Micro-nutrient Components of a Typical Nutrient Solution (Adapted from Cooper, 1988).

Nutrient	Concentration required (g/L)
Macro-nutrient	
$Ca(NO_3)_2$	0.435
KNO_3	0.323
NH_4NO_3	0.008
KH_2PO_4	0.109
K_2HPO_4	0.052
$MgSO_4/7H_2O$	0.185
NaCl	0.012
Micro-nutrient	
$(NH_4)_6Mo_7O_{24}/4H_2O$	5.0×10^{-5}
H_3BO_3	1.5×10^{-3}
$MnSO_4/4H_2O$	2.0×10^{-3}
$CuSO_4/5H_2O$	2.5×10^{-4}
$ZnSO_4/7H_2O$	1.0×10^{-3}
Sequestered Fe 136	10.0×10^{-3}

Source: Adapted from Cooper, 1988

In the soilless culture system, it is much easier and safer to improve the quality of produce by appropriately manipulating the composition of a nutrient solution as compared to the crops grown in soil. An increase in the EC above the upper threshold level in the root zone of soilless-grown vegetable crops in order to boost product quality is the most common example (Sonneveld and Voogt, 2009). High EC in tomatoes results in high dry matter, sugar, and titratable acidity in the fruit juice, while a moderate increase improves vitamin C and total carotene levels in tomato fruit (Schnitzler and Gruda, 2002).

4.7 RECENT PERSPECTIVES

Soilless culture systems are based on environmentally sustainable technology and can contribute to higher crop yields, even in areas with challenging growing conditions. There is a fast development in the use of soilless cultivation technique over the years, as it separates the crop from the soil and its disadvantages. In the soilless culture technique, the use of nutrient solution and growing media helps in increasing crop yields and improving product quality compared to soil-grown greenhouse crops. In addition, the transition from soil to a soilless cultivation leads to the reduction in the use of pesticides and other toxic agrochemicals, which are essential for soil-grown crops to sterilize the soil and control soil-borne pathogens. Ultimately, recycling the surplus nutrient solution that drains away after each irrigation will help to reduce nitrate and phosphate leaching into surface and groundwater resources. The countless potential given by this method is unquestionable, that includes productive, qualitative, and environmental benefits because of the efficient usage of water and nutrients.

Cultivation on high-quality, consumer-oriented expanding media with an emphasis on environmental concerns and sustainability will continue to be the common choice for soilless greenhouse production of edible vegetables, cut flowers, and seedlings in the near future. Over the years, the research on this culture system has primarily focused on the automation of nutrient and water supply, especially in closed systems where the additional nutrient solution is reused. Plant factories using soilless culture technology are being built with a lot of effort.

4.8 CONCLUSIONS

The soilless cultivation system is gaining more and more popularity with time, as it brings numerous advantages over soil-based cultivation. Despite of all the advantages, it helps in the reduction in the use of pesticides and other toxic agrochemicals, which are essential for soil-grown crops to sterilize the soil and control soil-borne pathogens. That is why soilless cultivation becomes even more important for all types of protected cultivation structures designed to enhance favorable climatic conditions. The farmers have started adopting the soilless cultivation systems keeping their financial status in view as well as the subsidy available on the initial cost of a protected structure.

KEYWORDS

- **soilless cultivation**
- **water culture**
- **nutrients**
- **substrate**
- **growing media**

REFERENCES

Adams, P. Nutritional Control in Hydroponics. In *Hydroponic Production of Vegetables and Ornamentals*; Savvas D., Passam H. C., Eds.; Athens, Greece: Embryo Publ., 2002; pp 211–261.

Buck, J. How Your Soilless Mix Can Improve Water and Nutrient Retention? 2018. https://www.greenhousecanada.com/inputs/growing-media/how-your-soilless-mix-can-improve-water-and-nutrient-retention-3236/(accessed Mar 3, 2019)

Cooper, A. 1. The System. 2. Operation of the System. In *The ABC of NFT. Nutrient Film Technique*; Grower Books Ed.; London, England, 1988; pp 3–123, ISBN 0901361224.

El-Kazzaz, K. A.; El-Kazzaz A. A. Soilless Agriculture a New and Advanced Method for Agriculture Development: An Introduction. *Agric. Res. Technol.* **2017,** *3* (2), 555610.

Gianquinto, G.; Orsini, F.; Michelon, N.; Ferreira da Silva, D.; Damasio de Faria, F. Improving Yield of Vegetables by using Soilless Micro-garden Technologies in Peri-urban Area of North-east Brazil. *Acta Hort.* **2006,** *747*, 57–65.

Gianquinto, G. P.; Muñoz, P.; Pardossi, A.; Ramazzotti, S.; Savvas, D. Soil Fertility and Plant Nutrition. In *Good Agricultural Practices for Greenhouse Vegetable Crops-Principles for Mediterranean Climate Areas*; Rome, Italy: FAO, Plant Production and Protection Paper 217, 2013, pp 205–269.

Grillas, S.; Lucas, M.; Bardopoulous, E.; Sarafopoulos, S. Perlite based Soilless Culture System: Current Commercial Applications and Prospects. *Acta Hort.* **2001,** *548*, 105–113.

Gruda, N.; Schnitzler, W. H. Suitability of Wood Fiber Substrates for Production of Vegetable Transplants. I. Physical Properties of Wood Fiber Substrates. *Sci. Hort.* **2004,** *100*, 309–322.

Gruda, N.; Caron, J.; Prasad, M.; Maher, M. J. Growing Media. In *Encyclopedia of Soil Sciences*; 3rd edn., Lal. R., ed; CRC Press, Taylor & Francis Group, 2016b; pp 1053–1058. https://doi.org/10.1081/E-ESS3-120053784.

Gruda, N.; Gianquinto, G.; Tüzel, Y.; Savvas, D. Culture Soil-less. In *Encyclopedia of Soil Sciences*; 3rd edn., Lal. R., ed; CRC Press, Taylor & Francis Group, 2016a, pp 533–537. https://doi.org/10.1081/E-ESS3-120053777

Gruda, N.; Qaryouti, M. M.; Leonardi, C. Growing Media. In *Good Agricultural Practices for Greenhouse Vegetable Crops–Principles for Mediterranean Climate Areas*; Rome, Italy: FAO, Plant Production and Protection Paper 217, 2013, pp 271–302.

Gruda, N.; Savvas, D.; Colla, G.; Rouphael, Y. Impacts of Genetic Material and Current Technologies on Product Quality of Selected Greenhouse Vegetables-A Review. *Europ. J. Hort. Sci.* **2018,** *83* (5), 319–328.

Hoagland, D. R.; Arnon, D. I. The Water-culture Method for Growing Plants without Soil. Calif. *Agric. Exp. St., Circ.* **1950,** *347* (Revised by D. I. Arnon).

Maher, M. J.; Thomson, D. Growth and Mn Content of Tomato (*Lycopersicon Esculentum*) Seedlings Grown in Sitka Spruce (*Picea Sitchensis* (Bong.) Carr.) Bark Substrate. *Sci. Hort.* **1991,** *48*, 223–231.

Nemati, M. R.; Simard, F.; Fortin, J.-P.; Beaudoin, J. Potential use of Biochar in Growing Media. *Vadose Zone J.* **2014,** *14*, 1–8.

Pardossi, A.; Malorgio, F.; Incrocci, L.; Tognoni, F. Hydroponic Technologies for Greenhouse Crops. In *Crops: Quality, Growth and Biotechnology*; Dris R. ed.; WFL Publisher, Helsinki, Finland, 2005, pp 360–378.

Postma, J.; van Os, E.; Bonants, P. J. M. Pathogen Detection and Management Strategies in Soilless Plant Growing Systems. In *Soilless Culture: Theory and Practice*; Raviv M.,

Lieth H. J. Eds.; Amsterdam: Elsevier, 2008, pp 425–457. https://doi.org/10.1016/B978-044452975-6.50012-5
Raviv, M. Composts in Growing Media: What's New and What's Next? *Acta Hort.* **2013,** *982*, 39–52.
Raviv, M.; Lieth, J. H. *Soilless Culture: Theory and practice*; Elsevier: Amsterdam, 2008.
Raviv, M.; Lieth, J. H. Significance of Soilless Cultivation in Agriculture. In Raviv, M.; Lieth J. H. Eds.; *Soilless Culture: Theory and Practice*; Academic Press: San Diego, USA, 2008, pp 111.
Savvas, D.; Gruda, N. Application of Soilless Culture Technologies in the Modern Greenhouse Industry-A Review. *Europ. J. Hort. Sci.* **2018,** *83* (5), 280–293
Savvas, D. Nutritional Management of Vegetables and Ornamental Plants in Hydroponics. In *Crop Management and Postharvest Handling of Horticultural Products, Vol. I: Quality Management*; Dris R., Niskanen R., Jain S. M., Eds.; Enfield, N.H., U.S.A.: Science Publishers, 2001, pp 37–87.
Savvas, D. Nutrient Solution Recycling. In *Hydroponic Production of Vegetables and Ornamentals*; Savvas D.; Passam H. C. Eds.; Embryo Publ.: Athens, Greece, 2002, pp. 299–343.
Savvas, D. Hydroponics: A Modern Technology Supporting the Application of Integrated Crop Management in Greenhouse. *J. Food Agric. Environ.* **2003,** *1*, 80–86.
Savvas, D.; Gianquinto, G. P.; Tüzel, Y.; Gruda, N. Soilless Culture. In *Good Agricultural Practices for Greenhouse Vegetable Crops-Principles for Mediterranean Climate Areas (Rome: FAO)*; Plant Production and Protection Paper 217, 2013, pp 303–354.
Savvas, D.; Oztekin, G. B.; Tepecik, M.; Ropokis, A.; Tüzel, Y.; Ntatsi, G.; Schwarz, D. Impact of Grafting and Rootstock on Nutrient to Water Uptake Ratios During the First Month after Planting of Hydroponically Grown Tomato. *J. Hort. Sci. Biotechnol.* **2017,** *92*, 294–302.
Schmilewski, G. Growing Medium Constituents used in the EU. *Acta Hort.* **2009,** *819*, 33–46.
Schnitzler, W. H.; Gruda, N. Hydroponics and Product Quality. In *Hydroponic Production of Vegetables and Ornamentals*; Savvas D., Passam H. C. Eds.; Embryo Publ.: Athens, Greece, 2002, pp 373–411.
Singh, M. C.; Singh, J. P.; Singh, K. G. Mathematical Modeling of Greenhouse Microclimate Under Vertically Trained Soilless Cropped Conditions. *Agric. Res.* **2021a,** *10.* https://doi.org/10.1007/s40003-021-00609-0
Singh, M. C.; Singh, J. P.; Singh, K. G.; Gupta, O. P.; Kumar, G. Computation of Vapour Pressure Deficit and Crop Transpiration Via Development of a Web Based Computer Module. *J. Agrometeorol.* **2021b,** *23* (4), 243–248.
Singh, M. C.; Singh, A.; Singh, J. P.; Singh, K. G. Economic Viability of Soilless Cucumber Cultivation Under Naturally Ventilated Greenhouse Conditions. *Ind. J. Hort.* **2020,** *77* (2), 315–321.
Singh, M. C.; Singh, J. P.; Singh, K. G. Development of a Microclimate Model for Prediction of Temperatures Inside a Naturally Ventilated Greenhouse Under Cucumber Crop in Soilless Media. *Comput. Electron. Agric.* **2018a,** *154*, 227–238.
Singh, M. C.; Singh, J. P.; Singh, K. G. Development of Mathematical Models for Predicting Vapour Pressure (Deficit) Inside a Greenhouse Independently from Internal and External Climate. *J. Agrometeorol.* **2018b,** *20* (3), 238–241.
Singh, M. C.; Singh, K. G.; Singh, J. P. Nutrient and Water use Efficiency of Cucumbers Grown in Soilless Media Under a Naturally Ventilated Greenhouse. *J. Agric. Sci. Technol.* **2019b,** *21* (3), 193–207.

Singh, M. C.; Singh, K. G.; Singh, J. P. Yield of Soilless Cucumbers Planted Under Partially Controlled Greenhouse Environment in Relation to Deficit Fertigation. *Indian J. Hort.* **2018c,** *75* (2), 259–264.

Singh, M. C.; Singh, K. G.; Singh, J. P.; Mahal, A. K. Performance of Soilless Cucumbers in Relation to Differential Fertigation Under Naturally Ventilated Greenhouse Conditions. *J. Plant Nutr.* **2019a,** *42* (11–12), 1316–1332.

Sonneveld, C. Effects of Salinity on Substrate Grown Vegetables and Ornamentals in Greenhouse Horticulture. Thesis (The Netherlands: Wageningen University), 2000, p 151. https://doi.org/10.1007/978-90-481-2532-6

Sonneveld, C. Composition of Nutrient Solutions. In *Hydroponic Production of Vegetables and Ornamentals*; Savvas D.; Passam H. C.; Eds.; Embryo Publ.: Athens, Greece, 2002, pp 179–210. https://doi.org/10.1080/01904169909365692.

Sonneveld, C.; Voogt, W. *Plant Nutrition of Greenhouse Crops*, Dordrecht, Heidelberg, London, New York: Springer, 2009; p 431.

Wohanka, W. Nutrient Solution Disinfection. In *Hydroponic Production of Vegetables and Ornamentals*; Savvas D., Passam H. C. Eds.; Embryo Publ.: Athens, Greece, 2002, pp 345–372.

CHAPTER 5

Greenhouse Utilization for Vegetable Cultivation

CHAUDHARI GANESH VASUDEO[1], PRATAPSINGH SURESH KHAPTE[2], GOPAL RAMDAS MAHAJAN[1], MATHALA JULIET GUPTA[1], R. RAMESH[1], and A. R. DESAI[1]

[1]ICAR-CCARI, Old Goa, Goa, India

[2]ICAR-NIASM, Baramati, Maharashtra, India

ABSTRACT

In the regular growing season, climatic vagaries reduce vegetable production under open-field conditions. The seasonal varieties certainly pose a limitation for off-season vegetable production, qualitative and quantitatively; however, the demand for fresh vegetables prevails year-round. The eye-appealing exterior is still an essential index of quality vegetable produce. Greenhouse cultivation of vegetables is not just a pro-off-season technology, but it has also shown potential for enhancing the yield of vegetables with better quality produce. The present chapter provides consolidated information regarding greenhouse utilization for "vegetable cultivation" through introductory background, global and Indian greenhouse vegetable production scenario, vegetable cropping sequences, and improved greenhouse cultivation techniques elaborated under eight significant topics for growing greenhouse vegetables. Other greenhouse usages like "vegetable nursery" and "research purpose" are briefed to highlight the importance of the protected structures.

Protected Cultivation: Structural Design, Crop Management, Modeling and Automation.
Mahesh Chand Singh and K. K. Sharma (Eds.)

5.1 INTRODUCTION

Protected cultivation is possibly the most intensive method of crop production, conserving water and land, and protecting the environment (Jensen, 2002). The first known example of protected cultivation of vegetables dates back to 42–37 B.C., when cucumbers were grown in carts. These were produced for the Roman Emperor, Tiberius. These carts were wheeled into sunlight during the day and back indoors during the night. The record reveals the voyage collection of European explorers from the 16th century who tried to domesticate the tropical plants in Europe and create greenhouses (Reddy, 2016). Since then, greenhouse technology has progressed from the wooden frame to iron frame, from glass covers to plastic glazing, and the various sophisticated additions and modern modifications. In a nutshell, the essence of greenhouse/ protected cultivation is improving crop microclimate by providing optimum temperature levels, light, preferably in the wavelength range of 400–700 nm known as photosynthetically active radiations (PAR), elevated CO_2 levels, and pest and disease-free environment. However, the greenhouses/protected structures are known with different terms namely, glasshouse, polyhouse, shade-net house, walk-in-tunnel, etc., based on the type of glazing material used, design and size of structure, advancements made, etc. Nearly 90% of the greenhouses are being constructed utilizing ultraviolet (U.V.)-stabilized polythene sheets as the glazing material (Reddy, 2016). Therefore, polyhouse has almost become a synonym for greenhouse in general.

Protected agriculture is being carried out in an area of 5,630,000 ha worldwide (Karaca, 2020). The greenhouse production of vegetables is an essential component of global protected agriculture. Under greenhouse cultivation, the area for vegetables was estimated to be 497,815 ha globally in 2018 (Anon, 2018). Therefore, it is not surprising to note that the global market of greenhouse horticulture (including vegetables) for 2019 recorded a value of 30 billion USD and projected an annual increase of around 9% for the next lustrum (Koukounaras, 2021). Horticulture crops, particularly the vegetables grown under open-field and stressed conditions (abiotic and biotic stress), are restricted to seasonal cultivation. Therefore, have a limitation for their off-season production under open fields. The climatic vagaries, even under a normal growing period, result in reduced yield. On the other hand, the demand for fresh vegetables prevails year-round. The primary objective of protected cultivation is to achieve the higher productivity coupled with improved quality of horticulture produce by overcoming the above mentioned limitations through use of: (1) structures suitable for a particular location's climate, (2) improved cultivation technologies, (3) appropriate management protocols, (4) suitable cropping sequences, etc. to retain a healthy crop with enhanced yield. Ultimately, this can extend the availability of fresh vegetables beyond their normal growing season and obtain the remunerative quantity and quality produce. Further, the low productivity of vegetables and luxurious utilization of natural resources in open-field cultivation can be addressed by adopting protected cultivation in resource-scarce regions. Production of vegetable crops under protected structures provides high input use efficiency under different agroclimatic conditions.

5.2 GREENHOUSE UTILIZATION

5.2.1 GLOBAL AND INDIAN CONTEXT

More than 100 countries in the world are producing greenhouse vegetables at the commercial level. The countries growing greenhouse vegetables in an area of 1000 ha and above are mentioned in Table 5.1. In India, the beginning of protected cultivation technology was made in the early 1990s (Chandra et al., 2002; Sabir and Singh, 2013). However, India's exposure to high-tech protected vegetable cultivation and high-value horticulture farming, in the real sense, came through the Indo-Israel project on greenhouse cultivation at IARI, New Delhi, during 1998–2003 (Reddy, 2016). The studies on greenhouse utilization are under continuous progress in India, which have helped identify greenhouse vegetable varieties, cropping patterns/sequences, improved cultivation techniques, and high-value vegetables and other crops. Like other countries, greenhouse cultivation has gained popularity in India also. It ensures the provision of a favorable microclimate, besides protecting crops from various biotic and abiotic factors, which cause considerable losses to productivity (Khapte et al., 2018).

Greenhouses are being constructed in the Ladakh area to widen the current growing season of vegetables from 3 to 8 months. In the northeastern region, fundamentally as rain shelters, protected structures are being built for better harvesting of vegetables. Vegetable seedlings are raised in greenhouses in the northern plains and the same is now adopted across the country. Andhra Pradesh, Gujarat, Maharashtra, Tamil Nadu, West Bengal, Haryana, and Punjab have consistently expanded areas under the protected cultivation during 2007–2012 (Indore et al., 2019). In the year 2012, the total area under protected cultivation in India was around 25,000 ha, while the area under greenhouse vegetable cultivation was around 2000 ha (Sabir and Singh, 2013). It is also noteworthy here that the greenhouse vegetable cultivation was just around 700 ha in the year 2011 (Hickman, 2011) registering almost three times increase in the area within a year. The major expansion of area under protected vegetable cultivation in India has taken place from the year 2011 and beyond.

TABLE 5.1 Worldwide Areas Under Greenhouse Vegetable Production (Adapted from Hickman, 2011).

Country	Area (ha)	Country	Area (ha)
Algeria	≈5,000	Lebanon	≈1,000
Argentina	≈2,200	Libya	≈1,000
Australia	≈1,300	Mexico	≈4,300
Belgium	≈1,600	Moldova	≈500
Brazil	≈1,000	Morocco	≈16,500
Bulgaria	≈1,100	Netherland	≈4,600
Canada	≈1,200	New Zealand	≈700
Chile	≈2,100	Palestine	≈3,300
China	>81,000	Poland	≈5,200
Columbia	≈1,200	Portugal	≈1,500
Ecuador	≈2,700	Romania	≈1,300
Egypt	≈1,200	Russia	≈1,400
France	≈10,000	Serbia	≈1,000
Germany	≈1,400	South Korea	≈47,000
Greece	≈5,000	Spain	≈70,400
Hungary	≈5,400	Syria	≈3,100
India	≈700	Tunisia	≈1,300
Iran	≈4,000	Turkey	≈33,500
Israel	≈4,000	UAE	≈800
Italy	≈25,000	Ukraine	≈2,700
Japan	≈36,000	United Kingdom	≈700
Jordan	≈2,000	USA	≈700

Source: Adapted from Hickman, 2011

5.3 OFF-SEASON VEGETABLE CULTIVATION AND CROPPING SEQUENCE STUDIES

Greenhouse cultivation of vegetables gives higher yield and better produce quality than the open-field cultivation of vegetables cultivated during the regular growing season (Singh et al., 2018c, 2019b). However, the actual greenhouse utilization for vegetable production is in off-season cultivation (Singh et al., 2018a, 2018b).

There is harsh winter during the December–January months under the north Indian plains. However, off-season cultivation of tomato, French bean, and cucurbitaceous vegetables namely, cucumber, summer squash, bottle gourd, etc., can be taken up during the winter season using different protected structures such as cloches and walk-in tunnels. Cloches and walk-in tunnels are temporary protected structures that can be used from December to mid-February for the off-season cultivation of vegetables. After the weather improves, cloches are dismantled and the glazing of the walk-in tunnels is removed. The walk-in tunnel is suitable for growing off-season vegetables under north Indian plains and lower hills (Table 5.2).

TABLE 5.2 Off-season Vegetable Crops Cultivated Under Walk-in Tunnels (Adapted from IARI, 2014).

Crop	Planting time	Availability	Yield (t/ha)	Net profit (Rupees in lakh/ha)
Bottle gourd	Oct–Nov	Feb–May	25.0–30.0	2.5–3.0
Cucumber	Oct–Nov	Dec–Mar	15.0–30.0	1.0–1.5
French bean	Oct–Nov	Dec–Mar	6.0–8.0	1.0–1.25
Pumpkin	Nov–Dec	Feb–May	30.0–40.0	1.5–2.0
Summer squash	Nov–Dec	Feb–Apr	40.0–50.0	2.0–2.5
Tomato	Oct–Nov	Feb–May	20.0–25.0	1.0–1.5

Source: Adapted from IARI, 2014

Various studies on off-season vegetable cultivation have been taken up inside polyhouses under AICRP on plasticulture engineering and technology (AICRP on PET). Some profitable cropping sequences are recommended for round-the-year cultivation for particular regions (Table 5.3) with their benefit–cost ratios (BCRs).

A naturally ventilated polyhouse is a structure made of G.I. pipes, insect-proof net or shade net material, and transparent plastic sheets. All four walls of such greenhouses are fitted with an insect-proof (40 mesh nylon) net or 50% shade net. Rollable plastic curtains are used over these nets to facilitate proper cross ventilation. An insect-proof nylon net is also used in place of roof ventilators for natural airflow and insect-free ventilation. For irrigation, a low-pressure drip irrigation system is used. Such kinds of polyhouses do not demand electricity for operational use. This type of structure is suitable for peri-urban areas where high-value vegetables like tomato, capsicum, parthenocarpic cucumber, etc. can be grown (IARI, 2014). Single crop (tomato/capsicum/cucumber) round-the-year cropping looks attractive with an alternate year of crop change (Table 5.4). The parthenocarpic cucumber has registered the benefit–cost ratio of 2.0 (Table 5.4). Singh et al. (2007) reported that under low cost naturally ventilated polyhouses in northern plains of India, the year-round cultivation of cucumber is most suitable and economical. It is observed that cultivation of cucumber year-round in a modified naturally ventilated greenhouse under the west coast ecosystem is economical (Gupta et al., 2017, 2021). Singh et al. (2020, 2021) have also reported soilless cucumber (parthenocarpic) cultivation to be economically viable when cultivated inside a naturally ventilated polyhouse.

TABLE 5.3 Remunerative Cropping Sequences Suggested for Year-round Cultivation Under Protected Structures at Different Locations (Adapted from Indore et al., 2019).

Crop rotation/sequence	Region/area	Benefit–cost ratio (B:C)
Capsicum–Capsicum	Palampur, Himachal Pradesh	4.2–5.2
Capsicum–Cucumber–French bean		≈5.0
Capsicum–Tomato–Tomato–Spinach	Hills of Uttarakhand	2.3
Summer squash–French bean–Tomato–Spinach		2.9
Capsicum–Tomato–Carrot–Lettuce	Northeastern region	1.4–2.6
Spinach–Broccoli–Capsicum	Punjab and Rajasthan	4.1

Source: Adapted from Indore et al., 2019

TABLE 5.4 Pattern of Cropping for Year-round Cultivation Inside a Naturally Ventilated Greenhouse or Polyhouse (Adapted from IARI, 2014).

Pattern of cropping	Duration	Varieties/hybrids	Benefit–cost ratio (B:C)
Capsicum	August to May	Bombi, Indra, Swarna, and Orobelle	≈2.6
Parthenocarpic cucumber	Mid-July to May*	Isatis and Kian	2.0
Tomato	August to May	GS-600	2.5

Three crops.

Source: Adapted from IARI, 2014

5.4 IMPROVED GREENHOUSE CULTIVATION TECHNIQUES

Tomato, capsicum, cucumber, and lettuce are primary greenhouse vegetables in different parts of the world; however, China grows tomatoes, capsicum, and eggplant as significant greenhouse vegetables (Hickman, 2011). Therefore, we can say that tomato, capsicum, and parthenocarpic cucumber are the principal greenhouse vegetables grown in different parts of the globe. The eight significant points related to the improved techniques of greenhouse cultivation (Figure 5.1) are discussed below that need to be considered for growing tomato, capsicum, and parthenocarpic cucumber inside the greenhouses.

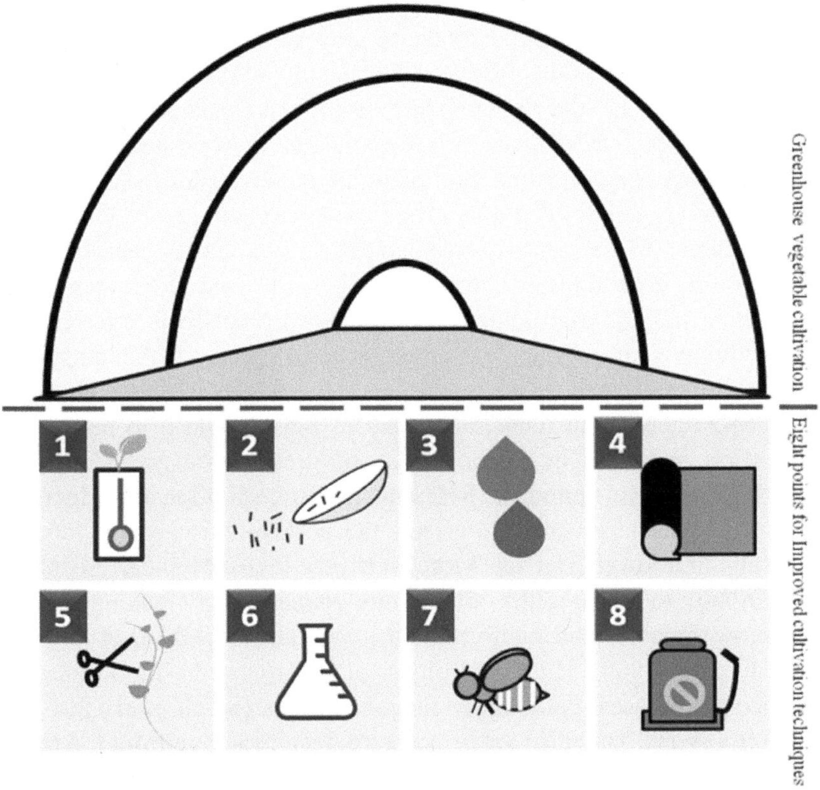

FIGURE 5.1 Eight points for improved greenhouse cultivation techniques for vegetable production.

5.4.1 SELECTION OF CROP VARIETIES AND GREENHOUSE ENVIRONMENT

The selection of crop varieties for greenhouse cultivation is an important decision that determines the response of all the other inputs, which enhance merely the inherent traits and qualities of the chosen combination. It is better to select those high-yielding tomato varieties that show indeterminate growth with two-three locular firm fruits and have a better shelf life for greenhouse cultivation. Indeterminate types of high-yielding capsicum varieties that produce firm fruits with thick pericarp are preferred under greenhouse cultivation. They are available in different fruit colors (green, yellow, red, purple, etc.). Since there is a high initial investment in greenhouse construction, growers cannot afford the wrong selection of cultivars, especially of very costly gynoecious parthenocarpic cucumber hybrids (Kumar et al., 2019). The parthenocarpic cucumber varieties do not need pollination, and therefore they are best suited for greenhouse cultivation.

Plant growth and development are strongly associated with the crop environment, which determines the yield and quality of produce (Table 5.5). Air temperature is a crucial microclimatic parameter (Wein, 1997; Singh et al., 2016, 2017b, 2018b) affecting crop production and quality. Greenhouse crops are chiefly of warm season adapted to air temperatures between 17 and 27°C, with the crucial lower and upper range of 10 and 35°C, respectively (Kittas et al., 2005). High temperature has been reported to interfere with the reproductive development of numerous plant species. If the temperature of tomato fruits exceeds 30°C, lycopene synthesis gets inhibited (Brandt et al., 2006). Higher night temperature causes blossom drop in bell pepper. Buckeye rot incidence in tomatoes may be greater in a greenhouse with warm, high relative humidity (RH) and low transpiration rate. Increased humidity inside the greenhouse affects the pollen dehiscence in tomatoes and results in a lower fruit set. Moreover, very high RH under greenhouse results in high turgor pressure, which may increase fruit cracking and light coloring in tomatoes and malformation in cucumber (Bakker et al., 1987; Leonardi et al., 2000). High humidity is also associated with an increased incidence of diseases, particularly fungal diseases (Singh et al., 2017a) in greenhouse crops. The ideal vapor pressure deficit ranging from 0.4 to 0.79 kPa is recommended for most of the greenhouse crops (Omafra, 2005). In the hot arid region of India, the performance of gynoecious cucumber (cv. Terminator) was assessed under protected structures namely, naturally ventilated polyhouse (NVP), insect-proof net house (INH), and shade net house (SNH). During the cropping period, the photosynthetically active radiation (PAR) inside these structures ranged from 154–842 $\mu mol\ m^{-2}\ s^{-1}$, much lower than the outside. Among the structures, the air temperature was 1.2 and 0.7°C lower, while relative humidity was 17 and 4% higher in NVP and SNH, respectively, compared to INH. Hence, it revealed that the NVP is the best low cost protected structure for successfully cultivating greenhouse cucumber in Indian hot arid regions (Khapte et al., 2021). For high humidity coupled with high rainfall and moderate to high temperatures prevailing in coastal regions of India, fan-ventilated greenhouses were found suitable (Gupta et al., 2016, 2020).

TABLE 5.5 Temperature and Humidity Requirements for Improved Yield and Quality of High-value Vegetable Crops (Adapted from Anon, 2021a, 2021b).

Parameter	Crops		
	Cucumber	**Sweet pepper**	**Tomato**
Day and night time average temperature (°C)	23 and 17–21	21–23 and 16–17	22 and 17–18
Relative humidity (%)	60–70	50–70	60–85

Source: Adapted from Anon, 2021a, 2021b

The studies revealed that most greenhouse crops show a positive response to increased carbon dioxide (CO_2) levels up to 1000–1500 ppm (Nelson, 2002). However, the response of crops to CO_2 was the highest in the lower concentrations. There is an upper limit for CO_2 enrichment beyond which there are no additional benefits, but the toxicity once it reaches an extreme limit. According to Nederhoff (1995), the most effective range of CO_2 concentration is that time when it is increased from 350 to 400 ppm. The CO_2 enrichment to the levels of 1000–1200 μmol mol^{-1} is a common practice for greenhouse cultivation to increase yield and earliness (Chalabi et al., 2002). Response of greenhouse tomato to CO_2 enrichment is strongly influenced by the duration of CO_2 enrichment and the greenhouse climate. The CO_2 enrichment at 800 μL L^{-1} concentration leads to an increase in bell pepper's early and total yield (Akilli et al., 2000). In the case of cucumber, CO_2 enrichment at 3000 v.p.m. resulted in better fruit yield (Enoch et al., 1976).

5.4.2 GROWING SUBSTRATE

The usual method of growing vegetables inside a greenhouse involves using existing soils as a growing substrate. Rigorous soil use demands soil maintenance by soil testing and treatment for fertility replenishment. Also, sanitization, solarization, mulching, and fumigation of soil are essential to maintain soil health and structure inside the closed structure (Sabir and Singh, 2013). For getting higher yield, soil solarization and other methods such as the incorporation of manure, compost, and organic amendments for improving soil structure and replenishing soil fertility are vital, which ultimately increase soil microbial activities and also reduce soil-borne diseases (Padmodaya and Reddy, 1999; Gelosmino et al., 2010; Maruo et al., 2002; Michel and Lazzeri, 2010; Ge Tida et al., 2011; Sabir and Singh, 2013). Apart from this, the research endeavors for soilless cultivation paved the way for the use of local and artificially composed chemically inert materials like sand, sawdust, coir, rockwool, cocopeat, perlite, vermiculite, etc. Although the soilless approach for vegetable cultivation in India is comparatively nascent, it is usually adopted on large scale by commercial growers. It has been widely adopted in developed countries with different advancements like hydroponics, aeroponics, aquaponics, etc. for intensive vegetable production.

The yield of different vegetables tends to be higher where plants are grown in various growing media and substrates, which can be mixed according to plant needs (Olle et al., 2012). In contrast to the benefits mentioned above of various growing media, Fandi et al. (2008) found increased yield where plants were grown in soil compared to sand substrates. Furthermore, the studies continuously reflect the benefit of mixing organic and inorganic media components for growing greenhouse vegetables with improved production performance (Tzortzakis and Economakis, 2005; Gao et al., 2010). Sabir and Singh (2013) aptly supported the idea of an integrated approach of cultural practices for more benefits in the long run. Indian greenhouse utilization, especially for vegetable cultivation, mainly integrates soil and soilless, inorganic–organic media components with integrated nutrient management (INM) approaches. Details of vegetable quality and productivity influenced by growing medium (Olle et al., 2012) are worth considering before preparing a media composition for producing a particular greenhouse crop.

5.4.3 IRRIGATION AND FERTIGATION

The varieties of greenhouse vegetables have continuous vegetative and reproductive growth, since balanced nutrition and precise irrigation are required for proper growth and development. Fertigation scheduling for the principal greenhouse vegetables is mentioned in Tables 5.6 and 5.7. Drip irrigation and fertigation require information such as plant needs, water quality, and soil properties. Fertilizer schedule also varies for different vegetables based on the crop stage, varieties, season, etc. (Sabir and Singh, 2013). Nutritional disorders may arise in soil or soilless media due to imbalanced fertilization and/or root environment limitation, posing severe yield and quality losses (Savvas et al., 2008).

TABLE 5.6 Fertigation Schedules of High-value Vegetable Crops Cultivated Under Protected Structures (Adapted from Open Access Source: Hasan et al., 2010).

Crop	Spacing (m)		Fertigation dose (ppm)*			Fertigation dose in kg/1000 m²			Yield (q/1000 m²)
	Plant to plant	Row to row	N	P	K	N	P	K	
Capsicum	0.30	0.40	100–200	50–100	100–220	32	15	36	40–50
Cucumber	0.30	0.40	120–220	60–80	120–240	10–16	4–7	11–18	30–40
Tomato	0.45	0.40	100–250	50–80	100–250	35	12	39	150–200

**Doses vary based on crop variety stage/season.*

Source: Adapted from Hasan et al. 2010

TABLE 5.7 Nutrient Solution Compositions for Some Greenhouse Vegetables (Adapted from Voogt, 2005).

Crop	Nutrient solution (mg/dl*)					
	Ca	Mg	K	SO_4	NH_4	NO_3
Capsicum	36	18	90	18	7.2	151.2
Cucumber	36	18	63	18	16.2	151.2
Tomato	36	27	90	27	7.2	169.2

**Deciliter.*

Source: Adapted from Voogt, 2005

5.4.4 MULCHING

In general, the use of polythene mulch in vegetable production is reported to have manifold benefits like improved hydrothermal regimes of soil, efficient utilization of resources, and excellent soil–water–plant relationship (Ashworth and Hurrison, 1983; Biswas et al., 2015). Mulching is very significant for conserving moisture (Kader, 2019) and suppressing weeds (Sánchez et al., 2008). Its sheet color variation also works for repelling some pests (Rathee et al., 2018). Mulching has been producing rewarding results under protected conditions also and contributes ultimately to crop yield. Singh et al. (2017) studied the effect of mulching under protected conditions in tomatoes. They found double shaded plastic mulch to be significantly superior for producing more fruits per plant, fruits per m^2, with the maximum net return. A study conducted by Dhaliwal et al. (2017) revealed that the yield performance of bell pepper in north Indian plains could be optimized if the crop is planted in a polyhouse under black polythene mulch at an appropriate planting date (around October 30). They recorded improved total yields with early bell pepper production in areas constrained by suboptimal temperature conditions. Rolaniya et al. (2018) found that black polythene mulch in cucumber helped produce significantly more fruits per vine, compared to straw mulch or the control.

5.4.5 TRAINING–PRUNING SYSTEMS

The vegetable varieties suitable for greenhouse are indeterminate in growth habit. Therefore, regular training–pruning is required from 1 to 2 weeks after transplanting. The vines are trained by plastic twine, loosely anchored at the base of the plant with the help of plastic clips to the overhead training wires. The yarn should be wrapped in a clockwise manner along with the plant after fixing the clips. Regular removing of auxiliary shoots in tomato needs to be done once a week. Otherwise, there will be an unnecessary waste of the nutrients applied to the plant, affecting fruiting. Regular and proper pruning of lower leaves helps in good air movement under the plant and facilitates plant management. Different training–pruning systems were studied for the utilization of vertical space in a greenhouse. In the single-stem system of tomato, the significantly higher values for quality parameters namely, lycopene content, total soluble solids (TSS), and vitamin A (Kumar and Patel, 2016), and significantly higher tomato yield per plant (Patel et al., 2016; Yadav et al., 2017) were reported. Pathan and Mathad (2017) found that cherry tomato cultivar (Cv.) HAT-121 responded better to spacing and training levels. From the yield point of view, 45 × 30 cm spacing with four stems training system was found better for the cherry tomato. Still, the wider spacing (45 × 60 cm) and double-stem training were found better concerning economy, quality, and export. Khoshkam et al. (2014) revealed the significantly maximum value of TSS, vitamin C, and acidity in tomato fruits where two-stem training was followed. In capsicum cultivation, four leaders or four-stem training systems were reported for achieving significantly higher fruit yield per plant (Shetty and Manohar, 2008; Thakur et al., 2018). A technique called 3G cutting for pruning cucurbitaceous crops like cucumber, bottle gourd, pumpkin, and muskmelon is being adopted for the varieties exhibiting a more comprehensive male:female flower-bearing ratio (Mourya et al., 2021). However, improved hybrids with gynoecious–parthenocarpic nature in cucumber provide only female flowers. Therefore, the cucumber crop being a vine, plants are trained upward, and trellising of the plant is done to avoid damage to flower buds (Sabir and Singh, 2013) which gives better fruit yield.

5.4.6 PLANT GROWTH REGULATORS

The precise concentration and stage of crop at which the particular plant growth regulator is being applied determine the success with its use. Ram et al. (2014) observed significant increase in the number of fruits per plant and fruit yield per ha with application of GA3 @ 60 ppm on tomato Cv. kashi vishesh in field conditions. Bharti et al. (2017) explained the importance of NAA @ 20 ppm when sprayed on capsicum, which gave better growth and yield along with higher B:C value. The TIBA, NAA, and MH (200–400 ppm) increased fruit size, fruit length, or fruit diameter in cucumber, whereas spray of NAA @ 150 ppm produced maximum cucumber fruit yield (Gosai et al., 2020).

5.4.7 POLLINATION SUPPLEMENTATION

The major vegetable crops grown under greenhouse are self-pollinated. However, the high humidity in naturally ventilated polyhouse hinders the pollen dehiscence. Although tomato is a self-pollinated crop, its exerted stigma type varieties get facilitated with pollination supplementation. Yankit et al. (2018) studied bumblebee pollination in tomatoes under protected conditions. They reported a significant increase in the number of fruits per cluster, fruits per plant, and seeds per fruit compared to the control treatment. Capsicum is an often cross-pollinated crop and responds to pollination supplementation. Significantly superior numbers of fruits per plant were reported by Putra et al. (2014) when *Apis cerena* was employed as a pollinator in *Capsicum annum*. Studies conducted by Rai et al. (2008) in *Cucumis sativus* exhibited that *Apis mellifera* pollination contributed significantly to harvesting higher fruits per vine; however, the present cucumber varieties being parthenocarpic do not need pollination for fruit set. In addition to this, battery-operated pollinators or vibrators can also facilitate pollination and improve the fruit sets.

5.4.8 PLANT PROTECTION TECHNIQUES

Successful greenhouse management can be done by preventing pests and diseases from entering the structure using an integrated approach, which is of utmost importance. It is imperative to implement various integrated pest management (IPM) (Radcliffe et al., 2009), integrated disease management

(IDM) (Gupta and Sharma, 2014), and exclusion strategies as well as cultural practices for possible management of insect pests (nematodes, whiteflies, aphids, thrips, leaf miners, fruit borers, etc.) and diseases (damping-off, wilt, leaf spot, early blight, late blight; other fungal, bacterial, and viral infections). Importantly, once pests and diseases enter the greenhouse, there are limited options to manage them. The most important aspect is to select insect pests and disease-resistant varieties wherever possible. One of the essential things in protected cultivation is to start with pest-free/disease-free planting material. When new planting material lands at the greenhouse, it must be examined for infestation before sowing/planting.

Similarly, new media need to be sterilized/treated for prevention against nematode and soil-borne disease inoculums. One of the critical components in managing soil-borne plant diseases is biocontrol formulations in the ground or media. Various other approaches like proper crop rotation, greenhouse ventilation, and use of clean tools or equipment are also equally important. In addition to the above, proper disposal of trash/plant debris, removal of indoor/immediate-outdoor weeds, and cleaning the greenhouse after every cultivation phase are crucial to trim down invasive agents that develop on outdoor weeds (Rathee et al., 2018).

The other strategies involve ultraviolet-blocking greenhouse cladding material to manipulate insects. Ultraviolet vision effectively prevents the immigration of a wide range of insect pests (whiteflies, aphids, thrips, and leaf miners) from the outdoor into the protected environment. The aphid, whiteflies, thrips, and leaf miner populations were observed to be less on tomatoes grown in a plastic house made of polythene treated to exclude ultraviolet wavelengths in ordinary plastic houses. The use of "ultraviolet-A films" helps reduce the number of insecticide applications by 50–80% to manage *Spodoptera sp.* (Nguyen et al., 2009; Rathee et al., 2018).

The fine mesh insect-proof nets act as a mechanical barrier to the greenhouse openings. Insect-proof screens of 10 × 20 and 10 × 22 μm give adequate exclusion of whiteflies, *Trialeurodes vaporariorum* and *Bemisia tabaci* without impeding their natural enemies (*Diglyphus idea* and *Eretmocerus erimicus*). Reflective silver/metalized mulches are also used under protected conditions to repel sucking pests and thereby the viral diseases. Metalized mulch was effective in reducing silverleaf whitefly entry by 90%. The blend of screening and metalized mulch used together will provide the most significant total reduction of whitefly entry. For whiteflies, aphids, thrips, and leaf miner adults, yellow sticky cards are an excellent supplement to record the pest incidence in the protected environment. Blue-colored sticky traps can also be used (Rathee et al., 2018) for trapping sucking pests. It is always suggested to use judiciously suitable agrochemicals only as a definitive resort, if necessary, with due consideration of the waiting period before harvesting the produce.

5.5 OTHER USES OF GREENHOUSES

5.5.1 VEGETABLE NURSERY

Nowadays, protected structures are also being used to raise nurseries of healthy vegetable seedlings on a commercial scale so that ready-to-transplant seedlings are made accessible to growers at a reasonable cost. The nurserymen use protected structures to produce the quality and genuine seedlings of popular, improved, and ruling varieties/hybrids, and make them available both during "off-season" and at commencement of the "regular growing season."

5.5.2 RESEARCH PURPOSE

Earlier use of protected structures was confined only for research purposes. The process of vegetable micropropagation, in vitro culturing, incubation room, and hardening of in vitro plantlets for transplanting to the open field is not new to science. However, such types of research studies on plants are not possible without the use of protected structures. Phytotron is an enclosed protected structure that provides controlled environmental conditions for growing plants intended for plant physiological and botanical research. Protected structures are also important for plant protection trials to be conducted in a controlled environment.

The protected structures are also being constructed for plant breeding programs for speedy generation advancement of breeding material, which otherwise needs a regular crop season under open-field conditions. Conversely, inside protected structures with manipulation of the photoperiodic and temperature exposure, "speed breeding" is being favored by breeders for bringing plant genetic improvement in hitherto not attainable time.

5.6 CONCLUSION

Limitation of climatic vagaries and seasonal varieties hinders open-field cultivation of vegetables; on the other hand, greenhouse vegetable cultivation offers increased quantity and quality of products even in the off-season.

Consolidated information regarding greenhouse utilization for "vegetable cultivation" is presented in the chapter through an introduction, a glance at global and Indian greenhouse vegetable, vegetable cropping sequences, and improved greenhouse cultivation techniques of principal greenhouse-grown vegetables structured in eight significant points. The other greenhouse utilizations like "vegetable nursery" and "research purpose usages" are also briefed to highlight the importance of the protected structures for vegetables.

On the part of challenges, refinement of technologies for the production of safe vegetables is of utmost importance from the consumers' health point of view, especially when vegetables are consumed in raw form as a salad. Cost-effective and remunerative protected cultivation technologies are to be equally focused. Rapid urbanization and soil sealing under civil construction have become new challenges of the era due to which the cultivable open fields are becoming dwindled year after year. But, the increasing demand for horticulture crops, especially vegetables continues to prevail in time to come. Greenhouse technology is the boon for enhancing the productivity per unit area utilizing the limited land resource for catering the ever increasing demand.

In Indian context, the challenges also include firstly, to increase temperature inside the greenhouse during harsh winter months under extreme north Indian plains and northern hill conditions; secondly, reducing the temperature inside the greenhouse during summer months under lower north Indian plains and central and peninsular India. Similarly, adoption of greenhouse technology in specific locations need refinements like strategy for reducing humidity inside the greenhouse and managing algal growth on greenhouse cladding material under high rainfall as well as humid regions of the coastal ecosystem and northeastern hills.

KEYWORDS

- **greenhouse**
- **vegetables**
- **protected cultivation**
- **controlled environment**
- **polyhouse**

REFERENCES

Akilli, M., O¨ zmerzi, A., Ercan, N. Effect of CO2 enrichment on yield of some vegetables grown in greenhouses. **2000**. *Acta Hort.* 534, 231–234.

Anonymous. International Greenhouse Vegetable Production Statistics, 2018. https://www.cuestaroble.com/(accessed April 24, 2021)

Anonymous 2021a. https://www.deruiterseeds.com(accessed April 24, 2021)

Anonymous 2021b. http://agriculture.infoagro.com(accessed April 24, 2021)

Ashworth, S. and Harrison, H. Evaluation of Mulches for Use in the Home Garden. Hort. Science, 1983, 18, 180–182.

Bakker, J. C.; Welles, G. H. M.; van Uffelen, J. A. M. The Effects of Day and Night Humidity on Yield and Quality of Glasshouse Cucumbers. *J. Hortic. Sci.* **1987,** *62*, 363–370.

Bharti N, Kumar S, Tank R, Desai K. *Green farming*. **2017**; *8*(4): 907–911

Biswas, S.K., Akanda, A.R., Rahman, M.S., Hossain M.A. Effect of drip irrigation and mulching on yield, water-use efficiency and economics of tomato Plant Soil Environ. **2015,** *61*(3): 97–102.

Brandt, S.; P'ek, Z.; Barna, E.; Lugasi, A.; Helyes, L. Lycopene Content and Colour of Ripening Tomatoes as Affected by Environmental Conditions. *J. Sci. Food. Agric.* **2006,** *86*, 568–572. DOI: 10.1002/jsfa.2390

Chalabi, Z. S.; Biro, A.; Bailey, B. J.; Aikman, D. P.; Cockshull, K. E. S.E.-Structures and Environment: Optimal Control Strategies for Carbon Dioxide Enrichment in Greenhouse Tomato Crops-Part 1: Using Pure Carbon Dioxide. *Biosys. Eng.* **2002,** *81* (4), 421–431.

Chandra, P.; Shrivastava, R.; Dogra, A. K., Gupta, M. J. Strategies for Development of Protected Cultivation in India. In *Approaches for Sustainable Development of Horticulture*; Singh H. P., Negi J. P., Samuel, J. C., Eds.; DAC, MOA, **2002**; pp 92–96.

Dhaliwal, M. S.; Sharma, S. P.; Jindal, S. K.; Dhaliwal, L. K.; Gaikwad, A. K. Growth and Yield of Bell Pepper as Influenced by the Growing Environment, Mulch, and Planting Date. *J. Crop Improv*. **2017,** *31* (6), 830–846.

Enoch, H. Z., Rylski, I.; Spiegelman, M. CO2 enrichment of strawberry and cucumber plants grown in unheated greenhouses in Israel. *Scientia Horticulturae*. **1976**. *5*(1): 33–41.

Fandi, M.; Al-Muhtaseb, J. A.; Hussein, M. A. Yield and Fruit Quality of Tomato as Affected by the Substrate in an Open Soilless Culture. *Jordan J. Agric. Sci.* **2008,** *4*, 65–72.

Gao, H. B.; Zhang, T. J.; Lv, G. Y.; Zhang, G. H.; Wu, X. L.; Li, J. R.; Gong, B. B. Effects of Different Compound Substrates on Growth, Yield and Fruit Quality of Cucumber. *Acta Hortic*. **2010,** *856*, 173–180.

Ge TiDa; Nie San'an; Wu JinShui; Shen JianLin; Xiao He'ai; Tong ChengLi; Huang DanFeng; Hong; Yun Iwasaki, K. Chemical Properties Microbial Biomass and Activity Differ Between Soils of Organic and Conventional Horticultural Systems Under Greenhouse and Open Field Management: A Case Study. *J. Soils Sediments* **2011,** *11* (1): 25–36.

Gelsomino, A.; Abenavoli, M. R.; Princi, G.; Attina, E.; Cacco, G.; Sorgona, A. Compost from Fresh Orange Waste: A Suitable Substrate for Nursery and Field Crops? *Compost Sci. Util.* **2010,** *18* (3), 201–210.

Gupta, M. J. Greenhouse Design for High Rainfall, High Humidity, Moderate to High Temperature Coastal Regions. *New Age Prot Cult.* **2016,** *2* (1), 23–24.

Gupta, M. J.; Paramesha, V.; Thangam, M.; Arunachalam, V. Energy Analysis of Different Cucumber Varieties in Protected Cultivation under Humid Tropic Condition of West Coast India. *Prog. Hort.* **2021,** *53* (1), 90–94.

Gupta, M. J.; Thangam, M.; Ramesh, R.; Maruthadurai, R.; Arunachalam, V. Modified Single Span Greenhouse for Small Farmers in the West Coastal Ecosystem. *New Age Prot Cult.* **2017,** 4 (1), 17–19.

Gupta, M. J.; Thangam, M.; Arunachalam, V. Microclimatic Studies in a Double-span Greenhouse under Wind Driven and Fan Ventilated Conditions in West Coast of India. *Acta Hort.* **2020,** *1271*, 227–234.

Gupta, S. K.; Sharma, M. *Approaches and Trends in Plant Disease Management*; Scientific Publishers: India, **2014;** pp 1–429. ISBN: 978-81-7233-887-9.

Gosai, S.; Adhikari, S.; Khanal, S.; Poudel, P. B. Effects of Plant Growth Regulators on Growth, Flowering, Fruiting and Fruit Yield of Cucumber (*Cucumis Sativus* L.): A Review. *Arch. Agric. Environ. Sci.* **2020,** *5* (3), 268–274.

Hasan, M.; Singh, B.; Singh, M. C.; Singh, A. K.; Kaore, S. V.; Naved Sabir T.; Tomar, B. S. Fertigation Scheduling for Horticultural Crops. *Tech Bulletin-ICN*, **2010,** 80.

Hickman, G. W. A Review of Current Data on International Production of Vegetables in Greenhouses, **2011,** 73.

Leonardi, C.; Guichard, S.; Berti, N. High Vapour Pressure Deficit Influences Growth, Transpiration and Quality of Tomato Fruits. *Sci. Hort.* **2000,** *88*, 285–296.

IARI. *Technological Options for Enhanced Productivity and Profit, Indian Agricultural Research Institute*; New Delhi-110 012, 2014; pp 258. ISBN: 978-93-83168-14-9.

Indore, N.; Kannujia, P.; Kale, S. J.; Panda, N.; Malathi, A. N. Offseason Vegetable Production Under Protected Cultivation. *Int. J. Eng. Sci. Res.* **2019,** *9* (5), 54–57.

Jensen, M. H. Controlled Environment Agriculture in Deserts Tropics and Temperate Regions-A World Review. *Acta Hort.* **2002,** *578*, 19–25.

Kader, M. A.; Singha, A.; Begum, M. A.; et al. Mulching as Water-saving Technique in Dryland Agriculture: Review Article. *Bull. Natl. Res. Cent.* **2019,** *43*, 147. DOI: https://doi.org/10.1186/s42269-019-0186-7

Karaca, C. Current Situation, Problems and Solution of Protected Cultivation in Antalya. In *Current Researches in Agriculture, Forestry and Aquaculture Scienc.*; Atılgan A., Saltuk, B., Eds.; Duvar Publishing: Turkey, 2020. ISBN: 978-625-7680-04-2.

Khapte, P. S.; Kumar P.; Saxena A.; Singh A. Performance Evaluation and Character Association Studies in arid Region Greenhouse Tomato Hybrids. *Ind. J. Hort.* **2018,** *75* (3), 457–462.

Khapte, P. S.; Meena, H. M.; Kumar, P., Burman, U.; Saxena, A.; Kumar, P. Influence of Different Protected Cultivation Structures on Performance of Cucumber (*Cucumis Sativus* L.) in Indian Hot Arid Region. *J. Agrometeorol.* **2021,** *23* (3), 265–271.

Khoshkam, S.; Seyed, Z; Aeen, A. The Impact of Different Plant Training Systems on Quantitative and Qualitative Parameters of Greenhouse Tomato Cultivars. *Int. J. Farming Allied Sci.* **2014,** *3* (6), 659–663.

Koukounaras, A. Advanced Greenhouse Horticulture: New Technologies and Cultivation Practices. *Hort.* **2021,** *7*, 1. DOI: https://doi.org/10.3390/horticulturae7010001.

Kumar, P., Khapte, P. S., Saxena, A. and Kumar P. Evaluation of gynoecious Cucumber (*Cucumis sativus*) Hybrids for Early-summer Greenhouse Production in Western Indian Arid Plains. *Ind. J. Agric. Sci.* **2019,** *89* (3), 545–550.

Olle, M.; Ngouajio, M.; Siomos, A. Vegetable Quality and Productivity as Influenced by Growing Medium: A Review. *Žemdirbystė-Agric.* **2012,** *99* (4), 399–408.

Pathan, A. K.; Mathad, J. C. Effect of Planting Geometry and Training Methods on Yield of Cherry Tomato Grown under Shade House. *J. Farm Sci.* **2017,** *30* (3), 435–436.

Nguyen, T. H. N.; Borgemeisterj, M.; Poehling, H. M. Manipulation of Ultraviolet Light Affects Immigration Behavior of *Ceratothripoides Claratris* (Thysanoptera: Thripidae). *J. Econ. Entomol.* **2009,** *102* (4), 1559–1566.

Sabir, N.; Singh, B. Protected Cultivation of Vegetables in Global Arena: A Review. *Indian J. Agric. Sci.*, **2013,** *83* (2), 123–135.

Sánchez, E.; Lamont Jr, W. J.; Orzolek, M. D. Newspaper Mulches for Suppressing Weeds for Organic High-tunnel Cucumber Production. *Hort. Technol.* **2008,** *18* (1), 154–157.

Singh, B.; Kumar, M.; Sirohi, N. P. S. Protected Cultivation of Cucurbits under Low-cost Protected Structures: A Sustainable Technology for Peri-urban Areas of Northern India. *Acta Hort.* **2007,** *731*, 267–272.

Tzortzakis, N. G.; Economakis, C. D. Shredded Maize Stems as an Alternative Substrate Medium: Effect on Growth, Flowering and Yield of Tomato in Soilless Culture. *J. Veg. Sci.* **2005,** *11*, 57–70.

Voogt, W. Composition of the Basic Nutrient Solution for Fertigation for Some Greenhouse Crops. In *International Symposium on Fertigation; Optimizing the Utilization of Water and Nutrients*; Beijing, September 20–24, 2005.

Reddy, P. P. *Protected Cultivation. Sustainable Crop Protection under Protected Cultivation*; Springer Science + Business Media Singapore, **2016**, pp 1–11. DOI: https://doi.org/10.1007/978-981-287-952-3_1

Kumar, S.; Patel, N. Effect of Spacing and Supportive Training on Quality Parameters in Tomato under Protected Culture. *Bioinfolet.* **2016,** *13* (2B), 420–424.

Kittas, C.; Karamanis, M.; Katsoulas, N. Air Temperature Regime in a Forced Ventilated Greenhouse with Rose Crop. *Energy Build.* **2005,** *37* (8), 807–812.

Leonardi, C.; Guichard, S.; Berti, N. High vapour pressure deficit influences growth, transpiration and quality of tomato fruits. *Sci. Hortic.* **2000,** *88*, 285–296.

Mourya D., Rani M.; Akhtar S.; Chattopadhyay T.; Baranwal S. Lauki me 3G Cutting Se Adhik Upaj. *Phal-Phool.* **2021,** *4* (July–August) issue, 44–45. https://icar.org.in/sites/default/files/Phal-Phool-JULY-AUG-2021.pdf

Omafra, S. *Growing Greenhouse Vegetables*; Guelph, Ontario, Canada: Ontario Ministry of Agriculture, Food, and Rural Affairs, 2005.

Patel, N.; Kumar, S.; Tank, R.; Patel, P.; Tiwari, J. Response of Tomato (*Solanum Lycopersicum* L.) to Varying Levels of Spacing and Training under Protected Conditions. *Green Farming Int. J.* **2016**, *7* (2), 438–441.

Putra, R. E.; Permana, A. D.; Kinasih, I. Application of Asiatic Honey Bees (*Apis Cerana*) and Stingless Bees (*Trigona Laeviceps*) as Pollinator Agents of Hot Pepper (*Capsicum Annuum* L.) at Local Indonesia Farm System. *Psyche* **2014,** 5. Article ID 687979, DOI. https://doi.org/10.1155/2014/687979

Rai, A.; Gracy, R.; Kumar, A.; Chaurasia, S.; Rai, M. Effect of *Apis Mellifera* Pollination on the Yield Attributing Characters and Yield of Cucumber (*Cucumis Sativus* L.). *Veg. Sci.* **2008,** *32* (2), 201–202.

Ram, R. B.; Prakash, J.; Meena, M. L. Growth, Flowering, Fruiting, Yield and Quality of Tomato (*Lycopersicon Esculentum* Mill.) as Influenced Plant Bio Regulators. *Inter. J. of Plant Sci.* **2014,** *9* (1), 67–71.

Rolaniya, O.; Verma, I.; Bhunia, S.; Choudhary, S. Effect of Irrigation Levels and Mulch on Growth and Yield of Cucumber (*Cucumis Sativus* L.) under Polyhouse. *Inter. J. Curr. Microbiol. App. Sci.* **2018,** *7* (3), 3748–3756.

Savvas, D.; Ntatsi, G.; Passam, H. C. Plant Nutrition and Physiological Disorders in Greenhouse Grown Tomato, Pepper and Eggplant. *The Eur. J. Plant Sci. Biotechnol.* **2008,** *2* (Special issue 1), 45–61.

Shetty, G.; Manohar, K. Influence of Pruning and Growth Regulators on Flowering, Fruit Set and Yield of Coloured Capsicum (*Capsicum Annuum* L.) cv. Orobelle under Naturally Ventilated Greenhouse. *The Asian J. Hort.* **2008,** *3* (2), 213–216.

Singh, H.; Sharma, P.; Kumar, P.; Dhillon, N.; Sekhon, B. Influence of Mulching on Growth and Yield of Tomato (*Solanum Lycopersicum* L.) under Protected Environment. *Int. J. Biotechnol.* **2017,** *19* (2), 1–6.

Singh, M. C.; Yousuf, A.; Singh, J. P. Greenhouse Microclimate Modeling under Cropped Conditions-A Review. *Res. Environ. Life Sci.* **2016,** *9* (12), 1552–1557.

Singh, M. C.; Singh, J. P.; Pandey, S. K.; Mahay, D.; Srivastava, V. Factors Affecting the Performance of Greenhouse Cucumber Cultivation-A Review. *Int. J. Curr. Microbiol. Appl. Sci.* **2017a,** *6* (10), 2304–2323.

Singh, M. C.; Singh, J. P.; Singh, K. G. Optimal Operating Microclimatic Conditions for Drip Fertigated Cucumbers in Soilless Media under a Naturally Ventilated Greenhouse. *Indian J. Ecol.* **2017b,** *44*, 821–826.

Singh, M. C.; Singh, J. P.; Pandey, S. K.; Cutting, N. G.; Sharma, P.; Shrivastav, V.; Sharma, P. A Review of Three Commonly used Techniques of Controlling Greenhouse Microclimate. *Int. J. Curr. Microbiol. Appl. Sci.* **2018**a, *7* (1), 3491–3505.

Singh, M. C.; Kachwaya, D. S.; Kalsi, K. Soilless Cucumber Cultivation under Protective Structures in Relation to Irrigation Coupled Fertigation Management, Economic Viability and Potential Benefits-A Review. *Int. J. Curr. Microbiol. Appl. Sci.* **2018b,** *7* (03), 2451–2468.

Singh, M. C.; Singh, K. G.; Singh, J. P. Performance of Soilless Cucumbers Planted under Partially Controlled Greenhouse Environment in Relation to Deficit Fertigation. *Ind. J. Hort.* **2018**c, *75* (2), 259–264.

Singh, M. C.; Singh, K. G.; Singh, J. P. Nutrient and Water Use Efficiency of Cucumbers Grown in Soilless Media under a Naturally Ventilated Greenhouse. *J. Agric. Sci. Technol.* **2019a,** *21* (3), 193–207.

Singh, M. C.; Singh, K. G.; Singh, J. P.; Mahal, A. K. Performance of Soilless Cucumbers in Relation to Differential Fertigation under Naturally Ventilated Greenhouse Conditions. *J. Plant Nutr.* **2019b,** *42* (11–12), 1316–1332.

Singh, M. C.; Singh, A.; Singh, J. P.; Singh, K. G. Economic Viability of Soilless Cucumber Cultivation under Naturally Ventilated Greenhouse Conditions. *Indian J. Hort.* **2020,** *77* (2), 315–321.

Singh, M. C. Status of Protected Cultivation in India, Challenges, Scope and Future Needs for Its Successful Adoption- Review. *Agric. Res.* **2021,** 10 (accepted article).

Thakur, G.; Singh, A.; Patel, P.; Maurya, P.; Kumar, U. J. Effect of Training Level on Growth and Yield of Capsicum (*Capsicum Annuum* L) Hybrid Buffalo under Natural Ventilated Polyhouse. *Journal Pharmacogn. Phytochem.* **2018,** S.P. I., 82–85.

Yadav, S.; Ameta, K. D.; Sharma, S. K.; Dubey, R. B.; Rathore, R. S.; Kumar, H.; Kapuriya, V. K. Effect of Spacing and Training on Vegetative Growth Characteristics and Yield of Tomato (*Solanum Lycopersicum* L.) Grown in Polyhouse. *Inter. J. Curr. Microb. App. Sci.* **2017,** *6* (5), 1969–1976.

Yankit, P.; Rana, K.; Sharma, H.; Thakur, M.; Thakur, R. K. Effect of Bumble Bee Pollination on Quality and Yield of Tomato (*Solanum Lycopersicum* Mill.) Grown under Protected Conditions. *Int. J. Curr. Microb. App. Sci.* **2018,** *7* (1), 257–263.

Maruo, N.; Furukawa, Y.; Taira, K.; Asano, T. Transition of Soil Nitrogen with High Levels of Compost Application under Greenhouse Culture [Japanese]. *Bull. Nara Pref. Agric. Exp. Stn.* **2002,** *33*, 35–37.

Michel VV and Lazzeri L. Green Manure and Organic Amendments to Control Corky Root of Tomato. *Acta Hortic.* **2010,** *883*, 287–294.

Nelson, P. V. *Greenhouse Operation and Management*, 6th ed.; Prentice Hall, Upper Saddle River: New Jersey, USA, **2002;** pp 692.

Nederhoff, E. Commercial Grower **1995,** *50* (10), 25–26.

Padmodaya, B.; Reddy, H. R. Effect of Organic Amendments on Seedling Disease of Tomato caused by *Fusarium Oxysporm* sp *Lycopersici. J. Mycol. Plant Pathol.* **1999,** *29* (1), 38–41.

Radcliffe, E. B.; Hutchison, W. D.; Cancelado, R. E. *Integrated Pest Management Concepts, Tactics, Strategies and Case Studies*; Cambridge University Press, 2009; pp 1–529. ISBN 978-0-521-69931-0.

Rathee, M.; Singh, N. V.; Dalal, P. K.; Mehra, S. Integrated Pest Management under Protected Cultivation: A Review. *J. Entomol. Zool. Stud.* **2018,** *6* (2), 1201–1208.

Wein, H. C. *The Physiology of Vegetable Crops*; CAB International: Wallingford, U.K., **1997;** p 662.

CHAPTER 6

Effect of Greenhouse Microclimate on Crop Performance

ALPANA JOSHI, AKANSHA SAXENA, and SUBRATA K. DAS

School of Biological Engineering and Life Sciences, Shobhit Institute of Engineering and Technology (Deemed-to-be-University), Meerut, India

ABSTRACT

The commercial production of crops in controlled greenhouse conditions has become popular because it provides a favorable environment for sustainable crop production throughout the year. The microclimate of a greenhouse depends on solar radiation, temperature distribution, and relative humidity. Greenhouse microclimate is not uniform across the greenhouse and is significantly affected by crop geometry and greenhouse management. Additionally, it is influenced by the material used, the shape of a greenhouse, its positioning, and the wind direction. Greenhouse microclimate is the main driving force that adversely affects plant metabolic activities, yield, and fruit quality. High summer temperature, vapor pressure deficit, and solar radiation are the major setbacks for successful greenhouse crop production, which directly increase stress in plants and decrease crop productivity and quality. The natural ventilation, shade cloths, cooling pads, fogging systems, and roof sprinklers are the most common methods used by growers to decrease solar radiation and air and leaf temperatures, which eventually improve crop growth and prevent disease development. During winter, the photosynthetic rates are reduced due to the loss of light which leads to a proportional loss in crop growth and productivity, so the main systems used for greenhouse climatic control are heating, ventilation, and CO_2 enrichment.

Protected Cultivation: Structural Design, Crop Management, Modeling and Automation.
Mahesh Chand Singh and K. K. Sharma (Eds.)

6.1 INTRODUCTION

Greenhouse is a closed structure that maintains a suitable environment for improved crop growth and higher production (Tiwari, 2003). Greenhouse systems have numerous applications like aquaculture, crop farming, crop drying, poultry, and soil solarization (Aritesty and Wulandani, 2014; Azaizia et al., 2017). Greenhouse cultivation of vegetable crops is also becoming popular and successful, especially in the hilly regions of India. There are about 115 countries using greenhouse farming system for the commercial production of vegetable crops (Sabir and Singh, 2013). Additionally, improved growth of vegetable crops in controlled environmental conditions gives a better product quality and increased production because of little use of chemical fertilizers and pesticides, which is not possible under open-field conditions (Hanan, 1998; Chauhan et al., 2017). The chances of insects and disease infestation is also higher in the open fields of tropics due to extreme solar radiation, high rainfall, and high humidity.

Greenhouse system can be classified on the basis of types of materials used (glass house, plastic film, and rigid panel greenhouse) and shapes of the cover (spherical dome, hyperbolic paraboloid, quonset, modified quonset (modified IARI model), gothic arch, mansard roof, even span, uneven span, vinery, and modified arch) (Tiwari, 2003). Multispan metal-framed modular greenhouses are the most dominant type of greenhouses, whereas quonset-shaped "polytunnels" and air-supported greenhouses utilize a lightweight and transparent plastic film as the cover material, and the rate of heat loss from a polyethylene-covered greenhouse is higher than the glass greenhouse (Figure 6.1). Although, the deployment of insulating screens or the use of multiple-paned greenhouses may reduce heat losses (Santamouris et al., 1996; Ghosal and Tiwari, 2006; Shukla et al., 2006; Kuehni et al., 2016).

Greenhouse microclimate is a combination of physical processes which are governed by environmental conditions like solar radiation, greenhouse structure, crop type temperature distribution, and relative humidity (Singh et al., 2017a, 2017b). The crop productivity is not only influenced by the genetic composition of plants but the microclimate also has a major role. There are two key components of the greenhouse microclimate, that is, greenhouse illumination (light) and temperature (Singh et al., 2018). The illumination is most important because light is a vital component for the photosynthesis process, and heat and light are required throughout the development stages of the crop. Moreover, the greenhouse location, cladding material, shape, and orientation also affect the inside microclimate of the greenhouse.

Temperature is the other essential component of the microclimate; higher temperature, solar radiation, and high vapor pressure deficit are the major concerns which indirectly decrease crop productivity and fruit quality by causing stress in plants (Ghani et al., 2019). Typically, the temperature of different parts of greenhouse is not even and keeps changing due to constant air circulation, but at the same time, ventilation is essential for heat transfer process. Hence, it is crucial to understand the greenhouse microclimate and its characteristics to maintain optimum operating conditions of greenhouse during various periods of plant growth and development (Table 6.1).

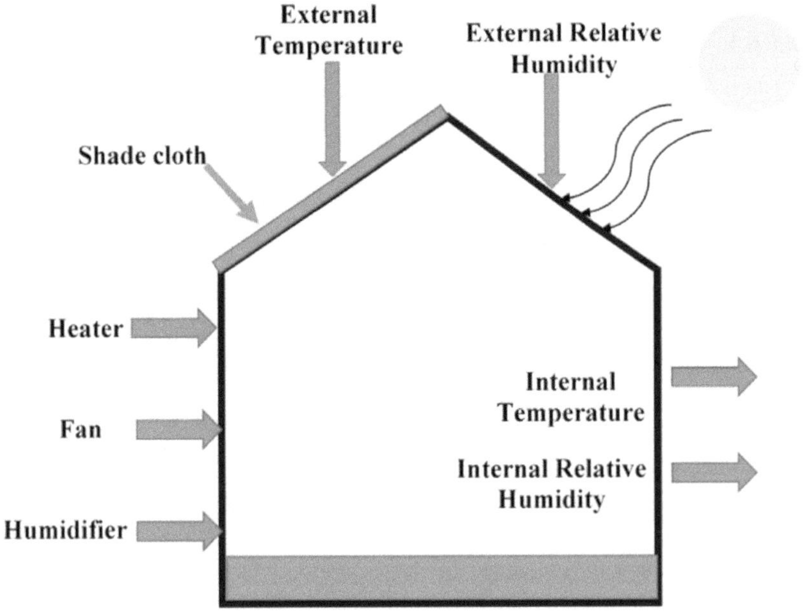

FIGURE 6.1 Basic design elements of greenhouse microclimate.

6.2 MICROCLIMATE CONTROL IN GREENHOUSE

Microclimate control in greenhouses is essential to achieve optimal growing environment, and the main objective of optimizing greenhouse microclimate is to attain maximum crop growth and production (Singh et al., 2016). The greenhouse contains automatic control system that monitors air temperature, electrical conductivity, soil temperature and moisture, relative humidity, and carbon dioxide concentrations (Radojevic et al., 2014).

Greenhouse materials allow shorter wavelength to pass through, but long wavelength radiation is stuck inside the greenhouse which eventually rises the temperature of greenhouse. Greenhouse overheating is the serious problem for plant growth and development, especially in arid and semiarid regions. Because of the greenhouse effect cooling is crucial to provide the optimum range of temperatures and humidity inside the greenhouse (Shamshiri et al., 2018; Choaba et al., 2019). Cooling systems may be generally classified into passive and active cooling systems. Passive cooling system does not use energy to cool the greenhouse, whereas the active cooling system uses energy to cool the greenhouse (Yang et al., 2013).

TABLE 6.1 Key Factors Affecting the Greenhouse Microclimate.

Sr. No.	Characteristics	Optimum climatic condition	Operational parameters
1	Temperature	Air temperature	Ventilation rate
		Wind speed insolation	Heating and cooling systems, shades and thermal screens
2	Photosynthetically active radiation intensity	Insolation	Presence of shades and additional lighting
3	Atmospheric elements	Levels of humidity, CO_2, NOx, SO_2	Ventilation rate, humidity control, CO_2 supplementation

There are different methods of passive cooling like natural ventilation, white washing, and shading nets, which are considered simple and affordable ways to reduce greenhouse heat accumulation (Sethi and Sharma, 2007; Garcia et al., 2011; McCartney et al., 2018; Aroca-Delgado et al., 2018; Singh et al., 2018). Natural ventilation is the simplest and cheapest way to reduce heat inside the greenhouse in the tropical climate. Some studies suggested that natural ventilation alone is not enough to reduce the temperature in tropical region (Romantchik et al., 2017). Therefore, additional cooling system is important for proper growth and crop production. The active cooling methods like evaporative pads, fogging, misting, and fan ventilation are being used successfully in tropical weather (Table 6.2). As far as the cooling efficiency is concerned, it is higher in temperate climates where the air temperature and humidity levels are at lower side (Vox et al., 2010; Ganguly et al., 2010; Giacomelli et al., 2012; Ishii et al., 2016; Romantchik et al., 2017; Rong et al., 2017; McCartney et al., 2018; Nikolaou et al., 2018).

TABLE 6.2 Different Types of Cooling Technologies and Performance of Greenhouse.

Sr. No.	Technology	Method	Location	Performance	Cooling
1	Air conditioning system using a cool storage water tank and three heat pumps	Steady-state simulation model	Yamagata, Tokyo and Kagoshima, Japan	62–100% of the greenhouse cooling demands	Lower air temperature to 15°C in night time cooling
2	Air conditioning system using ground water	Estimation of overall heat transfer coefficient	Borj Cedria, Tunisia	Reduce the greenhouse air temperature by 12°C	Maintain temperature to the required levels during summer period
3	Air conditioning system with heat storage system and a fine wire heat exchanger	Numerical matrix-based model	Wageningen, The Netherlands	Cooling capacity of 20 kW	Reduces air temperature up to 25°C
4	Different colored shade nets	Variation of relative shading intensities	Moravac near Aleksinac, Serbia	Enhancement of productivity by improving temperature extremes	Increase in efficiency
5	Evaporative cooling	Mathematical modeling	Saudi Arabia	Energy effectiveness up to 50.1%	Yes
			Delhi, India	0.5 as maximum low instantaneous loss efficiency	Reduce the air temperature by 6°C
6	Fan–pad evaporative cooling system	Complete randomize design (CRD) Wind tunnel measurements, constant water flow rate	Sudan Aarhus, Denmark	Cooling efficiency up to 90% Cooling efficiency up to 69.2%	Decrease in air temperature by 11.77% Decrease in air temperature by 10%
7	Fan–pad evaporative cooling system combined with shading mechanisms	Thermal modeling considering steady-state condition	Kolkata, India	0.88 of saturation efficiency of cooling pad	Maintains air temperature

TABLE 6.2 *(Continued)*

Sr. No.	Technology	Method	Location	Performance	Cooling
8	Fan–pad system supplied by photovoltaic panels	Mathematical model	Chapingo, Mexico	Can generate up to 10,533 kWh per year	Maintains temperature below 25°C
9	Fog-cooled greenhouse	Experimental study	Arta, Western Greece	Vapor pressure deficit was about 45% lower under fog condition	Maintains temperature below 26°C
10	Ground source evaporative cooling system	Psychometric and thermodynamic analysis	Karabuk, Turkey	Refreshment efficiency of 38%	Yes, can be maintained to the required levels
11	Heat pump and grid connected solar photovoltaic panels	Energy Plus software	Izmir, Turkey	Electricity generation can meet greenhouse demand	Yes
12	Heat pump air conditioner	Steady-state model	Bangkok, Thailand		Reduce air temperature up to 18°C
13	Hybrid low energy direct evaporating cooler	Climate assessment tool	Adelaide, Brisbane, Canberra, Darwin, Hobart, Melbourne, Perth, Sydney	Decline in energy consumption	It can be controlled to the required levels
14	Indirect–direct evaporative cooling	Estimation of the cooling loads of the greenhouse	Baghdad, Iraq	Higher cooling efficiency as compared to the direct evaporative cooler	Yes, temperature decrease from 12.1 to 21.6°C
15	Modified plastic net houses	Measurement of solar radiation transmittance	Riyadh, Saudi Arabia	Reduction of the costs for construction, water consumption and electric energy	Yes, depends on the porosity and shading color of the nets
16	Natural ventilation	Thermal model	China	Energy saving	Yes, can be controlled to the required levels

TABLE 6.2 *(Continued)*

Sr. No.	Technology	Method	Location	Performance	Cooling
17	Naturally ventilated greenhouse	Dynamic climate model	Zimbabwe		Yes, can lower air temperature by 10°C
18	Naturally ventilated greenhouse with a polypropylene mulch covering the soil	Semiempirical dynamic model	Almeria, Spain	Energy saving	Yes
19	Naturally ventilated greenhouse with a mist system	Thermal model	Volos, Eastern Greece	Evaporative cooling potential increased	Yes
20	Naturally ventilated greenhouse equipped with screens	Numerical study	Bologna, Italy	Not available	Reduces air temperature
21	Roof cooling system using direct evaporation from a porous layer	Experimental study	Manchester, United Kingdom	Relative evaporation rates up to 1.2	Reduces air temperature by 5°C
22	Rotative photovoltaic panels and reflective aluminum mirrors with ventilation	Energy balance equation	Lazio, Italy	Decrease of solar radiation	Maintain inside temperature at 26°C
23	Solar cooling plant	Dynamic simulation model	Bari, Southern Italy	Energy saving	Maintained air temperature to the required level
24	Solar hybrid power system	Four-parameter model	Kolkata, India	Can support the energy requirement of floriculture greenhouse	Yes
25	Surplus air heat pump system	Air thermal energy utilization and temperature control	South Korea	Energy conservation up to 76.4%	Maintained air temperature to the required level
26	Variable pressure fogging system	Quantitative feedback theory	Beit Dagan, Israel	The performance of the controllers was not affected by set points	Maintained air temperature to the required level

6.2.1 SHADING SYSTEM

In arid and semiarid regions, the air temperature exceeds most of the time due to the high levels of solar radiation. In such situation, appropriate cooling methods are required to create a favorable environment condition for crop growth and development (Abdel-Ghany et al., 2012, 2016). There are many methods used to control the excess heat generation in the greenhouse due to greenhouse effect. Avoiding solar radiation from entering the greenhouse is the most suitable preventive method for cooling greenhouses, especially in arid and semiarid regions (Hesham et al., 2016). Shading is one of the commonly used methods, which helps to keep optimum temperature and humidity required for crop production (Miguel et al., 1994; Hawa, 2006). Shading systems include the use of nets of different colors, liquid foams between the walls, reflective shade screens, shade cloths, and water film over the roof. The unique characteristic of shading systems is to limit the incoming amount of light energy inside the greenhouse and hence reduce the solar load of the greenhouse (Ghoulem et al., 2019). Effect of screening net with 35% shading rate was compared with other cooling systems, and humidity was coming out as a limiting factor among all microclimatic data like solar radiation, temperature, and photosynthetically active radiation (PAR) (Ilic et al., 2012).

Shading greenhouse is an effective and simple method to achieve a suitable environment throughout the crop growth, and to enhance crop productivity and quality in tropical region. Sometimes, shading has adverse effect on crop growth and production as the reduction in light below the acceptable levels reduces the photosynthesis rate resulting in low production. Although, increase in the leaf area index, number of fruits per plant, and no effect on total dry matter content was observed in tomato crop. In arid region, shading along with a cooling method is able to maintain greenhouse air temperature at 5–10°C lower, increase the relative humidity by 15–20%, and decrease the transmitted solar radiation by 30–50% as compared to a greenhouse without shading; whereas in cold regions, shading materials act as insulators which significantly reduce heat loss from greenhouses during night. In such condition, the air temperature of greenhouse can be kept 5°C higher than outside temperature along with 15–20% saving in energy used for heating greenhouse (Villarreal-Guerrero et al., 2012; Xu et al., 2015; Ishii et al., 2016; Davies et al., 2019; Yildirim and Bilir, 2017).

Whitewash shading is an old method of shading mostly used to modify the greenhouse microclimate. It is a temporary shading system and can get washed away with rainfall. The main purpose of whitewash shading is to reduce the intensity of solar radiation entering in the greenhouse, where the mixture of calcium oxide or calcium carbonate is used to paint the outer surface of glass/polyethylene of greenhouse (Baille et al., 2001; Garcia et al., 2011). The combination of whitewash shading or shading net with natural ventilation and/or evaporative cooling can lower the air temperature (approx. 10%), solar radiation (approx. 50%), increase relative humidity (approx. 20%), and enhance productivity and quality of crops (Ahemd et al., 2016). Moreover, hybrid system comprising rotating photovoltaic panels and reflective mirrors for shading in greenhouse is used to increase the production of crops (Marucci and Cappuccini, 2016). Excessive heat can cause the indoor temperature to rise giving detrimental effects to crop growth and production. It is crucial to have an accurate estimation of the solar radiation to build a viable physical model, and mass and heat transfer coefficients significantly impact the greenhouse energy and mass balance (Tiwari et al., 1998; Pieters and Deltour, 1999; Sharma and Tiwari, 1999; Sengar and Kothari, 2008; Taki et al., 2016; Su and Xu, 2017; Lopez-Cruz et al., 2018).

6.2.2 NATURAL VENTILATION

Natural ventilation is a very simple technique for greenhouse cooling of tropical regions which requires little or no external energy. It can be achieved by proper placing of roof and side wall opening of greenhouse (Ghoulem et al., 2019). Different types of design were proposed to develop a naturally ventilated greenhouse. Naturally ventilated five span greenhouse design was developed with 30° slope roof, side wall, and roof vents for each span. The study indicated that the proposed design could provide four times ventilation rates and proper temperature distribution as compared to parral-type greenhouse. The effect of vent openings on a multispan greenhouse microclimate was studied during the summer and winter. Different configurations of roof opening affect temperature, air flow, and relative humidity. The increasing size of roof vent also reduces the dehumidification time. Additionally, the first span of greenhouse microclimate is different than the last span due to roof opening configuration, which directly affects crop growth and development (Reyes-Rosas et al., 2017).

6.2.3 EVAPORATIVE COOLING

There are three kinds of evaporative cooling systems used in greenhouse: direct cooling, indirect cooling, and mixed mode cooling systems. Direct evaporative cooling is the commonly used method of cooling where water is directly applied, whereas in indirect cooling method, energy exchanges between the two air streams (Misra and Ghosh, 2018). Evaporative cooling technologies include fan pad and fog cooling, which are effectively used in hot and arid locations. Fan and pad system contained three main equipments like evaporative pad, exhaust fan, and circulation pumps. Performance study of greenhouse system which is surrounded by shallow water ponds and floating wetted surfaces demonstrated the reduction in temperature by 3–6°C (Misra and Ghosh, 2017). Uneven distribution of temperature across the greenhouse is the major drawback of the fan pad system. In contrast, fogging systems are an equally effective and uniform method of greenhouse cooling. The size of the fine droplet (10–20 μm) is useful, since it has a high water evaporation rate, and droplet disperses into the air quickly and evaporates early before they fall on plant leaves resulting in no wetting of leaves and fruits. Hence, the poorly managed fogging system can lead to increased pests and diseases incidence, and low product quality and yield (Arbel et al., 2003; Ahmed et al., 2011).

6.3 RESPONSE OF PLANT TO GREENHOUSE MICROCLIMATE

The growth of crop in field and in greenhouse is a function of various physiological reactions. Factors like light, temperature, carbon dioxide, soil/growth medium, relative humidity, and ventilation constitute the plant microclimate and effect the growth of plant in the greenhouse. The response of plant to greenhouse microclimate is discussed below.

6.3.1 LIGHT

The solar radiation consists of various wavelengths including visible spectrum of light, ultraviolet (UV), and infrared (IR). However, plants require the visible light, 400–700 nm PAR, for photosynthesis. The process of photosynthesis involves production of carbohydrates from carbon dioxide and water in the presence of chlorophyll and light energy. Light is fundamental for photosynthesis and photomorphogenesis (Santosh et al., 2017).

It is generally known that plant growth increases with increasing light intensity. However, recent studies showed that growth of plants being subjected to high intensity light is reduced due to stress of plants (Pearson et al., 1995). Earlier, it was thought that increase in light intensity will lead to increase in growth but this misconception was removed when experimental studies showed that high intensity light causes stress in plants (Wakchaure et al., 2017). Therefore, optimum and crop-specific light must be provided for plant growth and photosynthesis.

The light energy is transmitted and reflected back from greenhouse which is highly covered by the type of covering material, that is, glass, polyethylene, polyvinyl chloride, and acrylic panel. The coating material and latitude of plants also regulate the light intensity to which the plants are exposed. A study carried out by Yalcın and Erturk (2019) reported the effect of fluorescent (K_2SiF_6:Mn^{4+}) and scattering (TiO_2) pigmented greenhouse coatings on net photosynthesis of soybean at three different latitudes. It was found that fluorescent and scattering pigments did not increase the PAR transmitted to greenhouse. The plant growth was better at lower latitudes due to lesser solar intensity stress. Also, the scattering (TiO_2) pigments were better than fluorescent (K_2SiF_6:$Mn^{4+)}$ pigments to combat the stress. The TiO_2 particles are cheaper and more stable than K_2SiF_6:Mn^{4+}. The use of pigments in the coating reduces net daily photosynthesis for high latitudes.

The different shading screens also affect the light transmission inside the greenhouse. Holcman and Sentelhas in 2012 studied the effect of five different shading screens on the greenhouse covered with transparent low-density polyethylene (LDPE). The various screens used were thermoreflective screen, a control without screen, red screen, blue screen, and black screen on bromeliad (*Aechmea fasciata*). The parameters like incoming solar radiation (Rg), PAR, and net radiation (Rn) were estimated. The control and red screen treatments resulted in the highest PAR, net radiation, and solar radiation transmissivity, however, the black screen displayed opposite trend.

The revolution in greenhouse lightening started with introduction of light emitting diode which replaced the traditional lightening system. Light emitting diode (LED) is a semiconductor; the wavelength of light emitted depends on the properties of semiconductor material. The special property of LED is the wavelength emitted matches with plant photoreceptors, which allows for increased production and influences morphology of plant (Morrow, 2008). The LED light technology is energy efficient and allows for selection of specific lightening spectra. In earlier years, LED was costlier than traditional fluorescent or high-pressure sodium (HPS); although with advancement in technology, cheaper and better LEDs are being produced. Earlier researches strived to study the effect of combination lightening of LED with fluorescent (Lin et al., 2013) and LED with high-pressure sodium lamps (Pinho et al., 2007) on plant metabolism.

In greenhouse conditions, the source of light and position of light in relation to plant also affects the yield characteristics. Apart from the usual positioning of light in greenhouse, supplemental light applied in close contact of crop canopy yields higher crop yield and light use efficiency. For this purpose, LEDs (low bulb temperature) are used for interlightening as they do not cause heat damage to plants. In cucumber, use of blue/red LEDs interlightening improved leaf photosynthesis and dry mass of leaf, but no effect was recorded on fruit productivity (Trouwborst et al., 2010). In case of sweet pepper, plants interlightening led to increase in number of fruits and early maturation, thus leading to improvement in yield (Jokinen et al., 2012). A common thumb rule in greenhouse cultivation is assumed that a 1% increase in light results in a 1% yield increase. A survey indicated that for most of the crops a 1% light increment results in 0.5–1% increase in harvest. For vegetable 0.8–1%, for fruit 0.7–1%, for cut flowers 0.6–1%, for bulb flowers 0.25–1.25%, for flowering pot plants 0.5–1%, and 0.65% for nonflowering pot plants increase in harvestable product was observed.

6.3.2 TEMPERATURE

Internal greenhouse factors like temperature and light radiation are critical for crop cultivation (Singh et al., 2018). All the enzymatic reactions affecting plant physiology and morphology are directly affected by the surrounding temperature, whether the plants are in open field or closed area. The common rule is the rate of enzyme action doubles with increase of 60°C in temperature. Apart from this, the rate of transpiration and respiration is also a function of surrounding temperature. The greenhouse crops are usually grown at a day temperature, which is 3–60°C higher than the night temperature on a cloudy day and 80°C higher on a clear day. The optimum temperature inside the greenhouse for crop cultivation is species specific. Greenhouse overheating or under and uneven heating are common problems. The uniform temperature and humidity inside the greenhouse are essential for optimum productivity (Arbel et al., 2003; Singh et al., 2021).

Various studies have illustrated the effect of temperature on plant growth and development inside the greenhouse. A study was conducted (Liu et al., 2020) on the difference in cucumber yield in conventional solar greenhouses (SG) and sunken solar greenhouses (SSG). It was observed that inside the SSG, the daily minimum air temperature increased to more than 10°C, in spite of outside air temperature lower than 10°C. The warmer ambient temperature reduced low-temperature injury, increased growth, and showed higher cucumber fruit production.

The greenhouse temperature control system enables to maintain a uniform temperature gradient across the whole greenhouse. However, studies have shown that certain plant parts are more temperature-sensitive than others. In tomatoes, the most temperature-sensitive parts are flowers (Sato et al., 2000, 2002), followed by roots, shoot tips (Grimstad and Frimanslund, 1993), and developing fruits (Bertin, 2005). After this finding, local temperature control of the plant parts is practiced as it is more energy efficient than uniform heating and has no effect on the yield of the plants.

In greenhouse, several vegetables have been cultivated under local temperature control and studied for yield or yield attributing characters. A study conducted by Kawasaki et al. (2011) concluded that local heating of shoot tip on tomato leads to higher or similar yield as compared to conventional heating. Also, the local heating reduced the fuel consumption by 26%. Similarly, local heating of flower buds leads to improved pollen viability, better fruit set, and higher fruit yield in tomato (Kawasaki et al., 2010). In melon (*Cucumis melo* L.) and watermelon (*Citurullus lanatus*), sucrose accumulation was enhanced in fruits after local fruit heating (Kano et al., 2012; Matsumoto et al., 2012). The local temperature control of roots also affects various physiological functions and yield in plants. Root zone cooling in lettuce (*Lactuca sativa* L.) improved shoot dry weight (Ilahi et al., 2017). In cucumber, root zone cooling under warm air temperature alleviated fruit number and weight (Moon et al., 2007). In sweet pepper, higher calcium content in fruits was observed after root zone cooling treatment. Apart from cooling treatment, heating of root zone has also been carried out. In strawberry (*Fragaria* × *ananassa* Duch.), root zone heating at low air temperature promoted flower bud differentiation, flowering of axillary flower clusters, and higher commercial fruit yield (Kim et al., 2009). All the above studies were conducted in greenhouse in regulated temperature conditions. This proves that greenhouse technology (GT) can be used to study various plant physiological reactions under different temperature conditions.

The greenhouse heating mechanism has also served as a model to study climate change. In a study, *Silene capitata* Kom—a Korean endangered species was cultivated under (control) normal conditions in open and (treatment) under elevated carbon dioxide levels in a greenhouse. The experimental result stated that under high CO_2 concentration and temperature, stomata cannot control the photosynthetic rate. This environment suppresses excessive photosynthetic production by reducing chlorophyll content. It is proposed that in such condition, stomata are controlled for transpiration activity rather than for photosynthesis to prevent heat damage to the plant. It is estimated that *S. capitata* growth will be limited to areas of adequate water supply as the plant would fail to combat water stress (Park et al., 2019). Such kind of studies can serve as a model to predict crop growth pattern and behavior in future.

6.3.3 CARBON DIOXIDE

The carbon dioxide (CO_2) is an important greenhouse gas and a fundamental reactant in plant photosynthesis. A general statement is under elevated CO_2 concentrations, the plant growth is increased by absorbing more carbon dioxide (CO_2 fertilization effect) (Ainsworth, 2008). Under greenhouse conditions, several plant species have been studied with varying carbon dioxide concentrations, and their effect on plant yield and physiology has been determined. Zheng et al. (2018) had studied growth of three perennial grasses, tall fescue (*Festuca arundinacea* Schreb.), perennial ryegrass (*Lolium perenne* L.), and Kentucky bluegrass (*Poa pratensis* L.), under CO_2 concentrations at 400, 600, 800, 1000, and 1200 ppm. The optimal CO_2 concentration was found at 945, 915, and 1151 ppm for the above ground biomass of tall fescue, perennial ryegrass, and Kentucky bluegrass, respectively in greenhouse.

In cucumber crop, CO_2 enrichment was evaluated under greenhouse and average increase in fruit biomass production for both dry and fresh matter was noticed (Shanchez-Guerrero et al., 2005, 2009). Elevated CO_2 (eCO_2) is used as a fertilizer in greenhouse for pot plants, cut flower, and vegetable cultivation (Mortensen, 1987; Bisbis et al., 2018). Thus, GT has been used as a tool for determining optimum carbon dioxide levels for various vegetables and for efficient cultivation of plants.

6.3.4 VENTILATION

The greenhouse ventilation is responsible for air exchange and it affects temperature and carbon dioxide levels, therefore ventilation system indirectly affects the plant physiological processes and yield (Singh et al., 2021). Several experiments have been conducted to determine ventilation type for enhancing plant growth. The effect of natural ventilation and evaporative cooling (fan and pad cooling system) was evaluated in tropical climate in tomato, *Solanum lycopersicum*, cv FMTT260 (AVRDC, Shanhua,

Taiwan). The evaporative cooling (fan and pad cooling system) resulted in a better microclimate (air temperature and CO_2 concentration) but increased the incidence of fungal infection compared to natural ventilation (Ghosal et al., 2003; Jamaludin et al., 2014; Guan et al., 2015). In another study, two cooling systems (fan–pad evaporative as opposed to fan ventilation) were evaluated on greenhouse microclimate and soilless cucumber growth. The results suggested that fan ventilation cooling system prevented overheating in soilless cucumber crops in greenhouses without compromising yield (Nikolaou et al., 2019). The ventilation systems are an important part of greenhouse microclimate control. The ventilation systems are plant/crop specific. The optimum yield and better quality of crops, plants, vegetables or flowers in greenhouse is a function of several factors like light, temperature, carbon dioxide levels, and ventilation system, and their interactions (Santolini et al., 2018).

6.4 IMPORTANCE OF GT

The development of human civilization is a witness to problem addressing and problem solving. Each day technological advances are recorded in various fields, including agriculture. The journey from traditional agriculture to growing plants in space had seen many milestones in between like development of greenhouse technology, in vitro techniques, vertical farming, and hydroponics.

The key reasons behind development of all the above mentioned techniques are the rising human population, changes in consumption patterns, exploitation of natural resources, and food wastage (Foley et al., 2011). In current times, food security is one of the major targets around the globe (Hasan et al., 2018). The current food production is limited by availability of land and resources. It is estimated that in 2050, world production must increase by 70% to combat global hunger (FAO, 2009). In order to achieve this target, it is projected that 53% increase in the consumption of water resources and an increment of 38% of farmed land would be required (Fischer, 2007). The exploitation of land resources and conversion of land in farmland would lead to increased deforestation and damage to ecosystem (Garnett et al., 2013; Aznar-Sanchez et al., 2019).

The problems of feeding millions, maintaining ecological stability, and avoiding overexploitation of resources can be solved by employing GT (Tilman et al., 2011). The GT involves growing plants in a framed structure covered with transparent or translucent material under partial or fully controlled environmental conditions to get optimum growth and productivity. Initially, the GT was employed to cultivate vegetables in temperate conditions. However, with the advancement of technology it is now being presently used to grow flowers and commercial crops.

Presently, around 50 countries in the world actively use GT to enhance crop production and productivity, including India. In USA, 4000 ha area is under greenhouse mostly for flower production. In Spain and Italy, almost 25,000 and 18,500 ha areas are under greenhouses for vegetable production. Apart from this, Canada also produces most of its off-season vegetables and flowers under greenhouse. Netherlands, the leading country in flower export has 89,600 ha of flower producing area under greenhouse. The success of greenhouse is not limited to temperate countries but warmer countries like Israel and Turkey have 15,000 and 10,000 ha areas under GT for cut flower production. Egypt has about 1000 ha area under greenhouses. Japan has almost, 40,000 ha under greenhouse, out of which 7500 ha is solely for fruit production. In South Korea, almost 21,000 ha area is under greenhouses for cultivation of flowers and fruits. China has 48,000 ha of area under cover out of which 11,000 ha is under fruits production. Countries like Philippines also rely on GT for crop production due to rainy conditions (Pandey and Pandey, 2015).

In India, greenhouses were introduced in 1980s. GT is used in Ladakh region for cultivation of vegetables. In the northeast area, greenhouses are employed for protection from rain and to grow off-season vegetables. In northern India, seedlings of vegetables and flowers are germinated under greenhouses for quality improvement and for targeting market. Many states like Maharashtra, Tamil Nadu, and Karnataka use GT for floriculture for domestic and export purposes. The widespread usage of GT is popular around the globe.

Greenhouse farming has several advantages over traditional agriculture like enhanced crop production per year. The agriculture in controlled environment uses lesser water (70–80%) and the crops suffer lesser from plant pathogens, microbes, and pests (Despommier, 2013). The sustainable development is defined as the one that meets current needs without compromising the ability of future generations to meet their needs. Sustainable agriculture strives for a balance between agricultural production and preservation of nature (Chaplin-Kramer, 2015). The modern agriculture practices cause water scarcity, nutrient-depleted soils, and pollution (Magwaza et al., 2020). The pollution from agrichemicals like fertilizers or herbicides contaminates the environment affecting the health of the local populations (Tilman, 1999).

Several studies have indicated that agricultural practices and chemicals cause loss of biodiversity and harm the ecosystem (Castro et al., 2019). Lastly, the overexploitation of land resources and usage of chemical destroy the soil characteristics leading to soil erosion and loss of fertility (Foucher et al., 2014). Therefore, GT is important to achieve sustainability and productivity.

The greenhouse cultivation allows production of vegetables and flowers throughout the year in favorable microclimate. The GT not only provides appropriate environment to plant, but also promotes better growth and fruiting as compared to open field. On a commercial level, GT is used for hybrid seeds, high-value vegetables and ornamentals, cut flowers, and medicinal plants production. These are high-value and low-volume goods that fetch a good price in the market. The GT is also widely used for post-harvest drying of produce. In India, commonly sun drying is used, but the sundried product does not meet the international quality standard and fails to sell in the international markets (Tiris et al., 1994; Esper and Muhlbauer, 1998). It has been observed that greenhouse drying technology produces better quality product over open sun drying (Kumar et al., 2009). The heating inside the greenhouse is caused due to greenhouse effect. The long wave solar radiation enters the glass cover and is transmitted inside after reflection from glass. A part of transmitted radiation is reflected back in the form of short wave radiation from the crops, which is again transmitted to atmosphere. The remaining radiation is trapped inside the greenhouse, leading to heating up of inside of the glass cover. Unlike open sun drying, GT does not allow loss of long wave radiation to atmosphere leading to increased heat trapping inside greenhouse. The convection losses to surroundings are also minimized by glass cover which leads to increased greenhouse air temperature. Thus, due to the marvel of engineering, the greenhouse serves as a source of excellent and commercial crop drying.

The GT is a controlled system unlike open field and the various equipments inside the greenhouses are connected to microprocessors and computers leading to control over the microclimate inside the greenhouse (Shamshiri et al., 2018). In recent years, GT is powered with renewable energy which makes it more favorable for sustainable agriculture (Chel and Kaushik, 2011). Greenhouse system allows for usage of reused water, rainwater, and desalinated seawater, thus proving to be a useful tool in water-limited areas (Shekarchi and Shahnia, 2019). Newer, cheaper, and efficient covering material is being evolved each day to optimize production and yield (Petropoulos et al., 2018). The most beneficial aspect of GT is that it caters to the local market requirements.

6.5 CONCLUSIONS

Greenhouse climate control is one of the challenging tasks in precision agriculture. The potential benefits of closed greenhouses for increased crop production, reduction of CO_2 emissions, water savings, and improved crop protection are high. Closed and semiclosed greenhouses make it possible to combine more controlled cultivation, with high levels of production and savings in fossil fuel use.

The improved yield is not only influenced by the genetic structure of plants, but the temperature and humidity are also the main variables that have a direct association with crop yield. Temperature and humidity control, combined with high light intensity and CO_2 concentration in a closed greenhouse enhanced the crop production. Moreover, the required changes in cultivation strategy in a closed greenhouse have led to increases in production. However, sophisticated new technologies involving microprocessors, data loggers, automated irrigation, and fertigation system allow fully microclimatic and other input control. Further research on controlling greenhouse microclimate parameters in relation to increased crop production will help sustainable production of greenhouse crops.

KEYWORDS

- **greenhouse system**
- **microclimate**
- **crop yield**
- **photosynthesis rate**
- **solar radiation**
- **cooling system**

REFERENCES

Abdel-Ghany, A. M.; Al-Helal, I. M.; Alzahrani, S. M.; Alsadon, A. A.; Ali, I. M.; Elleithy, R. M. Covering Materials Incorporating Radiation–Preventing Techniques to Meet Greenhouse Cooling Challenges in Arid Regions: A Review. *Sci. World J.* **2012,** *2012*, 906360.

Abdel-Ghany, A. M.; Al-Helal, I. M.; Picuno, P.; Shady, M. R. Modified Plastic Net–Houses As Alternative Agricultural Structures for Saving Energy and Water in Hot and Sunny Regions. *Renew. Energy* **2016,** *93*, 332–339.

Ahemd, H. A.; Al-Faraj, A. A.; Abdel-Ghany, A. M. Shading Greenhouses to Improve the Microclimate, Energy and Water Saving in Hot Regions: A Review. *Sci. Hort.* **2016,** *201*, 36–45.

Ahmed, E. M.; Abaas, O.; Ahmed, M.; Ismail, M. R. Performance Evaluation of Three Different Types of Local Evaporative Cooling Pads in Greenhouses in Sudan. *Saudi. J. Biol. Sci.* **2011,** *18*, 45–51.

Ainsworth, E. A. Rice Production in a Changing Climate: A Meta-Analysis of Responses to Elevated Carbon Dioxide and Elevated Ozone Concentration. *Glob. Chang. Biol.* **2008,** *14*, 1642–1650.

Arbel, A.; Barak, M.; Shklyar, A. Combination of Forced Ventilation and Fogging Systems for Cooling Greenhouses. *Biosyst. Eng.* **2003,** *84*, 45–55.

Aritesty, E.; Wulandani, D. Performance of the Rack Type–Greenhouse Effect Solar Dryer for Wild Ginger (*Curcuma Xanthorizza* Roxb.) Drying. *Energy. Procedia.* **2014,** *47*, 94–100.

Aroca-Delgado, R.; Perez-Alonso, J.; Callejon-Ferre, A. J.; Velazquez-Marti, B. Compatibility between Crops and Solar Panels: An Overview from Shading Systems. *Sustainability* **2018,** *10*, 743.

Azaizia, Z.; Kooli, S.; Elkhadraoui, A.; Hamdi, I.; Guizani, A. Investigation of a New Solar Greenhouse Drying System for Peppers. *Int. J. Hydrog. Energy* **2017,** *42*, 8818–8826.

Aznar-Sanchez, J. A.; Piquer-Rodriguez, M.; Velasco-Munoz, J. F.; Manzano-Agugliaro, F. Worldwide Research Trends on Sustainable Land Use in Agriculture. *Land Use Policy* **2019,** *87*, 104069.

Baille, A.; Kittas, C.; Katsoulas, N. Influence of Whitening on Greenhouse Microclimate and Crop Energy Partitioning. *Agric. For. Meteorol.* **2001,** *107*, 293–306.

Bertin, N. Analysis of The Tomato Fruit Growth Response to Temperature and Plant Fruit Load In Relation to Cell Division, Cell Expansion and DNA Endo-Reduplication. *Ann. Bot.* **2005,** *95*, 439–447.

Bisbis, M. B.; Gruda, N.; Blanke, M. Potential Impacts of Climate Change on Vegetable Production and Product Quality–A Review. *J. Clean. Prod.* **2018,** *170*, 1602–1620.

Castro, A. J.; Lopez-Rodríguez, M. D.; Giagnocavo, C.; Gimenez, M.; Cespedes, L.; La Calle, A. Valera, D. L. Six Collective Challenges for Sustainability of Almeria Greenhouse Horticulture. *Int. J. Environ. Res. Public Health* **2019,** *16*, 4097.

Chaplin-Kramer, R.; Sharp, R. P.; Mandle, L.; Sim, S.; Johnson, J.; Butnar, I.; Kareiva, P. M. Spatial Patterns of Agricultural Expansion Determine Impacts on Biodiversity and Carbon Storage. *Proc. Natl. Acad. Sci.* **2015,** *112*, 7402–7407.

Chauhan, P. S.; Kumar, A.; Gupta, B. A Review on Thermal Models for 436 Greenhouse Dryers. *Renew. Sustain. Energy. Rev.* **2017,** *75*, 548–558.

Chel, A.; Kaushik, G. Renewable Energy for Sustainable Agriculture. *Agron. Sustain. Dev.* **2011,** *31*, 91–118.

Choab, N.; Allouhi, A.; El Maakoul, A.; Kousksou, T.; Saadeddine, S.; Jamil, A. Review on Greenhouse Microclimate and Application: Design Parameters, Thermal Modelling and Simulation, Climate Controlling Technologies. *Sol. Energy* **2019,** *191*, 109–137.

Chou, S. K.; Chua, K. J.; Ho, J. C.; Ooi, C. L. On The Study of an Energy–Efficient Greenhouse for Heating, Cooling and Dehumidification Applications. *Appl. Energy* **2004,** *77*, 355–373.

Davies, P. A.; Zaragoza, G. Ideal Performance of a Self–Cooling Greenhouse. *Appl. Therm. Eng.* **2019,** *149*, 502–511.

Despommier, D. Farming Up the City: The Rise of Urban Vertical Farms. *Trends. Biotechnol.* **2013,** *31*, 388–389.

Domis, M.; Papadopoulos, A. P.; Gosselin, A. Greenhouse Tomato Fruit Quality. *Hort. Rev.* **2002,** *26*, 239–349.

Esper, A.; Muhlbauer, W. Solar Drying–An Effective Means of Food Preservation. *Renew. Energy* **1998,** *15*, 95–100.

Fischer, G.; Tubiello, F. N.; Van Velthuizen, H.; Wiberg, D. A. Climate Change Impacts on Irrigation Water Requirements: Effects of Mitigation, 1990–2080. *Technol. Forecast. Soc. Change* **2007,** *74*, 1083–1107.

Foley, J. A.; Ramankutty, N.; Brauman, K. A.; Cassidy, E. S.; Gerber, J. S.; Johnston, M.; Mueller, N. D.; O'Connell, C.; Ray, D. K.; West, P. C.; Balzer, C.; Bennett, E. M.; Carpenter, S. R.; Hill, J.; Monfreda, C.; Polasky, S.; Rockstrom, J.; Sheehan, J.; Siebert, S.; Tilman, D.; Zaks, D. P. Solutions for a Cultivated Planet. *Nature* **2011,** *7369*, 337–342.

Food and Agriculture Organization of the United Nations. *High Level Expert Forum—How to Feed the World in 2050, Office of the Director*; Agricultural Development Economics Division: Rome, Italy, 2009.

Foucher, A.; Salvador-Blanes, S.; Evrard, O.; Simonneau, A.; Chapron, E.; Courp, T. Desmet, M. Increase in Soil Erosion after Agricultural Intensification: Evidence from a Lowland Basin in France. *Anthropocene* **2014,** *7*, 30–41.

Ganguly, A.; Misra, D.; Ghosh, S. Modelling and Analysis of Solar Photovoltaic–Electrolyser–Fuel Cell Hybrid Power System Integrated with a Floriculture Greenhouse. *Energy Build.* **2010,** *42*, 2036–2043.

Garcia, M. L.; Medrano, E.; Sanchez-Guerrero, M. C.; Lorenzo, P. Climatic Effects of Two Cooling Systems in Greenhouses in the Mediterranean Area: External Mobile Shading and Fog System. *Biosyst. Eng.* **2011,** *108*, 133–143.

Garnett, T.; Appleby, M. C.; Balmford, A.; Bateman, I. J.; Benton, T. G.; Bloomer, P.; Godfray, H. C. J. Sustainable Intensification in Agriculture: Premises and Policies. *Science* **2013,** *341*, 33–34.

Ghani, S.; Bakochristou, F.; ElBialy, E. M. A. A.; Gamaledin, S. M. A.; Rashwan, M. M.; Abdelhalim, A. M.; Ismail, S. M. Design Challenges of Agricultural Greenhouses in Hot and Arid Environments–A Review. *Eng. Agric. Environ. Food* **2019,** *12*, 48–70.

Ghosal, M. K.; Tiwari, G. N. Modelling and Parametric Studies for Thermal Performance of an Earth to Air Heat Exchanger Integrated with a Greenhouse. *Energy Convers. Manag.* **2006,** *47*, 1779–1798.

Ghosal, M. K.; Tiwari, G. N.; Srivastava, N. S. L. Modelling and Experimental Validation of a Greenhouse with Evaporative Cooling by Moving Water Film Over External Shade Cloth. *Energy Build.* **2003,** *35*, 843–850.

Ghoulem, M.; El Moueddeb, K.; Nehdi, E.; Boukhanouf, R.; Calautit, J. K. Greenhouse Design and Cooling Technologies for Sustainable Food Cultivation in Hot Climates: Review of Current Practice and Future Status. *Biosyst. Eng.* **2019,** *183*, 121–150.

Giacomelli, G.; Sase, S.; Cramer, R.; Hoogeboom, J.; MacKenzie, A.; Parbst, K.; Scarascia–Mugnozza, G.; Selina, P.; Sharp, D.; Voogt, J. Greenhouse Production Systems for People. *Acta. Hort.* **2012,** *927*, 23–38.

Grimstad, S. O.; Frimanslund, E. Effect of Different Day and Night Temperature Regimes on Greenhouse Cucumber Young Plant Production, Flower Bud Formation and Early Yield. *Sci. Hort.* **1993,** *53*, 191–204.

Guan, L.; Bennett, M.; Bell, J. Evaluating the Potential Use of Direct Evaporative Cooling in Australia. *Energy Build.* **2015,** *108*, 185–194.

Hanan, J. J. *Greenhouses, Advanced Technology for Protected Horticulture*; CRC: Florida, 1998.

Hasan, M. K.; Desiere, S.; D'Haese, M.; Kumar, L. Impact of Climate–Smart Agriculture Adoption on the Food Security of Coastal Farmers in Bangladesh. *Food. Secur.* **2018,** *10*, 1073–1088.

Hawa, Z. E. J. Carbon Dioxide Enriched Production Technology for Controlled Environment System in the Lowland Tropics. *Acta. Hort.* **2006,** *710.*

Hesham, A. A.; Al-Faraj, A. A.; Abdel-Ghany, A. M. Shading Greenhouses to Improve the Microclimate, Energy and Water Saving in Hot Regions: A Review. *Sci. Hort.* **2016,** *201*, 36–45.

Holcman, E.; Sentelhas, P. C. Microclimate under Different Shading Screens in Greenhouses Cultivated with Bromeliads. *Bras. de Eng. Agricola e Ambient* **2012,** *16*, 858–863.

Ilahi, W. F. F.; Ahmad, D.; Husain, M. C. Effects of Root Zone Cooling on Butterhead Lettuce Grown in Tropical Conditions in a Coir–Perlite Mixture. *Hort. Environ. Biotechnol.* **2017,** *58*, 1–4.

Ilic, Z. S.; Milenkovic, L.; Stanojevic, L.; Cvetkovic, D.; Fallik, E. Effects of the Modification of Light Intensity by Color Shade Nets on Yield and Quality of Tomato Fruits. *Sci. Hort.* **2012,** *139*, 90–95.

Ishii, M.; Sase, S.; Moriyama, H.; Okushima, L.; Ikeguchi, A.; Hayashi, M.; Giacomelli, G. A. Controlled Environment Agriculture for Effective Plant Production Systems in a Semiarid Greenhouse. *Jpn. Agric. Res. Q.* **2016,** *50*, 101–113.

Jamaludin, D.; Ahmad, D.; Kamaruddin, R.; Jaafar, H. Z. Microclimate Inside a Tropical Greenhouse Equipped with Evaporative Cooling Pads. *Pertanika. J. Sci. Technol.* **2014,** *22*, 255–272.

Kano, Y.; Matsumoto, J.; Aoki, Y. S.; Madachi, T. Effects of Heating Bearing Shoot Near Fruit on Cell Size, Sucrose Metabolizing Enzymes and Sugar Accumulation in Watermelon Fruit. *J. Jpn. Soc. Hort. Sci.* **2012,** *81*, 171–176.

Kawasaki, Y.; Suzuki, K.; Yasuba, K.; Kawashima, H.; Sasaki, H.; Takaichi, M. Effect of Local Heating Around the Tomato Shoot Apex and Flower Clusters on Plant Surface Temperature and Characteristics Related to Fruit Yield. *Hort. Res.* **2010,** *9*, 345–350.

Kawasaki, Y.; Suzuki, K.; Yasuba, K.; Takaichi, M. Effect of Local Air Heating by a Hanging Duct Near the Tomato Shoot Apex and Flower Clusters on Vertical Temperature Distribution, Fruit Yield and Fuel Consumption. *Hort. Res.* **2011,** *10*, 395–400.

Kim, Y.; Endo, M.; Kiriiwa, Y.; Chen, L.; Nukaya, A. Effects of Root Zone Heating During Daytime on The Flowering, Growth and Yield of Strawberry Akihime Grown in Substrate Culture. *Hort. Res.* **2009,** *8*, 193–199.

Kuehni, S. M. S.; Bou-Zeid, E.; Webb, C.; Shokri, N. Roof Cooling by Direct Evaporation from a Porous Layer. *Energy Build.* **2016,** *127*, 521–528.

Kumar, K. S.; Tiwari, K. N.; Jha, M. K. Design and Technology for Greenhouse Cooling in Tropical and Subtropical Regions: A Review. *Energy Build.* **2009,** *41*, 1269–1275.

Lin, K. H.; Huang, M. Y.; Huang, W. D.; Hsu, M. H.; Yang, Z. W.; Yang, C. M. The Effects of Red, Blue, and White Light–Emitting Diodes on the Growth, Development, and Edible Quality of Hydroponically Grown Lettuce (*Lactuca Sativa L.* var. Capitata). *Sci. Hort.* **2013,** *150*, 86–91.

Liu, H.; Yin, C.; Hu, X.; Tanny, J.; Tang, X. Microclimate Characteristics and Evapotranspiration Estimates of Cucumber Plants in a Newly Developed Sunken Solar Greenhouse. *Water* **2020,** *12*, 2275.

Lopez-Cruz, I. L.; Fitz-Rodriguez, E.; Salazar-Moreno, R.; Rojano-Aguilar, A.; Kacira, M. Development and Analysis of Dynamical Mathematical Models of Greenhouse Climate: A Review. *Eur. J. Hort. Sci.* **2018,** *83*, 269–279.

Magwaza, S. T.; Magwaza, L. S.; Odindo, A. O.; Mditshwa, A. Hydroponic Technology as Decentralised System for Domestic Wastewater Treatment and Vegetable Production in Urban Agriculture: A Review. *Sci. Total Environ.* **2020,** *698*, 134154.

Marucci, A.; Cappuccini, A. Dynamic Photovoltaic Greenhouse: Energy Balance in Completely Clear Sky Condition during the Hot Period. *Energy* **2016,** *102*, 302–312.

Matsumoto, J.; Kano, Y.; Madachi, T.; Aoki, Y. S. Heating Bearing Shoots near Fruits Promotes Sugar Accumulation in Melon Fruit. *Sci. Hort.* **2012,** *133*, 18–22.

McCartney, L.; Orsat, V.; Lefsrud, M. G. An Experimental Study of the Cooling Performance and Airflow Patterns in a Model Natural Ventilation Augmented Cooling (NVAC) Greenhouse. *Biosyst. Eng.* **2018,** *174*, 173–189.

Miguel, A. F.; Silva, A. M.; Rosa, R. Solar Irradiation inside a Single–Span Greenhouse with Shading Screens. *J. Agric. Eng. Res.* **1994,** *59*, 61–72.

Misra, D.; Ghosh, S. Evaporative Cooling Technologies for Greenhouses: A Comprehensive Review.

Agric. Eng. Int.: CIGR J. **2018,** *20*, 1–15.

Misra, D.; Ghosh, S. Performance Study of a Floricultural Greenhouse Surrounded by Shallow Water Ponds. *Int. J. Renew. Energy Res.* **2017,** *6*, 137.

Moon, J. H.; Kang, Y. K.; Suh, H. D. Effect of root–zone cooling on the growth and yield of cucumber at supraoptimal air temperature. In *XXVII International Horticultural Congress–IHC2006;* International Symposium on Advances in Environmental Control, Automation, 2006; pp 271–274.

Morrow, R. C. LED Lighting in Horticulture. *Hort. Sci.* **2008,** *43*, 1947–1950.

Mortensen, L. M. CO2 enrichment in Greenhouses. Crop Responses. *Sci. Hort.* **1987,** *33*, 1–25.

Nikolaou, G.; Neocleous, D.; Katsoulas, N.; Kittas, C. Dynamic Assessment of Whitewash Shading and Evaporative Cooling on the Greenhouse Microclimate and Cucumber Growth in a Mediterranean Climate. *Ital. J. Agrometeorol.* **2018,** *2*, 15–26.

Nikolaou, G.; Neocleous, D.; Katsoulas, N.; Kittas, C. Effects of Cooling Systems on Greenhouse Microclimate and Cucumber Growth under Mediterranean Climatic Conditions. *Agronomy* **2019,** *9*, 300.

Pandey, S.; Pandey, A. Greenhouse Technology. *Int. J. Res. Granthaalayah* **2015,** *3*, 1–3.

Park, J. H.; Han, Y. S.; Lee, E. P.; Lee, S. Y.; Jeong, H. M.; You, Y. H. Studies on the Changes in Phenological, Growth and Physiological Responses of Silene Capitata Kom; An Endangered Plant In Korea, Under Climate Change Treatment. *J. Ecol. Environ.* **2019,** *43*, 1–9.

Pearson, S.; Wheldon, A. E.; Hadley, P. Radiation Transmission and Fluorescence of Nine Greenhouse Cladding Materials. *J. Agric. Eng. Res.* **1995,** *62*, 61–69.

Petropoulos, S. A.; Fernandes, A.; Katsoulas, N.; Barros, L.; Ferreira, I. C. The Effect of Covering Material on the Yield, Quality and Chemical Composition of Greenhouse-Grown Tomato Fruit. *J. Sci. Food Agric.* **2019,** *99*, 3057–3068.

Pieters, J. G.; Deltour, J. M. Modelling Solar Energy Input in Greenhouses. *Sol. Energy* **1999,** *67*, 119–130.

Pinho, P.; Tahvonen, R.; Lukkala, R.; Sarkka, L.; Halonen, L.; Tetri, E. Evaluation of Lettuce Growth under Multi–Spectral–Component Supplemental Solid–State Lighting in Greenhouse Environment. *Acta. Hort.* **2007,** *711*, 111–119.

Radojevic, N.; Kostadinovic, D.; Vlajkovic, H.; Veg, E. Microclimate Control in Greenhouses. Trans Rotation Experiment in North China Using the EU Rotate N Simulation Model. *Env. Pollution* **2014,** *158*, 2218–2229.

Reyes-Rosas, A.; Molina-Aiz, F. D.; Valera, D. L.; Lopez, A.; Khamkure, S. Development of a Single Energy Balance Model for Prediction of Temperatures inside a Naturally Ventilated Greenhouse with Polypropylene Soil Mulch. *Comput. Electron. Agric.* **2017,** *142*, 9–28.

Romantchik, E.; Rios, E.; Sanchez, E.; Lopez, I.; Sanchez, J. R. Determination of Energy to be Supplied by Photovoltaic Systems for Fan–Pad Systems in Cooling Process of Greenhouses. *Appl. Therm. Eng.* **2017,** *114*, 1161–1168.

Rong, L.; Pedersen, P.; Jensen, T. L.; Morsing, S.; Zhang, G. Dynamic Performance of an Evaporative Cooling Pad Investigated in a Wind Tunnel for Application in Hot and Arid Climate. *Biosyst. Eng.* **2017,** *156*, 173–182.

Sabir, N.; Singh, B. Protected Cultivation of Vegetables in Global Arena: A Review. *Indian J. Agric. Sci.* **2013,** *83*, 123–35.

Sanchez-Guerrero, M. C.; Lorenzo, P.; Medrano, E.; Baille, A.; Castilla, N. Effects of EC–based Irrigation Scheduling and CO2 Enrichment on Water Use Efficiency of a Greenhouse Cucumber Crop. *Agric. Water Manag.* **2009,** *96*, 429–436.

Sanchez-Guerrero, M. C.; Lorenzo, P.; Medrano, E.; Castilla, N.; Soriano, T.; Baille, A. Effect of Variable CO_2 Enrichment on Greenhouse Production in Mild Winter Climates. *Agric. For. Meteorol.* **2005,** *132*, 244–252.

Santamouris, M.; Mihalakakou, G.; Balaras, C. A.; Lewis, J. O.; Vallindras, M.; Argiriou, A. Energy Conservation in Greenhouses with Buried Pipes. *Energy* **1996,** *21*, 353–360.

Santolini, E.; Pulvirenti, B.; Benni, S.; Barbaresi, L.; Torreggiani, D.; Tassinari, P. Numerical Study of Wind–Driven Natural Ventilation in a Greenhouse with Screens. *Comput. Electron. Agric.* **2018,** *149*, 41–53.

Santosh, D. T.; Tiwari, K. N.; Singh, V. K.; Reddy, A. R. G. Micro Climate Control in Greenhouse. *Int. J. Curr. Microbiol. Appl. Sci.* **2017,** *6*, 1730–1742.

Sato, S.; Peet, M. M.; Thomas, J. F. Determining Critical Pre- and Post-Anthesis Periods and Physiological Processes in *Lycopersicon esculentum* Mill. Exposed to Moderately Elevated Temperatures. *J. Exp. Bot.* **2002,** *53*, 1187–1195.

Sato, S.; Peet, M. M.; Thomas, J. F. Physiological Factors Limit Fruit Set of Tomato *(Lycopersicon esculentum* Mill.) Under Chronic, Mild Heat Stress. *Plant Cell Environ.* **2000,** *23*, 719–726.

Sengar, S. H.; Kothari, S. Thermal Modelling and Performance Evaluation of Arch Shape Greenhouse for Nursery Raising. *Afr. J. Math. Comput. Sci. Res.* **2008**, *1*, 1–9.

Sethi, V. P.; Sharma, S. K. Survey of Cooling Technologies for Worldwide Agricultural Greenhouse Applications. *Sol. Energy* **2007,** *81*, 1447–1459.

Shamshiri, R. R.; Jones, J. W.; Thorp, K. R.; Ahmad, D.; Che Man H.; Taheri S. Review of Optimum Temperature, Humidity, and Vapour Pressure Deficit for Microclimate Evaluation and Control in Greenhouse Cultivation of Tomato: A Review. *Int. Agrophycs.* **2018,** *32*, 2.

Sharma, P. K.; Tiwari, G. N. Parametric Study of a Greenhouse by Using Runge–Kutta Methods. *Energy Convers. Manag.* **1999,** *40*, 901–912.

Shekarchi, N.; Shahnia, F. A Comprehensive Review of Solar-Driven Desalination Technologies for Off-Grid Greenhouses. *Int. J. Energy Res.* **2019,** *43*, 1357–1386.

Shukla, A.; Tiwari, G. N.; Sodha, M. S. Thermal Modelling for Greenhouse Heating by using Thermal Curtain and an Earth–Air Heat Exchanger. *Build. Environ.* **2006,** *41*, 843–850.

Singh, M. C. Status of Protected Cultivation in India, Challenges, Scope and Future Needs for its Successful Adoption–Review. *Agric. Res.* **2021,** *10.*

Singh, M. C.; Singh, J. P.; Singh, K. G. Optimal Operating Microclimatic Conditions for Drip Fertigated Cucumbers in Soilless Media under a Naturally Ventilated Greenhouse. *Indian J. Ecol.* **2017a,** *44,* 821–26.

Singh, M. C.; Singh, J. P.; Pandey, S. K.; Cutting, N. G.; Sharma, P.; Shrivastav, V.; Sharma, P. A Review of Three Commonly used Techniques of Controlling Greenhouse Microclimate. *Int. J. Curr. Microbiol. Appl. Sci.* **2018,** *7,* 3491–3505.

Singh, M. C.; Singh, J. P.; Pandey, S. K.; Mahay, D.; Srivastava, V. Factors Affecting the Performance of Greenhouse Cucumber Cultivation–A Review. *Int. J. Curr. Microbiol. Appl. Sci.* **2017b,** *6,* 2304–23.

Singh, M. C.; Yousuf, A.; Singh, J. P. Greenhouse Microclimate Modelling under Cropped Conditions–A Review. *Research in Environ. Life Sci.* **2016,** *9,* 1552–1557.

Su, Y.; Xu, L. Towards Discrete Time Model for Greenhouse Climate Control. *Eng. Agric. Environ. Food.* **2017,** *10,* 157–170.

Taki, M.; Ajabshirchi, Y.; Ranjbar, S. F.; Rohani, A.; Matloobi, M. Modelling and Experimental Validation of Heat Transfer and Energy Consumption in an Innovative Greenhouse Structure. *Inf. Process. Agric.* **2016,** *3,* 157–174.

Tilman, D. Global Environmental Impacts of Agricultural Expansion: The Need for Sustainable and Efficient Practices. *Proc. Natl. Acad. Sci.* **1999,** *96,* 5995–6000.

Tilman, D.; Balzer, C.; Hill, J.; Befort, B. L. Global Food Demand and the Sustainable Intensification of Agriculture. *Proc. Natl. Acad. Sci.* **2011,** *108,* 20260–20264.

Tiris, C.; Ozbalta, N.; Tiris, M.; Dincer, I. Performance of a Solar Dryer. *Energy* **1994,** *19,* 993–997.

Tiwari, G. N. *Greenhouse Technology for Controlled Environment*; Narosa Publishing House: New Delhi, India, 2003.

Tiwari, G. N.; Sharma, P. K.; Goyal, R. K.; Sutar, R. F. Estimation of an Efficiency Factor for a Greenhouse: A Numerical and Experimental Study. *Energy Build.* **1998,** *28,* 241–250.

Trouwborst, G.; Oosterkamp, J.; Hogewoning, S. W.; Harbinson, J.; Van Ieperen, W. The Responses of Light Interception, Photosynthesis and Fruit Yield of Cucumber to LED-Lighting within the Canopy. *Physiol. Plant.* **2010,** *138,* 289–300.

Villarreal-Guerrero, F.; Kacira, M.; Fitz-Rodriguez, E.; Linker, R.; Kubota, C.; Giacomelli, G. A.; Arbel, A. Simulated Performance of a Greenhouse Cooling Control Strategy with Natural Ventilation and Fog Cooling. *Biosyst. Eng.* **2012,** *111,* 217–228.

Vox, G.; Teitel, M.; Pardossi, A.; Minuto, A.; Tinivella, F.; Schettini, E. *Sustainable Greenhouse Systems. Sustainable Agriculture: Technology, Planning and Management*; Nova Science Publishers, Inc.: New York, 2010, pp 1–79.

Wakchaure, G. C.; Minhas, P. S. Effect of Reduced PAR on Growth and Photosynthetic Efficiency of Soybean Genotypes. *J. Agrometeorol.* **2017,** *19,* 1–9.

Xu, J.; Li, Y.; Wang, R. Z.; Liu, W.; Zhou, P. Experimental Performance of Evaporative Cooling Pad Systems in Greenhouses in Humid Subtropical Climates. *Appl. Energy* **2015,** *138,* 291–301.

Yalcın, R. A.; Erturk, H. Improving Photosynthetic Efficiency using Greenhouse Coatings with Scattering and Fluorescent Pigments. *Mater. Express* **2019,** *6,* 085551.

Yang, S. H.; Joong, J. R. Utilization and Performance Evaluation of a Surplus Air Heat Pump System for Greenhouse Cooling and Heating. *Appl. Energy* **2013,** 244–251.

Yildirim, N.; Bilir, L. Evaluation of a Hybrid System for a Nearly Zero Energy Greenhouse. *Energy Convers. Manag.* **2017,** *148*, 1278–1290.

Zheng, Y.; Li, F.; Hao, L.; Shedayi, A. A.; Guo, L.; Ma, C.; Xu, M. The Optimal CO_2 Concentrations for the Growth of Three Perennial Grass Species. *BMC Plant Boil.* **2018,** *18*, 1–12.

CHAPTER 7

Irrigation and Fertigation Systems for Protected Cultivation

DILIP SINGH KACHWAYA[1], HARISH CHANDRA RATURI[1], ANUPREET SINGH TIWANA[2], and MOHD TUFAIL[3]

[1]*Department of Agriculture, Mata Gujri College, Fatehgarh Sahib, Punjab, India*

[2]*Department of Geography, Mata Gujri College, Fatehgarh Sahib, Punjab, India*

[3]*Department of Geography, Govt. P. G. College, Rajouri, Jammu and Kashmir, India*

ABSTRACT

Water (irrigation) is an utmost critical input in agriculture. However, freshwater resources are depleting at an alarming rate due to their overexploitation, particularly in the agricultural sector, where they are used without measurement. Thus, it becomes essential to apply measured quantities of water to the crops through proper scheduling, and improve the water use efficiency using the microirrigation techniques such as drip system. Using drip irrigation under different crops, the crop's water consumption may be reduced by 30–70% with an increased productivity of 20–90%. Drip irrigation tactics work well to increase crop yield by about thrice a similar amount of water utilized when compared with conventional methods. It is a practical method of applying fertilizers and nutrients (fertigation) exactly in the rooting zone of the plants. Thus, it has become imperative to conserve water, and improve water and nutrient use efficiencies (WNUE) by adopting a precise fertigation technique. Fertigation refers to supplying nutrients/

Protected Cultivation: Structural Design, Crop Management, Modeling and Automation.
Mahesh Chand Singh and K. K. Sharma (Eds.)

water-soluble fertilizers via microirrigation systems (drip irrigation, micro-sprinkler, and minisprinkler systems). Under open environmental conditions, the crop production has also become a challenging task due to the continued exposure to the adverse weather conditions. Thus, while deciding the rate of crop production, weather/climatic conditions should be strictly taken under consideration, particularly when grown in protected structures (greenhouses). Protected cultivation (greenhouse cultivation) is mainly intended to provide optimal growing conditions and protect crops from adverse weather and different insect-pests. Hence, the overall water and fertilizer use efficiencies significantly increase under microirrigation techniques, when the crops are subjected to controlled climatic conditions.

7.1 INTRODUCTION

At present, technological achievements in agriculture are highly needed to feed the huge mass of population, keeping in mind the nutritious and sound yield approach for longer sustainability, along with food security. Moreover, advancements in years require wiser efforts in terms of balance within horticulture production to combat massive rise in population, which is expected to become 10 billion in 2050. In horticultural crops, target can be achieved through protected cultivation by means of controlled environments and greenhouse conditions; wherein production succession through micro-climatic factors monitoring like light, temperature, humidity, and gaseous atmosphere is recently adopted (Rouphael et al., 2015). Advantageously, water scarcity is the foremost need to work upon under agricultural practices and is in real sense maintained under protected cultivation to a greater extent. (De Pascale et al., 2018). Being hit target, the primary motive to produce healthy food with least environmental damage is the action in demand. Also, in the last two decades, increased awareness of sustainable practices under protected cultivation has seen trended wisely to improve health of the population thus their longevity by these stated means (De Pascale et al., 2017).

In protected cultivation, irrigation and fertigation are the most important management factors through which farmers control fruit yield, quality, and plant development (Putra and Yuliando, 2015; Montesano et al., 2015; Van et al., 2002). Protected cultivation is defined as a technique in which the microclimate in the vicinity of the plant is regulated entirely or partially. Greenhouses help to conserve soil moisture and make efficient use of energy, particularly solar energy. It is required for year-round farming, higher returns, off-season production, assured output, improved crop quality, lower residues of pesticides, self-employment for educated rural youth in the agricultural sectors, weather vagaries, managed pollination, and easier conservation of plants. Protected cultivation technology aids in ensuring high yields and better quality, and financially viable returns (Maitra et al., 2020).

Fertilizers applied through broadcasting are not efficiently utilized by the plants, whereas fertigation permits a precise and uniform application of nutrients (Jat et al., 2011). Fertigation also increases the efficiency of applying the fertilizers, thus reducing the production cost and the possibility of groundwater pollution caused by the fertilizers' leaching.

The irrigation timing and volume are undoubtedly the most critical factors for efficacious irrigation management and preserving water, and these, in turn, also improve the yield both quantitatively and qualitatively (Zeng et al., 2009; Sezen et al., 2010).

7.2 GREENHOUSE TECHNOLOGY AND CROPS

Today, about 92% of crop plants are cultivated in open field/land conditions. Previously, starting from greenhouse technology adoption to protect the crop under temperature means resulted in good returns, which lends the basis for recently investigated other cultivation tactics under the headlines of protected cultivation. This is the achieved scientific way of providing favorable environment to crops. It safeguards the crops from harsh climatic disturbances resulting from cold climate, wind speed, rainfall, extreme solar radiation, temperature, insect-pests, and diseases. An ideal microclimate can be formed around the crop.

In the world, US has about 15,000 ha area occupied by greenhouses, mainly used for floriculture with annual income of greater than 3.4 billion USD. The area covered under greenhouse structures is expected to increase considerably, if the transportation cost of vegetables from neighboring countries increases.

Spain and Italy cultivate around 28,000 and 19,500 ha, respectively under vegetable crops like bean, cucumber, capsicum, tomato, and strawberry. Spain utilizes simple tunnel-type greenhouses that are embedded only with UV-stabilized polyethylene film. Similarly, Canada adopted the same greenhouse technology to grow flowers and off-season vegetable crops, mainly tomatoes, cucumbers, and capsicums. In another way, sensory likeness for hydroponically-grown greenhouse vegetables, in addition to the sound yield can be seen through revenue benefits by farmers on selling in Canada. The Netherlands adopted it since long back also for flowers and vegetables, thus counting under traditional exporters for these stuffs all over the globe. With 89,600 ha area, the greenhouse business of Dutch is undoubtedly the highly innovative in the world. The robust studies and developments retained the Dutch greenhouse industry at an advanced level. Gulf countries' hot temperature prevalence addresses the requirement for greenhouse and other protective cultivation techniques. Israel, with its leading exporter role for cut flowers with extensive cultivation under greenhouses (18,000 ha), and Turkey, with 12,000 ha area under cultivation of florets and vegetables, are the landmarks to recognize the value of this technology. In Saudi Arabia, cucumber and tomato crops have contribution of greater than 94% in the total greenhouse cultivation. The commonly used cooling technique deployed in these regions is cooling based on evaporation (i.e., evaporative cooling). Egypt occupies around 1400 ha area under greenhouse structures, comprising mainly of plastic-covered tunnel-type constructions. Natural ventilation is used for the regulation of temperature and humidity situations. Here, the main greenhouse crops are cucumbers, tomatoes, peppers, melons, and nursery plants. In Asian continent, China, followed by Japan are the major beneficiaries of greenhouse technology.

The way adopted and progression succeeded by China is appreciable than any other countries working in same direction all around the globe. In the late 70s, the area under greenhouses in China hit the level of 51,000 ha for flowers and 11,000 ha for fruit crops such as mango, lemon, grapes, loquat cherry, Japanese persimmon, and fig. Flexible plastic-based flexible packaging material is the only better means for frame and glazing aspect. Japan covers greater than 40,000 ha area under greenhouses, out of which about 7500 ha is dedicated to fruits only.

Japan produces a variety of vegetable and flower crops in greenhouses to meet a substantial share of demand for vegetables. Also in South Korea, area beyond 21,000 ha is covered by greenhouses for cultivating flowers and fruits. Interestingly, greenhouses authorize crop cultivation in severely cold climatic conditions of Russia and Canada. Furthermore, it allows production in areas with extremely hot climatic conditions of Kuwait, UAE, and Israel. In southeast Asia, mainly the Philippines, greenhouse cultivation has proved the crop production even with extreme rainfall. Therefore, protected/greenhouse crop growing is being adopted under all possible climate settings. While greenhouses have been there for more than 15 decades in different parts of the globe, India has started protected/greenhouse cultivation in 1980s, chiefly for research purposes. This may be because the priority was to achieve self-reliance in food production. But, in the recent past, as of globalization, emerging international trading, and the scarcity of food, a tremendous boost has been given to India's fresh harvest agricultural produce in export terms, which has uplifted a spurt in demand for the technology of greenhouse.

The greenhouse system may be one important component to sustain food for the growing Indian population/economy. As far as, India is concerned for development of greenhouse technology, a national committee on use of plastics in agriculture (NCPA) was established in 1982. The recommendations from this committee address location-specific trials for greenhouse technology for the approval in various regions of India. Greenhouses are being constructed in the Leh-Ladakh (UT) region for achieving extended growth periods (from 3 to 8 months) of the vegetable crops (Anon, 2013b). In northeast India, greenhouse structures are being built basically as rain shelters to facilitate off-season vegetable cultivation. In north India, flower and vegetable seedlings are being prepared in greenhouses to capture the early markets or improve quality of the seedlings. Propagation of difficult-to-root tree varieties has also been found to be very inspiring. Several commercial floriculture projects are coming up in Maharashtra, Tamil Nadu, and Karnataka states to fulfill the demands of both local as well as foreign markets.

The commercialization of greenhouse cultivation is in progress by 1988 forward, and today with amendments in the government's economic strategies and developing enterprises, numerous business houses have started to set up 100% export-oriented units. In a 4-years period only, with the enactment of the liberal strategies in 1991, about 103 schemes with FDI of more than Rs. 80 crores have been permitted to be set up at a projected cost of Rs. 1000 crores in the nearby areas of Pune, Bangalore, Hyderabad, and Delhi. Therefore, the area under the climatologically-organized greenhouses under these projects is projected to be about 300 ha. Some have started exports by now and acknowledged very favorable outcomes in accepting the quality in major markets of the foreign lands (Anon, 2013a).

7.2.1 GREENHOUSE CROPS

The selection of species and cultivars for protected cultivation is a critical factor that can have a considerable impact on the overall success and profitability of the operation (La Malfa and Leonardi, 2001). In winters, protected cultivation concentrates on cultivating vegetables which belong to the Solanaceae (peppers, tomatoes, eggplant, etc.) and Cucurbitaceous (cucumbers, melons, watermelons, summer squash, etc.) families. In several mediterranean countries, more than 80% area of these crops is cultivated under cold greenhouse conditions to meet local market requirements.

The selection of species and cultivars is critical in determining the long-term viability of protected cultivations. While the selecting process is complicated, some of the most important parts might be simplified. The decision is based on a step-by-step process that begins with a detailed examination of demands and priorities. This strategy is built on being aware of the available know-how as well as the ability to conduct experiments. The fruit and vegetable crops like capsicum, tomato, cucumber, lettuce, and strawberry; floriculture crops like orchid, gerbera, and rose are basically commercially cultivated in Indian greenhouses (Figure 7.6).

7.3 IRRIGATION SYSTEM FOR GREENHOUSE CROPS

Water has the most crucial involvement in human life, particularly in agri-practices. The pressure for agriculture's maximum effective water use intensifies with increased competition among various sectors. Therefore, improving the crop productivities per unit of water per unit land is the need of the hour. Hence, for the time being, sensible irrigation strategies are also a major consideration aspect for making significant improvements in the environment of crop production to a satisfactory count. Microirrigation resides among those strategies that primarily save unnecessary wastage of water and increase its use efficiently (Figure 7.1), specifically in agricultural practices (Zafari and Mohammadi, 2019). Lessening in water consumption by using drip irrigation fluctuates from 30 to 70% and productivity gains in the range of 20–90% for various crops, when compared with flood irrigation. In a nutshell, drip irrigation raises the yield potential of crops to a significant extent.

The irrigation requirement depends on the cultivated area, type of crop, weather condition, time of cultivation, the capacity of ventilation requirement, etc. Additionally, the type of soil or soil mix and type of container bed also affect the irrigation requirement. The crop quality could get hampered by the frequent application of water in the greenhouse. Therefore, the decision on irrigation scheduling should be taken by continuous inspection and understanding of the requirement. The irrigation systems used in the greenhouse are handwriting, perimeter watering, overhead sprinklers, boom watering, and drip irrigation (Debnath and Mohiuddin, 2020).

Surface drip and microsprinklers as microirrigation ways under greenhouses worked efficiently. Automation of irrigation becomes easier in protected structures/greenhouses, which is highly needed to improve water application efficiency and control of the environment by using computers and artificial intelligence techniques. Advantageously, the application of plant protection agents by way of irrigation, namely, herbicides, fungicides, insecticides, growth regulators, and biocontrol agents, has increased rapidly (Debnath and Mohiuddin, 2020), listed here in Table 7.1.

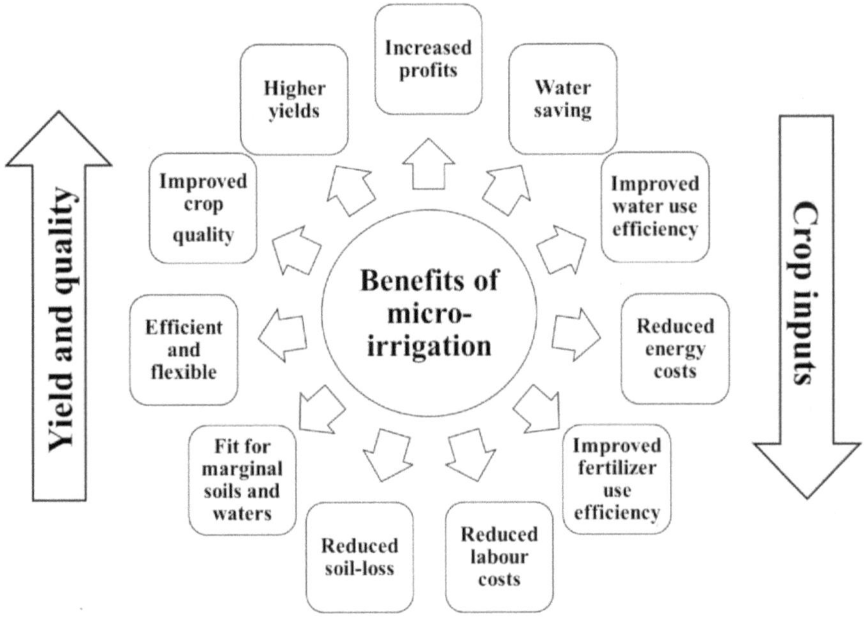

FIGURE 7.1 Benefits of microirrigation.

7.3.1 COMPONENTS OF MICROIRRIGATION

Microirrigation refers to a system that supplies water to the plants at lower discharge rate and pressure (<1.5–2 bar). This system, particularly the drip irrigation provides localized water distribution, near the rooting zone of the plants. The key parts of a microirrigation system are:

- Pump
- System of filters
- Control valves
- Delivery system:

TABLE 7.1 Suitable Fertilizers for Fertigation.

Name	Chemical formula	$N\text{-}P_2O_5\text{-}K_2O$ content (%)	Solubility (g/L 20°C)	Remarks
Ammonium nitrate	NH_4NO_3	34-0-0	1830	Incompatible with acids
Ammonium sulfate	$(NH4)_2SO_4$	21-0-0	760	Clogging with hard water
Phosphoric acid	H_3PO_4	0-52-0	457	Incompatible with calcium
Diammonium phosphate	$(NH4)_2HP_2O_5$	18-46-0	575	Contains phosphorus at high solubility
Potassium chloride	KCl	0-0-60	347	Chloride is toxic for some crops, cheapest K source
Potassium nitrate	KNO_3	13-0-44	316	Expensive, high nitrate
Potassium sulfate	K_2SO_4	0-0-50	110	Excellent source of sulfur, clogging with hard water
Urea	$CO(NH_2)_2$	46-0-0	1100	Water soluble

Mainlines, made of PVC or polyethylene (PE), as channels for water from source to the submains (PVC or PE) thence finally to drip lines. Drip lines composed of LDPE (low-density polyethylene) are placed parallel to the crop, installed with emitters at the end through which water percolates at root zone of the plant via dripping, bubbling, and microsprinkling (these can be of various types, suited to serve one or more plants).

Emitters/drippers can be categorized as online and inline:

- Online emitters are installed on laterals conveying water lengthwise in the plant rows. Such emitters can be installed on laterals of varying diameters allowing good operational flexibility. For most row crops, the quantity and spacing of the emitters can be modified to meet plant spacing. Drip lines can be hung or put directly on the ground.
- Inline emitters are the important parts of the lateral pipes. Such emitters offer higher resistance to emitter clogging up, and they are suitable for placing on the ground along the row of the crop or under the mulch.

7.3.2 MICROIRRIGATION SYSTEMS

Microirrigation systems can be grouped as:

- System operating with pressure and discharge range of 0.5–2.0 bar and 0.5–4 L/h, respectively.

- Systems operating with pressure and discharge range of 1–4 bar and 2–20 L/h, respectively.
- Systems operating with pressure and discharge range of 1–3 bar and 6–30 L/h, respectively.
- Systems operating with pressure and discharge range of 1–2.5 bar and 0.7–7 L/h, respectively.
- Systems using micro/minisprinklers.
- Sub-surface drip/trickle irrigation system.

The most recent classification of microirrigation system for greenhouses has been shown in Figure 7.2.

7.3.2.1 SYSTEMS WITH DRIP LINES

Perforated hoses, which consist of a thin polyethylene tube (usually 0.15–0.20 mm) with holes spaced at a fixed distance, fall under this category. Because the perforated tubes lack a flow rate control system, they cannot ensure uniformity of distribution, which is susceptible to becoming uneven due to frequent clogging of the pores. Light drip lines and dripper lines are two types of real drip lines (where the water flow is controlled). Light drip lines, which are commonly used for row-planted annual crops, have emitters that allow regular water distribution patterns. Drip lines with a dripper and adequate tube thickness last longer, give more consistent irrigation, and are appropriate for long-term crops. Drip lines with self-compensating drippers are also available for sloping ground or extended distances. There will be significant pressure changes at the beginning and end of the drip line in these circumstances, altering the flow rate of individual drippers. The self-compensating drippers can perform efficiently in these settings because they maintain constant flow rates at various pressures. A membrane of extremely elastic–plastic material that is distorted when water pressure is applied provides the self-compensating feature. The resulting expansion or contraction of the emitter's output section aids in the stabilization of the water flow. Many self-compensating drippers also have a labyrinth that self-cleans (anticlogging). The self-washing mechanism works on the idea that the membrane does not form up to around 0.5–0.7 bar, generating a turbulent flow in the tube (higher than the nominal flow rate) capable of eliminating particles that may build inside the dripper.

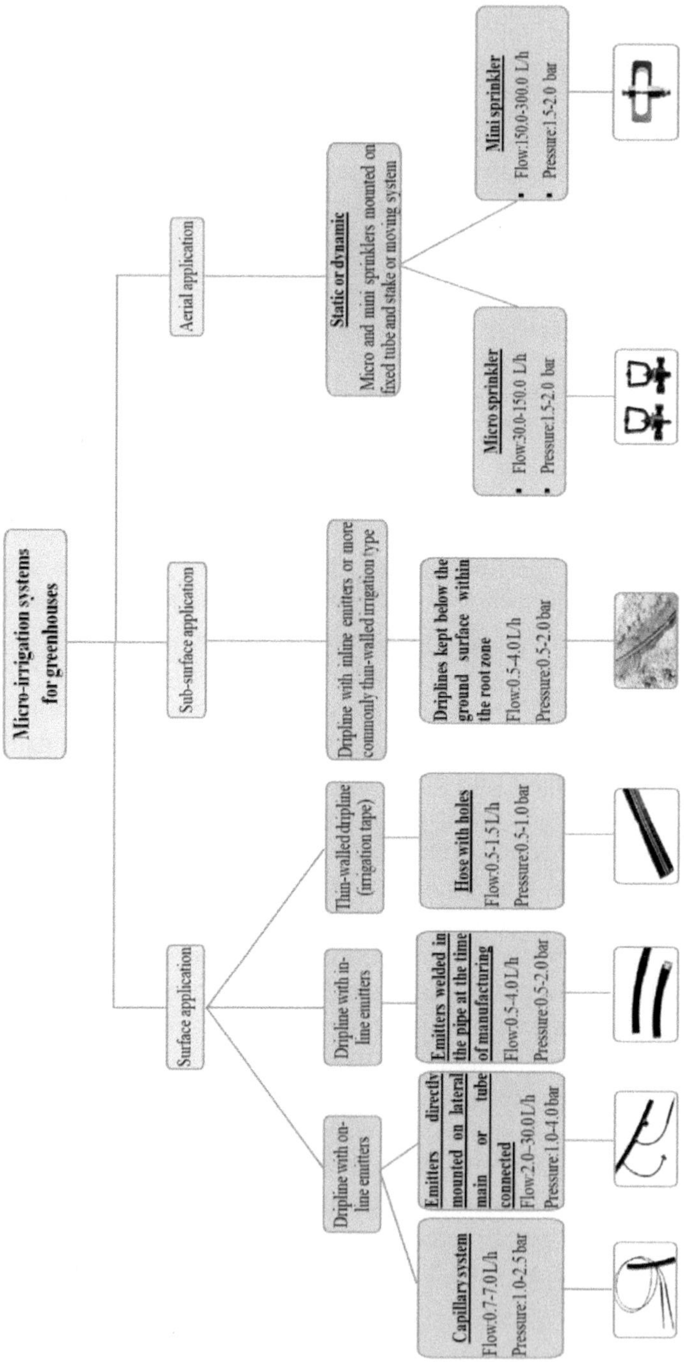

FIGURE 7.2 Microirrigation systems for protected cultivation.

Source: Adapted from Debnath and Pedda, 2020

7.3.2.2 SYSTEMS WITH DRIPPERS

Low-density polyethylene tubes with sizes ranging from 16 to 25 mm are used in these systems. Drippers with buttons or arrows can be used. Button-type drippers have a special labyrinth that enables the flow to be more uniform. Also, in this case, the self-pressure compensating models provide continuous supply of water irrespective of the variations in the working pressure.

These drippers can be mounted directly on the line (with a thin polyethylene tube delivering water to the plants) or at the end of short strands inserted on the main lines. Furthermore, special derivations enable single emitters to have multiple outputs (up to eight) when connected directly to the polyethylene tube, lowering overall irrigation expenses. The antidrain compensated no leakage dripper is one form of dripper (CNL). In this case, the dripper is equipped with a special membrane that completely closes the water leakage when the pressure falls below a predetermined level (0.3–1 bar, depending on the model), the line does not empty. This could be beneficial for soilless crops that require frequent and short irrigations. In this instance, there is a possibility of an excessive supply of water in the line's last sections for the line's emptying. The arrow-type drippers are labyrinth-style drippers that are connected to the main lines via tiny polyethylene tubes. By inverting the little tubes that connect the drippers to the major polyethylene lateral pipes, these may be readily inspected and cleaned.

7.3.2.3 SYSTEMS WITH INTERMITTENT EMITTERS

These systems are characterized by means of excessive unitary flow rates (6–30 L/h, with operating pressures of 1–3 bar). The essential advantages of those systems are the decreased incidence of clogging because of the higher working flow rate and the larger wet surface in comparison with systems with classic drippers. Overall, these features permit more uniform growth of the root systems.

7.3.2.4 SYSTEMS WITH CAPILLARY TUBES

These systems are becoming more popular because of their low cost and the possibility to have the lines preassembled by the manufacturer, resulting in significant labor cost savings. This system consists of a polyethylene tube with a diameter of 20–25 mm, which is inserted into capillary tubes with an internal diameter of 0.5–1.5 mm and sufficient length to reach the dispensing point. In these cases, the flow rate is directly proportional to the capillary diameter and inversely proportional to its length for a given pressure. The system includes rods connected to the capillary's end to enable for proper positioning close to the plants' roots. Because of pressure losses along the line, this type of system, although cost-effective, does not provide uniform water distribution. As a result, lines with a large number of drip points should be avoided. To give water under pressure to both ends of the main line with the capillaries, it is always advisable to use double-headed irrigation systems in a closed circuit with this type. As a result, the pressure gradient along the line is kept to a minimum.

7.3.2.5 SYSTEMS WITH MICRO/MINISPRINKLERS

Sprinklers are conventionally classified as follows:

- Microsprinklers with discharge of 30–150 L/h
- Minisprinklers with discharge of 150–350 L/h.

The microsprinklers need operating pressure of about 1.5–2.0 bar and emission holes calibrated between 0.08 and 0.23 cm. As a result, the flow rates and infiltration rates are significantly higher than with the drippers. The rate at which irrigation water may be absorbed by the soil is slightly lower. Sprinklers can also be standard or self-compensating, and they have the same features as the drippers. These systems can be divided as follows:

- Dynamic systems (also known as microsprinklers or minisprinklers, depending on flow rates) provide a circular wetting area with a greater radius than other systems, since they include moving parts during irrigation.
- Sprayers, also known as static (no moving parts) sprayers, are capable of having wetting areas of various shapes (circular, semicircular, and sectors).

Sprinklers can be installed directly on main or branch lines, on the tube, or on carrying rods placed into the ground. Different combinations correspond to different irrigation needs or farm management practices, but they can have a significant impact on irrigation quality and efficiency. If the irrigation system is automated, sprinklers can be used to combine irrigation and frost protection. This method is indicated in places where spring frost is moderate but harmful. Sprinklers that create large diameter droplets are more effective at transferring heat to the atmosphere and raising air temperature.

7.3.2.6 SUBSURFACE DRIP IRRIGATION

It discusses the use of underground drip tubes placed at a particular depth. This technology allows you to wet the underground root zone directly, reducing evaporative water loss at the soil surface and increasing water efficiency (Giancarlo and Albino, 2013).

7.4 FERTIGATION STRATEGIES ADOPTED UNDER PROTECTED STRUCTURE

Fertigation is the way wherein water-soluble fertilizers channelize to plants without any losses by dripping means. Nowadays, this is a widely adopted agronomically sound process of supplying water-soluble fertilizers within the lively root zone of the plants (Zafari and Mohammadi, 2019).

7.4.1 SUITABLE FERTILIZERS FOR FERTIGATION

7.4.1.1 FERTILIZERS SUITED BEST FOR FERTIGATION IN PROTECTED CULTIVATION

Fertilizers/nutrients used alone or in combination flourish the plants if utilized in appropriate concentration (s). These nutrients include N (150–200 ppm), P (50 ppm), K (200–400 ppm), Ca (150–200 ppm), Mg (50 ppm), B (0.2 ppm), Zn (0.1 ppm), Cu (0.1 ppm), Mn (1 ppm), Fe (5 ppm), and Mo (0.05 ppm). The fertilizers suitable for fertigation are listed in Table 7.1.

7.4.1.2 CHLORIDE-FREE FERTILIZERS

Such fertilizers are manufactured from urea, ammonium nitrate phosphate, and potassium nitrate as the elementary constituents, and are helpful for high-value crops and crops that are more sensitive to chloride injury.

7.4.1.3 LIQUID FERTILIZERS

Bulk fertilizers like urea, ammonium nitrate, phosphate and sulfate, potassium sulfate, chloride and nitrate, phosphoric acid, etc. are the main components for preparing liquid fertilizers. Such fertilizers are pure and prevent precipitation and blocking of emitters/drippers. They are normally acidic in nature with pH value in the range of 5.5–6.5 and aid in correcting the soil pH. A fertilizer solution having pH value of 7 or higher may possibly be utilized in soils with acidic nature.

7.4.1.4 NORMAL FERTILIZERS

These are formed by mixing urea, ammonium sulfate, nitrate and phosphate, potassium chloride, phosphoric acid, etc.

7.4.1.5 MICRONUTRIENTS

Micronutrients are usually supplied to the plants separately in majority of soils, as the application via fertigation would act in response with salts present in the water and result in blockage due to precipitation. The chelated micronutrients are extremely soluble in water, and the chances of blockage of emitters are least.

7.4.2 EFFICIENT MACHINERY AND FERTIGATION SELECTION METHODOLOGY

7.4.2.1 FERTIGATION TANK

A part of the irrigation water is directed from main supply line into a fertilizer solution tank in this system. The diverted water carries the fertilizer solution, and returns to the mainline through the exit assembly of the venturi. Fertilizer tank is required for a big irrigation system when crops require fertilizers in greater amount. However, with this system, the quantities of fertilizers incoming with water change continuously with time, with a very high concentration at the beginning. The fertilizer tanks exist in capacities of 90, 120, and 160 L (Figure 7.3).

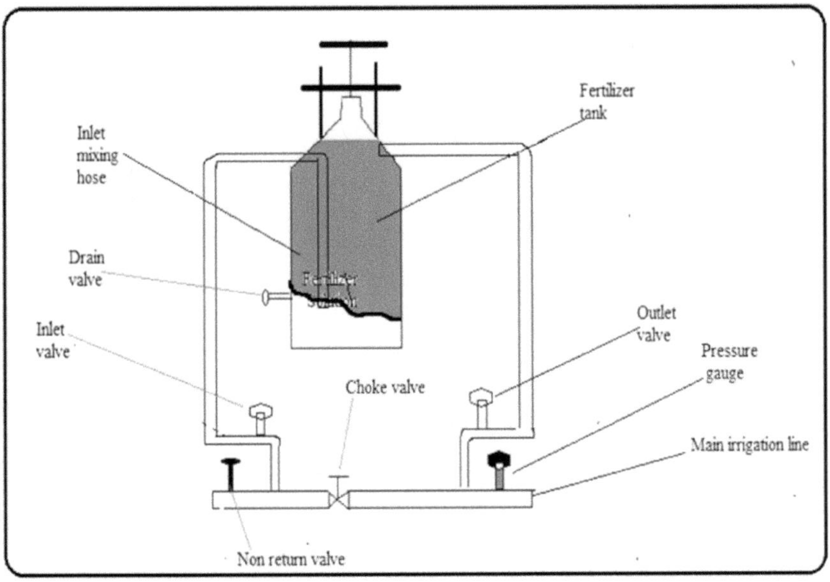

FIGURE 7.3 Fertigation tank.

Source: Adapted from Shirgure, 2013

7.4.2.2 VENTURI INJECTOR

In venturi injector, a vacuum is created by distracting a fraction of water to flow from main pipe via a venturi, which increases flow velocity. The created vacuum is therefore utilized to start drawing out fertilizer solution by suction pipe from a given tank. The venturi injector best suits the fertigation requirements of small growers, due to its low cost and ease in operation. But, the use of venturi may lead to a significant pressure drop in the system, resulting in uneven water and fertilizer distribution. The rate of suction of venturi varies as 30–120 L/h (Figure 7.4).

7.4.2.3 INJECTOR PUMPS

These are the diaphragm or piston pumps that are driven by the pressure created by flowing water through the irrigation system. The rate of injecting fertilizers is relative to the flow of water. A high degree of regulation over the fertilizer injection rate is possible without much pressure loss. The initial cost of the injector pump is high; however, it is appropriate for precise fertigation. The suction rate of this type of pump varies as 40–160 L/h (Figure 7.5).

FIGURE 7.4 Venturi injector.

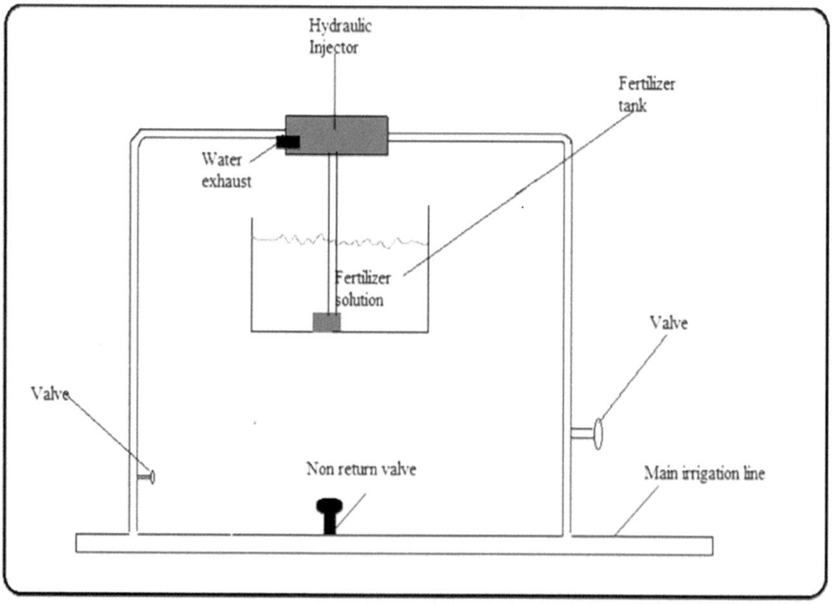

FIGURE 7.5 Fertilizer injector pump.

Source: Adapted from Shirgure, 2013

7.5 IRRIGATION AND FERTIGATION SCHEDULING IN GREENHOUSES

7.5.1 IRRIGATION SCHEDULING

Irrigation scheduling determines the amount of water applied to the crop (irrigation dose) and the application timing (irrigation frequency). The principal methods of irrigation scheduling in soil-bound crops and soilless growing systems are very similar. It includes water balance (determines crop water requirements from climatic data) and the use of soil or plant sensors. Irrigation scheduling based on measured plant water status and physiological response of a plant to drought using plant-based methods was extensively reviewed by several authors (Jones, 2004; Katsoulas et al., 2016). The different means of irrigation scheduling in greenhouses are summarized below (Table 7.2). The fertigation technique among soilless crops has been described by Singh et al. (2018). In soilless systems, we have just water as nutrient supply medium (fertilizers) to the crops (mixture of macro- and micronutrients). Application frequency for this nutrient solution may be varied either on daily basis or as a stock solution, that is, Hoagland solution (Hoagland and Arnon, 1950). The nutrient solution so formed should be monitored on regular basis for its pH and EC values, as well as the deficiency of micronutrients.

Alomran et al. (2013) reported that providing deficit irrigation at precise crop stages with 80% ET_c (i.e., lowering irrigation amount by about 40% ET_c) is the most suitable irrigation plan for obtaining improved production. If adopted for tomatoes under greenhouse roof, partial root drying saves water effectively to 50%. But, substantial saving of irrigation water can make the partial root drying viable in the water-scarce and costly regions (Zegbe, 2006).

Sensors are now recommended for irrigation scheduling in protected structures. There are two key problems when using sensors for irrigation scheduling (IS), namely replication (2–3 sensors per crop) and location (representative of the crop). Cost, convenience of use, preparation and maintenance requirements, technical support, data interpretation, availability of irrigation protocols, working language, and the user-friendliness of any software are all practical considerations. For both technical and commercial reasons, soil water sensors are a dynamic and rapidly evolving technology. While for decades, there have been practical on-farm applications, the technological developments of soil water sensors have generally been focused on scientific uses. Sensors are being used for automation of irrigation with automatic start and closure of water application to the crop (Thompson

TABLE 7.2 Methods for Scheduling Irrigation in Soil and Soilless-based Cultivation Systems in Greenhouses (Adapted from Nikolaou et al., 2019).

Scheduling irrigation	Basis	Method or tool/decision variable
Climate monitoring	Based on evapotranspiration (ET)	
	• Reference evapotranspiration (ET_o)	• Atmometer/ET_o
	• Crop evapotranspiration (ET_c)	• Class A Pan/ET_o
		• ET models/Crop water used
		• Lysimeters/ET_c
		• Reduce class A Pan/ET_o
	Based on solar radiation	Pyranometer/IF
Soil or substrate monitoring	Electrical conductivity (EC)	EC sensor/IF (soil and soilless-based cultivations)
	Drainage (%)	Mathematic formula and weighting devices/Irrigation volume and frequency-based (soilless cultivation)
	Physical properties	Mathematical formula/IF (soil and soilless-based cultivations)
	Water content (volumetric basis)	Dielectric sensors (e.g., TDR and FDR)/IF (soilless and soil-based cultivations)
	Water potential	• Tensiometer/IF
		• Electrical resistance sensors/IF (soil-based measurements) (e.g., gypsum blocks)
Phytosensing	Canopy temperature	Infrared thermometry/IT
	Crop/plant reflectance	Sensing equipment or indices for plant/crop reflectance (e.g., Normalized difference vegetation index (NDVI), photochemical reflectance index (PRI), etc.)/Detect the water stress
	Leaf water potential	Pressure chamber/Irrigation timing (IT)
	Stem diameter changes	Dentrometer/IT
	Stomata resistance	Diffusion porometer/IT
	Stem water flow	Heat balance technique for sap-flow measurement or sap-flow sensor/IT
Time clock	Based on time	Irrigation controllers/Irrigation frequency (IF)

Source: Adapted from Nikolaou et al., 2019

and Gallardo, 2005; Gallardo et al., 2013). Automatic irrigation control requires either (a) a triggering system as soon as the set-point reaches or (b) automatic observing of data of small intervals by use of sensors having quick reactions to the fluctuations in soil water status and an appropriate interface with a controller for irrigation. Appropriate protocols to determine the locations (in relation to emitter/plant placement) for installing sensors, depth for installation, repetitions, and mainly the protocols for interpreting the sensor data for automatic or manual irrigation are the challenges till date (Thompson and Gallardo, 2005; Thompson et al., 2007; Gallardo et al., 2013). Thompson et al. (2007) also presented detailed information about using static or dynamic protocols to interpret information gathered from capacitance-based sensors inside the protected structures. For commercial applications, affordable and robust sensors are required that measure soil water status continually and use wireless radio technology to transmit the data to the irrigation controller (Pardossi et al., 2009; Gallardo et al., 2013).

7.5.2 FERTIGATION SCHEDULING OF GREENHOUSE CROPS

Fertigation allows for a noteworthy improvement in fertilizer use effectiveness with better outcomes as compared to the other methods of applying fertilizers to the crops (40–45%) (Agostini et al., 2010; Solaimalai et al., 2005). The advantages of fertigation include applying fertilizers in small splits as per requirements, improved fertilizer distribution in the root zone, and the opportunity of preserving a low. Consequently, the use of fertigation has been found to decrease loss of transportable nutrients like nitrogen (up to 70%), as compared to the traditional methods of applying fertilizers (Solaimalai et al., 2005).

Fertigation in the greenhouse makes better use of water and nutrients, increases yield, and improves the final product's quality. Fertigation scheduling refers to the use of trickle irrigation to supply water and nutrients at the appropriate times for each crop stage. The most significant part of drip fertigation is fertilization scheduling, which involves knowing the correct amount of water and fertilizers to be administered by drip at regular intervals on a daily and monthly basis. Stages in horticulture crops differ with crop, such as the initial and final stages, vegetative stage, flowering and fruiting stages, and so on. Fertigation scheduling indicating the cropwise stages along with concentration of nutrients (N, P, and K) in terms of ppm (1 ppm N = 1 mg N/L of water) is given for horticultural crops (vegetables and flowers) under greenhouse (Tables 7.3 and 7.4). From the farmer's perspective,

TABLE 7.3 Fertigation Schedules of High-Value Crops Cultivated in Greenhouses (Adapted from open access source: Hasan et al., 2010).

Crop (plants/1000 m^2)	Scheduling period	Dosage (ppm)			Total dose (kg/1000 m^2)		
		N	P	K	N	P	K
Tomato and cherry tomato (2773)	August–September	100	50	100	35	12	39
	October	200	60	200			
	November–January	250	80	250			
	February	250	60	250			
	March	250	50	250			
	April	100	50	200			
Capsicum (4160)	August–September	100	60	100	32	15	36
	October	180	80	180			
	November	220	100	220			
	December	220	100	220			
	January–March	220	90	220			
	April	100	50	200			
Cucumber (4160)	**Cycle: August–October**						
	August	120	60	120	15	7	16
	September	180	80	200			
	September–October	180	80	200			
	Cycle: November–January						
	November	160	80	160	10	4	11
	December	220	80	240			
	December–January	220	80	240			
	Cycle: February–April						
	February	120	60	120	16	6	18
	March	18	60	200			
	March–April	180	60	200			

Source: Adapted from Hasan et al., 2010

TABLE 7.4 Fertigation Schedules of Flowers Cultivated in Greenhouses (Adapted from open access source: Hasan et al., 2010).

Crop (Plants/1000 m²)	Scheduling period	Dosage (ppm)			Total dose (kg/1000 m²)		
		N	P	K	N	P	K
Carnation (32,000)	September–October	50	30	40	14	8	13
	November–March	60	40	60			
	April–August	40	20	40			
Chrysanthemum (65,000)	September–October	80	50	60	21	13	19
	November–April	90	60	80			
	May–August	50	30	50			
Gerbera (16,000)	September–October	70	50	60	17	12	17
	November–April	80	60	80			
	May–August	40	24	40			
Lilium (32,000)	September–October	60	36	60	17	11	17
	November–March	80	50	80			
	April–August	50	30	50			
Rose (12,000)	September–October	80	50	60	28	17	25
	November–March	100	60	80			
	April–August	80	50	80			

Source: Adapted from Hasan et al., 2010

monthwise nutrient doses in horticultural crops are converted to 100% water-soluble fertilizers, urea phosphate (17:44:0), urea, and SOP in a user-friendly way. The proposed fertigation scheduling (Tables 7.3 and 7.4) is generally recommended for all types of soil. However, it is recommended to get the soil testing done and adjust the nutrient dosage accordingly. Fertigation scheduling for horticultural crops will benefit farmers in increasing the yield and quality of harvest through efficient use of water and nutrients (Hasan et al., 2010).

7.6 CASE STUDY: EFFECT OF IRRIGATION AND FERTIGATION IN GREENHOUSE CROPS

The phrase "nutriculture" refers to any method of cultivating plants in a medium other than natural soil. Water culture, aggregate culture, and the adsorbed-nutrient technique are all included. Hydroponics is a method of growing plants in which the nutrient solution is continuously circulated through the plant roots. The solution replaces soil as a source of water and nutrients for the plants. Tray agriculture and tank farming are other names of hydroponics system (Hogland and Arnon, 1950). We studied the effect of differential fertigation on the growth, yield, and quality of strawberries cultivated under a naturally ventilated polyhouse conditions in the winter season (Figure 7.7).

The temperature was recorded during the morning and afternoon using a psychometer from inside and outside of the structure to check whether the temperature is suitable for the crops or not; if not, then temperature was maintained with the help of a fan and pad cooling system. A hygrometer was used to measure relative humidity from 08:00 to 18:00 hours at 2-h intervals, and relative humidity was maintained using fogging.

It was found that fertigation in strawberries in greenhouse conditions gave better performance on to the selected traits like qualitative; fruit taste, color development, and size and quantitative; plant height, leaf area, and other vegetative parameters using copper solution with a concentration of 100% and their interaction with Camarosa variety.

7.7 AUTOMATION IN IRRIGATION SYSTEM

Water and nutrient supplies of crops are affected by several parameters like crop type, phase of development, and cultivation system. Though, a

Cucumber Capsicum

Lettuce

Tomato

FIGURE 7.6 Different crops under protected cultivation.

horticultural crop cultivated inside a protected structure flourishes well using water and fertilizers competitively than any other production schemes. It is essential to increase efficiency for which customized fertigation strategy should be made for automation of fertigation in greenhouses.

FIGURE 7.7 Strawberry in hydroponic system.

Since the late 1980s, tremendous progress has been made in the development, availability, and application of sensors in irrigation scheduling and automation, due to significant advances in electronics and information technology. Electrotensiometers are computer-based soil moisture tension recording sensors for controlling water application. Limiting drainage prevents nutrient leaching that can source pollution of aquifers. Nevertheless, for automatic irrigation control based on the soil matric potential (SMP) to be efficient, it becomes necessary to identify a suitable SMP value to begin irrigation depending on types of crops, stages of development, soil type (texture), and evaporative conditions (Hanson et al., 2000).

In conventional agricultural systems, irrigation automation includes microcontrollers, programmable logic controllers, and crop/sensor interfaces programmed to control specific tasks characterized by a sequential evolution (Rodriguez-Ortega et al., 2016). However, plant responses to their environment are characterized by nonlinearity, prompting the need for open and integrated architecture systems involving cooperation between different devices (Morimoto and Hashimoto, 1998; King and Sigrimis, 2001).

The internet of things and digital geographies offer the opportunity to improve PA in terms of productivity, food security, and sustainability in agriculture (Kamilaris et al., 2017; Zamora-Izquierdo et al., 2019). The IoT removes the gap between physical and cyber worlds by integrating sensor-based information, data communication, data cloud or fog-computing acquisition, and intelligent processing in real time (Goap et al., 2018; Mehra et al., 2018; Zamora-Izquierdo et al., 2019). Even though a British visionary, Kevin Ashton first mentioned the term "IoT" back in 1999, IoT and the platforms for agricultural systems are still at an early stage of development (Madakam et al., 2015; Tzounis et al., 2017; Karim et al., 2017). The IoT and cyber-physical systems describe hardware systems (platforms) that gather data from sensors, communicate with actuators, and cooperate to cover automation needs (Zamora-Izquierdo et al., 2019). At the same time, cloud-, fog-, and edge-computing software application stores can analyze collected data in real time (Tzounis et al., 2017). According to Khanna and Kaur (2019), the most common IoT-enabled technologies include:

- Bluetooth
- ZigBee
- Wireless sensor networks (WSNs)
- Radiofrequency identification
- Wireless fidelity
- Near-field communication

According to Shi et al. (2019), data were only analyzed and presented in a spreadsheet program such as Excel in past. Through expansion of cloud computation, decision making has become more reliant on artificial intelligence (AI) (King and Sigrimis, 2001; Mehraet et al., 2018). According to Kirtan et al. (2019) and Yong et al. (2018), the commonly adopted AI systems are:

- Fuzzy logic and neuro-fuzzy logic
- Artificial neural network (ANN)
- Bayesian networks
- Expert systems

Common to all these systems is the central processing unit (CPU) responsible for primary functions and using basic components such as microcontroller sand motherboards (Lakhiar et al., 2018):

- Arduino: A microcontroller motherboard with easy-to-use hardware and software, which can read analog input signals from different sensors. Two versions are Arduino Uno and Arduino Yun.
- Raspberry Pi: A small single-board computer with no storage onboard, which can run multiple programs at a time. It is easy to connect with internet.

Recently, a new concept, that is, the internet of underground things consists of differential sensors (e.g., sensor for indicating salinity level and moisture in soil) that are partially or fully implanted in soil (Vuran et al., 2018). An algorithm for irrigation automation of plants grown in pots on the basis of weighting type lysimeters has also been developed by Vera-Repulloa et al. (2015). In that system, each lysimeter is made up of metal to grip the pots, and four load cells are provided to measure changes in weight and drainage water. A communication network (Modbus/RS-485) receives data from load cells and processes them to control irrigation valves in real time. The software allowed for data acquisition, logging, and scheduling of irrigation on the basis of estimates of the real-time water balance obtained from the lysimeters. Helmer et al. (2005) explained the "crop assist system" for continuous monitoring of a vine crop grown inside a protected structure using 11 physiological parameters and irrigation on the basis of the real-time data received with pairs of load cells. According to SteidleNeto et al. (2014), the irrigation frequency could be managed through the "Hydra-Control" algorithm in tomato crop, which estimates the transpiration rate using the electrical conductivity (EC) of leachate under preset limits.

Sihombing et al. (2018) fabricated a system for regulating temperature and level of nutrient solution based on different sensors using a "logic if else" approach. In this, an Arduino Uno microcontroller with communication via low-cost wireless fidelity (Wi-Fi; ESP8266) was connected to an android smartphone. The system was used for the small-scale closed growing spinach. Likewise, Palande et al. (2018) developed the "Titan Smartronic" system for fully automating a hydroponic unit to eliminate the human contact after crop transplantation. The developed system utilized a number of sensors (e.g., EC and pH monitoring and a water temperature sensor) supported by a Raspberry Pi, Arduinos, and open-source software for maintaining the required threshold values. This system consists of an IoT-based network to monitor and control from android or Ios devices.

Mehra et al. (2018) developed a deep neural network-based system to manage tomato plants grown in a hydroponic tank using several parameters (e.g., pH, nutrition level, ambient temperature, humidity, and lighting) collected in the cloud in real time. The model was fitted using Raspberry Pi3, Arduino, and tensor-flow measurement. Zamora-Izquierdo et al. (2019) developed a multitier PA management platform based on edge and cloud computation. The IoT system was implemented in tomato hydroponic units in actual settings. The IoT application achieved water savings of about 30–80%. Likewise, Katsoulas et al. (2017) developed a web-based irrigation scheduling algorithm to optimize the irrigation interval rate and the amount of nutrient solution in each irrigation event in SCS. The crop transpiration rate was estimated from greenhouse climatic data (e.g., air temperature and relative humidity) and solar radiation values to determine irrigation rate. The amount of irrigation was estimated by taking into account substrate water content and water balance. The use of the algorithm resulted in a 100% increase in water and fertilizer use efficiency in two commercial hydroponic units compared to standard irrigation practices.

In order to optimize fertilizer use in hydroponics fertigation systems, Bueno-Delgado et al. (2016) developed an android application called "Ecofert." The quantities and types of proposed fertilizers for preparation of stock solutions were estimated according to crop type and specific nutrient needs, the available commercialized fertilizer type, and cost. "Ecofert" used linear programming and a cloud-based database of common fertilizers, which was updated daily. Pérez-Castro et al. (2017) also used a linear programming algorithm developed for android mobile devices (named "cFertigUAL") to determine optimal crop fertigation.

Today, advanced horticulturists pay great attention to the expanded set of climate indicators, including leaf temperature, leaf humidity, air temperature distribution along the vertical section of the greenhouse, and airspeed. A company named "Laboratory of Engineering System" has developed a computer climate system, which serves the basic microclimate parameters and allows controlling the previously named extended set of indicators. This system fully automatically supports control of all known types of organizational systems of the greenhouse by accurately maintaining the parameters set by the horticulturist.

7.8 FUTURE PROSPECTS

A possible future trend could include monitoring plant water stress indices on a real-time basis through remote sensors, utilizing nondestructive technology such as the use of reflectance indices. There is a need to develop a commercial irrigation controller unit to model and monitor soil-plant-atmosphere conditions utilizing artificial intelligence. In addition, the implementation of mathematical and empirical models, in combination with decision support systems, will be a beneficial tool for the better management of nutrient solutions in protected cultivated crops and better production.

7.9 CONCLUSIONS

Under the changing environmental conditions, the world is facing acute water shortages. Therefore, under such conditions, use of drip irrigation is quite useful and applicable for efficient water use. Drip irrigation reduces the cost of production apart from increasing the yield and quality in the greenhouse. With the help of drip fertigation in protected cultivation, the overall productivity of horticulture crops will be increased. Besides higher yield and higher resource use efficiency, a higher benefit–cost ratio is obtained in drip fertigation. Drip irrigation cum fertigation in greenhouse crops increases the yield due to higher nutrient uptake, improving water and fertilizer use efficiency. The fertilizers could be saved to the tune of 25–50% through drip fertigation.

Application of fertilizer through drip system is much more efficient than application in any of the conventional methods. Drip irrigation affords and indeed calls for a complete and total fertilizer application through the drip system, feeding the plants with small, frequent, and fully balanced nutrient diets; this in itself is a significant breakthrough in the nutrient application and is a continuation in further obtaining a considerable increase in crop production. Smart technologies have been developed to improve SCS using advanced sensors, communication, and big data technologies. This is important because in high technology greenhouses, where optimum climatic conditions can be achieved, crop production only becomes profitable when productivity is improved.

KEYWORDS

- **irrigation**
- **fertigation**
- **automation**
- **greenhouse**
- **scheduling**

REFERENCES

Agostini, F.; Tei, F.; Silgram, M.; Farneselli, M.; Benincasa, P.; Aller, M. Decreasing Nitrate Leaching in Vegetable Crops with Better N Management. In *Genetic Engineering, Bio-fertilisation, Soil Quality and Organic Farming*; Springer: Berlin, Germany, 2010; pp 147–200.

Alomran, A. M.; Louki, I. I.; Aly, A. A.; Nadeem, M. E. Impact of Deficit Irrigation on Soil Salinity and Cucumber Yield under Greenhouse Condition in an Arid Environment. *J. Agric. Sci. Technol.* **2013,** *15*, 1247–1259.

Anonymous[a], 2013. Indian Council of Agricultural Research (ICAR). Indian Agricultural Statistics Research Institute, 2013. http://ecoursesonline.iasri.res.in/mod/page/view.php?id=124111(accessed Oct 7, 2021).

Anonymous[b]. Design and Maintenance of Green House. Indian Council of Agricultural Research (ICAR). Indian Agricultural Statistics Research Institute, 2013. http://ecoursesonline.iasri.res. in/mod/page/view.php?id=124115 (accessed Oct 30, 2021).

Bueno-Delgado, M. V., Molina-Martínez, J. M., Correoso-Campillo, R.; Pavón-Mariño, P. Ecofert: An Android Application for the Optimization of Fertilizer Cost in Fertigation. *Comput. Electron. Agric.* **2016,** *121*, 32–42.

De Pascale S., Dalla Costa L., Vallone S., Barbieri G.; Maggio A. Increasing Water use Efficiency in Vegetable Crop Production: From Plant to Irrigation Systems Efficiency. *Hort Technol.* **2011,** *21*, 301–307.

De Pascale S., Rouphael Y., Gallardo M. and Thompson R. B. Water and Fertilization Management of Vegetables: State of the Art and Future Challenges. *Eur. J. Hort. Sci.* **2018,** *83* (5), 306–318.

De Pascale S.; Rouphael Y.; Pardossi A.; Gallardo M.; Thompson R. B. Recent Advances in Water and Nutrient Management of Soil-grown Crops in Mediterranean Greenhouses. *ActaHort.* **2017,** *1170*, 31–44.

De Swaef, T.; Steppe, K. Linking Stem Diameter Variations to Sap Flow, Turgor and Water Potential in Tomato. *Funct. Plant Biol.* **2010,** *37*, 429–438.

Debnath, S.; Pedda, G. P. S. K. Irrigation and Fertigation in Protected Cultivation Irrigation and Fertigation in Protected Cultivation. In *Protected Cultivation and Smart Agriculture*; Maitra S., Gaikwad D. J., Shankar T., Eds.; New Delhi Publishers: New Delhi, 2020; pp 40–54. ISBN: 978-81-948993-2-7, DOI: 10.30954/NDP-PCSA.2020.5

Gallardo, M.; Thompson, R. B.; Fernández, M. D. Water Requirements and Irrigation Management in Mediterranean Greenhouses: The Case of the South-east Coast of Spain. In *Good Agricultural Practices for Greenhouse Vegetable Crops*; Plant Production and Protection Paper 217; FAO: Rome, Italy, 2013; pp 109–136.

Giancarlo B; Albino M. Micro Irrigation. FAO-Plant Production and Protection Paper, **2013,** *217*, 149–167

Goap, A.; Sharma, D.; Shukla, A K.; Rama Krishna, C. An IoT based Smart Irrigation Management System using Machine Learning and Open Source Technologies. *Comput. Electron. Agric.* **2018,** *155*, 41–49.

Hanson, B. R.; Orloff, S.; Peters, D. Monitoring Soil Moisture Helps Refine Irrigation Management. *Calif. Agric.* **2000,** *54*, 38–42.

Hasan, M.; Singh, B.; Singh, M. C.; Singh, A. K.; Kaore, S. V.; Tarunendu,; Sabir, N.; Tomar, BS; Fertigation Scheduling for Horticultural Crops. Tech. Bull. No. TB-ICN:80/2010, Indian Agricultural Research Institute, New Delhi-110012 India, Sept 2010, pp 1–44.

Helmer, T.; Ehret, D. L.; Bittman, S. CropAssist, an Automated System for Direct Measurement of Greenhouse Tomato Growth and Water Use. *Comput. Electron. Agric.* **2005,** *48* (3), 198–215.

Hoagland, D. R.; Arnon, D. I. The Water Culture Method for Growing Plants Without Soil. *Agric. California Agricultural Experiment Station Circulation* **1950,** *347*, 32.

Jat A. R.; Wani P. S.; Sharawat K. L.; Singh P.; Dhaka B. L. Fertigation in Vegetable Crops for Higher Productivity and Resource use Efficiency. *Ind. J. Fert.* **2011,** *7*, 22–37.

Jones, H. G. Irrigation Scheduling: Advantages and Pitfalls of Plant-based Methods. *J. Exp. Bot.* **2004,** *55*, 2427–2436.

Kamilaris, A.; Kartakoullis, A.; Prenafeta-Boldú, F. X. A Review on the Practice of Big Data Analysis in Agriculture. *Comput. Electron. Agric.* **2017,** *143*, 23–37.

Karim, F.; Karim, F.; Frihida, A. Monitoring System using Web of Things in Precision Agriculture. *Procedia Comput. Sci.* **2017,** *110*, 402–409.

Katsoulas, N.; Bartzanas, T.; Kittas, C. Online Professional Irrigation Scheduling System for Greenhouse Crops. *Acta Hort.* **2017,** *1154*, 221–228.

Katsoulas, N.; Elvanidi, A.; Ferentinos, K. P.; Kacira, M.; Bartzanas, T.; Kittas, C. Crop Reflectance Monitoring as a Tool for Water Stress Detection in Greenhouses: A Review. *Biosyst. Eng.* **2016,** *151*, 374–398.

Khanna, A; Kaur, S. Evolution of Internet of Things (IoT) and its Significant Impact in the Field of Precision Agriculture. *Comput. Electron. Agric.* **2019,** *157*, 218–231.

Kim, M.; Kim, S.; Kim, Y.; Choi, Y.; Seo, M. Infrared Estimation of Canopy Temperature as Crop Water Stress Indicator. *Korean J. Soil Sci. Fert.* **2015,** *48*, 499–504.

King, R. E.; Sigrimis, N. Editorial. Computational Intelligence in Crop Production. *Comput. Electron. Agric.* **2001,** *31* (1), 1–3.

Kirtan, J., Aalap, D., Poojan, P. I.; Manan, S. A Comprehensive Review on Automation in Agriculture Using Artificial Intelligence. *Artif. Intell. Agric.* **2019,** *2*, 1–12.

La Malfa, G.; Leonardi, C. Crop Practices and Techniques: Trends and Needs. *Acta Hort.* **2001,** *559*, 31–42.

Lakhiar, I. A.; Jianmin, G.; Syed, T. N.; Chandio, F. A.; Buttar, N. A.; Qureshi, W. A. Monitoring and Control Systems in Agriculture using Intelligent Sensor Techniques: A Review of the Aeroponic System. *J. Sens.* **2018,** *2018*, 1–18.

Madakam, S.; Ramaswamy, R.; Tripathi, S. Internet of Things (IoT): A Literature Review. *J. Comput. Commun.* **2015,** *3* (5), 164–173.

Maitra, S.; Shankar, T.; Sairam, M.; Pine, S. Evaluation of Gerbera (Gerbera Jamesonii L.) Cultivars for Growth, Yield, and Flower Quality under Protected Cultivation. *Indian Journal of Natural Sciences* **2020,** *10* (60), 20271–20276.

Mehra, M.; Saxena, S.; Sankaranarayanan, S.; Tom, R. J.; Veeramanikandan, M. IoT based Hydroponics System using Deep Neural Networks. *Comput. Electron. Agric.* **2018,** *155*, 473–486.

Montesano, F. F.; Serio, F.; Mininni, C.; Signore, A.; Parente, A.; Santamaria, P. Tensiometer-Based Irrigation Management of Subirrigated Soilless Tomato: Effects of Substrate Matric Potential Control on Crop Performance. *Front. Plant Sci.* **2015,** *6*, 1–11.

Morimoto, T.; Hashimoto, Y. An Intelligent Control Technique Based on Fuzzy Controls, Neural Networks and Genetic Algorithms for Greenhouse Automation. *IFAC Artif. Intell. Agric.* **1998,** *31*, 61–66.

Nikolaou, G.; Neocleous, D.; Katsoulas, N.; Kittas, C. Irrigation of Greenhouse Crops. *Hort.* **2019,** *5* (1), 7.

Palande, V.; Zaheer, A.; George, K. Fully Automated Hydroponic System for Indoor Plant Growth. *Procedia Comput. Sci.* **2018,** *129*, 482–488.

Pardossi, A.; Incrocci, L.; Incrocci, G.; Fernando, M.; Bacci, L.; Rapi, B.; Marzialetti, P.; Hemming, J.; Balendonck, J. Root Zone Sensors for Irrigation Management in Intensive Agriculture. *Sensors* **2009,** *9*, 2809–2835.

Pérez-Castro, A.; Sánchez-Molina, J. A.; Castilla, M.; Sánchez-Moreno, J.; Moreno Úbedac, J. C; Magán, J. J. Fertig UAL: A Fertigation Management App for Greenhouse Vegetable Crops. *Agric. Water Manag.* **2017,** *183*, 186–193.

Putra, P. A.; Yuliando, H. Soilless Culture System to Support Water Use Efficiency and Product Quality: A Review. *Agric. Agric. Sci. Procedia.* **2015,** *3*, 283–288.

Rodriguez-Ortega, W. M.; Martinez, V.; Rivero, R. M.; Camara-Zapata, J. M.; Mestre, T.; Garcia-Sanchez, F. Use of a Smart Irrigation System to Study the Effects of Irrigation Management on the Agronomic and Physiological Responses of Tomato Plants Grown under Different Temperatures Regimes. *Agric. Water Manag.* **2016,** *183*, 158–168.

Rouphael Y.; Franken P.; Schneider C.; Schwarz D.; Giovanetti M.; Agnolucci M.; De Pascale S.; Bonini P.; Colla G. Arbuscularmycorrhizal Fungi Act as Biostimulants in Horticultural Crops. *Scientia Horticulturae* **2015,** *196*, 91–108.

Sezen, S. M.; Celikel, G.; Yazar, A.; Tekin, S.; Kapur, B. Effect of Irrigation Management on Yield and Quality of Tomatoes Grown in Different Soilless Media in a Glasshouse. *Sci. Res. Essays* **2010,** *5*, 41–48.

Shi, X.; An, X.; Zhao, Q.; Liu, H.; Xia, L.; Sun, X.; Guo, Y. State-of-the-Art Internet of Things in Protected Agriculture. *Sensors* **2019,** *19* (8), 1833.

Shirgure, P. S. Citrus Fertigation-a Technology of Water and Fertilizers Saving. *Scientific Journal of Crop Science* **2013,** *2* (5), 56–66.

Sihombing, P.; Karina, N. A.; Tarigan, J. T.; Syarif, M. I. Automated Hydroponics Nutrition Plant Systems Using Arduinouno Microcontroller based on Android. *J. Phys.: Conf. Ser.* **2018,** *978* (1), 012014.

Singh, M. C.; Kachwaya, D. S.; Kalsi, K. Soilless Cucumber Cultivation under Protective Structures in Relation to Irrigation Coupled Fertigation Management, Economic Viability and Potential Benefits - A Review. *Int. J. Curr. Microbiol. App. Sci.* **2018,** *7* (3), 2451–2468.

Solaimalai, A.; Baskar, M.; Sadasakthi, A.; Subburamu, K. Fertigation in High-value Crops—A Review. *Agric. Rev.* **2005,** *1*, 1–13.

SteidleNeto, A. J.; Zolnier, S.; de Carvalho Lopes, D. Development and Evaluation of an Automated System for Fertigation Control in Soilless Tomato Production. *Comput. Electron. Agric.* **2014,** *103*, 17–25.

Thompson, R. B.; Gallardo, M. Use of Soil Sensors for Irrigation Scheduling. In *Improvement of Water Use Efficiency in Protected Crops*; Fernández-Fernández, M. P., Lorenzo-Minguez, Mª. I. Cuadrado Gómez Eds.; Dirección General de Investigación y FormaciónAgraria de la Junta de Andalucía, Seville: Spain, 2005; pp 351–376.

Thompson, R. B.; Gallardo, M.; Valdez, L. C.; Fernández, M. D. Using Plant Water Status to Define Threshold Values for Irrigation Management of Vegetable Crops using Soil Moisture Sensors. *Agric. Water Manag.* **2007,** *88*, 147–158.

Tzounis, A.; Katsoulas, N., Bartzanas, T.; Kittas, C. Internet of Things in Agriculture, Recent Advances and Future Challenges. *Biosyst. Eng.* **2017,** *164*, 31–48.

Van Os, E. A.; Gieling, T. H.; Ruijs, M. N. A. Equipment for Hydroponic Installations. In *Hydroponic Production of Vegetables and Ornamental*; Savvas, D., Passam, H., Eds.; Embryo Publications: Athens, Greece, 2002; pp 104–140. ISBN 960-8002-12-5.

Vera-Repulloa, J. A.; Ruiz-Peñalverb, L.; Jiménez-Buendíaa, M.; Rosillob, J. J.; Molina-Martínez, J. M. Software for the Automatic Control of Irrigation using Weighing-drainage Lysimeters. *Agric. Water Manag.* **2015,** *151*, 4–12.

Vuran, M. C.; Salam, A.; Wong, R.; Irmak, S. Internet of Underground Things in Precision Agriculture: Architecture and Technology Aspects. *Ad Hoc Netw.* **2018,** *81*, 160–173.

Yong, W.; Shuaishuai, L.; Li, L.; Minzan, L.; Ming, L.; Arvanitis, K. G.; Georgieva, C.; Sigrimis, S. Smart Sensors from Ground to Cloud and Web Intelligence. *IFAC Papers Online* **2018,** *17*, 31–38.

Zafari Janbaz K.; Mohammadi N. K. A. Review on Drip Fertigation on Field Crops. *International Journal of Engineering Research & Technology* **2019,** *8* (11), 722–729.

Zamora-Izquierdo, M. A.; Santa, J.; Martínez, J. A.; Martínez, V.; Skarmeta, A. F. Smart Farming IoT Platform based on Edge and Cloud Computing. *Biosyst. Eng.* **2019,** *177*, 4–17.

Zegbe, J. A.; Behboudian, M. H.; Clothier, B. E. Yield and Fruit Quality in Processing Tomato under Partial Rootzone Drying. *Eur. J. Hortic. Sci.* **2006,** *71*, 252–258.

Zeng, C. Z.; Bie, Z. L.; Yuan, B. Z. Determination of Optimum Irrigation Water Amount for Drip-irrigated Muskmelon (Cucumis Melo L.) in Plastic Greenhouse. *Agric. Water Manag.* **2009,** *96*, 595–602.

CHAPTER 8

Recent Advances in Management of Diseases and Pests of Common Occurrence Under Protected Cultivation

SHIDEH MOJERLOU

Department of Horticulture and Plant Protection, Faculty of Agriculture, Shahrood University of Technology, Shahrood, Iran

ABSTRACT

Protected cultivations are more susceptible to pests and diseases compared to open fields and orchards due to continuous cropping, which provides them ample of food and a favorable environment inside the greenhouses. Along with implementation of modern technologies in agricultural systems, advanced disease and pest management approaches have been applied in the greenhouses and other protected cultivations. Also, these new intelligence technologies using large-scale datasets will help farmers, inspectors, and researchers to identify pests and diseases at early stages of disease establishment and development, and give them noteworthy insights to implement accurate, effective, and timely control methods. Hence, they are considered as promising approaches in future for plant protection in protected cultivations like open fields, which will improve quality and efficiency of agricultural products. Computer vision technologies, image processing including thermal imaging, hyperspectral imaging, chlorophyll florescence imaging, internet of things, and system approaches are some of the modern technologies which have been used for plant pest and disease management in protected cultivations and are discussed here.

Protected Cultivation: Structural Design, Crop Management, Modeling and Automation.
Mahesh Chand Singh and K. K. Sharma (Eds.)

8.1 INTRODUCTION

Plant diseases are considered as one of the major factors influencing food production and social developments all around the world (Palmgren et al., 2015; He et al., 2016). In order to achieve sustainable plant disease management, integrative alliance of different branches of science including biological science (e.g., plant pathology, breeding, agronomy, and soil science), environmental science, social science, and economic science is needed (He et al., 2016). Rapid and accurate diagnosis, identification, and forecasting of the plant diseases are always substantial, and related technologies such as molecular methods have been well established. Although their application is still changeable among farmers (Miller et al., 2009; He et al., 2016).

In the past decades, protected cultivation has been used as a powerful facility for agriculture products worldwide. In this condition, plants are protected from unfavorable outdoor environment and pests, and are grown in the optimal condition with regard to both quality and quantity (van Henten et al., 2013). In other words, protected horticulture refers to any approaches which protect crops from any meteoric events like rain, snow, high temperature, radiation, light intensity, CO_2 concentration, and so on. Generally, greenhouses, plastic shelters, shading houses, and low and walk-in tunnels are considered as protected cultivation systems (Shishido, 2011). Since this method of cultivation needs high enterprise and running costs, it has been utilized for high-value fruits and vegetables (i.e., tomatoes, sweet pepper, and cucumber), flowers (roses, chrysanthemums, and gerberas), and other potted plants (van Henten et al., 2013). One of the main steps in the greenhouse production process is plant maintenance operations (PMOs) (van Henten et al., 2013). Hence, diseases and pests management is considered as a level of the PMOs.

8.2 DISEASES AND PESTS MANAGEMENT IN PROTECTED CULTIVATION

Plant diseases develop differently in protected cultivation compared to open-field conditions, especially because of high temperature and high humidity. Hence, disease and pest management has become one of the major limiting factors in protected cultivation. Some of the common pests and diseases in the greenhouse condition are listed in Tables 8.1 and 8.2.

The conventional methods for the management of plant pests and diseases in the greenhouse condition include chemical control, cultural control (solarization, environmental control, and resistant cultivars), biological

TABLE 8.1 Some Common Diseases Occurring in Protected Cultivation.

Disease	Agent (s)	Host (s)
Bacterial disease		
Galls	*Agrobacterium tumefaciens*	Wide host range
Soft rot	*Pectobacterium carotovorum*	Cucumber, eggplant, melon, pepper, and tomato
Fungal disease		
Damping-off	*Phytophthora* spp., *Pythium* spp., *Rhizoctonia solani*	Cucumber, eggplant, melon, pepper, tomato, and gerbera
Downey mildew	*Bremia lactucae*	Lettuce
Downey mildew	*Pseudoperonospora cubensis*	Cucurbits
Fusarium wilt	*Fusarium oxysporum* f.sp. *lycopersici*	Tomato
Grey mold	*Botrytis cinerea*	Cucumber, eggplant, melon, pepper, tomato, and ornamental plants
Late blight	*Phytophthora infestans*	Tomato
Leaf mold	*Passalora fulva*	Tomato
Leaf spot	*Alternaria* spp., *Cercospora* sp., *Septoria* sp., *Gloeosporium* sp.	Several crops
Powdery mildew	*Podosphaera xanthii Erisyphe cichoracearum*	Cucurbits
Powdery mildew	*Sphaerotheca pannosae*	Roses
White rust	*Puccinia horiana*	Chrysanthemum
Viral disease		
Tomato spotted wilt virus (TSWV)	-	Gerbera, pepper, tomato, and ornamental plants
Tobacco mosaic virus (TMV)	-	Pepper, tomato, and ornamental plants
Potato virus Y (PVY)	-	Pepper
Cucumber mosaic virus (CMV)	-	Cucumber, lettuce, pepper, tomato, carnation, and gerbera

Source: Adapted from Fletcher (1984), Elad (1999), Jarvis and Jarvis (1992).

TABLE 8.2 Major Crops of Protected Cultivation and Their Common Pests.

Crop (s)	Common pest	Scientific name
Capsicum, cucumber, and lettuce	Aphids	*Aphis gossypii*
Chrysanthemum		*Macrosiphoniella sanborni*
Strawberry		*Myzus ascalonicus*
Capsicum, cucumber, gerbera, lettuce, and pepper		*Myzus persicae*
Orchid		*Toxoptera aurantii*
Capsicum, carnation, and tomato	Caterpillar (s)	*Helicoverpa armigera*
Capsicum, cucumber, rose, and tomato		*Spodoptera litura*
Tomato, chrysanthemum, cucumber, and gerbera	Leaf miner (s)	*Liriomyza trifolii*
Tomato	Mite (s)	*Aceria lycopersici*
Capsicum		*Polyphagotarsonemus latus*
Strawberry		*Stenotarsonemus fragariae*
Carnation		*Tetranychus cinnabarinus*
Cucumber, pepper, and tomato		*Tetranychus urticae*
Rose	Thrip (s)	*Scritothrips dorsalis*
Gerbera		*Thrips palmi*
Gerbera and capsicum	Whiteflies	*Bemisia tabaci*
Tomato, beans, capsicum, cucumber, and gerbera		*Trialeurodes vaporariorum*

Source: Adapted from Shipp et al. (1991), Sabir and Singh (2013).

control, natural enemies, and integrated pest management (IPM). The first rational way of pest and disease management is chemical control because of its prompt effect. Chemical control with protective fungicides has to be done before pathogen establishment. Due to intensive growth of plants in the greenhouses, it has to be applied at short intervals. However, chemical control has negative impact on human health, environment, disturbs natural balance, and leads to appearance of resistant pests and pathogens populations (Yucel et al., 2013). Also, inappropriate environmental management decreases the chemicals effectiveness and induces tolerant pathogens, and subsequently wrecks the crop production systems (Shishido, 2011). On the other hand, controlling the environmental factors is more feasible in protected environments than in open fields. For example, using appropriate means to manipulate pathogens' favorable humidity and temperature, and applying proper sanitation practices are more practical in these conditions (Eshenaur and Anderson, 2010). In Turkey, IPM method has been adopted in protected tomato cultivation since 1990, and comparisons were made between IPM-performed and nonperformed greenhouses. It has been revealed that the expenses and the number of pesticide applications were reduced by half, where IPM was considered (Yucel et al., 2013).

Over the past decades, standard disease control measurements have been used in greenhouse cropping routinely. Reduction of chemical applications has become a trend in recent years. Therefore, using biocontrol products in greenhouse crops have been developed extensively. For instance, *Trichoderma* spp. is considered as a promising biocontrol agent for many fungal diseases. Albeit, modern technologies have provided potent disease management methods in protected horticulture systems (Paulitz and Belanger, 2001; Calpas, 2010). Several machine vision systems have been developed to detect plant diseases at an early stage of infection, among which thermographic and fluorescent measuring approaches were more promising than multispectral and hyperspectral sensors (Chaerle and Van Der Straeten, 2000). In this chapter, we are discussing modern technologies which have been used in plant disease management in protected cultivations such as greenhouses, including computer vision-based technologies, thermal imaging, the internet of things (IoT), and systems approaches.

8.3 COMPUTER VISION TECHNOLOGIES

In recent years, monitoring plant diseases based on computer vision systems is considered as an important tool in agriculture, and its implementation increased rapidly (Gomes and Leta, 2012; Rico-Fernandez et al., 2019). Computer vision-based technologies are becoming more popular in agricultural automation in order to increase agriculture production and efficiency (Foflia and Reina, 2006). Graphic processing units (GPUs), deep belief networks (DBNs), and artificial intelligence are some examples of these advanced technologies (Mochida et al., 2019), which are capable of applying to the field of agricultural automation (Tian et al., 2020). Poor accuracy, precision, and timing are some issues of conventional plant disease management, which are based on manual inspections (Ramcharan et al., 2019). Hence, by application of computer vision technologies, prevention and disease control measures have been improved. Timely prevention and control of diseases, pests, and weeds can reduce damages, enhance efficiency, and contribute to sustainable agriculture (Akram et al., 2017; Tian et al., 2020).

Computer vision approaches have been used for leaf disease detection in several plants like banana, cotton, grape, lemon, beans, rose, and hydrangea. This method is separated into two sections, including supervised and unsupervised disease detection. The first step is used for system training of disease features (e.g., using artificial neural network for system learning) and the second part uses disease features for disease identification (Rout and Parida, 2019). Story and Kacira (2015) studied lettuce plants growing features and overall health in hydroponic system using computer vision monitoring system. Panoramic images were captured and analyzed using different types of cameras such as color, near-infrared (NIR), and thermal.

Osama et al. (2020) detected tomato diseases in greenhouse model using an automated intelligence system. They studied common pest and diseases in greenhouses including spider mites, early blight, late blight, leaf mold, target spot, mosaic virus, septoria spot, bacterial spot, and yellow leaf curl virus. Results showed that deep learning system could differentiate mentioned diseases with 94.8% accuracy. They also implemented a web application to monitor growth situation and possible diseases that may occur in the greenhouse. Identification of different diseases in tomato plants using various image processing techniques has been done by researchers (Sabrol and Satish, 2016; Shijie et al., 2017; Lu et al., 2018). Wspanialy and Moussa (Wspanialy and Moussa, 2020) introduced a general disease model based on computer vision system to predict several tomato plant diseases in greenhouse. Interestingly, this system can evaluate previously unseen leaf diseases and disease severity. They also introduced an annotated proportional disease severity dataset and associated estimation model, which are suitable for different plant diseases such as fungal and bacterial diseases.

Brahimi et al. (2017) implemented deep learning and convolutional neural network (CNN) approaches to classify diseases. Interestingly, they did not use preprocessing phase for image preparation, and used the images directly. They also used a large dataset that includes 14,828 images of nine diseases in tomato. Using these methods, they could detect infected leaves and localize symptoms in the leaves with the accuracy of 99.18%.

Electronic noses (ENs) system, which was introduced in 1982, has been used in wide aspects including security, medicine, agriculture, food industry, and environmental problems (Hines et al., 1999; Llobet et al., 1999; Dutta et al., 2003, 2004). Ghaffari et al. (2010) implemented automated system for diagnosis of diseases in tomato. Several algorithms based on artificial neural network (ANN), such as multilayer perceptron (MLP), learning vector quantization (LVQ), and radial basis function (RBF) were used. Results showed that RBF, MLP, and LVQ approaches could identify healthy plants, powdery mildew, and spider mite infected plants by 94%, 96%, and 98% accuracy, respectively. Hence, ENs are promising tools for differentiation between healthy and infected plants at early stages of the disease development in greenhouses.

8.4 THERMAL IMAGING

Application of imaging techniques for plant phenotyping is a relevant method for quantification of plants' appearance and efficiency under different environmental circumstances. Thus, plant phenotyping is considered as a valuable technique in precision agriculture to diagnose and detect plant stress, even in absence of symptoms (Mahlein, 2016). Optical sensors have been used in different aspects of plant pathology including study of plants responses to pathogens, pests, and abiotic stresses; identification of primary disease spots; evaluation of resistance or susceptibility reaction of plant genotypes; indexing of disease symptoms severity; and assessing of plant biomass and yield (Mahlein, 2016). Stomata and its activity are considered as an important physiological trait in plant growth and development, and play a crucial role in photosynthesis and transpiration by controlling carbon–water balances (Jones, 2014). For this reason, there is a correlation between stomatal conductance to water (g_s) and yield, tolerance of environmental stresses (Prashar et al., 2013), and leaf temperature (Scarth et al., 1948; Milthorpe and Spencer, 1957; Fuchs and Tanner, 1966; Jones, 1999). Thermography using thermal long-wave infrared (TIR) cameras has been applied widely to agriculture, although it has limited capacity for diagnosis. Indeed, increasing of the temperature may occur due to stomatal closure, which can be a consequence of unspecific response of plants to biotic and abiotic stresses. It also happens in low capacity of water evaporation or loss of vegetation (Prashar et al., 2013). On the other hand, some stresses are like pathogens reduce the leaf temperature temporally. They interfere in the movements of stomata in order to facilitate their entry into mesophyll (Zeng et al., 2010).

Stomata are natural entry of most pathogens into host tissues, and regulation of its movement has an important role in plant defense against pathogens (Sawinski et al., 2016). In order to overcome stomatal closure as a plant defense response, some pathogens are capable to repeal signaling pathways and activate stomatal aperture (Agurla and Raghavendra, 2016). Also, pathogens can disturb plant physiological progress like leaf development, cell wall, and leaf cuticle integrity or components, changing plant metabolism, and causing tissue necrosis (Grimmer et al., 2012; Baron et al., 2016; Smigaj et al., 2019).

Thermography has been used widely for detection of stress from lab to field scale by proximal or remote sensing approaches. Proximal sensing has been used mainly in growth chambers and greenhouses, and refers to imaging single leaf or entire plant from a close distance. In this method, static stands or robots (medium or small sizes) can be used for mounting the cameras (Pineda et al., 2021). Due to complex working circumstances, precision agriculture approaches like robotics face challenges in the protected cultivation. For example, it should be economically viable, and also it has disseminated market due to cultivated crops varieties (Van Henten et al., 2013). Proximal sensing has been used for detection of different biotic stresses in protected cultivations such as growth chambers and greenhouses by researchers.

8.4.1 VIRAL DISEASES

Thermography is often an useful approach to track viral infection in plants, even if they do not cause changes in g_s or leaf temperature (Montero et al., 2016). Many researches showed the increases in leaf temperature between +0.5 and +1.5K, and some revealed the decreases of −0.5K, compared to healthy plants (Pineda et al., 2021). Hypersensitive reaction of tobacco plants to *tobacco mosaic virus* (TMV) is visualized by local increased temperature before symptom's appearance on leaves, while no thermal reaction was visible in susceptible plants (Chaerle et al., 2004). In the case of *Nicotiana benthamiana* plants, which were infected by *pepper middle mottle virus* (PMMoV), asymptomatic leaves could be identified based on increasing the temperature of tissues around the main veins expanding to the neighbor tissues. It is noteworthy that thermal symptoms caused by severe strains were detected before those of the less virulent ones (Chaerle et al., 2006). Cucumber plants infected with *cucumber mosaic virus* (CMV) showed higher temperature in the whole leaf, compared to control. On the other hand, cucumber plants infected with *cucumber green mottle mosaic virus* (CGMMV) displayed different temperature pattern containing cold spots at infection sites. However, none of these changes were detectable before symptom's appearance in plants (Berdugo et al., 2014). Sometimes, coinfection of plants by different viruses leads to higher increase in temperature, compared to single-virus plant infection. For example, this phenomenon is detected in sweet potato plants infected with both *sweet potato feathery mottle virus* and *sweet potato chlorotic stunt virus* (Wang et al., 2019). Zhu et al. (2018) evaluated tomato leaves infected by virus using infrared thermal imaging process (Figure 8.1). The results showed that mock plants exhibited a uniform blue color and the average temperature was 23.2°C. After 3 days of inoculation, the leaves showed yellowish color and the average temperature increased about 0.6°C and reached to 24.1°C. The red-orange area appeared at the leaf tip after 6 days. In fact, this area was the first lesion which was confirmed after 10 days. Based on the results, at the beginning of the infection process, the location of lesion on the leaves can be identified using thermogram according to the increased temperature of the susceptible leaf.

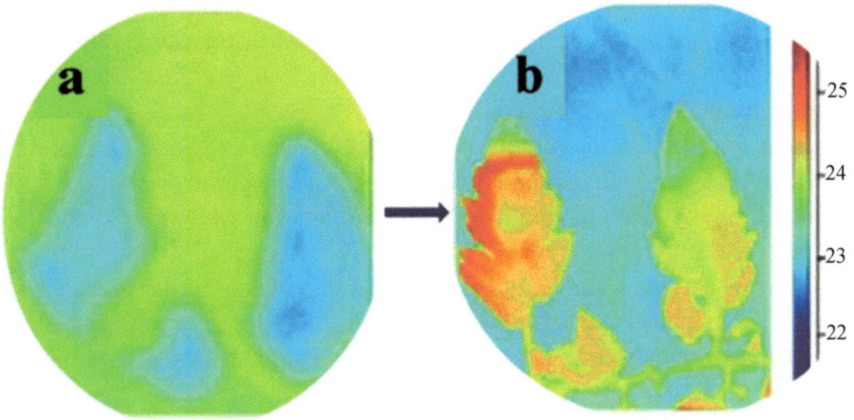

FIGURE 8.1 Infrared thermal imaging (a) before and (b) 9 days after inoculation of tomato plants by virus.

Source: Adapted from Zhu et al. (2018).

8.4.2 BACTERIAL DISEASES

Based on literatures, temperature has been changed in bacterial infected plants between −3.3 and +3K, compared to healthy plants. Galls produced by *Agrobacterium tumefaciens* in castor oil plants was the coldest part, due to lack of protective cuticle against water loss (Schurr et al., 1996). Decreased temperature at the infection sites could be due to virulence factors, which interfere with signaling pathways leading to opening of the stomata (Zheng et al., 2012). Researchers could identify hypersensitive reaction (HR) in *Pseudomonas syringae* pv. tomato DC3000-bean pathosystem before symptom's appearance (1 h post infection) using thermography. This method also was useful to recognize systemic infection in bean plants infected by *P. syringae* pv.

phaseolicola 1448A, 2 h after infection (Perez-Bueno et al., 2015). Moreover, TIR and multicolor fluorescent images have been used to provide algorithm for differentiation of infected melon and zucchini plants from healthy ones in bacterial infection (Perez-Bueno et al., 2016; Pineda et al., 2017, 2018).

8.4.3 FUNGAL AND FUNGAL-LIKE DISEASES

In the case of fungal pathogens and oomycetes which affect plants transpiration, thermography has been used widely to detect presymptomatic infections. Firstly, these pathogens can induce reduction or increase in leaf temperature at infection sites or whole leaves. While after disease development, temperature falls due to cell death and water loss. TIR studies revealed that biotrophic fungi and oomycetes can decrease temperature ranging from 0.1 to −2.5K; while increases from +0.6 to +2K have been recorded. In addition, hemibiotrophic pathogens can drop leaf temperature between −0.6 and −2.2K, and in contrast, raise it from +0.4 to +7.5K. Eventually, necrotrophic fungi decrease temperature between −2 and 5K, and increases are varied from +0.3 to +9K (Pineda et al., 2021).

Many researchers studied the detection of fungal and fungal-like diseases using TIR in indoor and outdoor conditions. Regarding cucumber plants infected by downey and powdery mildews, local lesions showed lower temperature in comparison with surrounding healthy tissues because of unusual stomatal opening. In the case of powdery mildew infection, TIR approach could not detect infection before symptom's appearance, while it was capable of early diagnosis of downy mildew (Oerke et al., 2006; Lindenthal et al., 2005; Berdugo et al., 2014; Wen et al., 2019). Although, *Pseudoperonospora cubensis*, causal agent of downy mildew in cucumber, is a biotrophic oomycete, it causes necrotrophic lesions at the late phase of infection and increases leaf temperature (Oerke et al., 2006; Lindenthal et al., 2014). Jafari et al. (2017) studied roses plants infected by powdery mildew in the greenhouse using TIR method.

Also, TIR imaging was used to describe disease severity in hemibiotrophic fungi, and showed correlation with infection stage and host plant resistance (Oerke, 2011; Belin et al., 2013). Finally, in the case of necrotrophic fungi, *B. cinerea* infected roses exhibited a temperature reduction at initially formed local lesions, pursued by an increase in temperature after the appearance of necrosis (Jafari et al., 2017). In addition, results revealed that median temperature and maximum temperature differences (MTD) index could diagnose the presence and fungal growth in flowers, at least 1 day before appearance of visible symptoms. Although, some patterns such as temperature frequency distribution and thermal histogram shape are more efficient than above-mentioned indices (Jafari et al., 2016).

8.5 HYPERSPECTRAL IMAGING

Along with the technology advancements, agriculture like other sciences has benefited from these improvements. Optical sensor is one of them, which can be used for plant disease detection and identification on different scales from tissues in laboratory to a single plant in a greenhouse or climate chambers in fields. It has numerous privileges compared to other methods, such as molecular methods. For instance, it is not invasive, therefore researchers can use it for time series assessments and it reduces the required samples. In addition, it can be applied in automated measurement systems, and leads to decreasing the workload, economic, and ecological costs compared to manual measurement systems (Mahlein, 2016; Mahlein et al., 2017; Virlet et al., 2017; Walter et al., 2017). Diseases make changes in plant physiological process, including tissue color, leaf shape, plants morphology, and density. Hence, researchers used hyperspectral imaging method to figure out plant diseases on various scales in different pathosystems (West et al., 2003; Sankaran et al., 2010; Mahlein, 2016). Fungal pathogens influence the spectral spectrum of the plants based on their interaction type, nutrition style (e.g., biotrophic, hemibiotrophic, and necrotrophic), growth type, developmental stage, and disease severity. These changes in photosynthesis are detectable between 500 and 680 nm wavelength (Thomas et al., 2018). Using hyperspectral method, it is possible to specify the causal agent, which is not possible by other methods like thermography and chlorophyll fluorescence (Bravo et al., 2004; Mahlein et al., 2010; Hillnhutter et al., 2012). Essential element of hyperspectral sensor method is similar to RGB cameras, and is the light amount which reaches out to the sensors. But, unlike RGB cameras in which only three bands of visual parts of electromagnetic spectrum are measured, in this method, several hundred bands are detected by sensors (Thomas et al., 2018). Accordingly, images were achieved in this method containing two dimensions, spatial and spectral information (Jensen, 2006); due to huge amount of data, it needs advanced data analysis methods to extract relevant information from the images (Behmann et al., 2015). Three different sensor types have been implemented in hyperspectral imaging method, including the push broom/line scanner, the filter-based sensor setup, and a whisk broom setup (Figure 8.2). Push broom cameras are line scanners and full scan can be completed by moving the sensors or using mirror. Filter-based system evaluates the sample by different filters, which splits the scan ray or uses combined filter pattern. Eventually, a spectrometer is moved over the region of interest in whisk broom sensors and spectral pixels are collected. All these sensors result in a 3D hypercube, which has two spatial axes and one spectral axis (Paulus and Mahlein, 2020).

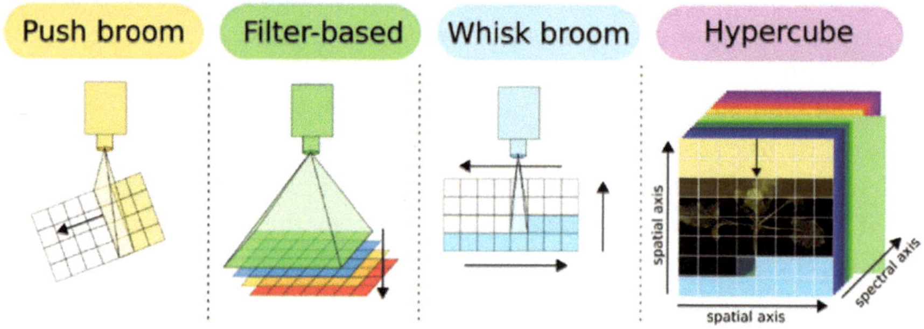

FIGURE 8.2 Overview of hyperspectral sensor imaging techniques.

Source: Adapted from Paulus and Mahlein (2020). http://creativecommons.org/licenses/by/4.0/

Mahlein et al. (2017) applied a hyperspectral imaging method to detect barley powdery mildew disease on the microscopic and canopy scale under a protected cultivation system, and sugar beet with different foliar diseases on different scales. Results showed that this is a potential and promising method for plant phenotyping, precision agriculture, and plant pathogen interactions on different scales, which has some supplementary tasks such as calibration of camera systems and using combination of hyperspectral and 3D measuring methods to get more accurate results (Mahlein et al., 2012; Behmann et al., 2015, 2016; Wahabjada et al., 2015; Roscher et al., 2016). Furthermore, in this method, spatial resolution and mixed pixel's quantity are important, and depend on the distance between the objects and the sensors. For detection of single symptom or infected leaves and plants, sensors with about 1-m spatial resolution are less suitable and in these cases, proximal sensors are premier (Oerke et al., 2014). Also, spectral reflectance method has been used to investigate changes in wheat plants, which were infected by greenbug on canopy scale in greenhouse and field. Results showed that hyperspectral imaging is capable to detect greenbug infections on wheat canopies (Mirik et al., 2006).

Plant phenotype is affected by different factors such as environment, stress, and genotype (Figure 8.3). For biological explanation of reflectance spectrum, physiological and histological factors have been used by researchers. Gene expression profile and enzyme activity's role on spectral reflectance patterns need more studies (Kuska and Mahlein, 2018).

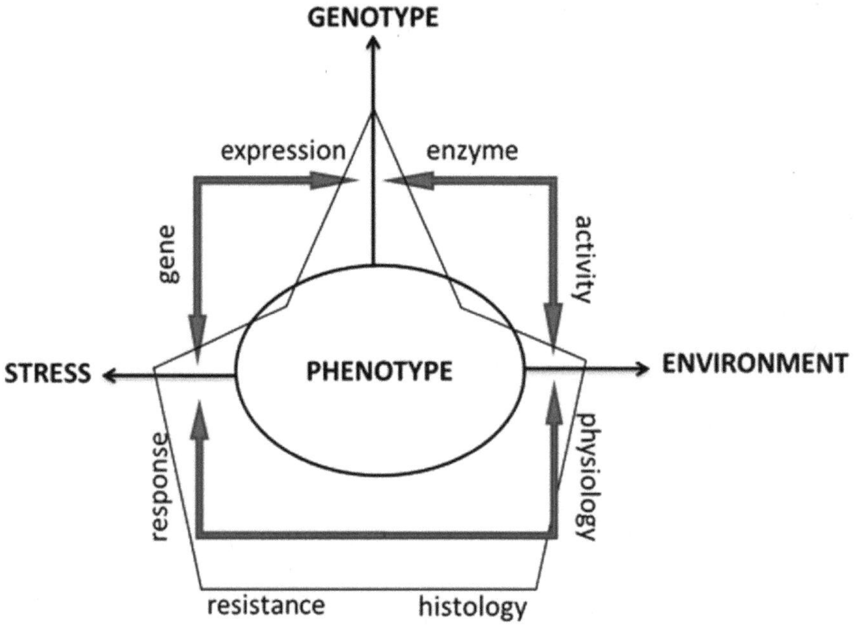

FIGURE 8.3 Factors affecting plant phenotype.
Source: Adapted from Kuska and Mahlein (2018).

8.6 CHLOROPHYLL FLUORESCENCE IMAGING

Plants absorb light energy and it distributes in photosynthesis, thermal dissipation, and chlorophyll fluorescence emission (Buscmann, 1999). Chlorophyll fluorescence imaging (Chl-FI) is a potent method to measure photosynthesis at cellular, leaf, and whole plant scale for plant phenotyping (Perez-Bueno et al., 2019). There is a reverse correlation between Chl-FI and photosynthesis and/or thermal dissipation. This method has been used for different plant pathogen systems, including viral and fungal diseases (Chaerle and Van Der Straeten, 2000; Chou et al., 2000; Lohaus et al., 2000). Based on the researches, inhibition of metabolic reactions which are not linked to photosynthesis can have impact on chlorophyll fluorescence emission (Barbagallo et al., 2003). Chaerle et al. (2004) studied Chl-FI system for two different plant-pathogen interactions (i.e., fungal and viral pathogens) in comparison with thermal imaging system. Results showed that Chl-FI differentiations at early disease stages (presymptomatic stages) are less, compared to thermal imaging system. As a result, utilization of Chl-FI and/or thermal imaging in combination with automatic systems leads to accelerate screening of plant population against diseases at early stages. Occasionally, Chl-F cannot differentiate between healthy and infected tissues clearly or do so at late stage of the disease; in these cases, other methods have to be applied (Perez-Bueno et al., 2019). Differences in Chl-F emission intensity have been revealed in zucchini cotyledons inoculated with *cucumber mosaic virus* (CMV) before symptom's appearance (Técsi et al., 1994). Therefore, it has been proved that Chl-FI is a promising method for on time detecting of viral movement in plants. Ch-FI in combination with multicolor fluorescence imaging and thermography was used to distinguish between healthy and infected plants in melon plants infected by *Dickeya dadantii* (Pineda et al., 2018).

8.7 THE INTERNET OF THINGS (IOT)

The internet of things (IoT) has a broad spectrum meaning which was developed in 1999, and has changed along with advances in information technologies. It was firstly applied in healthcare, smart city, and industrial production. Also, agriculture can be a proper candidate of utilizing IoT solutions because it includes monitoring of vast areas (Islam et al., 2015; Asghari et al., 2018; Shi et al., 2019). IoT is a potent linkage of physical instruments, vehicles, home facilities, items related to electronics, software, sensor, and network which attach and make connection among mentioned items. Indeed, it provides a direct connection between physical world and systems based on computer (Yakkundimath et al., 2018). In the protected agriculture or controllable agriculture, artificial techniques are used to change climate factors like light, temperature, and humidity to prepare suitable condition for growth of plants to a certain extent. Thus, IoT technology is capable of applying in protected agriculture, and some of the elaborated IoT solutions can be transferred into protected agriculture (Jensen and Malter, 1995). This system helps farmers to control soil condition, climate change, and plant health (Dlodlo and Kalezhi, 2015). It also can monitor environmental factors like temperature, humidity, and CO_2 concentration (Peng et al., 2015; Lu et al., 2018). On the other hand, IoT system can be used in the protected cultivation. Indeed, cameras can trap diseases and pests in the greenhouse and help farmers to protect the crops in timely manner (Ma et al., 2015).

Various types of IoT sensors have been used in greenhouse conditions to inform farmers about environmental situation, location, plant growth, and pest and disease situation. They helped farmers to monitor pest and diseases in the greenhouse. After that, all of the captured images and raw data are sent to cloud centers to be processed and analyzed using different mathematical models and algorithms with regards to each disease and pest (Pixia and Xiangdong, 2013; Ahmed et al., 2016). Eventually, cloud centers give farmers information about identified pests and diseases. Also, they predict and warn on possible tragedies and give recommendation of expert systems (Shi et al., 2019).

ZigBee technology has been used by researchers to design a pest trap automatic monitoring system. This system can assess pest's density at different sites and inform farmers when it goes above the set value (Tirelli et al., 2011). Other researchers used ultrasound to propose a pest monitoring system which helped them to prepare instructions for early detection and eradication of the pests (Ahouandjinou et al., 2017). Foughali et al. (2018) used IoT approach to propose a system for prevention and making decision about potato late blight disease. Currently, pests and diseases warning systems are based on historical data, and are more focused on plant protection instead of timeliness. Hence, researchers should concentrate more on online monitoring, early disease detection, and warning systems at the early stages of plant infection in the future (Shi et al., 2019).

Thorat et al. (2017) applied IoT system to detect leaf diseases such as black spot, botrytis blight, leaf spot, powdery mildew, and rust on different crops. The results showed that image processing could detect leaf diseases successfully. Huddar et al. (2012) proposed an algorithm to identify pests in greenhouses and farms using image processing. The algorithm was used to identify whiteflies with 96% accuracy. Gavaskar and Sumithra (2017) developed a monitoring system using IoT for pest control.

Khan et al. (2020) studied IoT as a tool for monitoring and managing environmental conditions, water irrigation, and detection of leaf diseases in a greenhouse. They implemented IoT technologies such as image processing tools, single-board microcontroller, temperature and humidity measuring sensor, and a testing platform.

A model named "IoTOMATO" has been used by researchers using chatbot. It was an IoT-based system for monitoring tomato using various sensors. In addition, deep learning has been applied to detect diseases of leaves. Eventually, AlexNet- and SqueeseNet-based convolutional neural network models were used for training and testing, in order to achieve a suitable deep learning algorithm (Yosuke et al., 2017).

8.8 SYSTEMS APPROACHES

Pests and disease management in horticultural nurseries is very challenging due to plant material and spatial and temporal heterogeneity. Traditional pest and disease prevention methods are based on classification, end-point surveys, and quarantine, and are not suitable for unknown pathogens or pests. A new method called hazard analysis of critical control points (HACCP) has been used by US food and drug administration officers in 1972. This method includes a series of logical steps to identify, evaluate, and correct sources of hazard. HACCP approach has been modified and implemented by researchers in Oregon to reduce the damage caused by *Phytophthora* species in nursery plants (Parke and Grunwald, 2012). Since these approaches are focused on production of healthy plants, nurseries which use these methods can find their way to national and international markets. In fact, farmers learn how to treat the root of contamination, not only the symptoms. In this study, they introduced the critical control points (CCPs) for *Phytophthora* species contamination management in nurseries and best management practices at each step. This method can be developed for every pest and pathogen. It is noteworthy that HACCP-based methods are promising means of controlling pests and pathogens in nursery trade. It is an active system that needs to be improved routinely by learning and reassessment of CCPs to implement new and/or revised existing control strategies (Parke and Grunwald, 2012).

8.9 CONCLUSIONS

Farmers are always concerned about pest and diseases management in protected cultivations such as greenhouses more than open fields, due to its proper and special condition for disease occurrence and development. Recently, advanced technologies have been used in agriculture like other sciences in order to increase agriculture products and help farmers to protect crops against pests and diseases in a timely manner. Based on the above-mentioned approaches, it can be concluded that computer vision intelligence technology using large-scale datasets will be used in different aspects of agriculture such as pest and disease managements in protected cultivations like open fields in future. Also, by implementing cutting-edge technologies like IoT, deep learning, image processing, and spectral analysis technologies, efficiency and quality of the agricultural products will be improved (Coech and Langat, 2018; Rehman et al., 2019) that will help farmers to identify pests and diseases at the early stages of the infection process, and give them valuable suggestion and insights to make effective decision and take timely action.

KEYWORDS

- **plant protection**
- **greenhouse crops**
- **computer vision technology**
- **imaging system**
- **Internet of Things**

REFERENCES

Agurla, S.; Raghavendra, A. S. Convergence and Divergence of Signaling Events in Guard Cells during Stomatal Closure by Plant Hormones or Microbial Elicitors. *Front. Plant Sci.* **2016,** *7,* 1332. Available from: https://www.frontiersin.org/article/10.3389/fpls.2016.01332

Ahmed, R. Z.; Biradar, R. C.; Chaudhari, S. S. Cluster-Based Data Aggregation for Pest Identification in Coffee Plantations Using Wireless Sensor Networks. *Comput. Electr. Eng.* **2016,** *56* (C), 591–607.

Ahouandjinou, A. S. R. M.; Kiki, P. M. A. F.; Assogba, K. In *Smart Environment Monitoring System by using Sensors Ultrasonic Detection of Farm Pests.* 2nd International Conference on Bio-engineering for Smart Technologies (BioSMART), 2017; pp 1–5.

Akram, T.; Naqvi, S. R.; Haider, S. A.; Kamran, M. Towards Real-time Crops Surveillance for Disease Classification: Exploiting Parallelism in Computer Vision. *Comput. Electr. Eng.* **2017,** *59,* 15–26.

Asghari, P.; Rahmani, A.; Haj, Seyyed, Javadi, H. Internet of Things Applications: A Systematic Review. *Comput. Networks* **2018,** 148.

Barbagallo, R. P.; Oxborough, K.; Pallett, K. E.; Baker, N. R. Rapid, Noninvasive Screening for Perturbations of Metabolism and Plant Growth using Chlorophyll Fluorescence Imaging. *Plant Physiol.* **2003,** *132* (2), 485–493.

Baron, M.; Pineda, M.; Perez-Bueno, M. L. Picturing Pathogen Infection in Plants. *Z. Naturforsch. C.* **2016,** *71* (9–10), 355–368. Available from: http://europepmc.org/abstract/MED/27626766

Behmann, J.; Mahlein, A.-K.; Paulus, S.; Kuhlmann, H.; Oerke, E.-C.; Plumer, L. Calibration of Hyperspectral Close-range Pushbroom Cameras for Plant Phenotyping. *ISPRS J. Photogramm. Remote Sens.* **2015,** *106,* 172–182.

Behmann, J.; Mahlein, A.-K.; Paulus, S.; Dupuis, J.; Kuhlmann, H.; Oerke, E.-C., et al. Generation and Application of Hyperspectral 3D Plant Models: Methods and Challenges. *Mach. Vis. Appl.* **2016,** *27* (5), 611–624.

Belin, E.; Rousseau, D.; Boureau, T.; Caffier, V. Thermography Versus Chlorophyll Fluorescence Imaging for Detection and Quantification of Apple Scab. *Comput. Electron. Agric.* **2013,** *90*, 159–163.

Berdugo, C. A.; Zito, R.; Paulus, S.; Mahlein, A.-K. Fusion of Sensor Data for the Detection and Differentiation of Plant Diseases in Cucumber. *Plant Pathol.* **2014,** *63* (6), 1344–1356. Available from: https://doi.org/10.1111/ppa.12219

Brahimi, M.; Boukhalfa, K.; Moussaoui, A. Deep Learning for Tomato Diseases: Classification and Symptoms Visualization. *Appl. Artif. Intell.* **2017,** *31* (4), 299–315.

Bravo, C.; Moshou, D.; Oberti, R.; West, J.; McCartney, A.; Bodria, L., et al. Foliar Disease Detection in the Field using Optical Sensor Fusion. 2004.

Buschmann, C. Thermal Dissipation during Photosynthetic Induction and Subsequent Dark Recovery as Measured by Photoacoustic Signals. *Photosynthetica* **1999,** *36* (1), 149–161.

Calpas, J. *Pest and disease management in commercial greenhouses*; Agriculture and Rural Development, Government of Alberta; 2010.

Chaerle, L.; Van-Der-Straeten, D. Imaging Techniques and the Early Detection of Plant Stress. *Trends Plant Sci.* **2000,** *5* (11), 495–501.

Chaerle, L.; Hagenbeek, D.; De-Bruyne, E.; Valcke, R.; Van-Der-Straeten, D. Thermal and Chlorophyll-fluorescence Imaging Distinguish Plant-pathogen Interactions at an Early Stage. *Plant Cell Physiol.* **2004,** *45* (7), 887–896.

Chaerle, L.; Pineda, M.; Romero-Aranda, R.; Van-Der-Straeten, D.; Baron, M. Robotized Thermal and Chlorophyll Fluorescence Imaging of Pepper Mild Mottle Virus Infection in Nicotiana Benthamiana. *Plant Cell Physiol.* **2006,** *47* (9), 1323–1336.

Chou, H.; Bundock, N.; Rolfe, S. A.; Scholes, J. D. Infection of Arabidopsis Thaliana Leaves with Albugo Candida (White Blister Rust) Causes a Reprogramming of Host Metabolism. *Mol. Plant Pathol.* **2000,** *1* (2), 99–113.

Dlodlo, N.; Kalezhi J. In *The Internet of Things in Agriculture for Sustainable Rural Development*. International Conference on Emerging Trends in Networks and Computer Communications (ETNCC), 2015; pp 13–18.

Dutta, R.; Hines, E. L.; Gardner, J. W.; Udrea, D. D.; Boilot, P. Non-destructive Egg Freshness Determination: An Electronic Nose based Approach. *Meas. Sci. Technol.* **2003,** *14* (2), 190.

Dutta, R.; Gardner, J. W.; Hines, E. L. Classification of Ear, Nose, and Throat Bacteria using a Neural-network-based Electronic Nose. *MRS Bull.* **2004,** *29* (10), 709–713.

Elad, Y. Plant Diseases in Greenhouses. *Ecosyst. World* **1999,** 191–212.

Eshenaur, B.; Anderson, R. Managing the Greenhouse Environment to Control Plant Diseases. 2010.

Fletcher, J. T. *Diseases of Greenhouse Plants.* Longman; 1984.

Foughali, K.; Fathallah, K.; Frihida, A. Using Cloud IOT for Disease Prevention in Precision Agriculture. *Procedia. Comput. Sci.* **2018,** *130*, 575–582.

Foglia, M. M.; Reina, G. Agricultural Robot for Radicchio Harvesting. *J F Robot [Internet].* **2006,** *23* (6-7), 363–377.

Fuchs, M.; Tanner, C. B. Infrared Thermometry of Vegetation 1. *Agron. J.* **1966,** *58* (6), 597–601.

Gavaskar, S.; Sumithra, A. Design and Development of Pest Monitoring System for Implementing Precision Agriculture using IOT. *Int. J. Sci. Technol. Eng.* **2017,** *3*, 46–48.

Ghaffari, R.; Zhang, F.; Iliescu, D.; Hines, E.; Leeson, M.; Napier, R., et al. In *Early Detection of Diseases in Tomato Crops: An Electronic Nose and Intelligent Systems Approach.* The 2010 International Joint Conference on Neural Networks (IJCNN), IEEE; 2010. pp 1–6.

Gomes, J. F. S.; Leta, F. R. Applications of Computer Vision Techniques in the Agriculture and Food Industry: A Review. *Eur. Food Res. Technol.* **2012,** *235* (6), 989–1000.
Grimmer, M. K.; John-Foulkes, M.; Paveley, N. D. Foliar Pathogenesis and Plant Water Relations: A Review. *J. Exp. Bot.* **2012,** *63* (12), 4321–4331.
Islam, S. M. R.; Kwak, D.; Kabir, M. D. H.; Hossain, M.; Kwak, K.-S. The Internet of Things for Health Care: A Comprehensive Survey. *IEEE Access* **2015,** *3*, 678–708.
He, D.; Zhan, J.; Xie, L. Problems, Challenges and Future of Plant Disease Management: From an Ecological Point of View. *J. Integr. Agric.* **2016,** *15* (4), 705–715.
Hillnhutter, C.; Mahlein, A.-K.; Sikora, R. A.; Oerke, E.-C. Use of Imaging Spectroscopy to Discriminate Symptoms Caused by Heterodera Schachtii and Rhizoctonia Solani on Sugar Beet. *Precis. Agric.* **2012,** *13* (1), 17–32.
Hines, E. L.; Llobet, E.; Gardner, J. W. Neural Network based Electronic Nose for Apple Ripeness Determination. *Electron. Lett.* **1999,** *35* (10), 821–823.
Huddar, S. R.; Gowri, S.; Keerthana, K.; Vasanthi, S.; Rupanagudi, S. Novel Algorithm for Segmentation and Automatic Identification of Pests on Plants using Image Processing. *Proc. Int. Conf. Comp. Comm. & Netw. Tech.* 2012; pp 1–5.
Jafari, M.; Minaei, S.; Safaie, N.; Torkamani-Azar, F.; Sadeghi, M. Classification using Radial-basis Neural Networks based on Thermographic Assessment of Botrytis Cinerea Infected Cut Rose Flowers Treated with Methyl Jasmonate. *J. Crop. Prot.* **2016,** *5* (4), 591–602.
Jafari, M.; Minaei, S.; Safaie, N. Detection of Pre-symptomatic Rose Powdery-mildew and Gray-mold Diseases based on Thermal Vision. *Infrared Phys. Technol.* **2017,** *85*, 170–183.
Jarvis, W. R.; Jarvis, W. R. *Managing diseases in greenhouse crops,* Vol. 288; Aps press St. Paul; 1992.
Jensen, M. H.; Malter, A. J. *Protected Agriculture: A Global Review*; World Bank Publications: Washington, DC, USA; 1995.
Jensen, J. R. Remote Sensing of the Environment an Earth Resource Perspective: Low Price Edition Pearson Education. New Delhi: India, 2006.
Jones, H. G. Use of Thermography for Quantitative Studies of Spatial and Temporal Variation of Stomatal Conductance Over Leaf Surfaces. *Plant Cell. Environ.* **1999,** *22* (9), 1043–1055.
Jones, H. *Plants and Microclimate: A Quantitative Approach to Environmental Plant Physiology*, 3rd ed; Cambridge University Press: Cambridge, UK; 2014.
Khan, F. A.; Ibrahim, A. A.; Zeki, A. M. Environmental Monitoring and Disease Detection of Plants in Smart Greenhouse using Internet of Things. *J. Phys. Commun.* **2020,** *4* (5), 55008.
Koech, R.; Langat, P. Improving Irrigation Water use Efficiency: A Review of Advances, Challenges and Opportunities in the Australian Context. *Water* **2018,** *10* (12), 1771.
Kuska, M. T.; Mahlein, A.-K. Aiming at Decision Making in Plant Disease Protection and Phenotyping by the use of Optical Sensors. *Eur. J. Plant Pathol.* **2018,** *152* (4), 987–992.
Lindenthal, M.; Steiner, U.; Dehne, H.-W.; Oerke, E.-C. Effect of Downy Mildew Development on Transpiration of Cucumber Leaves Visualized by Digital Infrared Thermography. *Phytopathology* **2005,** *95* (3), 233–240.
Liu, P.; Li, B.; Yu X. In *Intelligent Control System of Cucumber Production in the Greenhouse Based on Internet of Things*. International Conference on Cloud Computing and Security, 2018; pp 395–406.
Llobet, E.; Hines, E. L.; Gardner, J. W.; Franco, S. Non-destructive Banana Ripeness Determination using a Neural Network-based Electronic Nose. *Meas. Sci. Technol.* **1999,** *10* (6), 538.
Lohaus, G.; Heldt, H. W.; Osmond, C. B. Infection with Phloem Limited Abutilon Mosaic Virus Causes Localized Carbohydrate Accumulation in Leaves of Abutilon Striatum:

Relationships to Symptom Development and Effects on Chlorophyll Fluorescence Quenching During Photosynthetic Induction. *Plant Biol.* **2000,** *2* (2), 161–167.
Lu, J.; Zhou, M.; Gao, Y.; Jiang, H. Using Hyperspectral Imaging to Discriminate Yellow Leaf Curl Disease in Tomato Leaves. *Precis. Agric.* **2018,** *19* (3), 379–394.
Ma, J.; Li, X.; Wen, H.; Fu, Z.; Zhang, L. A Key Frame Extraction Method for Processing Greenhouse Vegetables Production Monitoring Video. *Comput. Electron. Agric.* **2015,** *111*, 92–102.
Mahlein, A.-K.; Steiner, U.; Dehne, H.-W.; Oerke, E.-C. Spectral Signatures of Sugar Beet Leaves for the Detection and Differentiation of Diseases. *Precis. Agric.* **2010,** *11* (4), 413–431.
Mahlein, A.-K.; Steiner, U.; Hillnhutter, C.; Dehne, H.-W.; Oerke, E.-C. Hyperspectral Imaging for Small-scale Analysis of Symptoms Caused by Different Sugar Beet Diseases. *Plant Methods* **2012,** *8* (1), 1–13.
Mahlein, A.-K. Plant Disease Detection by Imaging Sensors–Parallels and Specific Demands for Precision Agriculture and Plant Phenotyping. *Plant Dis.* **2016,** *100* (2), 241–251.
Mahlein, A. K.; Kuska, M. T.; Thomas, S.; Bohnenkamp, D.; Alisaac, E.; Behmann, J.; Wahabzada, M.; Kersting, K. Plant Disease Detection by Hyperspectral Imaging: From the Lab to the Field. *Adv. Anim. Biosci.* **2017,** *8* (2), 238.
Miller, S. A.; Beed, F. D.; Harmon, C. L. Plant Disease Diagnostic Capabilities and Networks. *Annu. Rev. Phytopathol.* **2009,** *47*, 15–38.
Milthorpe, F. L.; Spencer, E. J. Experimental Studies of the Factors Controlling Transpiration: III. The Interrelations Between Transpiration Rate, Stomatal Movement, and Leaf-water Content. *J. Exp. Bot.* **1957,** *8* (3), 413–437.
Mirik, M.; Michels-Jr, G. J.; Kassymzhanova-Mirik, S.; Elliott, N. C, Bowling, R. Hyperspectral Spectrometry as a Means to Differentiate Uninfested and Infested Winter Wheat by Greenbug (Hemiptera: Aphididae). *J. Econ. Entomol.* **2006,** *99* (5), 1682–1690.
Mochida, K.; Koda, S.; Inoue, K.; Hirayama, T.; Tanaka, S.; Nishii, R., et al. Computer Vision-based Phenotyping for Improvement of Plant Productivity: A Machine Learning Perspective. *Gigascience* **2019,** *8* (1), 153.
Montero, R.; Perez-Bueno, M. L.; Baron, M.; Florez-Sarasa, I.; Tohge, T.; Fernie, A. R., et al. Alterations in Primary and Secondary Metabolism in Vitis vinifera 'Malvasía de Banyalbufar' Upon Infection with Grapevine Leafroll-associated Virus 3. *Physiol. Plant* **2016,** *157* (4), 442–452.
Oerke, E. C.; Steiner, U.; Dehne, H. W.; Lindenthal, M. Thermal Imaging of Cucumber Leaves Affected by Downy Mildew and Environmental Conditions. *J. Exp. Bot.* **2006,** *57* (9), 2121–2132.
Oerke, E.-C.; Frohling, P.; Steiner, U. Thermographic Assessment of Scab Disease on Apple Leaves. *Precis. Agric.* **2011,** *12* (5), 699–715.
Oerke, E.-C.; Mahlein, A.-K.; Steiner, U. Proximal Sensing of Plant Diseases. In *Detection and Diagnostics of Plant Pathogens;* Springer, 2014; pp 55–68.
Osama, R.; Ashraf, N. E. H.; Yasser, A.; Abdel Fatah, S.; Masry, N. El.; Abdel Raouf, A. In: *Detecting Plant's Diseases in Greenhouse using Deep Learning.* 2020 2nd Novel Intelligent and Leading Emerging Sciences Conference (NILES), 2020; pp 75–80.
Palmgren, M. G.; Edenbrandt, A. K.; Vedel, S. E.; Andersen, M. M.; Landes, X.; Østerberg, J. T.; Falhof, J.; Olsen, L. I.; Christensen, S. B.; SandØe, P.; Gamborg, C.; Kappel, K.; Thorsen, B. J.; Pagh, P. Are We Ready for Back-to-nature Crop Breeding? *Trends Plant Sci.* **2015,** *20* (3),155–164.

Pang, Z.; Chen, Q.; Han, W.; Zheng, L. Value-centric Design of the Internet-of-things Solution for Food Supply Chain: Value Creation, Sensor Portfolio and Information Fusion. *Inf. Syst. Front.* **2015,** *17* (2), 289–319.

Parke, J. L.; Grunwald, N. J. A Systems Approach for Management of Pests and Pathogens of Nursery Crops. *Plant Dis.* **2012,** *96* (9), 1236–1244.

Paulitz, T. C.; Bélanger, R. R. Biological Control in Greenhouse Systems. *Annu. Rev. Phytopathol.* **2001,** *39* (1), 103–133.

Paulus, S.; Mahlein, A.-K. Technical Workflows for Hyperspectral Plant Image Assessment and Processing on the Greenhouse and Laboratory Scale. *Gigascience* **2020,** *9* (8), giaa090.

Perez-Bueno, M. L.; Ciscato, M.; VandeVen, M.; Garcia-Luque, I.; Valcke, R.; Baron, M. Imaging Viral Infection: Studies on Nicotiana Benthamiana Plants Infected with the Pepper Mild Mottle Tobamovirus. *Photosynth. Res.* **2006,** *90* (2), 111–123.

Perez-Bueno, M. L.; Pineda, M.; Diaz-Casado, E.; Baron, M. Spatial and Temporal Dynamics of Primary and Secondary Metabolism in Phaseolus Vulgaris Challenged by Pseudomonas Syringae. *Physiol. Plant* **2015,** *153* (1), 161–174.

Perez-Bueno, M. L.; Pineda, M.; Cabeza, F. M.; Baron, M. Multicolor Fluorescence Imaging as a Candidate for Disease Detection in Plant Phenotyping. *Front. Plant Sci.* **2016,** *7*, 1790.

Perez-Bueno, M. L.; Pineda, M.; Baron, M. Phenotyping Plant Responses to Biotic Stress by Chlorophyll Fluorescence Imaging. *Front. Plant Sci.* **2019,** 10, 1135. https://www.frontiersin.org/article/10.3389/fpls.2019.01135

Pineda, M.; Olejnickova, J.; Csefalvay, L.; Baron, M. Tracking Viral Movement in Plants by Means of Chlorophyll Fluorescence Imaging. *J. Plant Physiol.* **2011,** *168* (17), 2035–2040.

Pineda, M.; Perez-Bueno, M. L.; Paredes, V.; Baron, M. Use of Multicolour Fluorescence Imaging for Diagnosis of Bacterial and Fungal Infection on Zucchini by Implementing Machine Learning. *Funct. Plant Biol.* **2017,** *44* (6), 563–572.

Pineda, M.; Perez-Bueno, M. L.; Baron, M. Detection of Bacterial Infection in Melon Plants by Classification Methods based on Imaging Data. *Front. Plant Sci.* **2018,** *9*, 164.

Pineda, M.; Baron, M.; Perez-Bueno, M.-L. Thermal Imaging for Plant Stress Detection and Phenotyping. *Remote Sens.* **2021,** *13* (1), 68.

Pixia, D.; Xiangdong, W. Recognition of Greenhouse Cucumber Disease based on Image Processing Technology. *Open J. Appl. Sci.* **2013,** *3* (01), 27.

Prashar, A.; Yildiz, J.; McNicol, J. W.; Bryan, G. J.; Jones, H. G. Infra-red Thermography for High Throughput Field Phenotyping in Solanum Tuberosum. *PLoS One* **2013,** *8* (6), 65816.

Ramcharan, A.; McCloskey, P.; Baranowski, K.; Mbilinyi, N.; Mrisho, L.; Ndalahwa, M., et al. A Mobile-based Deep Learning Model for Cassava Disease Diagnosis. *Front. Plant Sci.* **2019,** *10*, 272.

Rehman, T. U.; Mahmud, M. S.; Chang, Y. K.; Jin, J.; Shin, J. Current and Future Applications of Statistical Machine Learning Algorithms for Agricultural Machine Vision Systems. *Comput. Electron. Agric.* **2019,** *156*, 585–605.

Rico-Fernández, M. P.; Rios-Cabrera, R.; Castelan, M.; Guerrero-Reyes, H. I.; Juarez-Maldonado, A. A. Contextualized Approach for Segmentation of Foliage in Different Crop Species. *Comput. Electron. Agric.* **2019,** *156*, 378–386.

Roscher, R.; Behmann, J.; Mahlein, A.-K.; Dupuis, J.; Kuhlmann, H.; Plumer, L. Detection of Disease Symptoms on Hyperspectral 3d Plant Models. *ISPRS Ann. Photogramm. Remote Sens. Spat. Inf. Sci.* **2016,** *3*(7).

Rout, R.; Parida, P. In *A Review on Leaf Disease Detection Using Computer Vision Approach.* International Conference on Innovation in Modern Science and Technology, Springer, 2019; pp 863–871.

Sabir, N.; Singh, B. Protected Cultivation of Vegetables in Global Arena: A Review. *Indian J. Agric. Sci.* **2013,** *83* (2), 123–135.

Sabrol, H.; Satish, K. In *Tomato Plant Disease Classification in Digital Images using Classification Tree.* 2016 International Conference on Communication and Signal Processing (ICCSP), IEEE, 2016; pp 1242–1246.

Sankaran, S.; Mishra, A.; Ehsani, R.; Davis, C. A Review of Advanced Techniques for Detecting Plant Diseases. *Comput. Electron. Agric.* **2010,** *72* (1), 1–13.

Sawinski, K.; Mersmann, S.; Robatzek, S.; Bohmer, M. Guarding the Green: Pathways to Stomatal Immunity. *Mol. Plant-Microbe. Interact.* **2013,** *26* (6), 626–632.

Scarth, G. W.; Loewy, A.; Shaw, M. Use of the Infrared Total Absorption Method for Estimating the Time Course of Photosynthesis and Transpiration. *Can. J. Res.* **1948,** *26* (1), 94–107.

Schurr, U.; Schuberth, B.; Aloni, R.; Pradel, K. S.; Schmundt, D.; Jahne, B., et al. Structural and Functional Evidence for Xylem-mediated Water Transport and High Transpiration in Agrobacterium Tumefaciens-induced Tumors of Ricinus Communis. *Bot. Acta* **1996,** *109* (5), 405–411.

Shi, X.; An, X.; Zhao, Q.; Liu, H.; Xia, L.; Sun, X., et al. State-of-the-art Internet of Things in Protected Agriculture. *Sensors* **2019,** *19* (8), 1833.

Shijie, J.; Peiyi, J.; Siping, H. Automatic Detection of Tomato Diseases and Pests based on Leaf Images. In *2017 Chinese Automation Congress (CAC)*; IEEE, 2017; pp 2510–2537.

Shipp, J. L.; Boland, G. J.; Shaw, L. A. Integrated Pest Management of Disease and Arthropod Pests of Greenhouse Vegetable Crops in Ontario: Current Status and Future Possibilities. *Can. J. Plant Sci.* **1991,** *71* (3), 887–914.

Shishido, M. Plant Disease Management in Protected Horticulture. *Hort. Res.* **2011,** *65*, 7–18.

Smigaj, M.; Gaulton, R.; Suarez, J. C.; Barr, S. L. Canopy Temperature from an Unmanned Aerial Vehicle as an Indicator of Tree Stress Associated with Red Band Needle Blight Severity. *For. Ecol. Manage.* **2019,** *433*, 699–708.

Story, D.; Kacira, M. Design and Implementation of a Computer Vision-guided Greenhouse Crop Diagnostics System. *Mach. Vis. Appl.* **2015,** *26* (4), 495–506.

Tecsi, L. I.; Maule, A. J.; Smith, A. M.; Leegood, R. C. Complex, Localized Changes in CO_2 Assimilation and Starch Content Associated with the Susceptible Interaction Between Cucumber Mosaic Virus and a Cucurbit Host. *Plant J.* **1994,** *5* (6), 837–847.

Thomas, S.; Kuska, M. T.; Bohnenkamp, D.; Brugger, A.; Alisaac, E.; Wahabzada, M., et al. Benefits of Hyperspectral Imaging for Plant Disease Detection and Plant Protection: A Technical Perspective. *J. Plant Dis. Prot.* **2018,** *125* (1), 5–20.

Thorat, A.; Kumari, S.; Valakunde, N. D. In *An IoT based Smart Solution for Leaf Disease Detection.* 2017 International Conference on Big Data, IoT and Data Science (BID), 2017; pp 193–198.

Tian, H.; Wang, T.; Liu, Y.; Qiao, X.; Li, Y. Computer Vision Technology in Agricultural Automation—A Review. *Inf. Process. Agric.* **2020,** *7* (1), 1–19.

Tirelli, P.; Borghese, N. A.; Pedersini, F.; Galassi, G.; Oberti R. In *Automatic Monitoring of Pest Insects Traps by Zigbee-based Wireless Networking of Image Sensors.* IEEE International Instrumentation and Measurement Technology Conference, 2011; pp 1–5.

van Henten, E. J.; Bac, C. W.; Hemming, J.; Edan Y. In *Robotics in Protected Cultivation.* 4th IFAC Conference on Modelling and Control in Agriculture, Horticulture and Post-Harvest Industry, 2013; pp 27–30.

Virlet, N.; Sabermanesh, K.; Sadeghi-Tehran, P.; Hawkesford, M. J. Field Scanalyzer: An Automated Robotic Field Phenotyping Platform for Detailed Crop Monitoring. *Funct. Plant Biol.* **2017,** *44* (1), 143–153.

Wahabzada, M.; Paulus, S.; Kersting, K.; Mahlein, A.-K. Automated Interpretation of 3D Laserscanned Point Clouds for Plant Organ Segmentation. *BMC Bioinformatics* **2015,** *16* (1), 1–11.

Walter, A.; Liebisch, F.; Hund, A. Plant Phenotyping: From Bean Weighing to Image Analysis. *Plant Methods* **2015,** *11* (1), 1–11.

Wang, L.; Poque, S.; Valkonen, J. P. T. Phenotyping Viral Infection in Sweetpotato using a High-throughput Chlorophyll Fluorescence and Thermal Imaging Platform. *Plant Methods* **2019,** *15* (1), 1–14.

Wen, D.-M.; Chen, M.-X.; Zhao, L.; Ji, T.; Li, M.; Yang, X.-T. Use of Thermal Imaging and Fourier Transform Infrared Spectroscopy for the Pre-symptomatic Detection of Cucumber Downy Mildew. *Eur. J. Plant Pathol.* **2019,** *155* (2), 405–416.

West, J. S.; Bravo, C.; Oberti, R.; Lemaire, D.; Moshou, D.; McCartney, H. A. The Potential of Optical Canopy Measurement for Targeted Control of Field Crop Diseases. *Annu. Rev. Phytopathol.* **2003,** *41* (1), 593–614.

Wspanialy, P.; Moussa, M. A Detection and Severity Estimation System for Generic Diseases of Tomato Greenhouse Plants. *Comput. Electron. Agric [Internet].* **2020,** *178*, 105701. Available from: https://www.sciencedirect.com/science/article/pii/S0168169920302O5

Yakkundimath, R.; Saunshi, G.; Kamatar, V. Plant Disease Detection using IoT. *Glob. J. Comput. Sci. Technol.* **2018,** *8* (9), 18902–18906.

Yosuke, K.; Masahiro, I.; Nobuhiro, O.; Hiroshi, I. Development of "IoTOMATO" that is an Agriculture Monitoring System by using Chat Bot. *FIT2017, proceeding IPSJ IEICE* **2017,** *4*, 229–230.

Yucel, S.; Kececi, M.; Yurtmen, M.; Yildiz, R. C.; Ozarslandan, A.; Can, C. Integrated Pest Management of Protected Vegetable Cultivation in Turkey. *Eur. J. Plant Sci. Biotechnol.* **2013,** *7* (1), 7–13.

Zeng, W.; Melotto, M.; He, S. Y. Plant Stomata: A Checkpoint of Host Immunity and Pathogen Virulence. *Curr. Opin. Biotechnol.* **2010,** *21* (5), 599–603.

Zheng, X.; Spivey, N. W.; Zeng, W.; Liu, P.-P.; Fu, Z. Q.; Klessig, D. F., et al. Coronatine Promotes Pseudomonas Syringae Virulence in Plants by Activating a Signaling Cascade that Inhibits Salicylic Acid Accumulation. *Cell Host. Microbe.* **2012,** *11* (6), 587–596.

Zhu, W.; Chen, H.; Ciechanowska, I.; Spaner, D. Application of Infrared Thermal Imaging for the Rapid Diagnosis of Crop Disease. *IFAC-Papers OnLine* **2018,** *51* (17), 424–430.

CHAPTER 9

Insect-Pests and Diseases in Greenhouse Cultivation and Their Biological Control

ANJANI KHULBE[1] and POONAM BATRA[2]

[1]*Indian Council of Agricultural Research, Krishi Bhawan, New Delhi, India*

[2]*Regional Plant Quarantine Station, Ministry of Agriculture and Farmers Welfare, Rangpuri, New Delhi, India*

ABSTRACT

Global Agriculture industry is under significant pressure due to climate change, continuously decreasing resources and increasing population. In this modern agriculture era, the growers are turning toward advanced technologies to increase production efficiency and crop resilience. Vegetables in India are mainly cultivated according to the seasonal need of a region due to wide range of diverse agro-climatic conditions of the country. Therefore, greenhouse technology for production of high value horticultural crops is gaining a pace everywhere in India, and a trend of gradual increase in acreage of protected farming has been observed. This area has been recorded to be up to 40,000 ha including both horticulture and ornamental crops. The ideal and stable environment for plants under greenhouse also provides a congenial environment for plant growth one hand and an excellent platform for the development of insect-pest and diseases on the other. Many fungal, bacterial, viral, and nematode diseases and various pests can be seen in the greenhouse conditions if managed inadequately that may affect the crop(s) to a great extent under protected environment. The mode of managing the insect-pest and disease problem under green house is still chemical treatment, but seeing the increase in occurrence of pesticide resistance and pesticide residues,

Protected Cultivation: Structural Design, Crop Management, Modeling and Automation.
Mahesh Chand Singh and K. K. Sharma (Eds.)

people are moving toward safe option of biological control. The chapter is being discussed below and it briefly covers the insect-pests and diseases that can be commonly observed under protected cultivation along with the biological control measures which can be useful in their management. Various biological control agents like *Trichoderma harzianum* Strain T-22, Strain T-39; *Streptomyces griseoviridis* Strain K6; *Bacillus firmus* BioNem WP, *Verticillium lecanii Bacillus thuringiensis,* etc., are commercially available and widely used by the farmers but there are several other bio-control agents (BCAs) that need to be popularized by ensuring their availability to the farmers. An account of such bioagents has been given in order to widen the arena of non-chemical options that are available worldwide. In a country like India, where we find huge diversity of plants, where it is needed now to explore, identify, and develop its own BCAs and reduce the hurdles in the development and use of new bio-pesticides. The chapter gives a brief account of challenges being faced in development, commercialization and use of bio-agents, in order to bring into light, the workable areas for researchers and policy makers to get the real potential of biological control agents especially under protected cultivation.

9.1 INTRODUCTION

The greenhouse technology which started about 200 years ago in European countries has now been adopted by more than 50 countries on a commercial scale. The world's total greenhouse area is more than 3,07,000 ha which includes both plastic and glass greenhouses (Beniwal, 2015). In India, the area under protected cultivation is about 40,000 ha which includes both vegetable and ornamental crops (Singh, 2014). Some of the major states having cultivation of crops under greenhouse are Maharashtra, Uttarakhand, Karnataka, and Jammu & Kashmir (Beniwal, 2015). In past few decades, the area under greenhouse has increased manifolds with steep escalation in cultivation of high value crops.

Lot of pests and diseases occur in protected cultivation similar to field conditions and the method of control of these is still with chemicals. But seeing the increased interest in biological control of pests and diseases in protected cultivation in some European countries, America and Japan, other countries are also moving toward this safe option (Bueno et al., 2010). The increase in occurrence of pesticide resistance and pesticide residues apart from concern related to legislative restrictions and safety of the workers is continuously motivating toward adoption of biological control methods as an alternative (Sharma, 2018; Sharma et al., 2019). In European countries, over 80% of the biological control is used in protected crops to manage insect-pests in tomato, cucumber, and sweet pepper; and around 90% of all commercialized natural enemies are used for all the horticultural crops. However, an increase in the usage of biocontrol in greenhouses has been reported in some cut flower crops namely rose, gerbera, chrysanthemum, orchid, etc., and specific potted plants like poinsettia, anthurium, etc., since the year 1990 (Bueno et al., 2010). Biological control measures can be developed by using a combination of potential natural enemies of indigenous (native) and exotic occurrence and, it may play a decisive role when native bioagents alone become useless against the pest (van Lanteren, 2003).

Till 1960s, the pesticides were able to control the greenhouse insect-pests easily, but during 1970s, the problems of pest resistance against pesticides in greenhouse increased rapidly. Today, majority of the key pests are characterized by high level of resistance to various pesticides. Due to pesticide resistance and extensive trade of plant material internationally, many herbivorous insects and mites have reached to global pest status. Therefore, by considering pesticide resistance together with consumer and environmental factors, the growers are replacing simple pesticides-based control programs with broader integrated control systems (Brødsgaard et al., 1999) in which biological control measures are playing an important role.

9.2 DISEASES AND INSECT-PESTS UNDER PROTECTED CULTIVATION

In open field, the plants are subjected to several uncontrolled environmental factors like temperature, humidity, heat, air, etc., which cannot be managed. When plants are grown in controlled environment or under protected cultivation, all the environmental parameters are more manageable. Such conditions are produced that are optimum for the growth of a healthy plant. The temperature, humidity, and light conditions are programmed based on the stage of growth of the plant (Singh et al., 2017b, 2018a,b). Soil and air moisture are kept at its optimum for maintaining good health of the plants. The ideal and stable environment with warm, humid, and abundant food under green house also provides an excellent platform for the development of diseases and insect-pests more than field conditions.

9.3 COMMON OCCURRING DISEASES UNDER PROTECTED CULTIVATION

Growth rate of plant depends on temperature, and any fluctuation in temperature condition causes the plants to undergo stress which may lead to diseases (Singh et al., 2017a). Similarly, humidity level also affects the plant growth and when the humidity is higher there are more chances of plant diseases (Singh et al., 2017a). For example, the appropriate humidity for greenhouse plants in the daytime is 25–70% and the humidity levels at night rise generally to 90–100%. During rainy season, the humidity may remain 100% for several days (Shishido, 2011). Overcrowding in plants can also be the reason of high humidity which increases the risk of diseases in plants. Under protected cultivation, the different optimal temperature and humidity conditions required by the pathogens are given in Table 9.1.

9.3.1 COMMON FUNGAL DISEASES UNDER PROTECTED ENVIRONMENT

9.3.1.1 POWDERY MILDEW

Most of the floral and foliage plants in the green house are found to be susceptible to powdery mildew disease. It is caused by *Oidium violae, Oidiopsis sicula, Sphaerotheca fuliginea, and S. humuli* (Shishido, 2011) *Erisyphae chicoracearum* and its growth can be seen usually on the upper surface of the leaves and sometimes may be seen on the lower surface also. Older leaves show more prominent symptoms as compared to lower leaves (Malathrakiset and Goumas, 1999).

9.3.1.2 GRAY MOLD

The symptoms of gray mold appear as blighting, leaf spot, damping off, and stem canker. Temperature of 17–27°C and high humidity are favorable conditions for disease development. Infected plant part shows fluffy, grayish to brown fungal growth in high humidity conditions. Tulip fire blight is caused by *Botrytis tulipae* and *Botrytis gladiolorum* damages gladiolus and some other members of Iridaceae family. Apart from causing losses in green

TABLE 9.1 Optimal Temperature and Humidity Conditions of Vegetable Diseases Common in Protected Horticulture Systems.

Crop (s): vegetable/fruit	Disease	Pathogen (s)	Favorable temperature (°C)	Humidity condition
Bell pepper, cucumber, eggplant, strawberry and tomato	Gray Mold	*Botrytis cinerea*	≈20	dry < moist
Tomato	Powdery Mildew	*Oidium violae*/*Oidiopsis sicula*	23–25	dry > moist
Eggplant		*Sphaerotheca fuliginea*/*Oidium* sp./ *Oidiopsis sicula*		
Bell pepper		*Oidiopsis sicula*		
Cucumber and melon		*Sphaerotheca fuliginea*		
Strawberry		*Sphaerotheca humuli*		
Tomato	Fusarium Wilt	*Fusarium oxysporum* f.sp. *lycopersici*	24–28	NA
Cucumber		*Fusarium oxysporum* f.sp. *cucumerinum*		
Melon		*Fusarium oxysporum* f.sp. *melonis*		
Strawberry		*Fusarium oxysporum* f.sp. *fragariae*		
Eggplant and tomato	Verticillium Wilt	*Verticillium dahlia*	23–28	NA
Strawberry			20–25	NA
Bell pepper, eggplant, and tomato	Bacterial Wilt	*Ralstonia solanacearum*	≈30	NA
Cucumber and melon	Downy Mildew	*Pseudoperonospora cubensis*	20–25	dry < moist
Cucumber and melon	Gummy Stem Blight	*Didymella bryoniae*	20–24	dry < moist
Tomato	Crown (foot) and Root Rot	*Fusarium oxysporum* f. sp. *radicis-lycopersici*	10–20	NA
Tomato	Brown Root Rot/Corky root	*Pyrenochaeta lycopersici*	13–18	NA
Tomato	Late Blight	*Phytophthora infestans*	≈20	dry < moist

TABLE 9.1 *(Continued)*

Crop (s): vegetable/fruit	Disease	Pathogen (s)	Favorable temperature (°C)	Humidity condition
Tomato	Leaf Mold	*Cladosporium fulva*	20–23	dry < moist
Eggplant	Sclerotinia Stem Rot	*Sclerotinia sclerotiorum*	15–24	dry < moist
Bell pepper	Phytophthora Blight	*Phytophthora capsici*	28–30	dry < moist
Cucumber	Phytophthora Rot	*Phytophthora melonis/P. nicotianae*	28–30	dry < moist
Cucumber	Scab	*Cladosporium cucumerinum*	≈17	dry < moist
Cucumber	Corynespora Leaf Spot	*Corynespora cassiicola*	≈28	dry < moist
Cucumber	Sclerotinia Stem Rot	*Sclerotinia sclerotiorum*	18–20	dry < moist
Cucumber	Bacterial Spot/Angular Leaf Spot	*Pseudomonas syringae*pv. *lachrymans*	24–28	dry < moist
Melon	Monosporascus Root Rot	*Monosporascus cannonballus*	25–30	NA
Strawberry	Anthracnose/Crown Rot	*Colletotrichum acutatum/C. fragariae/ Glomerella cingulate*	≈30	NA

NA, information is not available.

Source: Adapted from Shishido (2011).

house, *Botrytis* spp. may also cause severe post-harvest losses (Shishido, 2011; Goldammer, 2019).

9.3.1.3 ROOT ROTS

Plant may get affected by root rot at any stage of growth. Symptoms initially appear as nutrient deficiency and later reduction in growth, wilting, and ultimately death of plant can be seen. Roots get discolored and become brown to black and later may get rotten. It is a very common disease and most of the plants in greenhouse are prone to it. Various *Pythium* spp. and *Phytophthora* spp. are a major cause of root rots (Tompkins and Middleton, 1950). Some other root rots are corky root rot of tomato caused by *Pyrenochaeta lycopersici*, root rot and crown rot of tomato caused by *Fusarium oxysporum* and black root rot of Cucurbits is caused by *Phomopsis sclerotioides* (Malathrakiset and Goumas, 1999; Goldammer, 2019).

9.3.1.4 DOWNY MILDEWS

Downy mildew scan often by seen in ornamental crops, particularly basil, coleus, and snapdragon and also in vegetable crops. Symptoms are sometimes confused with nutrient imbalances or bacterial diseases. The disease occurs on undersurface of the leaves where fungal sporulation occurs in which it appears as indistinct, whitish or grayish to purplish growth. Shishido in 2011 wrote and mentioned that these downy mildew pathogens are relatives of *Pythium* and *Phytophthora.*

9.3.1.5 WILT

Wilt in greenhouse may be caused by *Fusarium* or *Verticillium* species. Foliage on the affected stems show wilting and usually become yellow. Commonly, greenhouse tomato becomes diseased with Fusarium wilts (*F. oxysporum* f. sp. *lycopersici*), cucumber (*Fusarium oxysporum* f. sp. *cucumerinum*), melon (*Fusarium oxysporum* f. sp. *melonis*), gladiolus (*Fusarium oxysporum* f. sp. *gladioli*), carnation (*Fusarium oxysporum* f. sp. *dianthi*), and chrysanthemum (*Fusarium oxysporum* f. sp. *chrysanthemi*). The causal organism of *Verticillium phialophora* wilt may be *Verticillium dahliae*, *Verticilium albo-atrum*, and *Phialophora cinerescens* (*Verticillium cinerescens*). These fungi are known to infect a huge number of plants and among them the majority of the plants grown in greenhouses. Members of Solanaceae family such as tomato, eggplant, and pepper are more prone to the disease. Symptoms of Fusarium wilt and Verticillium wilt are very similar according to Malathrakiset and Goumas, 1999.

9.3.1.6 SCLEROTINIA ROT

Sclerotinia rot is a common greenhouse disease caused by *Sclerotinia sclerotiorum* that damages lettuce, eggplant, tomato, cucumber, pepper, etc. Symptoms appear along stem, leaves, flowers, and fruits. Stem infection is more severe. Leaves above the infection area become yellow, wither, and die. The pathogen produces sclerotia that are small bean seed size or smaller that fall on the soil where they can survive for several years. High relative humidity (RH) and moderate temperature are required for infection (Malathrakiset and Goumas, 1999; Shishido, 2011).

9.3.1.7 DIDYMELLA DISEASES

Tomato and eggplant diseases like stem rot or canker caused by *Didymella lycopersici* and Gummy stem blight disease of cucurbits caused by *Didymell abryoniae* are very severe diseases found in greenhouses. During cool weather and high relative humidity, all aerial plant parts of hosts are affected in greenhouses. Collar and root system of the plant may also get affected causing yellowing and withering and the plant may die later. Cankerous growth can be commonly seen along the stem and the petioles. In case of aforesaid diseases, parts of the plant lying beyond the cankers may show death. The large spots on the foliage caused by these two diseases may cover the whole surface of leaf (Malathrakiset and Goumas, 1999).

9.3.1.8 CLADOSPORIUM DISEASES

This disease attacks cucumber, squash, melon, etc., and the symptoms appear as circular or angular spots on the leaves, which are water-soaked. It also affects fruit which is more serious. Diseases such as tomato leaf mold (*Cladosporium fulvum*) and cucurbit scab (*Cladosporium cucumerinum*) are two examples (Fletcher, 1984; Goldammer, 2019).

9.3.2 COMMON BACTERIAL DISEASES UNDER PROTECTED CULTIVATION

9.3.2.1 SOFT-ROTS

The wounds on the base of the plants those are propagated through cuttings are highly prone to attack by soft rot bacterium like *Erwinia*. Temperature more than 30°C and plenty of water make the conditions congenial for disease development. The affected portion becomes soft and slimy (Malathrakiset and Goumas, 1999).

9.3.2.2 BACTERIAL BLIGHT

These diseases are caused by *Xanthomonas campestris*. Plants get wilted but will not show root rot or stem canker symptoms unless severe. One or two leaves may become soft and droop. Disease caused by *X. c* pv. *pelargoni* in Geranium sometimes cause leaf spots. Systemic wilt of geranium is caused by *Ralstonia* (*Pseudomonas*) *solanacearum*. *Xanthomonas campestris pv. pelargoni* is known to infect only geranium but species of *Ralstonia* infect a number of vegetable such as tomato and potato, and different ornamental plants (Fletcher, 1984; Malathrakiset and Goumas, 1999).

9.3.2.3 BACTERIAL LEAF SPOTS

Pseudomonas cichorii and *P. syringae* are known to attack on many ornamental plants like chrysanthemum, geranium, and many more which are responsible to cause leaf spots.

9.3.2.4 WILTS

Some of the bacterial diseases that appear under protected cultivation are tomato bacterial canker (*Clavibacter michiganensis.* ssp. *michiganensis*), Carnation stunt due to bacterium *Erwinia chrysanthemi* pv. *dianthicola* (Malathrakiset and Goumas, 1999), bacterial wilt of solanaceous crops caused by *Ralstonia solanacearum* (Shishido, 2011).

9.3.2.5 ROTS

The diseases, that is, tomato soft rots (*Erwinia carotovora* ssp. *carotovora, Erwinia carotovora* ssp. *atroseptica, Pseudomonas viridiflava*) and tomato pith necrosis (*Pseudomonas corrugate, P. viridiflava, P. cichorii*) are few examples (Fletcher, 1984; Malathrakiset and Goumas, 1999).

9.3.2.6 LEAF AND STEM SPOTS

Some of the leaf and stem spots caused by bacteria are tomato speck (*Pseudomonas syringae*), bacterial spot occur on pepper and tomato caused by *Xanthomonas vesicatoria* and angular leaf spot of cucurbits (*Pseudomonas syringae*) (Malathrakiset and Goumas, 1999).

9.3.3 COMMON VIRAL DISEASES UNDER PROTECTED CULTIVATION

Many viruses of different types are known to infect the plants in greenhouses which can be transmitted by insects, mites, nematodes, and cultural practices. Symptoms of viral infection appear on leaves, stem, flowers, and other parts which gradually increase in size. Various types of symptoms may be observed which depend on host, prevailing environment, strain of the virus, and that period for how long the infection occurs. Usually, a range of hypoplastic symptoms like common, yellow and yellow vein mosaic, spots with necrotic areas and discolored rings, distortion of leaves or flowers, and rarely pattern of stripe or streaking may visible and characterized the infection caused by the virus. Tobacco mosaic virus (TMV), tomato yellow leaf curl virus (TYLCV), tomato spotted wilt virus (TSWV), cucumber mosaic virus (CMV) can be seen as commonly occurring viruses under greenhouse cultivation (Pundt, 2010). According to previous reports, it is reported that insects such as aphids, whiteflies, leafhoppers, thrips, mites, and also fungi and nematodes are found capable to transmit over 80% of viruses pathogenic to plants. Propagative plant materials like cuttings, bulbs, seeds, etc., are frequently associated to spread of plant viruses. However, some can also be spread through mechanical means like rubbing, abrasion, or by handling. Unlike Tobbaco Mosaic Virus, maximum number of phytoviruses of horticulture crops prevalent under protected cultivation can last only in their animate plants and insect hosts for short a period (Shishido, 2011).

9.3.4 NEMATODE-DISEASES UNDER PROTECTED CULTIVATION

Nematodes enter into the greenhouse through cuttings and soil. Different symptoms are produced on different host. Infestation with root-knot nematode (RKN) (*Meloidogyne* spp.), a very important genus of plant–parasitic nematodes (PPN) causes poor growth, reduction in qualitative and quantitative yield of the crop, and also decrease in resistance to other stresses like drought, etc. Moreover, complete loss of the crop may occur due to extensive damage. Four species of *Meloidogyne* (*M. incognita, M. javanica, M. hapla,* and *M. arenaria*) which are known to be major pests which attack on ornamental and vegetable crops, field crops, and also on pasture and grasses. All of the aforesaid species of RKN are associated to cause around 90% damage caused by nematodes (Kepenekcia et al., 2018). Among the most prevalent species of foliar plant pathogenic nematodes discovered in ornamentals are *Aphelenchoides fragariae and A. ritzemabosi*. The foliar nematodes on *Begonia* develop reddish coloration on leaves before developing dry patches and on chrysanthemum produce distinct black lesions bordered by the leaf veins. Necrotic patches are formed on the bottom of the leaves of African violet, and newly developing leaves become distorted and the leaves of Easter lily get twisted (Singh, 2014).

9.4 COMMON GREENHOUSE INSECT AND MITE PESTS

The common insect and mite pests of greenhouse crops include whiteflies, aphids, thrips, caterpillars, leaf miner, and mites. Some of the insects not only damage the crop, but also serve as vectors for many phytoviruses which are responsible to cause numerous diseases in plants. Damage on different crops caused by insect-pests under protected cultivation varies from 15 to 37% (Dhandapani et al., 2003; Rathee et al., 2018). Aphids can cause up to 38.85% damage on chili crops (Chinniah et al., 2016) whereas losses reported from whitefly varies among different vegetables crops 13.60% in capsicum (Singh and Joshi, 2020), 26% in cucumber (Ghongade, 2020; Dhandapani et al., 2003) reported 54% damage in okra. The list of major insect pests and mites causing major damage in the greenhouse is given in Table 9.2.

The description of some of the major insect-pests and mites is as under:

TABLE 9.2 Indian Scenario of Insect-Pests in Protected Cultivation.

Insect		**Host**
Common name	**Scientific name**	**Vegetable/fruit/ornamental crop (s)**
Aphid	*Aphis gossypii*	Capsicum
	Myzuses calonicus	Strawberry
	Myzus persicae	Capsicum and gerbera
	Toxoptera aurantii, Macrosiphum luteum	Orchid
	Macrosiphoniella sanborni	Chrysanthemum
	Helicoverpa armigera	Capsicum, carnation, and tomato
Caterpillar	*Spodoptera litura*	Capsicum, cucumber, tomato, and rose
	Liriomyza trifolii	Cucumber, tomato, and ornamentals (chrysanthemum, gerbera, etc.)
Leaf-miner	*Acerialycopersici*	Tomato
	Tetranychus neocalidonicus	Cucumber
	Polyphagotarsonemus latus	Capsicum
Mite	*Stenotarsonemus fragariae*	Strawberry
	Tetranychus cinnabarinu,	Carnation
	Tetranychus urticae(Spider mite)	Capsicum, cucumber, tomato, and ornamentals (carnation, gerbera, etc.)
	Scritothrips dorsalis	Rose
Thrip	*Thrips palmi, Thrips tabaci*	Gerbera
	Bemisia tabaci	Capsicum and gerbera
Whiteflies	*Trialeurodes vaporariorum*	Beans, capsicum, cucumber, tomato, and ornamentals (gerbera plus more than 30 hosts)

Source: Adapted from Sabir and Singh (2013) and Singh (2015).

9.4.1 APHIDS

Aphids are usually the number one insect pests in greenhouse. They are the major pests of the Solanaceae family and other vegetables crops. The most devastating insect pests of crops grown under protected settings around the world are *Aphis gossypii* and *Myzus persicae*. They can transfer over 100 different viruses to plants, causing fruit abortion, chlorosis, necrosis, and reduced development, among other problems (Perdikis et al., 2008). Cucumber mosaic virus (CMV) is an example of highly important and harmful virus of vegetable crops transmitted by *M. persicae*. In India, it reduced the output of greenhouse vegetable crops by 19.43% (Singh and Joshi, 2020). *M. persicae* is recognized to harm causing death a variety of host plants across the globe, and has been reported in Indian states, e.g., Himachal Pradesh, Maharashtra, and Punjab on different crops such as tomato, capsicum, cucumber, cut flower, etc. (Singh et al., 2004; Weintraub, 2007; Vashisth, 2009; Ibrahim et al., 2011; Sood et al., 2012; Sabir and Singh, 2013; Gavkare et al., 2014; Javed et al., 2019). Aphids are difficult pests to control because they relatively develop resistance to insecticides very easily (Foster et al., 2000; Criniti et al., 2008) and this was the prime cause to encourage pest control by adopting non-chemical biological methods (Devonshire, 1988).

9.4.2 WHITEFLIES

Whiteflies are also among the major pests in greenhouse crops. They are usually oval shaped, with yellow body covering with white wax on their wings. They mostly suck the sap from lower surface of leaf excreting honeydew in huge quantity which serves as a substratum to grow fungal mold and reduces photosynthetic activity leading to lower fruit. Population of whiteflies is high under protected cultivation, which is responsible for leaf yellowing and their shrinkage, as well as affecting the flower quality and overall health of the plant. These are one of the most challenging pests which are very difficult to manage in a greenhouse. Whitefly of sweet potato (*Bemisia tabaci* (Gennadius) and greenhouse whitefly (*Trialeurodes vaporariorum*) are the two most common whitefly species observed in greenhouse (Westwood). The whitefly of tobacco is known for having its potential vector quality for several phytoviruses which is the riskiest thing related to it (Gerling et al., 2001).

9.4.3 LEAFMINER

In a greenhouse, leaf miners can be also a difficult pest to control. The adult is a small black to yellowish insect with a yellow abdomen. This insect makes a little incision on the upper surface of the leaf and lays small white eggs. Larva of the leaf miner mines into the leaf after hatching, forming a typical serpentine tunnel inside the leaf. The larvae continue to make tunnel through the leaf by eating its tissues between the upper and lower surfaces. The larva becomes protected and may escape from anything that is sprayed on it once it is within the leaf. Leaf miners pupate in the soil; however, if the larva falls onto a hard surface or a concrete floor, it will die and not able to complete its life cycle. Problem of leaf miner occurrence is very less or rare in hydroponic system of greenhouses with concrete flooring than with alternative flooring materials. There are beneficial parasites which can control and eliminate leaf miners when properly introduced.

9.4.4 THRIPS

Thrips are tiny, elongated insects with an adult of small size <2.5 mm length and life stages comprising of the egg, larval stage with two instars, the pre-pupal, three pupal stage, and the adult. The eggs are kidney-shaped which are placed beneath the surface of the leaf. The larva feeds on plant tissue which can be leaf, flowers, or fruits. The damage is caused by adult and larval instars. Thrips attack on foliage, blooms, buds, and immature fruits. Their feeding is characterized by rasping the surface of plants and sucking up the sap. Leaves show mottling or silvery look that have been seriously affected. Some specific species may act as vectors and transmit plant viruses. The major pest species of concern to vegetable crops in protected cultivation are *Thrips palmi, Thrips tabaci*, and *Frankliniella occidentalis.*

9.4.5 TOBACCO CATERPILLAR

The tobacco caterpillar's adult moth (*Spodoptera litura*) is brown and a pattern of greyish-brown color with wavy lines of gray color. Each female may lay approximately 300 eggs in brown hair-covered clusters. The fully grown larva is silky black and up to 5 cm long which has yellowish green stripes dorsally and white bands laterally and their pupation occurs in the ground. They primarily feed on leaves, but they do occasionally harm fruits.

Female of the moth that has found their way inside the greenhouse lays eggs in huge quantities. The attacked leaves are skeletonized due to voracious eating by gregarious early instars and subsequent instars dispersed around the field. During the day hours, the larvae hide themselves inside the cracks and gaps of the field, and then emerge at night to feast voraciously on various parts of the plant. Isolated places inside the greenhouse may develop heavy infestation, where early spot treatments at accurate time can give effective management.

9.4.6 TOMATO FRUIT BORER

One of the major insect-pests occurs commonly on crops which are grown in open field is tomato fruit borer. They are usually not frequent in greenhouses, but if they get established, may responsible to cause significant loss. The caterpillars attack on fruits targeting a variety of other crops and feed directly. The wholly developed caterpillar appears green or brown striped with stout-body, and 3–4 cm in length. The light brown colored moth has a tiny brownish spot on its forewings. The female moth lays her eggs once at a time on foliage, buds, and blooms. The immature larva feeds on leaves while the adult bores into floral bud to eat them. Tomato fruit borer eats blossoms and young tomato fruits. Heavily infested crop may be ruined by these caterpillars if early and necessary action is not taken to control them.

9.4.7 MITES

Mites are tiny insect-pests penetrate into the plant tissue and sucking up the cell sap with their mouthparts which cause havoc on a variety of greenhouse crops. Two-spotted spider mite (*Tetranychus urticae*) and the broad mite (*Polyphagotarsonemus latus*) are the main issues in India for the crops grown inside greenhouse and poly-net limiting crop output. The typical damaging symptoms of broad mite comprise curling, twisting of foliage downside, and tanning of terminal foliage. Affected leaf under severe infestation, become golden yellow and purplish tinge underneath and finally fall off. Such plants show rosette look which dies subsequently. Conversely, two-spotted spider mite produces white speckling with yellow tinge and netting under severe infestation leads to reduced photosynthesis. In case of such infestation rather than attempting to address, the problem with insecticides, it is often preferable to destroy affected plants or parts.

9.5 BIOLOGICAL CONTROL OF DISEASES AND INSECT-PESTS UNDER PROTECTED CULTIVATION

9.5.1 BIODIVERSITY ENRICHMENT

By selecting the crop monoculture mankind has reduced the natural biodiversity. The insect-pests and weeds that appear along with the crops are kept away with the help of various artificial chemicals. The species enjoys mutually beneficial relationship and cooperation exists among the communities. With this beneficial relationship, the productivity of crop increases with intercropping together with reducing the ability of locating the crop by the pest. The incidence of white flies and thrips also known to be reduced when mulching is practiced.

The placement of organic materials like crop residue, hay, leaves, and twigs growing along the walls of the greenhouse enrich the number of resident spiders (*Lycosa pseudoannulata*) and ground beetles (Coleoptera: Carabidae). The population of all insect-pests, together with aphids (*Brevicoryne brassicae*), leafminers (*Phytomyza nigricornis*), and semiloopers (*Autographa nigrisigna* Walker) decreases in those fields where organic mulching was done as compare to unmulched one (Xu et al., 2008).

Due to excess absorption of nitrogen, the plants become susceptible to several fungal pathogens. Intercropping plays an important role by absorbing the excess amount of accessible nitrogen by the plant from soil. Xu et al. (2008) described that the planting of brassica leafy vegetable before transplantation of tomato seedling, absorbs the surplus nitrogen. As a result, the tomato plants showed fewer incidences of leaf blight and higher fruit growth.

9.5.2 USE OF BOTANICALS

Higher plants are used as biopesticides as they are recognized for their valuable pesticidal properties. Extracts of onion, garlic, eucalyptus, tobacco, and neem are known to control many plant–pathogenic fungi (Sharma et al., 2013; Sharma and Sandhu, 2021) and insects. Pyrethrum is known to be one of the oldest insecticides. Dried and finely crushed, chrysanthemum flowers possess an active ingredient known as pyrethrin and were marketed as insecticides because it acts rapidly on insects and paralyzed them.The ginkgo leaves extract repels adult whiteflies (80%) and killed nymphs (65%) and it is safe to natural enemies. Likewise, in similar way, the spray of neem is found efficacious to manage fungi, thrips, aphids, caterpillars, beetles, leaf miners, and others and has no adverse effect on frog (*Hylaarborea japonica*), spider (*Lycosa pseudoannulata*), ladybird beetle (*Hippodamia convergens*), and parasitoid wasp (*Eretmocerus sp.* nr. *californicus*) (Xu et al., 2008). The neem seed kernel extract (NSKE) shows significantly good results against more than 100 pest species of different crops, comprising aphids, Gypsy moth, confused flour beetle, Mexican bean beetle, citrus mealy bug, tobacco budworm, etc. (Chandrashekara et al., 2012). Pyrethrum extract are also known to repel adult whiteflies and with the application of *Zanthoxylum* on plants, the aphids are not killed but are only knocked out and fell off in soil (Xu et al., 2008).

9.5.3 BIOLOGICAL CONTROL OF COMMON GREENHOUSE DISEASES

The disease management strategies of greenhouse can be suitably altered as per the requirement due to its protected nature and its microclimatic condition. There is lack of specific chemicals that are particularly suitable for the greenhouse conditions. The chemicals that are used, pose great health hazards to the workers because of fumes in the closed environment. As the breakdown, weathering, and wash-off of chemicals on the leaves or in substrates are lower in greenhouses as compared to the field, due to which fungicides may have a longer residual activity. The limited choice of fungicides has also led to development of fungicide resistance in the pathogens. Thus, the greenhouse and the protected structures offer a unique opportunity for development and experimenting with biological control agents.

Several formulations of either of fungi *Gliocladium/Trichoderma* or the bacteria *Pseudomonas* and *Bacillus* have been widely used as bioagents against soil-borne pathogens. These products are registered as biofungicides but are also used as plant strengtheners (Beniwal, 2015; Sharma, 2018). Brief account of biocontrol agents available for green house is as follows:

9.5.3.1 BIOLOGICAL CONTROL OF SOIL-BORNE PATHOGENS

As mentioned by Butt et al. (2001), there are over 80 products for biocontrol of soil-borne pathogens worldwide and most of them are formulations either of the fungi *Gliocladium/Trichoderma* or the bacteria *Pseudomonas* and *Bacillus*.

Coniothyrium minitans: This mycoparasite is known to destroy *Sclerotinia sclerotiorum* and *S. minor* sclerotia. It is used for the control of Sclerotinia wilt of lettuce in greenhouse (Budge et al., 1995).

Gliocladium virens (=*Trichoderma virens*): It was isolated in the late 1980s from the soil in Maryland and its presence has been reported throughout the world. It was developed for control of *Pythium ultimum* and *Rhizoctonia solani* in soilless mixes (Lumsden and Locke, 1989). This fungus produces two fungitoxic compounds such as glioviren and gliotoxin (Beniwal, 2015).

Trichoderma harzianum Strain T-22: This strain of *T. harzianum* has been extensively tested in greenhouse and field trials (Harman, 2000). Being a rhizosphere competent strain, this fungus can colonize all parts of the root system and remain there for a longer duration when applied as a seed treatment, greenhouse soil drench or granules, or in-furrow drench or granules. This strain has shown to reduce Fusarium root rot and Crown rot of tomato when grown in potting mix containing T-22 and transplanted into the field. The T-22 strain controlled *R. solani* in poinsettia, geraniums and *Catharanthus*, and *Pythium* on impatiens, geraniums, and petunias in greenhouse trails. It must be applied as a preventive measure before occurrence of the disease (Datnoff et al., 1995; Nemec et al., 1996). This fungus inhibits the pathogens by mycoparasitism via production of chitinases, beta 1–3 glucanases and beta 1–4 glucanases, antibiotics, competition, induced resistance, solubilization of inorganic plant nutrients, and inactivation of the pathogen's enzymes that play significant role during infection and invasion (Sharma, 2013; Beniwal, 2015; Sharma, 2018).

Streptomyces griseoviridis Strain K61: This fungus was originally isolated from sphagnum peat and was first tested as a biocontrol agent against Fusarium wilt of carnations in commercial greenhouses. This strain reduced the spread of disease and increased yield (Lahdenper, 1987). *S. griseoviridis* is marketed under the name Mycostop in Europe and the USA (Beniwal, 2015).

Gliocladium catenulatum Strain J1446: It is mainly effective against fungus that cause damping-off, root rot, seed rot, and wilt. It is used for greenhouse as well as indoor plants. It is available as wettable powder formulation that is suitable for soil, roots, or foliage application (McQuilken and Mohammadi, 1997). Incorporation of this strain into the growing mix or drenching in ornamental bedding plants reduced damping-off caused by *Pythium* and *Rhizoctonia.* In some cases, *G. catenulatum* has been found to be as effective as the fungicides, propamocarb, or tolclofos (Beniwal, 2015).

Nonpathogenic *Fusarium oxysporum* Strain Fo47: This strain is effective against Fusarium wilt diseases on carnation (Postma and Rattink, 1992), tomato (Fuchs et al., 1999) and Fusarium root rot and crown rot on tomato (Lemanceau and Alabouvette, 1991).

To get rid of soil-borne pathogens in the greenhouse, the properly composted aged manure should be used. This manure should be treated with consortia of bioagents for proper mineralization as well as the multiplier effect of bioagents for effective control of soil-borne pathogens. The manure should be covered and kept away from garbage contamination by pests. It should be stored in a clean, dry location, separated from pesticides, and properly labeled to reduce the risk of environment contamination. All the equipment used should be clean. Avoid the manure application if produce is nearing maturity or being harvested and fertilizer should be applied as per need of crop which is related with existing nutrition in soil (Sabir et al., 2010).

9.5.3.2 BIOLOGICAL CONTROL OF FOLIAR PATHOGENS

Ampelomyces quisqualis: This is the first organism reported to be a hyperparasite of powdery mildews (Yarwood, 1932) and known to antagonize fungal species in the orders Erysiphales, Mucorales, and Perisporiales. This antagonist is effective only under very high humidity.

Trichoderma harzianum Strain T-39: The strain T-39 of *T. harzianum*, is targeted at *Botrytis cinerea*. The formulation Trichodex of this strain has been reported to effectively control diseases caused by *Botrytis* sp. affecting greenhouse crops in Israel (Elad et al., 1993) and many other nations (Elad et al., 1995) where it is registered for agricultural use. It is known to be the first commercially available greenhouse product. Like other bio-control agents, environmental conditions affect the efficacy of *T. harzianum* T-39. As mentioned by Paulitz and Belanger, 2001 the integration of this bio-agent with other disease control practices has saved 60% of the chemical sprays, which were replaced by Trichodex.

Bacillus subtilis Strain QST713: The strain QST713 of *B. subtilis* is currently available in the formulation of wettable powder. This is effective against about 40 plant diseases including common greenhouse diseases such as damping-off (*P. ultimum* and *R. solani*), gray mold (*B. cinerea*), and powdery mildews (Paulitz and Belanger, 2001).

Ulocladium atrum: Application of the spore suspension of *U. atrum* reduces the infection of *B. cinerea* on greenhouse tomatoes and other greenhouse crops such as geranium and roses (Kohl et al., 2000; Gerlagh et al., 2001).

9.5.3.3 BIOLOGICAL CONTROL OF NEMATODES

Some of the entomo-pathogenic nematodes (EPNs) are known to be effective against *Meloidogyne* spp. Most of them show some level of reduction in RKN populations, infection rates, or reproduction (El-Ashry et al., 2018).

Purpureocillium lilacinum (also known as Paecilomyces lilacinus), a soil-dwelling fungus that can attack and parasitize females of nematode, their eggs and juveniles, and thus found to decrease population density of RKN in the soil. Population of cotton aphid *(Gossypium hirsutum)* is also reduced when it lives endophytically inside the cotton plant. Numerous countries have developed the bioformulation of this fungus strains as a nematicide for biocontrol of nematodes (Kepenekcia et al., 2018). BioNematon (*P. lilacinum*) is a commercial product developed by the T. Stones company (Tamil Nadu, India) and was introduced to Turkey to control RKN of vegetable crops. BioNem WP is an another bioproduct of Agro-Green (Minrav, Ashdod, Israel) containing bacterium *Bacillus firmus* which is effective against *M. incognita, M. hapla, Tylenchulus semipenetrans, Heterodera* spp., *Ditylenchus dipsaci*, and *Xiphinema index* (Keren-Zur et al., 2000). Terefe et al. (2009) in Ethiopia, BioNem was evaluated against RKN of tomato under greenhouse and laboratory where results exhibited that BioNem was able to reduce number of galls and eggs in each egg mass, and final population of nematode too; moreover, it increased the plant height, fresh and dry shoot weight.

It was also observed that integration of biological control agent *P. lilacinum* and plant growth promoting rhizobacteria, *Pseudomonas fluorescens* along with farm yard manure reduces the incidence of root-knot disease of okra and also increase plant height (20%) and enhance marketable fruit yield (71%) as compared to individual treatments and/or control (Singh et al., 2014).

9.5.3.4 BIOLOGICAL CONTROL OF GREENHOUSE DISEASES IN INDIA

In India, the research in the field of Integrated Pest Management could not reach up to the progress and problems of plant protection at large and greenhouse cultivation in particular. *Trichoderma* spp. and *Pseudomonas* spp. are the most common bioagents used against *Pythium* spp, *Fusarium oxysporum, Rhizoctonia solani, Phytophthora* spp., *Sclerotium rolfsii*, and nematodes. Major diseases of greenhouse crops and their biocontrol in India are compiled in Table 9.3.

TABLE 9.3 Biocontrol Agents (Spray Application) for Important Diseases/Pathogens of Greenhouse Crops in India.

Origin	Biopesticide	Biocontrol organism	Vegetable/fruit/ornamental crop (s)	Target disease/pathogen
Spore powder	AQ 10	*Ampelomyces quisqualis*	Cucurbits, grapes, strawberries, tomatoes, and ornamentals	Powdery Mildew
	Ketomium	*Colletotrichum globosum and Colletotrichum cupreum*	Vegetables	Leaf Spot
	Plant Shield, Top shield and Trichodex	*Trichoderma. Harzianum*	Vegetables and ornamentals	Gray Mold and Fusarium Wilt
	Serenade Cercospora	*Bacillus subtillus*	Cucurbits, grapes, vegetables, and ornamentals	Powdery Mildew, Downy Mildew, Leaf Spot, early and Late Blight
	Sporodex	*Pseudozyma flocculosa*	Cucumber, tomato, and rose	Powdery Mildew and Fusarium Wilt
Spore suspension	Cease	*Bacillus subtilis*	Vegetables and Ornamentals	Anthracnose, Powdery Mildew, Leaf Spot, Gray Mold (Xanthomonas, Pseudomonas)
Pellets: bacterial cells	Bio-save	*Pseudomonas syringae*	Vegetables and ornamentals	Gray Mold
Wettable powder: bacterial cells	Blight Ban	*Pseudomonas fluorescens and Erwinia amylovora*	Tomato and strawberry	Russet inducing bacteria

Source: Adapted from Beniwal (2015).

9.5.4 BIOONTROL AGENTS (BCA) FOR COMMON GREENHOUSE INSECT-PESTS

BCAs are of utmost importance and widely used against several insect pests under protected cultivation. Some of the vital bioagents utilized to manage insect-pests in greenhouse are described below and listed in Table 9.4.

TABLE 9.4 Major Biocontrol Agents of Key Pests of Greenhouses.

Pest	Agent	
Mite(s)	**Predator**	*Neoseiulis cucumeris* *Orius laevigatus* *Phytoseiulis persimilis*
Thrip(s)		*Neoseiulis cucumeris* *Orius laevigatus*
Whiteflies		*Chrysoperla spp* *Orius laevigatus*
Aphid(s)		*Apidoletes aphidomyza* *Chrysoperla spp.* *Oriuslae vigatus*
Caterpillar(s) (*Spodoptera sp.*)		*Chrysoperla carnea*
Whiteflies	**Parasitoid**	*Eretmocerus mundus Encarsia formosa*
Leafminer		*Dacnusas iberica* *Diglyphus isaea*
Aphid(s)		*Aphidus colemani* *Aphidus matricarae*
Caterpillar(s) (*Spodoptera sp.*)		*Trichogramma spp.*
Whiteflies	**Entomopathogens**	*Beauveria bassiana* *Verticillium lecanii*
Leafminer		*Bacillus thuringiensis*
Caterpillar(s) (*Spodoptera sp.*)		*Bacillus thuringiensis, Sl*NPV, *Ha*NPV

Source: Adapted from Rathee et al. (2018).

9.5.4.1 BIOCONTROL AGENTS FOR WHITEFLIES

Biological control of whiteflies in greenhouse environment is mainly based on the use of parasitoids, predators, and pathogens.

9.5.4.1.1 Parasitoids

Encarsia formosa: The chalcidoid wasp *Encarsia formosa* is the well-known natural enemy of whitefly. It is among those natural enemies which are commercialized maximally for biological control of whiteflies on ornamental and vegetable crops in greenhouses and nurseries. Adult is dark headed with yellow abdomen and black thorax measuring less than 1 mm in length. It is effective mostly at higher temp (>70F). Females of the adult wasp feed on all juvenile stages of the insect with the help of their ovipositors and consume the exuding blood. Periodic introductions and "pest in first" approaches are used for introducing *E. formosa* inside greenhouse. Most suitable time to release the wasps is when the third juvenile stage of whitefly has developed on the plants or alternatively they can be periodically released after the establishment of crop. To achieve successful control, there may need 10 introductions of the wasp. The wasps can also be released when first adult of the whitefly is observed on the plants. This is the most adopted technique for releasing the wasps. During cultural practice like removal of lower foliage, extra caution should be taken as pupation wasps usually progresses here which may otherwise be thrown out along with discarded lower leaves. Hence, removal of the lower leaves of greenhouse crop should be done only after emergence of parasitoids.

Eretmocerus spp.: *Eretmocerus mundus* Mercet is a solitary ecto-endo-parasitoid of whitefly nymphs. Female lays eggs between the surface of the leaves and the host nymph. It is an effective parasitoid against greenhouse whitefly and has shown to give good control of *B. tabaci* in greenhouse tomatoes in Spain (Arno et al., 2005). *Encarsia spp.* could invade the greenhouses where the parasitoid *E. mundus* had been released, and in some cases became the predominant species (Weintraub et al., 2002).

9.5.4.1.2 Predators

Amblyseius swirskii: It is an useful predatory mite commonly found in the Eastern Mediterranean region. The egg and too early instar of whitefly are consumed by it and highly effective for biological control in the greenhouse and nursery crops. *Amblyseius swirskii* is pear shaped and measures 5.0 mm long. The eggs are round and transparent and measure up to 0.14 mm in diameter. These predatory mites are used to manage the whiteflies and thrips by many growers (Messelink et al., 2008). These mites also feed on spider mites but juvenile stages of whiteflies and thrips are preferred to prey (Xuenong and Enkegaard, 2009).

Delphastus catalinae: Delphastus catalinae Horn. (earlier, *D. pusillus* (LeConte) (Coleoptera: Coccinellidae) is an effective predator of whitefly and feeds on its eggs and nymphs which effectively reduces localized high population density of whiteflies. Temperature ranges from 22 to 26°C shows better results when it feeds on *B. tabaci* (Legaspi et al., 2008). The optimal temperature ranges from 25 to 30°C (Kutut and Yigit, 2007) and relative humidity around 95% (Simmons et al., 2008) are found suitable for mass rearing of this predator.

Orius laevigatus: Predatory genera *Orius* (Hemiptera: Anthocoridae) are commonly occurring on vegetables. The predator, *Orius laevigatus* is a most extensively used for the biocontrol of thrips and female bugs can consume about 12 thrips a day. The bug feeds on non-prey food such as pollen and can reproduce which helps them to build up populations before the presence of the pests. There is huge possibility for their contribution to manage *B. tabaci* with co-occurrence of thrips (Arnó et al., 2008).

Nesidiocoris tenuis: It is another voracious predatory insect belongs to miridae family which has been showed its effectiveness to manage population of whiteflies on tomatoes under protected environment. This mirid feeds on whiteflies, leafminers, spider mites, and premature juveniles of Lepidopteran pests (Torreno, 1994; Carnero et al., 2000; Urbaneja et al., 2005). When its prey is not available, it starts to feed on plants, inhabiting naturally on tomato crops which help significantly in controlling the whitefly (Calvo and Urbaneja, 2003; Sanchez et al., 2003).

9.5.4.1.3 Fungal Entomopathogens

Fungi that parasitize the insects have been recognized as important natural enemies of insect pests which can be used as effective biological control agents against agricultural pests. It shows huge prospective to manage white-fies in greenhouse conditions (Faria and Wraight, 2001). Entomopathogenic fungus penetrates the host through the cuticle and kills the host by releasing toxins and deficiency of nutrients which are soluble in hemolymph. Therefore, entomopathogenic fungi are potential candidates to control insects. Some species of *Lecanicillium* (=*Verticillium*), *Isaria* (=Paecilomyces), and *Beauveria* are known to use as for the control of whitefly in greenhouses. BotaniGard (BioWorks Inc), a commercial product of *Beauveria bassiana* has been formulated to use on conventional crops, and another product Mycotrol OR has also been approved for organic production. *Isaria fumosoroseus* is available commercially as PFR-97 for controlling a number of arthropod pests including whitefly which has been now registered to use in greenhouses

(Certis USA, 2012). The application of these biopesticides can be done with equipment used to spray any foliar pesticide. However, there are some special considerations while using BCAs. These products and formulations are found incompatible with some synthetic fungicides and are damaged by numerous products of botanical origin having antifungal or fungicidal property. Slow mode of action is the main issue with these EPF though efficacy may be very high. Additionally, they may also be slower to kill the insect as compare to chemicals. These products are usually thought to be compatible with majority of natural enemies; however, the time for the release of natural enemy may be crucial after spray which may also depends on host (Mahr et al., 2001).

9.5.4.2 BIOCONTROL AGENTS FOR APHIDS

Parasitoids of aphids fall in Aphidiidae and Aphelinidae (Hymenoptera). Even though, several species of parasitoids are reported from open field which attack on aphids, however, very less among them exhibited their effectiveness under green-house.

9.5.4.2.1 Parasitoids

Aphidius colemani: This wasp specifically parasitizes the aphids of melon or cotton (*Aphis gossypii*) and the green peach aphid (*Myzus persicae*). This is extensively used in ornamental and greenhouse vegetable crops. It is also effective through open rearing techniques (banker plants) using maize, wheat, and ryegrass that are infested with those species of aphids that do not attack on the vegetables. This strategy has a high prospective for the introduction of these parasitoid wasps under glasshouse conditions on cucumber crop (Jacobson and Croft, 1998) and release of 400–2000 adult parasitoids is suggested for one acre area (Rathee et al., 2018).

Aphelinus abdominalis: These parasitoid wasps are able to attack on more than 200 species of aphids (Holler and Haardt, 1993). The egg laying occurs once into the nymphs of targeted aphid of middle age which are hatched after 2–3 days and the hatched larvae of parasitoid wasp start to feed on the host aphid without killing it. After 7 days, the larva pupates at 20°C and turns it into a black "mummy". Such types of mummyfied larvae appear first after 2–3 weeks of introduction of the parasitoid. It is recommended for the release of two to four adults of this parasitoid wasp for every 10^2 feet at weekly intervals or till 80–90% aphids' parasitization.

9.5.4.2.2 Predators

Aphidoletes aphidimyza: It is widely applied in vegetable crops for the biological control of aphids but rarely used in ornamental crops due to development of huge population of aphids on plants before fruitful biological control (Parrella, 2017). Its larva feeds more than 80 species of aphids (Yukawa et al., 1998) and feeding is done with the help of their mouth parts by penetrating the body of aphids. After piercing, body fluid of aphid is drawn out via sucking and aphid is paralyzed due to all of this. Efficacy of *Aphidoletes aphidimyza* has been confirmed to control *M. persicae* on vegetables such as tomato, brinjal, sweet pepper, etc. (Markkula et al., 1979; Meadow et al., 1985). In greenhouse with plastic covered floor, establishment of predator was slow because it acts as obstacle to pupate the larvae. Due to aforesaid reason, another material like peat may be used (Gilkeson, 1990a).

Green lacewing (*Chrysoperla carnea*): The green lacewing (*Chrysoperla carnea*), a voracious predator is an important natural enemy of several pest species as it is also tolerant to several pesticides. Its larval stage feeds on aphids while pollens and honey dew are consumed by adult stages. The larva is occasionally called as aphid lion which eats hundreds of (100–600) aphids.

Ladybird Beetle (*Adalia bipunctata*): It is a predatory ladybird beetle which is commonly occurring and distributed in Central Asian, European, and North American continents. They are one of the main beetles whose adults and the larvae are predacious on aphids. Though, predation on aphids is essential for beetle to produce the eggs (Sundby, 1968). Mainly the last larval stage during development, it is observed that hundreds of aphids are consumed by a single larva. Release of coccinellids may be more suitable during that stage when aphid infestation on the plant is observed in huge quantity, and such type of situation may occur when other control measures become fail (Hodek and Honk, 1996).

9.5.4.2.3 Entomopathogenic Fungi

Several entomopathogenic fungi are known to infect and kill aphids, but *Isaria fumosoroseus* is the only one that was developed as a commercial biological control agent with the brand name of PFR-97 20% WDG. This formulated product has been registered for the greenhouse crops like ornamentals, vegetables, melons, and strawberries, even including the greenhouse nursery plants those are transplanted in open fields.

9.5.4.3 BIOCONTROL AGENTS FOR LEAF-MINER

Parasitoids are considered the most common natural enemies of leaf-miners. The commercial species of parasitoids to control leaf-miner are *Dacnus asibirica* Telenga (Hymenoptera: Braconidae), *Diglyphus isaea* Walker (Hymenoptera: Eulophidae), and *Opius palipes* Wesmael (Hymenoptera: Braconidae). Being an ectoparasitoid, egg laying by *D. isaea* is observed beside the larvae of leaf-miner in the mine where its larvae can be evidently noticed by backlighting. The body of first and second instars larvae of leaf-miner is pierced by female parasitoids through their ovipositors and the excreted fluid is consumed as food (Ode and Heinz, 2002).

9.5.4.4 BIOLOGICAL CONTROL AGENTS OF THRIPS

The main biological control agents of thrips under protected cultivation are predatory mites, heteropteran bugs, and entomopathogenic fungi, but also nematodes and parasitic wasps. Phytoseiidae is the family of predatory mites which is well known for biological control of thrips. The phytoseiid predators were primarily considered to be mostly related to spider mites but reported its effectiveness to manage thrips. The reasons for their effectiveness relate to their potential to utilize alternative food sources and their numerical response to the prey population (Sabelis, 2008). A major biological attribute that makes phytoseiids useful in biological control is their utilization of pollen. Their phytophagous habits enable these predators to maintain populations on the plants in the absence of thrips, and therefore, existing populations can control thrips at the initial phase of crop colonization. For proper establishment, release of predators is done within sachets in the greenhouse, which are rearing units, where stored product mites are available as food for the predators. This technique enabled the use of the predators on plants, such as eggplant or cucumber, which are less efficient pollen producers.

9.5.4.5 BIOLOGICAL CONTROL AGENTS OF MITES

Biocontrol of spider mite is mainly based on the use of predatory phytoseiid mites, the most important of which are, *Phytoseiulus persimilis* and *Neoseiulus californicus* McGregor. *Phytoseiulus persimilis* is commonly used against *T.*

urticae worldwide on cucumber, tomato, and sweet pepper (van Lenteren and Woets, 1988). *Phytoseiulus persimilis* is capable of reducing large mite populations very rapidly, but it cannot survive for longer duration in greenhouses (Casey and Parrella, 2005) and hence, regular release of this mite is required. However, *M. occidentalis* is able to control mite population for longer time even at comparatively low density, and therefore, one inoculative release of this predatory mite supposed to be sufficient. Other predatory phytoseiid mites that can be used in the control of spider mites are *Neoseiulus barkeri*, *N. californicus*, and *N. fallacis*.

9.5.4.6 BIOLOGICAL CONTROL AGENTS OF CATERPILLAR

9.5.4.6.1 Parasite

Trichogramma wasp attacks the eggs of more than 200 species of moths and butterflies. These microscopic wasps prevent the crop damage by killing their host before the pest can cause the damage to plant. But in greenhouse condition extent of infestations are minimal and sporadic and hence egg parasites are not very effective. However, if large caterpillar infestations seen on regular basis, then *Trichogramma* release can be done at the first sign of moth flight through peak egg laying can provide control (Mahr et al., 2001; Kalyanasundaram and Kamala, 2016).

9.5.4.6.2 Pathogen

Bacillus thuringiensis can provide a good control of common greenhouse caterpillar pests (Kalyanasundaram and Kamala, 2016). *Commonly known as* Bt, it is a soil inhabiting bacterium that is highly selective and active mainly against caterpillar. The variety *kurstaki* and *azawi* are available in commercial formulations for control of caterpillars (Mahr et al., 2001).

9.5.4.6.3 Predator

Larvae of commercially available green lacewing *Chrysoperla (=Chrysopa) carnea* and *Chrysopa rufilabiris* are voracious predator that will feed on moth eggs and very small caterpillars (Mahr et al., 2001).

9.6 CHALLENGES AND FUTURE OF BIOLOGICAL CONTROL UNDER PROTECTED CULTIVATION

Biological control has been a widely acceptable green approach to pest and disease management system in greenhouses and fields. Greenhouses account for 0.02% area used in agriculture (Beniwal, 2015), whereas the successful use of biological control of greenhouse insect-pests and diseases in different crops has documented for several years (Paulitz et al., 2001; Pilkington et al., 2010) which has showed its effectiveness and reliability. Despite several advantages of biological control, its use is still limited and not increasing. According to Van Lenteren (2011), despite of 230 natural enemies which are commercially available, guiding principles of quality control, shiping, and releasing methods, suitable guidance for growers, and a very fruitful track record, there is a diminutive area under biological control which does not seem to be expending. The adoption of biocontrol strategy in protected farming is more complicated due to the maintenance of high aesthetic value and satisfying the quarantine requirements. Most of the factors limiting biological control in greenhouses are general and related to all types of biological control in field conditions.

The biological control agents have slow action and unlike chemical control lack the immediate effect (Kok, 1999; Chandrashekara et al., 2012). By the time, bioagent controls the pest populations, the pests may grow to an intolerable level that can make the introduction of biological control insignificant for a grower (van Emden, 2004). The intense use of pesticides is the only way to overcome risk of invasive species of pests or pathogen which creates an environment that is lethal for other arthropod natural enemies. Globally, it is found that till date there is no natural enemy which has been approved for invasive pest. The quantitative number of new natural enemies which may probably use in biological control practices has also been restricted due to constraints limiting the worldwide movement of natural enemies. The natural enemies of pests and other biological agents are affected by adverse environmental conditions. The fungal and bacterial bioagents require specific temperature and humidity conditions to establish themselves in crops. The controlled environment under protected cultivation can be an advantage for the establishment of biocontrol agents. The bioagents lack quick mode of action as they need time to establish themselves and if there is a sudden surge of pests it is hard to manage them with biological control, even with advanced monitoring systems. Moreover, multiple pests and diseases may attack simultaneously on many different crops typically grown in the greenhouse. All of this makes biological control to be more difficult since numerous different biocontrol agents have to be released which may increase the costs unlike the use of single and broad-spectrum category of pesticide which may control the several pests by spraying it once. Several natural enemies may also show competition and may interfere with other and/or intra-guild predation, this also makes biological control more complex under protected cultivation (Messelink and Jansen, 2014).

Different crops, their pests, and natural enemies may interact diversely on different plant cultivars or species, and their interactions seem to be more intricate as plant diversity increases. There is no alternative method apart from the use of pesticide to control harmful pest among the pest complex against which bioagents are ineffective, although it disrupts the biological control phenomenon. Natural enemies are particularly vulnerable to broad-spectrum category of insecticides with a period of longer residual toxicity whereas certain pesticides on the other hand, having short residual toxicity, may show potential of compatibility and work well with natural enemies. The growers when provided the details of safety and specificity of new molecules switch back again to the pesticide program. The distributor influence on growers toward the use of chemicals is mainly due to its reduced cost and good margin from the sale. This also reduces the prospect of biological control in the greenhouse industry.

There are also challenges of introduction of pests in new areas through expanding trade and invasiveness due to climate change. However, in the future, biological control agents should be mostly searched for among indigenous natural species especially in the case of natural enemies, due to the potentially negative effects to the native fauna if exotic natural enemies are released in a new area. Conservation biological control is another perspective as several natural enemies occur within the green agro-ecosystems. The exploitation of beneficial organisms would be worthwhile due to the low cost of application and the enhancement of sustainability, which is a major topic in modern agriculture strategies. Research focusing on new species may become a major topic, particularly due to the use of biological control in new areas with variable climatic conditions. Investigations of new biotypes of already commercialized bioagents will continue further so as to provide improvements in the efficacy of biological control programs. The effort to reduce the cost of application of biological control through the development of more efficient mass production methods of bioagents will be also the main objective. Biological control will continue to provide solutions to pest problems in greenhouse crops in order to serve public awareness of enhanced sustainability in food production (Perdikis et al., 2008). To get best output of commercial natural enemies applied in protected high-value crops, the greenhouse climate should be kept optimal to achieve their actual functioning (van Lenteren, 2012) and also favors the management of diseases with biological control agents (Beniwal, 2015) as the conditions in the greenhouse can be adjusted for biocontrol agents.

The key for efficient use of biological control agents is to educate the user, farmers or extension people for proper storing and application of biological control agents to achieve adequate disease/pest control. The use of biological control agents requires more specific methodology to manage pest than chemical control. At all the steps information are needed related to the pest's biology, the type of injury they cause, and best time to apply treatment, together with biology and environmental need of the biocontrol agent, and its compatibility with currently available production systems. Biological control has proved mostly effective against many of the serious insect-pests of greenhouse crops including aphids, silver leaf whiteflies, thrips, and mites which are potential vectors of many phytoviruses and diseases caused by *Pythium, Phytophthora, Rhizoctonia, Fusarium* sp., etc.

India having vast diversity of plants requires new biopesticides which can be used for the management of pests and diseases. In current scenario of India, it is needed and appropriate time for the development of indigenous biological control agents (BCA) because they will be cost-effective and environment friendly. The process of commercialization is a key impediment behind the development and application of innovative biopesticides in India. Farmers are hesitant to employ the new items due to their high price and lack of practical understanding (Tripathi et al., 2020). Presently, BCAs are not comparable to chemical pesticides in meeting efficacy which is needed for market expectations, but still these have a promising future if knowledge and advance methods of biotechnology are utilized. The availability of recent advance molecular technologies has facilitated for identification of new candidate agents and helped to interpret the modes of action after field applications. The new technologies like proteomics and functional genomics will provide new possibilities for insights into ecological constraints and will help to predict the unseen possibilities to determine the physiology and expression of crucial genes present within the biocontrol agents during mass production, storage, formulation, and application. Based on the insight of available literature, there is tremendous potential for expansion of biological control. However, all the biological products have slow mode of action and may take considerable time for killing pests as compare to conventional pesticides, but these can give an outstanding and long-lasting control (Parrella and Lewis, 2017). Although there are some promising new BCAs, but still restricted work is done related to extension at academic institutional and/or cooperative level to explore and advance these options.

9.7 CONCLUSIONS

The changes in legislation at national level, development of new policies, adoption of organic farming, and reduction in chemical pesticides are also increasing the scope of biocontrol agents. The researchers worldwide have also been supported to discover new biocontrol agents to strengthen the industry. Therefore, there seems to be great potential for future of biological control in IPM modules for greenhouse and nursery crops. Biological control shows advantages in comparison with conventional chemical control, such as practically zero resistance development, zero risks of pesticide residues in the produce, and environmental protection (Perdikis et al., 2008; Sharma, 2018). With the considerable increase in the adoption of biological control worldwide, it is expected that pest control in greenhouse crops will increasingly rely on biological control.

KEYWORDS

- **biological control**
- **diseases and pests**
- **entomopathogens**
- **natural enemies**
- **protected cultivation**

REFERENCES

Alomar, O.; Gabarra, R.; Castañe, C. The Aphid Parasitoid *Aphelinus Abdominalis* (Hym.: Aphelinidae) for Biological Control of *Macrosiphum Euphorbiae* on Tomatoes Grown in Unheated Plastic Greenhouses. *Bulletin IOBC/WPRS* **1997,** *20* (4), 203–205.

Arnó J.; Matas, M.; Marti, M.; Arino, J.; Roig, J.; Gabarra, R. Coexistence between *Trialeurodes Vaporariorum* and *Bemisia Tabaci* and Impact of Natural Enemies in Tomato Crops Under Mediterranean Conditions. *Bulletin IOBC/WPRS* **2005,** *28* (1), 1–4.

Arnó, J.; Roig, J.; Riudavets, J. Evaluation of *Orius Majusculus* and *O. Laevigatus*as Predators of *Bemisia Tabaci* and Estimation of Their Prey Preference. *Biol. Control* **2008,** *44*, 1–6.
Beniwal, R. S. Diseases Management of Crops in Protected Cultivations/Greenhouses. In *Protected Cultivation of Horticultural Crops*; Singh et al. Eds.; CCS HAU: Hisar, 2015; pp 61–72.
Brødsgaard, H. F.; Albajes, R. Insect and Mite Pests. In *Integrated Pest and Disease Management in Greenhouse Crops*; Albajes et al. Eds.; Kluwer Academic Publishers: Netherland, 1999; pp 34–47.
Budge, S. P.; McQuilken, M. P.; Fenlon, J. S.; Whipps, J. M. Use of *Coniothyrium Minitans* and *Gliocladium Virens* for Biological Control of *Sclerotinia Sclerotiorum* in Glasshouse Lettuce. *Biol. Control* **1995,** *5*, 513–22.
Bueno, V. H. P.; van Lenteren, J. C. Biological Control of Pests in Protected Cultivation: Implementation in Latin America and Successes in Europe, *Memorias, XXXVII Congreso Sociedad Colombiana de Entomologia*; Bogota, D.C., 2010; pp 261–269.
Burger, J. M. S.; Reijen, T. M.; van Lenteren, J. C.; Vet, L. M. Host Feeding in Insect Parasitoids: Why Destructively Feed Upon a Host that Excretes an Alternative? *Entomol. Exp. Appl.* **2004,** *112*, 207–215.
Butt, T.; Jackson, C.; Magan, N. Eds. *Fungi as Biocontrol Agents-Progress, Problems and Potential*; CABI Publishing: Wallingford, UK, 2001; p 416.
Calvo, J.; Urbaneja, A. Nesidiocoris Tenuis Reu. (Het.: Miridae) en to Mate: Amigo o enemigo? *Almeria en Verde* **2003,** *4*, 21–23.
Carnero, A.; Díaz, S.; Amador, M.; Hernández, M.; Hernández, E. Impact of *Nesidiocoris Tenuis* Reuter (Hemiptera: Miridae) on Whitefly Populations in Protected Tomato Crops. *Bulletin IOBC/WPRS* **2000,** *23*, 259.
Casey, C. A.; Parrella, M. P. Evaluation of a Mechanical Dispenser and Interplant Bridges on the Dispersal and Efficacy of the Predator *Phytoseiulus Persimilis* (Acari: Phytoseiidae) in Greenhouse Cut Roses. *Biol. Control* **2005,** *32*, 130–136.
Chandrashekara, K. N.; Manivannan, S.; Chandrashekara, S. C.; Chakravarthi, M. Biological Control of Plant Diseases. In *Ecofriendly Innovative Approaches in Plant Disease Management*; Singh, V. K., Singh, Y., Singh A. Eds.; International Book Distributors Publication: New Delhi, 2012; pp 147–166.
Chinniah, C.; Ravikumar, A.; Kalyanasundaram, M.; Parthiban, P. Management of Sucking Pests, by Integration of Organic Sources of Amendments and Foliar Application of Entomopathogenic Fungi on Chili. *J. Biopest.* **2016,** *9* (1), 34–40.
Criniti, A.; Mazzoni, E.; Cassanelli, S.; Cravedi, P.; Tondelli, A.; Bizzaro, D.; Manicardi, G. C. Biochemical and Molecular Diagnosis of Insecticide Resistance Conferred by Esterase, MACE, kdr and Super-kdr based Mechanisms in Italian Strains of the Peach Potato Aphid, *Myzus Persicae* (Sulzer). *Pestic. Biochem. Physiol.* **2008,** *90*, 168–174.
Datnoff, L. E.; Nemec, S.; Pernezny, K. Biological Control of Fusarium Crown and Root Rot of Tomato in Florida using *Trichoderma Harzianum* and *Glomus Intraradices*. *Biol. Control* **1995,** *5*, 427–31.
Denmark, H. A. Broad Mite, *Polyphagotarsonemus Latus* (Banks). FDACS-DPI Bureau of Entomology, Circular No. 213, 1980, p 2.
Devonshire, A. L. Resisance of Aphids to Insecticides. In *Aphids: Their Biology, Natural Enemies and Control* (Vol 2C); Minks, A. K., Har rewijn, P. Eds.; Elsevier: New York, Amsterdam, 1988; pp 123–135.
Dhandapani, N.; Shelkar, U. R.; Murugan, M. Bio-Intensive Pest Management (BIPM) in Major Vegetable Crops. *J. FoodAgric. Environ.* **2003,** *1* (2), 333–339.

Elad, Y.; Gullino, M. L. Shtienberg, D. Aloi, C. Coping with Tomato Gray Mold under Mediterranean Conditions. *Crop Prot.* **1995,** *14*,105–9.

Elad, Y.; Zimand, G.; Zaqs, Y.; Zuriel, S.; Chet, I. Use of *Trichoderma Harzianum* in Combination or Alternation with Fungicides to Control Cucumber Gray Mold (*Botrytis Cinerea*) under Commercial Greenhouse Conditions. *Plant Pathol.* **1993,** *42*, 324–32.

El-Ashry, R. M.; Eldeeb, A. M.; El-Marzoky, A. M.; Mahrous, M. E. Suppression of the Root-knot Nematode, *Meloidogyne Incognita* in Tomato Plants by Application of Certain Entomopathogenic Nematode Species under Greenhouse Conditions Egypt. *J. Agronematol.* **2018,** *17* (1), 25–42.

Faria, M.; Wraight, S. P. Biological Control of *Bemisia Tabaci* with Fungi. *Crop Prot.* **2001,** *20*, 767–778.

Fletcher, J. T. *Diseases of Greenhouse Plants*; Longman, UK., 1984, p. 351.

Foster, S. P.; Denholm, I.; Devonshire, A. L. The Ups and Downs of Insecticide Resistance in Peach-potato Aphids (*Myzus Persicae*) in the UK. *Crop Prot.* **2000,** *19*, 873–879.

Fuchs, J. G.; Möenne-Loccoz, J. G.; Défago, G. Ability of Non-pathogenic *Fusarium Oxysporum* Fo47 to Protect Tomato against Fusarium Wilt. *Biol. Control* **1999,** *14*, 105–10.

Gavkare, O.; Surjeet, K.; Yuvraj, S.; Bhopale, S. P. Abundance of the *Myzus Persicae* (Sulzer) under Protected Environment in Himachal Pradesh, India. *J. Ind. Pollut. Control* **2014,** *30* (2), 285–287.

Gerlagh, M.; Amsing, J. J.; Molhoek, W. M. L.; Bosker-van Zessen, A. I.; Lombaers-van der Plas, C. H.; Köhl, J. The effect of Treatment with *Ulocladium Atrum* on *Botrytis Cinerea* Attack of Geranium (*Pelargonium Zonale*) Stock Plants and Cuttings. *Eur. J. Plant Pathol.* **2001,** *107* (4), 377–86.

Gerling, D.; Alomar, O.; Arnó, J. Biological Control of *Bemisia Tabaci* using Predators and Parasitoids. *Crop Prot.* **2001,** *20*, 779–799.

Ghongade, D. S. Population Abundance and Management of Whitefly and Red Spider Mite on Parthenocarpic Cucumber Grown under Protected Conditions. Ph.D Dissertation, Punjab Agricultural University, 2020; p 154.

Gilkeson, L. A. Biological Control of Aphids in Greenhouse Sweet Peppers and Tomatoes. *Bulletin IOBC/WPRS* **1990,** *13* (5), 64–70.

Goldammer, T. *Greenhouse Management: A Guide to Operations and Technology*; Apex Publishers: USA, 2019; p 406.

Harman, G. E. Myths and Dogmas of Biocontrol: Changes in Perceptions Derived from Research on *Trichoderma Harzianum* T-22. *Plant Dis.* **2000,** *84*, 377–93.

Heimpel, G. E.; Collier, T. R. The Evolution of Host-feeding Behavior in Insect Parasitoids. *Biol. Rev.* **1996,** *71*, 373–400.

Heinz, K. M. Nelson, J. M. Interspecific Interactions among Natural Enemies of *Bemisia* in an Inundative Biological Control Program. *Biol. Control* **1996,** *6*, 305–318.

Hodek, I.; Honk, A. *Ecology of Coccinellidae*; Kluwer Academic Publishers: Dordrecht, 1996, p 480.

Ibrahim, L.; Hamieh, A.; Ghanem, H.; Ibrahim, S. K. Pathogenicity of Entomopathogenic Fungi from Lebanese Soils Against Aphid, Whitefly and Non-target Beneficial Insects. *Int. J. Agr. Sci.* **2011,** *3* (3), 156–164.

Jacobson, R. J.; Croft, P. Strategies for the Control of *Aphis Gossypii* Glover (Hom.: Aphididae) with *Aphidius Colemani* Viereck (Hym.: Braconidae) in Protected Cucumbers. *Biocontrol Sci. Technol.* **1998,** *8*, 377–387.

Javed, K.; Javed, H.; Mukhtar, T.; Qiu, D. Pathogenicity of Some Entomopathogenic Fungal Strains to Green Peach Aphid, *Myzus Persicae* Sulzer (Homoptera: Aphididae). *Egypt J. Biol. Pest Control* **2019,** *29* (92), 1–7.

Jones, W. A.; Greenberg, S. M. Suitability of *Bemisia Argentifolii* (Homop Tera: Aleyrodidae) Instars for the Parasitoid *Eretmocerus Mundus* (Hymenop Tera: Aphelinidae). *Environ. Entomol.* **1998,** *27*, 1569–1573.

Kalyanasundaram, M.; Kamala, I. M. Parasitoids. In: *Ecofriendly Pest Management for Food Security*, Omkar Ed.; Academic Press: USA, 2016; pp 109–138. http://dx.doi.org/10.1016/B978-0-12-803265-7.00004-X.

Kepenekcia, I.; Hazirb, S.; Oksalc, E.; Lewisd, E. E. Application Methods of *Steinernema Feltiae, Xenorhabdus Bovienii* and *Purpureocillium Lilacinum* to Control Root-knot Nematodes in Greenhouse Tomato Systems. *Crop Prot.* **2018,** *108*, 31–38.

Keren-Zur, M.; Antonov, J.; Bercovitz, A.; Feldman, K.; Husid, A.; Kenan, G.; Markov, N.; Rebhum, M. In Baccillus Firmus *Formulation for the Safe Control of Root Knot Nematodes*. Proceedings of the Brighton Crop Protection Conference on Pests and Diseases 2A., **2000,** pp. 47–52.

Kohl, J.; Gerlagh, M.; Grit, G. Biocontrol of *Botrytis Cinerea* by *Ulocladium Atrum* in Different Production Systems of Cyclamen. *Plant Dis.* **2000,** *84*, 569–73.

Kok, L. T.; Kok, V. *Biological Control for the Public*, Publications: Virginia Cooperative Extension (VCE), Virginia Tech, and Virginia State University, 1999. http://hdl.handle.net/10919/18720

Kutuk, H.; Yigit, A. Life Table of *Delphastus Catalinae* (Horn) (Coleop Tera: Coccinellidae) on Cotton Whitefly, *Bemisia Tabaci* (Genn.) (Hornoptera: Aleyrodidae) as prey. *J. Plant Dis. Prot.* **2007,** *114*, 20–25.

Lahdenperä, M. L. The Control of Fusarium Wilt on Carnation with a *Streptomyces* Preparation. *Acta Hort.* **1987,** *216*, 85–92.

Legaspi, J. C.; Legaspi, B. C. Jr.; Simmons, A. M.; Soumare, M. Life Table Analysis for Immature and Female Adults of the Predatory Beetle, *Delphastus Catalinae*, Feeding on Whiteflies under Three Constant Temperatures. *J. Insect Sci.* **2008,** *8*, 07.

Lemanceau, P.; Alabouvette, C. Biological Control of Fusarium Diseases by Fluorescent Pseudomonasand Nonpathogenic *Fusarium*. *Crop Prot.* **1991,** *10* (4), 279–86.

van Lenteren, J. C.; Bueno, V. H. P. Augmentative Biological Control of Arthropods in Latin America. *Biocontrol* **2003,** *48*, 123–139.

Lumsden, R. D.; Locke, J. C. Biological Control of Damping-off Caused by *Pythium Ultimum* and *Rhizoctonia Solani* with *Gliocladium Virens* in Soilless Mix. *Phytopathol.* **1989,** *79*, 361–66.

Mahr, S. E. R.; Cloyd, R. A.; Mahr, D. L.; Sadof, C. S. *Biological Control of Insects and Other Pests of Greenhouse Crops*, WI.: North Central Regional Publication, Cooprative Extenson Publishing, 2001; p 100.

Malathrakis, N. E.; Goumas, D. E. Fungal and Bacterial Diseases. In *Integrated Pest and Disease Management in Greenhouse Crops,* Albajes et al. Eds.; Kluwer Academic Publishers: Netherland, 1999; pp 34–47.

Markkula, M.; Tiitanen, K.; Hamalainen, M.; Forsberg, A. The Aphid Midge *Aphidoletes Aphidimyza* (Diptera, Cecidomyiidae) and its use in Biological Control of Aphids. *Ann. Zool. Fenn.* **1979,** *45*, 89–98.

McQuilken, M. P.; Mohammadi, O. Evaluation of a Commercial Formulation of *Gliocladium Catenulatum* (J1446) for Biocontrol of Damping-off in Bedding Plants. *Meded. Rijksfac. Landbouwwet. Univ. Gent.* **1997,** *62*, 987–92.

Meadow, R. H.; Kelly, W. C.; Shelton, A. M. Evaluation of *Aphidoletes Aphidimyza* (Dip.: Cecidomyiidae) for Control of *Myzus Persicae* (Hom.: Aphididae) in Greenhouse and Field Experiments in the United States. *Entomophaga* **1985,** *30*, 385–392.

Mehta, P.; Wyman, J. A.; Nakhla, M. K.; Maxwel, D. P. Transmission of Tomato Yellow Leaf Curl Geminivirus by *Bemisia Tabaci* (Homoptera: Aleyrodidae). *J. Econ. Entomol.* **1994,** *87*, 1291–1297.

Messelink, G. J.; van Maanen, R.; van Steenpaal, S. E. F.; Janssen, A. Biological Control of Thrips and Whiteflies by a Shared Predator: Two Pests are better than one. *Biol. Control* **2008,** *44*, 372–379.

Messelink, G. J.; Jansen A. Increased Control of Thrips and Aphids in Greenhouses with Two Species of Generalist Predatory Bugs Involved in Intraguild Predation. *Biol. Control* **2014,** *79*, 1–7.

Minuto, A.; Minuto, G.; Migheli, Q.; Mocioni, M.; Gullino, M. L. Effect of Antagonistic *Fusarium* spp. and of Different Commercial Biofungicide Formulations on Fusarium Wilt of Basil (*Ocimum Basilicum* L.). *Crop Prot.* **1997,** *16*, 765–69.

Nemec, S.; Datnoff, L. E.; Strandberg, J. Efficacy of Biocontrol Agents in Planting Mixes to Colonize Plant Roots and Control Root Diseases of Vegetables and Citrus. *Crop. Prot.* **1996,** *15*, 735–42.

New, T. R. Neuroptera. In: *Aphids: Their Biology, Natural Enemies and Control (Vol 2B)*, Minks A. K., Harrewijn, P. Eds.; Elsevier: New York, Amsterdam, 1988; pp 249–258.

Ode, P. J.; Heinz, K. M. Host-size-dependent Sex Ratio Theory and Improving Mass-reared Parasitoid Sex Ratios. *Biol. Control* **2002,** *24*, 31–41.

Parrella, M. P.; Lewis, E. Biological Control in Greenhouse and Nursery Production: Present Status and Future Directions. *Am. Entomol.* **2017,** *63* (4), 237–250.

Paulitz, T. C.; Belanger, R. R. Biological Control of Greenhouse System. *Annu. Rev. Phytopathol.* **2001,** *39*, 103–133.

Perdikis, D.; Kapaxidi, E.; Papadoulis, G. Biological Control of Insect and Mite Pests in Greenhouse Solanaceous Crops. *Eur. J. Plant Sci. Biotechnol.* **2008,** *2* (S1), 125–144.

Pilkington, L. J.; Messelink, G.; van Lenteren, J. C.; Le Mottee, K. Protected Biological Control-Biological Pest Management in the Greenhouse Industry. *Biol. Control* **2010,** *52*, 216–220.

Postma, J.; Rattink, H. Biological Control of Fusarium Wilt of Carnation with a Non-pathogenic Isolate of *Fusarium Oxysporum*. *Can. J. Bot.* **1992,** *70*, 1199–205.

Pundt, L. *Some Virus Diseases of Greenhouse Crops*, Integrated Pest Management, University of Connecticut, 2010; http://www.hort.uconn.edu/ipm/greenhs/htms/gcrpvirus.htm

Rathee, M.; Singh, N. V.; Dalal, P. K.; Mehra, S. Integrated Pest Management under Protected Cultivation: A Review. *J. Entomol. Zool. Stud.* **2018,** *6* (2), 1201–1208.

Rattink, H. Biological Control of *Fusarium* Wilt Disease of Carnation by a Non-pathogenic Isolate of *Fusarium Oxysporum*. *Acta Hortic.* **1992,** *307*, 37–42.

Reddy, P. P. *Sustainable Crop Protection under Protected Cultivation*, Springer: Singapore, 2016; p 434.

Sabelis, M. W.; Janssen, A.; Lesna, I.; Aratchige, N. S.; Nomikou, M.; van Rijn, P. J. C. Developments in the use of Predatory Mites for Biological Pest Control. *Bulletin IOBC/WPRS* **2008,** *32*, 187–199.

Sabir, N.; Singh, B. Protected Cultivation of Vegetables in Global Arena: A Review. *Indian J. Agric. Sci.* **2013,** *83* (2), 123–135.

Sabir, N.; Singh, B.; Hasan, M.; Sumitha, R.; Deka, S.; Tanwar, R. K.; Ahuja, D. B.; Tomar, B. S.; Bambawale, O. M.; Khah, E. M. *Good Agricultural Practices (GAP) for IPM in*

Protected Cultivation, Tech. Bull. No. 23, National Centre for Integrated Pest Management: New Delhi, 2010; p 16.

Sanchez, J. A. Zoophytophagy in the Plantbug *Nesidiocoris tenuis*. *Agric. For. Entomol.* **2008,** *10*, 75–80.

Sanchez, J. A.; Gillespie, D. R.; McGregor, R. R. The Effects of Mullein Plants (*Verbascum Thapsus*) on the Population Dynamics of *Dicyphus Hesperus* (Heteroptera: Miridae) in Tomato Greenhouses. *Biol. Control* **2003,** *28*, 313–319.

Scopes, N. E. A. The Potential of *Chrysopa Carnea* as a Biological Control Agent of *Myzus Persicae* on Glasshouse Chrysanthemums. *Ann. Appl. Biol.* **1969,** *64*, 433–439.

Sharma K. K. Growth Promotion of Paddy Influenced by Biological Seed Treatment using Fungal Antagonist. *Trichoderma. Int. J. Plant Prot.* **2013,** *6* (1), 51–55.

Sharma K. K. Trichoderma in Agriculture: An Overview of Global Scenario on Research and its Application. *Int. J. Curr. Microbiol. Appl. Sci.* **2018,** *7* (8), 1922–1933.

Sharma L.; Sharma K. K.; Sinha A. P. Potential Application of Botanicals, Essential Oils and Natural Products against *Sarocladium Oryzae*. *Ann. Plant Prot. Sci.* **2013,** *21* (1), 109–113.

Sharma L.; Shukla S. K.; Jaiswal V. P.; Sharma K. K. Characterization of Cultural Beneficial Bacterial Diversity for Plant Growth Promoting Attributes Associated with Rice Rhizosphere. *J. Environ. Biol.* **2019,** *40*, 1180–1187.

Sharma K. K.; Sandhu P. S. Efficacy of Fungicides and Botanicals against Alternaria Blight of Taramira (*Eruca Sativa* Mill.) in Sub-mountainous Punjab. *Agric. Res. J.* **2021,** *8* (1), 150–153.

Shishido, M. Plant Disease Management in Protected Horticulture. *Hort. Research* **2011,** *65*, 7–18.

Simmons, A. M.; Legaspi, J. C.; Legaspi, B. C. Responses of *Delphastus Catalinae* (Coleoptera: Coccinellidae), a Predator of Whiteflies (Hemiptera: Aleyrodidae), to Relative Humidity: Oviposition, Hatch, and Immature Survival. *Ann. Entomol. Soc. Am.* **2008,** *101*, 378–383.

Singh, H.; Joshi, N. Management of the Aphid, *Myzus Persicae* (Sulzer) and the Whitefly, *Bemisia Tabaci* (Gennadius), using Biorational on Capsicum under Protected Cultivation in India. *Egypt J. Biol. Pest Control* **2020,** *30*, 1–9.

Singh, B. In *Protected Cultivation of Horticultural Crops in India: Challenges and Opportunities*. 3rd International Conference on Agriculture & Horticulture, Hyderabad International Convention Centre, India, Oct 27–29, 2014; http://dx.doi.org/10.4172/2168-9881.S1.010

Singh, B. Integrated Pest Management in Greenhouses. In *Protected Cultivation of Horticultural Crops*, Singh et al. Eds.; CCS HAU: Hisar, 2015; pp 55–60.

Singh, M. C.; Singh, J. P.; Pandey, S. K.; Mahay, D.; Srivastava, V. Factors Affecting the Performance of Greenhouse Cucumber Cultivation - A Review. *Int. J. Curr. Microbiol. Appl. Sci.* **2017a,** *6* (10), 2304–2323.

Singh, M. C.; Singh, J. P.; Singh, K. G. Optimal Operating Microclimatic Conditions for Drip Fertigated Cucumbers in Soilless Media under a Naturally Ventilated Greenhouse. *Indian J. Ecol.* **2017b,** *44*, 821–826.

Singh, M. C.; Singh, J. P.; Pandey, S. K.; Cutting, N. G.; Sharma, P.; Shrivastav, V.; Sharma, P. A Review of Three Commonly used Techniques of Controlling Greenhouse Microclimate. *Int. J. Curr. Microbiol. Appl. Sci.* **2018a,** *7* (1), 3491–3505.

Singh, M. C.; Kachwaya, D. S.; Kalsi, K. Soilless Cucumber Cultivation under Protective Structures in Relation to Irrigation Coupled Fertigation Management, Economic Viability and Potential Benefits - A Review. *Int. J. Curr. Microbiol. Appl. Sci.* **2018b,** *7* (03), 2451–2468.

Singh, S.; Singh, B.; Singh, A. P. Integrated Management of Root-knot Disease of Okra Caused by Root-knot Nematode. *Meloidogyne incognita. Indian J. Nematol.* **2014,** *44* (2), 172–178.

Sood, A. K.; David, B. V. The Greenhouse Whitefly, *Trialeurodes Vaporarriorum* (Westwood). In *The Whitefly or Mealywing Bugs*, David, B. V. Ed.; LAP Lambert Academic Publishing Gmbh & Co. KG: Germany, 2012; pp 147–169.

Sundby, R. A. Some Factors Influencing the Reproduction and Longevity of the Predator *Coccinella Septempunctata* L. (Coleoptera: Coccinellidae). *Entomophaga* **1968,** *13*, 197–202.

Terefe, M.; Tefera, T.; Sakhuja, P. K. Effect of a Formulation of *Bacillus Firmus* on Root-knot Nematode *Meloidogyne Incognita* Infestation and the Growth of Tomato Plants in the Greenhouse and Nursery. *J. Invertebr. Pathol.* **2009,** *100*, 94–99.

Tompkins, C. M.; Middleton, J. T. Etiology and Control of Poinsettia Root and Stem Rot Caused by *Pythium* spp. and *Rhioctonia Solani. Hilgardia* **1950,** *20*, 171–182.

Torreno, H. S. Predation Behavior and Efficiency of the Bug *Cyrtopeltis Tenuis* (Hemiptera: Miridae) against the Cutworm, *Spodoptera Litura* (F.). *Philipp. Entomol.* **1994,** *9*, 426–434.

Tripathi, Y. N.; Divyanshu, K.; Kumar, S.; Jaiswal, L. K.; Khan, A.; Birla, H.; Gupta, A.; Singh, S. P.; Upadhyay R. S. Biopesticides: Current Status and Future Prospects in India. In *Bioeconomy for Sustainable Development*, Keswani C. Eds.; Springer: Singapore, 2020; pp 79–109. https://doi.org/10.1007/978-981-13-9431-7_6.

Urbaneja, A.; Tapia, G.; Stansly, P. Influence of Host Plant and Prey Availability on Developmental Time and Survivorship of *Nesidiocoris Tenuis* (Het.: Miridae). *Biocontrol Sci. Technol.* **2005,** *15*, 513–518.

van Emden, H. F.; Service, M. W. *Pest and Vector Control*, Cambridge University Press: UK, **2004,** p 362.

van Lenteren, J. C. The State of Commercial Augmentative Biological Control: Plenty of Natural Enemies, but a Frustrating Lack of Uptake. *Biocontrol* **2012,** *57*, 1–20.

van Lenteren, J. C.; Woets, J. Biological and Integrated Control in Green Houses. *Annu. Rev. Entomol.* **1988,** *33*, 239–269.

van Lenteren, J. C. The State of Commercial Augmentative Biological Control: Plenty of Natural Enemies, but a Frustrating Lack of Uptake. *Biocontrol* **2011,** *57* (1), 1–20. http://link.springer.com/article/10.1007%2Fs10526-011-9395-1

Vashisth, S. *Insect and Nematode Complex Associated with Some Polyhouse Crops*, CSK Himachal Pradesh Krishi Vishvavidyalaya: Palampur (Unpublished), MSc Thesis, 2009.

Weintraub, P.; Sapira, N.; Chiel, E.; Steinberg, S. Effect of Various Release Schedules of *Eretmocerus Mundus* on the Control of *Bemisia Tabaci* in Organic Greenhouse Peppers, in Israel-preliminary Results. *Bulletin IOBC/WPRS* **2002,** *25* (1), 301–304.

Weintraub, P. G. Review Integrated Control of Pests in Tropical and Subtropical Sweet Pepper Production. *Pest Manag. Sci.* **2007,** *63*, 753–760.

Xu, H.; Xu, R.; Qin, F.; Ma, G.; Yu, Y.; Shah, S. K. Biological Pest and Disease Control in Greenhouse Vegetable Production. *Acta Hort.* **2008**, *767*, 229–237.

Xuenong, X.; Enkegaard A. Prey Preference of the Predatory Mite *Amblyseius Swirskii* between First Instar Western Flower Thrips (*Frankliniella Occidentalis*) and Nymphs of the Two Spotted Spider Mite (*Tetranychus Urticae*). *J. Insect Sci.* **2009**, *10*, 1–11.

Yarwood, C. E. *Ampelomyces Quisqualis* on Clover Mildew. *Phytopathology* **1932,** *22*, 31.

Yukawa, J.; Yamaguchi, D.; Mizota, K.; Setokuchi, O. Distribution and Host Range of an Aphidophagous Species of Cecidomyiidae, *Aphidoletes Aphidimyza* (Diptera), in Japan. *Appl. Entomol. Zool. (Jpn.)* **1998,** *33*, 185–193.

CHAPTER 10

Conventional and Recent Approaches of Integrated Pest Management in Greenhouse Cultivation

SUNAULLAH BHAT and SANDEEP KUMAR

Department of Zoology, Kumaun University-SSJ Campus, Almora, Uttarakhand, India

ABSTRACT

In protected cultivation, natural environment is modified for optimum growth and yield of the crop without any stress to plant. Over the last decade, protected cultivation of vegetables has shown remarkable potential. However, insect-pests are one of the major threats in greenhouse production of the crops at the global level. Due to the presence of hot and humid environment along with abundance of food, greenhouses provide suitable habitat and atmosphere for the development of insect-pest. Often, the natural enemies of the various insect-pests that keep them under check in the natural environment are deprived off under protected atmosphere. For these reasons, insect-pest situations frequently become alarming in the protected cultivation than the outside natural environment. Therefore, crop damage due to arthropod pests in the greenhouses differs from season to season, crop to crop, and pest to pest. Insect-pests like aphids, *Helicoverpa armigera* and *Spodoptera litura*, and whiteflies have become serious concerns in greenhouse production of the crops.

Integrated pest management (IPM) is the systematic technique to overcome insect-pests. It has a range of tactics and approaches to either reduce pest populations or decrease their economic impact on the concerned crop. It is a site-specific approach for controlling pests that depends on correct

Protected Cultivation: Structural Design, Crop Management, Modeling and Automation.
Mahesh Chand Singh and K. K. Sharma (Eds.)

identification of the pests, their relationship with host plant, and pest biology. With a lifelong perspective, one can easily invest on integrated pest management and get a cleaner environment in greenhouse crop production.

10.1 INTRODUCTION

Globally, the vegetable and horticulture production has significantly increased during the last few decades. However, a large gap still exists as many developing countries did not produce sufficient fruit and vegetables to satisfy the nutritional need of their inhabitants. The predicted global population of 9.1 billion will need a 60% surge in agricultural and horticultural production. The main concern due to the change in diet nature during the last few years is that the fruits and vegetables demand is expected to continue (Alexandratos and Bruinsma, 2012). Vegetable and horticulture crop production faces numerous problems, including biotic (insect-pests, diseases, and weeds) and abiotic stresses (drought, floods, and poor soil fertility). As these crops depend on climatic conditions, this leads to visible seasonal variations to supply fruit and vegetables. Due to population upsurge, agriculture land is getting reduced day by day, which favors artificialization of fruit and vegetable cropping system to minimize the effect of biotic and abiotic factors.

The term "protected cultivation" in agriculture covers different methods having similar objective to modify the natural environment to accomplish plant growth (Singh et al., 2017a, 2017b). Protected cultivation of cash crops has shown incredible potential during the last few decades. As far as fruit and vegetable crops under protected cultivation are concerned, China ranks first occupying an area of 2.76 mha, while India stands at seventh position with 25,000 ha of area worldwide (Nair and Barche, 2014). With the advance of new agricultural technologies, such as the cultivation of crops in a controlled environment inside greenhouses, protected cultivation gets boosted worldwide, which is being adopted to achieve maximum production as well as a regular supply of vegetables and fruits, including off-season availability. By implementing protected cultivation technology, the farmer can look forward for better remuneration for high quality produce (Singh et al., 2018a, 2018b).

As the greenhouse provides favorable microclimate for the pest development, the crops grown inside the protected environment are vulnerable to numerous disease and pest attacks, thereby limiting the success of crop production. For decades, this issue has been the subject of active studies, with the biggest obstacle being the inherent trade-off between insect control, financial costs, human health consequences, and environmental risks. As a result, the issue of pest management must be approached holistically, which has led to the emergence of Integrated Pest Management (IPM).

The concept of IPM was introduced in the late 1950s and widely adopted during the 1970s and 1980s. IPM is a long-term management approach which combines biological, cultural, and chemical strategies (Figure 10.1) to combat pest up to tolerable level, with grower spending minimum cost and also safeguarding our environment. IPM also offers applied techniques for the effective control of pests in the greenhouses. With the implementation of comprehensive cultural techniques and monitoring methods, precise identification of problem, and timely application as well as assessment of suitable management approaches, farmers can improve crop production while reducing their dependence on monotonous pesticide applications. The PAMS approach, that is, Prevention, Avoidance, Monitoring, and Suppression is reported to be vital for IPM programs (NCIPMC, 2010). In this context, IPM has a four-tier approach as follows.

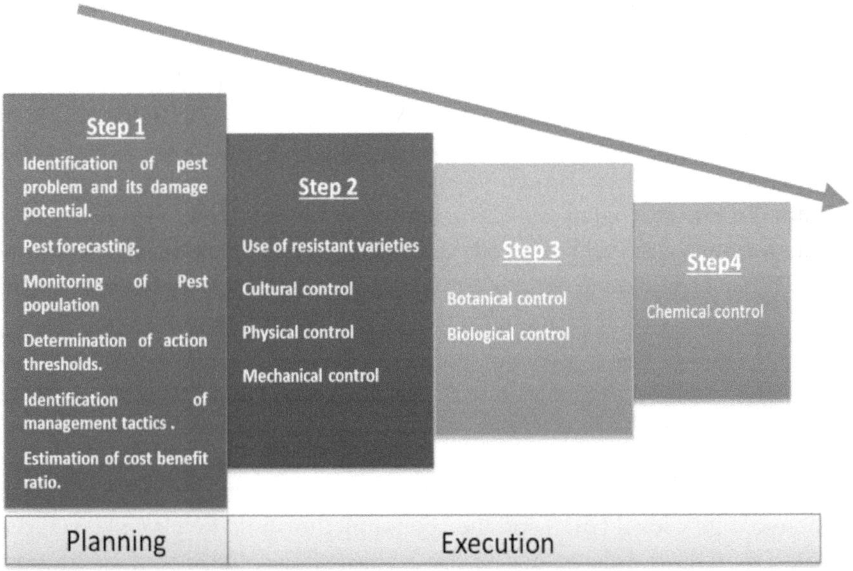

FIGURE 10.1 IPM for protected cultivation.

10.2 PEST IDENTIFICATION AND MONITORING

The first and foremost approach of any IPM is to identify the pest rightly and monitor its damage. Identification of pest by observing the symptoms and confirming it with the associated pest is more important. The damage of the pest is to be monitored at regular intervals, and thus its damage is under scan. Pest surveillance and monitoring is a basic prerequisite in IPM programs (Jayaraj and Rabindra, 1994). Monitoring can be visual or by trapping (McCravy, 2018), which is most preferred to monitor the pest population in protected cultivation, consisting of pitfall or sticky traps. Traps used for monitoring may be chromotropic (i.e., colored traps) or with attractants, for example, food baits or pheromone traps (i.e., baited traps). Another way to monitor pest population relies on active sampling (Fischnaller et al., 2017). Both these methods require manpower, and are therefore associated with high labor cost. Many researchers suggested the use of electronic devices for monitoring, for example, in 1980s for the first time optical sensors were used to monitor the tobacco budworm, *Heliothis virescens* and cabbage looper, *Trichoplusia ni* Hübner (Hendricks, 1985). Remote transmission technology fitted with image sensors is used to monitor insect-pests inside the greenhouses (Tirelli et al., 2011). For example, Rassati et al. (2016) used camera sensor device coupled with multifunnel trap for monitoring of bark beetles (Coleoptera: Scolytinae) and longhorn beetles (Coleoptera: Cerambycidae). Using camera for monitoring pest incidence needs high resolution and sharp imaging, especially for small insect-pests like whiteflies. For example, Doitsidis et al. (2017) selected a 2-megapixel camera and Shaked et al. (2018) selected a 5-megapixel camera (Omni Vision OV5647 NOIR Rasp Pi) to monitor *Bactrocera oleae* and *Ceratitis capitata* respectively (Figure 10.3). The data from these can be directly accessed with the help of internet. Ünlü et al. (2019) accessed the camera captures via internet due to the presence of a modern GPRS which permitted remote data transmission.

10.3 SETTING ACTION THRESHOLDS

A threshold level has to be set before taking any action in IPM, especially the economic injury level (EIL) and the economic threshold levels (ETL) (FAO, 1966). The EIL is defined as the lowest population density of insect-pest that will cause economic damage. Pedigo et al. (1986) defined "injury" as the impact of insect-pest on the host physiology; "damage" as the quantifiable loss of host's utility, most often including yield quantity or quality or aesthetics. Most IPM strategies demand the reduction of pest population to a level that does not cause economic damage. Therefore, a certain level of injury to the plant may not produce damage. ETL is defined as the amount of injury that will justify the cost of control (Stern et al., 1959). More logically,

ETL is the pest density at which management measures should be taken to prevent the pest from reaching a level that causes an economic injury.

$$\text{EIL} = \frac{\text{Cost of management}}{\text{Market value of produce} \times \text{Crop injury} \times \text{Damage yield loss} \times \text{Reduction of injury}}$$

Cost of management (Rs/ha); Market value per unit yield (Rs/tonne); Crop injury per pest (% defoliation or percent fruit change); Damage or yield loss per unit injury (% defoliation); proportional reduction in injury from pesticide use (Prabhuraj, 2016).

10.4 PREVENTION

Prevention is considered for the development of cropping systems that are potentially less likely to suffer major economic losses due to the prevalence of pests. Prevention can be done before the pest attack, if only we know that the pest may attack the crop by our previous experiences. In that cases, use of resistant varieties and making the conditions nonfavorable for the pest to survive or attack may be followed. But when the pest starts to attack the crop/produce, we can still prevent it from attaining the economic status by changing the conditions, that is, habitat manipulation (Nilsson et al., 2016). A wide range of methods may be applied which indirectly help to prevent pest attack. Preventive methods such as crop rotation, soil nutrient management, etc. along with other cultural practices are line of defense against insect-pests. Crop rotation is hardly utilized in protected cultivation; most of the greenhouses practiced continuous cropping with or without crop-free interval. However, crop rotations were found to be effective in reducing pest in crops (Mohler and Johnson, 2009).

10.5 MANAGEMENT

Management of pest is done by utilizing different tactics such as cultural, mechanical, behavioral, biological, and chemical. Apart from these, use of host resistance and biotechnological approaches like genetically-modified crops can also be included in IPM program (Katan, 2010). Preventive management strategies like resistance varieties of plants that are not preferred by the pest will harbor the pest and escape the damage (Stenberg and Muola, 2017).

10.6 IPM STRATEGIES

10.6.1 PREVENTIVE STRATEGIES

10.6.1.1 EXCLUSION

10.6.1.1.1 Physical Barriers

Protective cultivation relies on methods which associate knowledge of insect-pest ecology and modern technology for its control along with traditional farming practices. Prohibiting harmful insects from entering the greenhouse by means of physical barriers such as insect-proof net is considered as the first step in the IPM for protected cultivation. The prohibition of insect-pest is achieved by fitting fine-mesh screen, which acts as a mechanical barrier to the harmful insect (Bethke and Paine, 1991). These methods are discussed as under (Figure 10.2).

FIGURE 10.2 Different management measures as components of IPM.

10.6.1.1.1.1 Use of Insect-proof Nets

Insects may enter the greenhouse from outside through the ventilation openings. Installation of screens (net) on the ventilation openings will prevent or reduce the entry of pests. Insect-proof net (Figure 10.4) having pore size less than 200 μ is used in protected cultivation to prevent pests like whiteflies, aphids, leaf miner, and thrips. It is also effective method to control fruit borer like *Helicoverpa armigera*, the main pest of tomato fruit in greenhouse. Nets of 10 × 20 and 10 × 22 μ are satisfactory to exclude the whiteflies (*Trialeurodes vaporariorum* and *Bemisia tabaci*) without hampering the natural enemies movement (Hanafi et al., 2006). *Bemisia tabaci*, a small insect is the vector of tomato yellow leaf curl virus and has become a limiting factor for vegetable production in protected cultivation system.

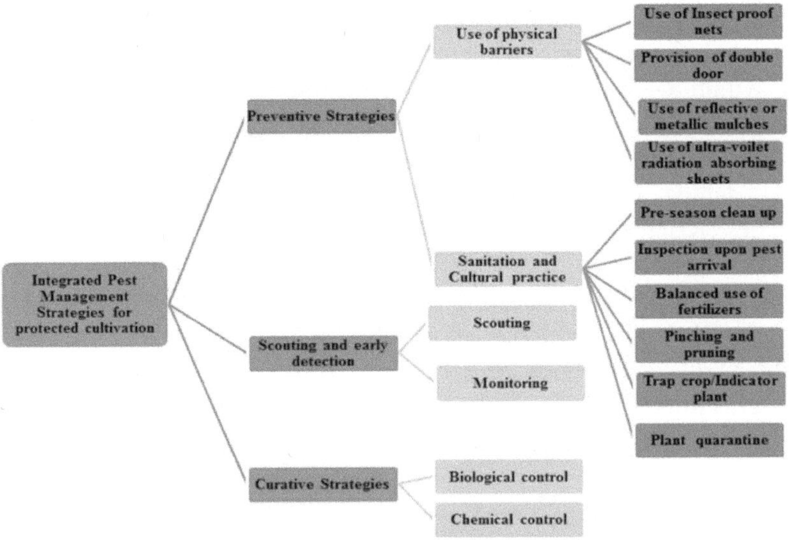

FIGURE 10.3 IPM strategies for protected cultivation.

FIGURE 10.4 Insect-proof net in protected cultivation.

10.6.1.1.1.2 Use of Reflective Mulches

Highly reflective or metalized plastic mulches (Figure 10.5) have been used in agriculture for many purposes, and are particularly effective in reducing the entry of whiteflies and thrips. These were found to be effective in forbidding aphids from plant and reduced the watermelon and zucchini mosaic virus within 3–6 weeks (Stapleton and Summers, 2002). Reflective ground cover around the crops has been used against colonization of insect-pests such as Asian citrus psyllid *Diaphorina citri* in citrus (Croxton and Stansly, 2014), aphids, leafhopper, leaf-mining flies, and thrips in peppers and tomatoes (Kring and Schuster, 1992; Csizinszky et al., 1995). Many pests like Asian citrus psyllid have been prohibited from colonizing the crops by planting solar light reflecting ground covers around them (Croxton and Stansly, 2014). The combined effect of insect-proof and metalized mulches together reduces the entry of harmful insect-pests. In addition, maximum light reflecting by aluminum mulches and reflective screens prevents insect infestations in both open fields and covered crops (Greer and Dole, 2003).

FIGURE 10.5 Reflective mulches in field.

10.6.1.1.1.3 Provision of Double Door

Restricted access is beneficial for protected cultivation as many insects may enter through clothes or sweep by the wind. Building a separated foyer to create a double-door entry (Figure 10.6) partially solves the problem of wind-carried insect-pests. The holes or tears anywhere on the net and door must be repaired immediately. It is essential that all greenhouses be fitted with a double-door or safety access system (SAS). An airlock double-door entrance for the nursery or any protected system prevents the easy entry of harmful insect-pests. The SAS is fitted usually with yellow sticky traps along both the sides (Rathee et al., 2018).

FIGURE 10.6 Double-door system in greenhouse.

10.6.1.1.1.4 Ultraviolet (UV) Radiation-Absorbing Sheets

These are considered to safeguard the crops from various insect-pests and vector-borne viral diseases by modifying insect behavior. Unfortunately, these problems are also faced by pollinators. To manipulate the UV vision of insects, greenhouses are made using UV-blocking cladding materials that have shown effectiveness to prevent the immigration of various insect-pests like aphids, thrips, leaf miners, and whiteflies from the external environment into the protected crop area (Papaioannou et al., 2012).

10.6.1.1.1.5 Colored Sticky Traps

As insects are attracted to specific color, this can be also used to control pests in the greenhouses. Colored sticky traps are simple, effective, and low-cost techniques for monitoring the presence of insect-pests along with their numbers. For instance, yellow sticky traps (Figure 10.7) attract numerous insects, and are often used to capture winged aphids, leaf miners, and adult whiteflies. *Thrips tabaci* is attracted toward blue (Otieno et al., 2018), yellow (Teulon and Penman, 2012), and white (Kahrer, 1992) sticky traps depending on the crop, however, blue sticky traps are most effective to attract thrips (Otieno et al., 2018). Colored sticky traps are commercially used in many countries for the control of insect-pests. However, sticky traps are a bit risky as they may also attract nontarget insect species such as pollinators, parasitoids, and predatory insects, thereby reducing their numbers (Pobozniak et al., 2020). Therefore, sticky traps should be selected based on knowledge of color preference of insect-pest. For instance, *Aeolothrips intermedius* predator of thrips and aphids shows less attraction for white and blue sticky traps, compared to yellow sticky traps (Trdan et al., 2005a).

FIGURE 10.7 Colored sticky traps for pest control.

10.6.1.2 AVOIDANCE

10.6.1.2.1 Sanitation of the Greenhouse

Sanitation is the most effective and cheapest method to escape pest or disease outbreak in the greenhouse. It is essential to keep the interior as well as the exterior area around the greenhouse free from weeds to minimize the pest attack. Residues of previous crops must be immediately destroyed after harvest. The eradication of infected leaves or plant parts along with key source of pest infestation, followed by disinfection of surfaces comes under sanitation procedure. It includes various techniques like cleaning of ventilation, plant materials, soil tools, etc. (Rathee et al., 2018).

10.6.1.2.2 Cultural Practice

There are extensive techniques of cultural practices with variable significances, but having a common objective to maximize economic returns from enhanced crop production under greenhouse environment. Cultural control encourages cobenefits within protected system, which includes pest suppression as well as conservation of natural predators. This can be accomplished by manipulating the environment and creating adverse biotic conditions, which do not allow the pests to colonize the crop, reducing the survival of the pest, minimizing injury, and supporting the natural enemies (Ferr, 2018). The preseason cleanup, inspection upon pest arrival, balanced use of fertilizers, pinching and pruning of infected plants, and trap crop also helps in the reduction of insect-pest population (Pimentel and Perkins, 2019). Balanced use of fertilizers for intake of nutrients should be followed. Nitrogenous fertilizers should be applied as per need for optimal growth of crops because heavy application of nitrogenous fertilizer causes unnecessary and excessive plant growth which favors growth of insect-pests like aphids. While, potassium application at a preferred dose has been found to decrease the insect-pest incidence (Rathee et al., 2018).

Predominant cultural practices such as intercropping and use of trap crops divert the feeding of insect-pests as well as preference for oviposition. In addition, trap crops also provide forage for the beneficial insects (Sarkar et al., 2018). Trap crops are good alternative to reduce pest incidence. Holden et al. (2012) found that trap crops have been revealed to efficiently control stink bugs, *Halyomorpha halys* (Hemiptera: Pentatomidae). Marigold has been tested as a trap crop targeting *Helicoverpa armigera* (Lepidoptera: Noctuidae) in tomato crops, and thrips in roses (Sarkar et al., 2018). An effective trap crop should have more attraction for pest than the main crop (Holden et al., 2012). Other cultural methods like date of sowing, irrigation practice, and cover crops also impede pest outbreaks (Kruidhof et al., 2020).

10.6.2 SCOUTING AND EARLY DETECTION

Scouting and early detection are essential to accomplish the insect infestation in protected cultivation. Pest monitoring is a regular practice to systematically manage the pest problem. Initial detection of pests has a vital role in pest monitoring techniques, for example, inspection of plants with a hand lens and yellow sticky traps for monitoring the pest. Crop surveillance on a regular basis detects conditions whether pests are absent or at levels far below economic damage, which prevents excessive application of control measures and expenses (Papaioannou et al., 2012).

Some indicator plants can be used for the detection of pests in protected cultivation for the early detection of a particular pest. For example, faba beans (*Vicia faba* L.) and some petunia plants detect the presence of thrips carrying TSMV. If these plants are infected with virus carrying thrips, symptoms are developed within a week of infestation. Some plants like peppers, tomatoes, and eggplants are infested with aphids every year. These plants can be monitored for early symptoms of aphids on young leaves, such as detection of honeydew on upper surface of leaves. When these plants are grown near flowering plants, peppers, tomatoes, and eggplants show the signs of thrips population like distorted young leaves along with silvery-flecked scars (Sarkar et al., 2018).

Thermal imaging has become a potential tool in agriculture for monitoring or detection of the pests in greenhouses and open fields (Varith et al., 2003). Thermal imaging is used in greenhouses to collect thermal images (IR imaging) of the pests. Due to implementation of low-cost embedded system, the advancement of modern agriculture techniques has become easier and faster. One of the challenging tasks is the use of wireless imaging in protected cultivation for pest detection. Certain systems are established for wireless imaging to monitor insect-pest by automatically assembling sticky paper traps images using RGB cameras fitted at various locations inside the greenhouses (Zhong et al., 2018). RGB cameras collect the images on sticky traps and are counted through image processing algorithms. Similarly, Martin et al. (2008) established a network of video camera for early insect-pest monitoring that senses the number of insect-pests on sticky paper traps with the help of real-time video analysis. However, little research has been done on the integration of wireless imaging and environmental sensing, specifically for insect-pest monitoring which could have enormous utility in many IPM applications.

10.6.3 CURATIVE MEASURES

10.6.3.1 NATURAL ENEMIES OF INSECT-PESTS (BIOLOGICAL CONTROL)

Most insects and other arthropods found in the agroecosystems do not feed on or harm plants. Some of them feed on and destroy insect-pests only. In most cases, the biological activity of these beneficial organisms can prevent or greatly limit pest problems. These farmer's friends can be categorized broadly as predators, parasites, and parasitoids (Figure 10.8). In the field of entomology, biological control is described as the utilization of insect predator, parasitoids, entomopathogenic nematodes or microbial pathogens to overcome the populations of different insect-pests (Flint and Dreistadt,

1998; Michaud, 2018). Biological control is more unpredictable, time taking, and somewhat less effective than pesticides, that is, chemical control. Though biological control is a safer means of pest control, but its consequences are to be studied properly before employing it (Van, 2000).

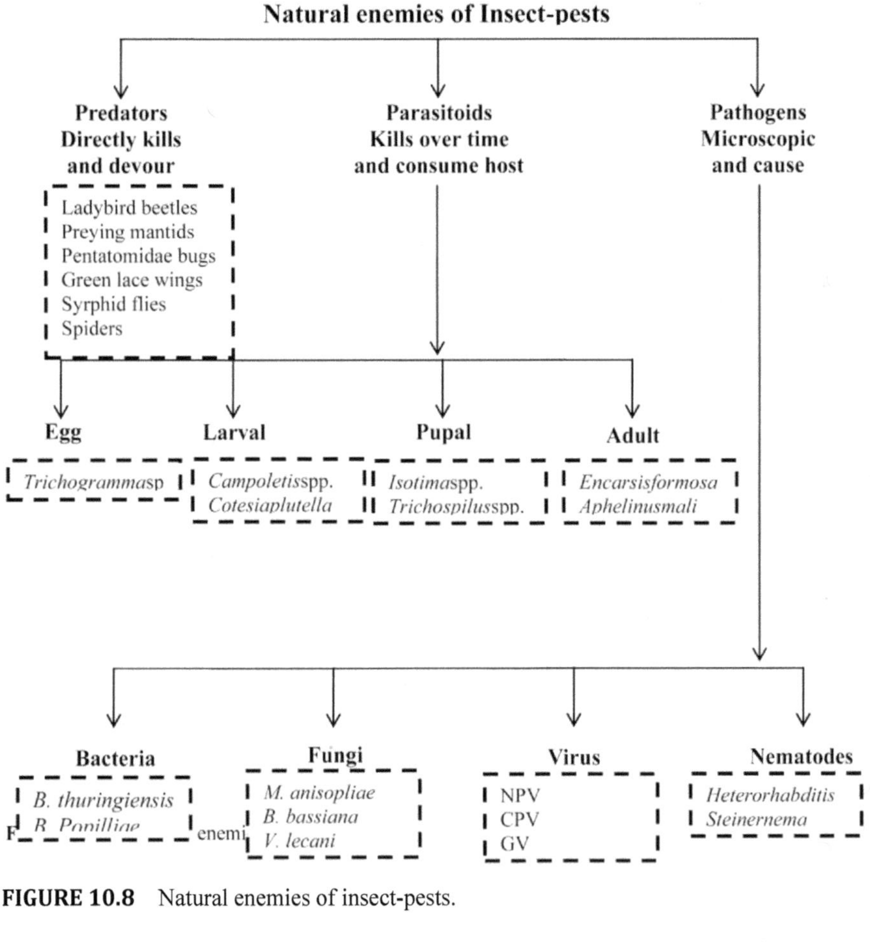

FIGURE 10.8 Natural enemies of insect-pests.

In greenhouses, the economically important pest species attacking vegetable crops are *Aphis gossypii*, *Macrosiphum euphorbiae*, *Aulacorthum solani*, and *Myzus persicae* (Blümel et al., 2004); those attacking ornamentals are *A. gossypii*, *M. euphorbiae*, *A. solani*, *M. persicae*, *Brachycaudus helichrysi* (Kaltenbach), *Macrosiphoniella sanborni* (Gillette), and *Macrosiphum rosae* L. (Chau et al., 2004). The efficacy of ladybirds as aphid predators in greenhouses has been reviewed (Powell et al., 2007). *Coccinella septempunctata* and *Hippodamia convergens* are two highly effective ladybird beetles found to control aphid population. The aphid gall midge, *Aphidoletes aphidimyza*, serves as an effective control for the outbreak of aphids. The midges are reported for their effectiveness in controlling aphid infestations, killing more aphids than they actually consumed. The larvae of midge established in aphid population rapidly eradicate the adult aphids (Hayes et al., 2019). *Tuta absoluta* (the tomato leaf miner) is one of the key insect-pests of tomato in greenhouses. Its larvae feed both on foliage and tomato fruit. Numerous predatory species, like *Macrolophus pygmaeus*, *Dichypus bolivari*, *Dicyphus errans*, *Nabis pseudoferus*, and *Nesidiocoris tenuis* have been evaluated as the natural enemies of *Tuta absoluta* (Ferracini et al., 2019). Among parasitoids, *Bracon nigricans* as tested in lab condition illustrates good behavioral plasticity to parasitize the *T. absoluta*. *Necremnus tutae*, a eulophid parasitoid was determined to show high mortality of *T. absoluta* larvae (Bodino et al., 2018).

10.6.3.2 CHEMICAL CONTROL (PESTICIDES)

Chemical control involves the application of synthetic, inorganic, or organic compounds that have a killing, inhibiting or repulsive consequence on organisms, which are also harmful to cultivated plants sometimes. Though chemicals have many side effects, the benefit of low cost, easy to use, and immediate preventive action makes chemical control as the most important pest control method (Kumar and Poehling, 2014). In India, Uttar Pradesh, Maharashtra, Andhra Pradesh, Haryana, and Punjab are the states which account for 70% of total pesticide consumption. The majority areas in NE region use low levels of pesticides. States like Jammu and Kashmir, Punjab, Haryana, and West Bengal consume >600 g/ha of pesticides whereas all the NE states, Rajasthan, and Madhya Pradesh consume <100 g/ha of pesticides (Devi et al., 2017).

Chemicals that kill the pests or inhibit their development are called pesticides. In addition, it also encompasses chemosterilants as well as growth regulators, which are used to interfere with the normal reproductive functions or developmental processes of pests. Insecticides can be classified as contact poisons, stomach poisons, and fumigants based on the mode of entry into the body of the insect. In general, chewing insects like caterpillars are managed using contact poisons and sucking pests are managed using systemic poisons. Based on the mode of action, insecticides may be physical poisons, protoplasmic poisons, respiratory poisons, nerve poisons, and growth regulators.

In general, pesticides are the last resort in IPM and used at the stage when the other management measures are not much helpful in pest management. However, proper pesticide at proper dose and at proper stage itself is an effective management of pests. Improper use of pesticides may result in pest resurgence, pest resistance, and pesticide residue, besides polluting the environment (Sabir et al., 2011). Pesticide application is also to be done at proper dose with proper mechanism (seed treatment, spray, dusting, and drenching), proper instrument, and a trained personnel. In protected cultivation, *T. absoluta* is a serious pest showing resistance to numerous pesticides such as abamectin, deltamethrin, and metamidophos in Argentina and to organophosphorus insecticides in Bolivia (Abbes et al., 2012). The insecticides permitted to control this pest are Voliam Targo (abamectin 18 g/L + chlorantraniliprole 45 g/L) 0.08%, Affirm (emamectin benzoate 0.95%) 0.15%, and Alverde (metaflumizone 240 g/L) 0.1% (Costache et al., 2018).

10.7 CONSTRAINTS AND FUTURE PROSPECTS

The IPM is complicated and time consuming. Farmers cannot afford to do series of operations to combat pest problems through IPM tactics. Being a quick action, pesticides are more impressive than IPM. Less IPM specialists are available in the country, as everyone is specialized in their own subject.

The application of IPM technique is a series of laborious and complex methods to implement as compared to conventional system, but strategic and timely application of IPM ensures successful management of pests with increased return to the farmers. Therefore, crop and region-specific IPM packages should be developed for protected cultivation, and farmers should be acknowledged about the right strategies of IPM. Pest monitoring is the basic pillar in IPM. In order to diminish crop loss, scouting and monitoring using suitable methods and tools for pests should be done timely. The advanced tools based on artificial intelligence (AI) such as various sensors and forecasting systems are indispensable for supervisory decision-making in IPM. So, future research and technology is being focused on these aspects. Statistical or mathematical models for pest population dynamics should be applied in IPM and broadly in agriculture for developing forecast models for efficient management of pests under greenhouses.

10.8 CONCLUSIONS

Pest management in protected cultivation is an obvious event of agricultural production system to sustain productivity. Several pest management strategies have already been developed against prevailing insect-pests problems. A wise combination of all these strategies forms the components of successful and ecofriendly integrated management of pest problems in the greenhouses. It is also noted that IPM strategies are specific to location, existing climatic conditions, pest, and other local socioeconomic factors. This would result in a total security and stability in both agricultural production and the ecosystem. Integrated approaches involving biological control agents, microbial pathogens, and botanicals are ecologically safer than insecticides to nontarget species like pollinators; however, it must be developed and reproduced at a large scale.

KEYWORDS

- **integrated pest management**
- **protected cultivation**
- ***Spodoptera litura***
- **greenhouse**
- **insect-pests**

REFERENCES

Abbes, K.; Harbi, A.; Chermiti, B. The Tomato Leafminer *Tuta absoluta* (Meyrick) in Tunisia: Current Status and Management Strategies. *EPPO Bull.* **2012,** *42* (2), 226–233.

Alexandratos, N.; Bruinsma, J. World Agriculture Towards 2030/2050: The 2012 Revision, 2012.

Bethke, J. A.; Paine, T. D. Screen Hole Size and Barriers for Exclusion of Insect Pests of Glasshouse Crops. *J. Entomol. Sci.* **1991,** *26*, 169–177.

Blümel, S. Biological Control of Aphids on Vegetable Crops. In *Biological Control in Protected Culture*; Heinz, K. M., Van Driesche, R. G.; Parrella, M. P., Eds.; Ball Publ.: Batavia, IL, USA, 2004; p 297.

Bodino, N.; Ferracini, C.; Tavella, L. Functional Response and Age-Specific Foraging Behavior of *Necremnus tutae* and *N. cosmopterix*, Native Natural Enemies of the Invasive Pest *Tuta absoluta* in Mediterranean Area. *J. Pest. Sci.* **2018,** *92* (4), 1467–1478.

Chau, A.; Heinz, K. M. Biological Control of Aphids on Ornamental Crops. In *Biological Control in Protected Culture*; Heinz, K. M., Van Driesche, R. G., Parrella, M. P., Eds.; Ball Publ.: Batavia, IL, USA, 2004; p 277.

Croxton, S. D.; Stansly, P. A. Metalized Polyethylene Mulch to Repel Asian Citrus Psyllid, Slow Spread of Huanglongbing and Improve Growth of New Citrus Plantings. *Pest Manag. Sci.* **2014,** *70* (2), 318–323.

Csizinszky, A. A.; Schuster, D. J.; Kring, J. B. Color Mulches Influence Yield and Insect Pest Populations in Tomatoes. *J. Am. Soc. Hortic. Sci.* **1995,** *120* (5), 778–784.

Devi, P. I.; Thomas, J.; Raju, R. K. Pesticide Consumption in India: A Spatiotemporal Analysis. *Agric. Econ. Res. Rev.* **2017,** *30* (347-2017-2047), 163–172.

Doitsidis, L.; Fouskitakis, G. N.; Varikou, K. N.; Rigakis II; Chatzichristofis, S. A.; Papafilippaki A. K., Birouraki A. E. Remote Monitoring of the *Bactrocera oleae* (Gmelin) (Diptera: Tephritidae) Population using an Automated McPhail trap. *Comput. Electron. Agric.* **2017,** *137*, 69–78.

FAO. In *Summary of Session and Discussion*, Proceedings of FAO Symposium on Integrated Pest Control; Food and Agriculture Organization: Rome, 1966; vol 2, p 87.

Ferracini, C.; Bueno, V. H. P.; Dindo, M. L.; Ingegno, B. L.; Luna, M. G.; Salas Gervassio, N. G.; Sanchez, N. E.; Siscaro, G.; van Lenteren, J. C.; Zappala, L.; Tavella, L. Natural Enemies of *Tuta absoluta* in the Mediterranean Basin, Europe and South Africa. *Biocontrol. Sci. Technol.* **2019,** *29* (6), 578–609

Ferrarezi, R. S.; Weiss, S. A.; Geiger, T. C.; Beamer K. P. Edible-Pod Peas as High-Value Crops in the US Virgin Islands. *Hort. Technol.* **2016,** *26* (5), 683–689.

Fischnaller, S.; Parth, M.; Messner, M.; Stocker, R.; Kerschbamer, C.; Reyes-Dominguez, Y.; Janik, K. Occurrence of Different *Cacopsylla* Species in Apple Orchards in South Tyrol (Italy) and Detection of Apple Proliferation Phytoplasma in *Cacopsylla melanoneura* and *Cacopsylla picta*. *Cicadina* **2017,** *17*, 37–51

Flint, M. L.; Dreistadt, S. H. Natural Enemies' Handbook: The Illustrated Guide to Biological Pest Control, Vol. 3386; University of California Press, 1998.

Greer, L.; Dole, J. M. Aluminum Foil, Aluminum-Painted, Plastic and Degradable Mulches Increase Yields and Decrease Insect Vectored Viral Diseases of Vegetables. *Hort. Technol.* **2003,** *13* (2), 276–284.

Hanafi, A.; Bouharroud, R.; Amouat, S.; Miftah. S. In *Efficiency of Insect Nets in Excluding Whiteflies and their Impact on Some Natural Biological Control Agents*, VIII International Symposium on Protected Cultivation in Mild Winter Climates: Advances in Soil and Soilless Cultivation under 747, 2006; pp 383–388.

Hayes, D. S.; Jordon-Thaden, I. E.; Cantley, J. T.; McDonnell, A. J.; Martine, C. T. Integrated Pest Management in the Academic Small Greenhouse Setting: A Case Study using *Solanum* spp.(Solanaceae). *A.P.P.S.* **2019,** *7* (8), 11281.

Hendricks, D. E. Electronic System for Detecting Trapped Boll Weevils in the Field and Transferring Incident Information to a Computer. *Southwest Entomol.* **1990,** *15* (1), 39–48.

Holden, M. H.; Ellner, S. P.; Lee, D. H.; Nyrop, J. P.; Sanderson, J. P. Designing an Effective Trap Cropping Strategy: The Effects of Attraction, Retention and Plant Spatial Distribution. *J. Appl. Ecol.* **2012,** *49*, 715–722.

Jayaraj, S.; Rabindra, R. J. Integrated Pest Management. In: *Agroclimatology and Sustainable Agriculture in Stressed Environments*; Virmani, et al., Eds.; Oxford and IBH Publishing: New Delhi, 1994; pp 103–113.

Kahrer, A. Monitoring the Timing of Peak Flight Activity of *Thrips tabaci* in Cabbage Fields. *Bull. OILB/SROP* **1992,** *15*, 28–35

Katan, J. Cultural Approaches for Disease Management: Present Status and Future Prospects. *Plant Pathol.* **2010,** *92* (4), S4.7–S4.9.

Kring, J. B.; Schuster, D. J. Management of Insects on Pepper and Tomato with UV-Reflective Mulches. *Fla. Entomol.* **1992,** *75* (1), 119–129.

Kruidhof, H. M.; Elmer, W. H. Cultural Methods for Greenhouse Pest and Disease Management. In: *Integrated Pest and Disease Management in Greenhouse Crops*; Springer: Cham, 2020; pp 285–330.

Kumar, P.; Poehling, H. M. Effects of Azadirachtin, Abamectin, and Spinosad on Sweet Potato Whitefly (Homoptera: Aleyrodidae) on Tomato Plants under Laboratory and Greenhouse Conditions in the Humid Tropics. *J. Econom. Entom.* **2014,** *100* (2), 411–420.

Marroni, M. V.; Teulon, D. A. J.; Scott, R. R.; Jaspers, M. V. Abundance of *Thrips obscuratus* (Crawford) (Thysanoptera: Thripidae) during Wine Grape Flowering in Different Regions in New Zealand. *N Z Entomol.* **2012,** *35* (1), 10–13.

Martin, V.; Moisan, S.; Paris, B.; Nicolas, O. In *Towards a Video Camera Network for Early Pest Detection in Greenhouses, ENDURE International Conference*; La Grande-Motte: France, 2008; pp 1–5.

McCravy, K. W. A Review of Sampling and Monitoring Methods for Beneficial Arthropods in Agroecosystems. *Insects* **2018,** *9*, 170.

Michaud, J. P. Challenges to Conservation Biological Control on the High Plains: 150 years of Evolutionary Rescue. *Biol. Control.* **2018,** *125*, 65–73.

Mohler, C. L.; Johnson, S. E. In *Crop Rotation on Organic Farms: A Planning Manual*; Natural Resource, Agriculture, and Engineering Service: Ithaca, NY, USA, 2009; p 156.

Nair, R.; Barche, S. Protected Cultivation of Vegetables-Present Status and Future Prospects in India. *Indian J. Appl. Res.* **2014,** *4*, 245–247.

Nilsson, U.; Porcel, M.; Świergiel, W.; Wivstad, M. Habitat Manipulation: As a Pest Management Tool in Vegetable and Fruit Cropping Systems, with the Focus on Insects and Mites, SLU, EPOK-Centre for Organic Food & Farming, 2016.

Nottingham, L. B.; Dively, G. P.; Schultz, P. B.; Herbert, D. A.; Kuhar, T. P. Natural History, Ecology, and Management of the Mexican Bean Beetle (Coleoptera: Coccinellidae) in the United States. *J. Integr. Pest Manag.* **2016,** *7* (1), 2.

Otieno, J. A.; Stukenberg, N.; Weller, J.; Poehling, H. M. Efficacy of LED-Enhanced Blue Sticky Traps Combined with the Synthetic Lure Lurem-TR for Trapping of Western Flower Thrips (*Frankliniella occidentalis*). *J. Pest. Sci.* **2018,** *91* (4), 1301–1314.

Papaioannou, C.; Katsoulas, N.; Maletsika, P.; Siomos, A.; Kittas, C. Effects of a UV-Absorbing Greenhouse Covering Film on Tomato Yield and Quality. *Span. J. Agric. Res.* **2012,** 4, 959–966.

Pedigo, L. P.; Hutchins, S. H.; Higley, L. G. Economic Injury Levels in Theory and Practice. *Annu. Rev. Entomol.* **1986,** *31*, 341–368.

Pimentel, D.; Perkins, J. H. Pest Control: Cultural and Environmental Aspects; CRC Press, 2019.

Pobozniak, M.; Tokarz, K.; Musynov, K. Evaluation of Sticky Trap Colour for Thrips (Thysanoptera) Monitoring in Pea Crops (*Pisum sativum* L.). *J. Plant. Dis. Prot.* **2020,** *127* (3), 307–321.

Powell, W.; Pell, J. K. Biological Control. In *Aphids as Crop Pests*; van Emden et al., Eds.; CAB International: Oxfordshire, UK, 2007; pp 469–513.

Prabhuraj, A. Pest, Category - Concept of ETL and EIL, 2016. http://www.agriglance.com/entomology/wpcontent/uploads/2017/11/ag_Pest_categories_EIL_ETL_lect31.pdf.

Rassati, D.; Faccoli, M.; Chinellato, F.; Hardwick, S.; Suckling, D. M.; Battisti, A. Web-Based Automatic Traps for Early Detection of Alien Wood-Boring Beetles. *Entomol. Exp. Appl.* **2016,** *160* (1), 91–95.

Rathee, M.; Singh, N. V.; Dalal, P. K.; Mehra, S. Integrated Pest Management under Protected Cultivation: A Eeview. *J. Entomol. Zool. Stud.* **2018,** *6*, 1201–1208.

Sabir, N.; Deka, S.; Singh, B.; Sumitha, R.; Hasan, M.; Kumar, M.; Bambawale, O. M. Integrated Pest Management for Greenhouse Cucumber: A Validation under North Indian Plains. *Indian J. Hort.* **2011,** *68* (3), 357–363.

Sarkar, S. C.; Wang, E.; Wu, S.; Lei, Z. Application of Trap Cropping as Companion Plants for the Management of Agricultural Pests: A Review. *Insects* **2018,** *9* (4), 128.

Shaked. B.; Amore, A.; Ioannou, C.; Valdés, F.; Alorda, B.; Papanastasiou, S.; Goldshtein, E.; Shenderey, C.; Leza, M.; Pontikakos, C.; Perdikis, D.; Tsiligiridis, T.; Tabilio, M. R.; Sciarretta, A.; Barceló, C.; Athanassiou, C.; Miranda, M. A.; Alchanatis, V.; Papadopoulos, N.; Nestel, D. Electronic Traps for Detection and Population Monitoring of Adult Fruit Flies (Diptera: Tephritidae). *J. Appl. Entomol.* **2018,** *142* (1-2), 43–51.

Singh, M. C.; Kachwaya, D. C., Kalsi, K. Soilless Cucumber Cultivation under Protective Structures in Relation to Irrigation Coupled Fertigation Management, Economic Viability and Potential Benefits-A Review. *Int. J. Curr. Microbiol. App. Sci.* **2018b,** *7* (03), 2451–2468.

Singh, M. C.; Singh, J. P.; Pandey, S. K.; Cutting, N. G.; Sharma, P.; Shrivastav, V.; Sharma, P. A Review of Three Commonly used Techniques of Controlling Greenhouse Microclimate. *Int. J. Curr. Microbiol. App. Sci.* **2018a,** *7* (1), 3491–3505.

Singh, M. C.; Singh, J. P.; Pandey, S. K.; Mahay, D.; Srivastava, V. Factors Affecting the Performance of Greenhouse Cucumber Cultivation-A Review. *Int. J. Curr. Microbiol. App. Sci.* **2017a,** *6* (10), 2304–2323.

Singh, M. C.; Singh, J. P.; Singh, K. G. Optimal Operating Microclimatic Conditions for Drip Fertigated Cucumbers in Soilless Media Under a Naturally Ventilated Greenhouse. *Ind. J. Ecol.* **2017b,** *44*, 821–826.

Stapleton, J. J.; Summers, C. G. Reflective Mulches for Management of Aphids and Aphid-Borne Virus Diseases in Late-Season Cantaloupe (*Cucumis melo* L. var. *cantalupensis*). *J. Crop Prot.* **2002,** *21*, 891–898.

Stenberg, J. A.; Muola, A. How Should Plant Resistance to Herbivores be Measured? *Front. Plant Sci.* **2017,** *8*, 663.

Stern, V. M.; Smith, R. F.; Van-den-Bosch, R.; Hagen, K. S. The Integration of Chemical and Biological Control of the Spotted Alfalfa Aphid: The Integrated Control Concept. *Hilgardia* **1959,** *29*, 81–101.

Teulon, D. A. J.; Penman, D. R. Colour Preferences of New Zealand Thrips (Terebrantia: Thysanoptera). *N.Z. Entomol.* **2012,** *15* (1), 8–13.

Tirelli, P.; Borghese, N. A.; Pedersini, F.; Galassi, G.; Oberti, R. In *Automatic Monitoring of Pest Insects Traps by Zigbee-Based Wireless Networking of Image Sensors*, Proceedings of the IEEE Instrumentation and Measurement Technology Conference; Hangzhou: China, 2011; pp 1–5.

Trdan, S.; Andjus, L.; Raspudić, E.; Kač, M. Distribution of *Aeolothrips intermedius* Bagnall (Thysanoptera: Aeolothripidae) and its Potential Prey Thysanoptera Species on Different Cultivated Host Plants. *J. Pest Sci.* **2005a,** *78* (4), 217–226.

Ünlü, L.; Akdemir, B.; Ögür, E.; Şahin, İ. Remote Monitoring of European Grapevine Moth, *Lobesia botrana* (Lepidoptera: Tortricidae) Population using Camera-Based Pheromone Traps in Vineyards. *Turkish J. A. F. Sci. Tech.* **2019,** *7* (4), 652–657.

Van-Lenteren, J. C. Success in Biological Control of Arthropods by Augmentation of Natural Enemies. In *Biological Control: Measures of Success*; Springer: Dordrecht, 2000; pp 77–103.

Varith, J.; Hyde, G. M.; Baritelle, A. L.; Fellman, J. K; Sattabongkot, T. Non-Contact Bruise Detection in Apples by Thermal Imaging. *I.F.S.E.T.* **2003,** *4*, 211–218.

Zhang, J.; Zhu, Y.; Zhang, X.; Ye, M.; Yang, J. Developing a Long Short-Term Memory (LSTM) Based Model for Predicting Water Table Depth in Agricultural Areas. *J. Hydrol.* **2018,** *561*, 918–929.

CHAPTER 11

Geometric and Aerodynamic Characterization of Insect-Proof Screens Used in Greenhouses: Aerodynamic Models

A. LÓPEZ-MARTÍNEZ[1], F.D. MOLINA-AIZ[1], D.L. VALERA[1], and H. FATNASSI[2]

[1]*CIAMBITAL (Centro de Investigación en Agrosistemas Intensivos Mediterráneos y Biotecnología Agroalimentaria), University of Almería, Almería, Spain*

[2]*The International Center for Biosaline Agriculture (ICBA), Academic City, Dubai, United Arab Emirates*

ABSTRACT

Insect-proof screens are a physical barrier against the passage of insect-pests in greenhouses. These nets are placed on all the vent openings of the greenhouses, mostly in areas with warm climate, such as the Mediterranean. With the use of insect-proof screens, it is possible to decrease the population of insect-pests inside the greenhouses, reducing the need to apply chemical protection treatments. In addition, the use of this protection technique prevents the insects beneficial to crops from leaving the greenhouses (pollinating insects or auxiliary insects used for the biological control of pests).

However, installation of these protective screens on the vents of greenhouses produces a significant reduction in their natural ventilation capacity, which significantly affects the indoor microclimate. There are several studies in which it has been shown that the use of these screens produces an increase

Protected Cultivation: Structural Design, Crop Management, Modeling and Automation.
Mahesh Chand Singh and K. K. Sharma (Eds.)

in the temperature and humidity of the air, a reduction in the speed with which air moves inside the greenhouse, a reduction in the kinetic energy of the air inside, an increase in vertical temperature gradient, etc. These effects on the microclimate can significantly affect the development and growth of crops.

The installation of these insect-proof screens in the vents of the greenhouses produces an additional resistance to airflow. The pressure drop generated depends on the geometric and aerodynamic characteristics of the insect-proof screen.

In this book chapter, we will talk about the geometric and aerodynamic characterization of these insect-proof screens used in greenhouses. We will also present different models that allow estimating the aerodynamic characteristics from some geometric parameters of the insect-proof screens.

11.1 INTRODUCTION

11.1.1 INSECT-PROOF SCREENS AND THE PROTECTION OF THE CROPS IN GREENHOUSES

The placement of insect-proof screens in the vents of greenhouses is the principal physical technique to avoid the entry of pest-insects in greenhouses; its use is essential in the greenhouses in regions where climatic conditions favor the spread of the insect-pests.

These screens are intended to reduce the entry of pest-insects into greenhouses, so the geometry of the pores will be defined based on the smallest pest insect. The small size of the pores of the meshes makes beneficial insects unable to leave the greenhouse, as Teitel (2007) indicated.

Some of the most damaging insects for crops in greenhouse are aphids (*Myzus persicae* Sulzev y *Aphis gossipii* Glover), leaf miners (*Liriomyza* sp.), whiteflies (*Trialeurodes vaporariorum* Westwood and *Bemisia tabaci* Gennadius), thrips (*Frankliniella occidentalis* Pergande and *Thrips tabaci* Lindeman), and currently the tomato moths (*Tuta absoluta* Meyrick).

These pest-insects, in addition to causing direct damage to crops due to the activity of insects themselves, can be transmitting vectors of highly damaging phytopathogenic viruses for crops. For example, the whitefly (Figure 11.1a) transmits *Tomato Yellow Leaf Curl Virus* (TYLCV) and the thrips (Figure 11.1b) are a vector of the *Tomato Spotted Wilt Virus* (TSWV).

Manufacture of insect-proof screens must be carried out in such a way that the limiting dimension of the pores (maximum width) should be less than the size of the smallest insect to be excluded. The capacity of nets to exclude insects also depends on the ability of each species to adapt its body to the dimensions of the pores. Bethke and Paine (1991) determined that *Bemisia tabaci* passed through insect-proof screens in which the pore width was at least twice the width of the insect's thorax, while *Frankliniella occidentalis* passed through insect-proof screens with pores slightly larger than its thorax, a sign of its ability.

FIGURE 11.1 Example of whitefly (a) and thrips (b) on samples of insect-proof screen (10 × 20 threads cm^{-2}).

Source: Reprinted from López, 2011

Nucifora (1994), in laboratory tests, observed that the 10 × 20 threads cm^{-2} insect-proof screens prevented 100% of the passage of *Bemisia tabaci* (pore width of 0.28 and 0.32 mm). In the field tests, the 10 × 20 threads cm^{-2} insect-proof screens significantly reduced the presence of *Bemisia tabaci* in comparison with a greenhouse with vents without nets.

Berlinger et al. (2002) studied eight insect-proof screens in a laboratory. With a thread density of 10 × 20 threads cm^{-2} (pore width of 290 μm), the penetration percentage of *Bemisia tabaci* was only 6.2%. They also observed that the most effective nets in the laboratory were also effective in field trials.

Ticó et al. (2002) compared, in the field, the efficacy of a 10 × 20 threads cm^{-2} (pore width 270 μm) and a 6 × 8 threads cm^{-2} insect-proof screens. The first reduced the number of *Bemisia tabaci* individuals inside the greenhouse and thus the incidence of the TYLCV virus. However, both screens were completely ineffective against *Frankliniella occidentalis*.

From the results obtained by the different authors, it can be deduced that current screens can be quite effective against *Bemisia tabaci*, while *Frankliniella occidentalis*, due to its smaller size and greater ability, can penetrate more easily through commercial insect-proof screens. There is a wide variety of commercial screens, but currently the information provided by manufacturers regarding their characteristics is very scarce. It is usually limited to indicating the density of threads and in some cases the porosity. This lack of information can confuse the farmer. Insect-proof screens with equal thread density, but made with distinct threads' diameters will have different pore lengths and therefore, their effectiveness as a physical barrier will vary.

A study published in 2016 indicated that 95.4% of greenhouse growers in the southern Spain (Almería) used insect-proof screens to protect their greenhouses (Valera et al., 2016). In the same study, the type of nets (thread density) that was usually used in side vents was indicated: 15 × 30 threads cm^{-2} (58.3%) and 10 × 20 threads cm^{-2} (25.6%); for roof vents: 15 × 30 threads cm^{-2} (56.0%) and 10 × 20 threads cm^{-2} (22.56%) (Valera et al., 2016).

It should also be taken into account that the use of insect-proof screens should not be considered as an isolated protection method, but must be combined with other alternative protection techniques (such as biological control). In this way, the set of methods should contribute to a decrease in the use of phytosanitary products, whose use should be limited for environmental reasons to reduce the risk of people who apply such products to improve the quality of the final product and the image in front of the consumer. In addition, by reducing the number of phytosanitary treatments, farmers can comply with international standards and controls (Teitel, 2007).

Thanks to the use of insect-proof screens, the population levels of pests can be reduced, reducing the rate of plant damages and infections transmitted by insect vectors. Use of insect-proof screens results in less need for the application of phytosanitary products (Baker and Jones, 1989; Berlinger et al., 1991, 1992; Taylor et al., 2001; Möller et al., 2004; Teitel, 2007; Castellano et al., 2008; Teitel and Wenger, 2014), having a significant impact on the economy of farms. Taylor et al. (2001) showed that *Bemisia tabaci* caused, from 1980 to 1990, economic losses between 15 and 32 million dollars, which could have gone much below with the use of insect-proof screens.

11.1.2 INSECT-PROOF SCREENS AND NATURAL VENTILATION IN GREENHOUSES

Insect-proof screens must be installed in all greenhouse vent openings. As the main drawback, the use of insect-proof screens has a very negative effect on the natural ventilation and the microclimate of the greenhouse (López et al., 2014, 2015, 2016). There are several authors who have analyzed the physical characteristics of the insect-proof screens, both from the point of view of its geometry and aerodynamics. The effect of the screens on the natural ventilation capacity of greenhouses and on their microclimate has also been studied in depth. First, low speed wind tunnels have been used to analyze the aerodynamic behavior of the insect-proof screens (Miguel et al., 1997; Dierickx, 1998; Klose and Tantau, 2004; Valera et al., 2006; Álvarez, 2010; Pérez-Vega et al., 2016). Other authors used CFD simulations to study the aerodynamic properties of the insect-proof screens (Teitel and Wenger, 2014). Results can vary significantly if the simulations are performed on the design of the actual mesh structure or assuming the mesh as a porous structure (Teitel, 2010).

In addition to the analysis of the aerodynamic properties of the insect-proof screens using CFD simulations (Teitel, 2010; Teitel and Wenger, 2014), many authors have used CFD models to study the effect of these insect-proof screens on the microclimate of greenhouses with natural ventilation (Muñoz et al., 1999; Miguel and Silva, 2000; Bartzanas et al., 2002; Kittas et al., 2002; Fatnassi et al., 2002, 2003, 2006; Molina-Aiz et al., 2004b; Soni et al., 2005; Harmanto et al., 2006; Katsoulas et al., 2006; Kittas et al., 2008; Baeza et al., 2009; López et al., 2014, 2016).

The climatic parameters in the greenhouse are conditioned by the characteristics of insect-proof screens, the geometry of vents, and outside wind characteristics (Miguel and Silva, 2000). The insect-proof screens reduce the natural ventilation capacity of the greenhouses and the inside air velocity (Kittas et al., 2008). By reducing the porosity of the screens, temperature and humidity inside the greenhouse increase as a consequence of reduction in ventilation (Fatnassi et al., 2002, 2003). Bartzanas et al. (2002) observed that a insect-proof screen with a porosity of 53% reduced 50% of the ventilation rate, increasing 4°C inside temperature. In the same way, Baeza et al. (2009) showed that a screen with a porosity of 28% could reduce the ventilation rate up to 77–87%. The nets affect the vertical gradient of inside air temperature in the greenhouse. Insect-proof screens with low porosity can generate a significant vertical air temperature gradient, increasing the air temperature with the height (Soni et al., 2005).

All these adverse facets of the insect-proof screens could be increased by the accumulation of dirt on the screens. Linker et al. (2002) observed that 2 months after the installation of an insect-proof screen in a greenhouse of 50 m^2, the pressure loss coefficient of the nets went from 12 to 200. López et al. (2013) observed that the accumulation of dirt in the screens can reduce the porosity of the insect-proof screens up to 26.1%. Besides, this accumulation of small particles in the screens can increase their pressure drop coefficient up to 61.2% (López et al., 2018).

Work should continue on the development of new insect-proof screens that achieves greater reductions in pore size, but maintain similar or higher levels of porosity, either through the use of threads of lesser diameter or by other means. It must also be taken into account that the use of the insect-proof screens is not an isolated method, but must be combined with other methods of struggle that allow not to limit the pore size of the insect-proof screens so much, being able to work in favor of the greenhouse microclimate with even higher porosity meshes.

11.2 GEOMETRIC CHARACTERIZATION

11.2.1 BIDIMENSIONAL CHARACTERISTICS

This chapter explains a numerical analysis of microscope digital images to determine the bidimensional geometric characteristics of the insect-proof screens. Commercial image analysis softwares or specific softwares (as used for data presented in this section, *Euclides* v1.4) can be used to obtain numerical values of properties of the screens as important as porosity, thread density, and average pore surface, among others.

To perform the analysis in a representative way, at least three samples from each insect-proof screen must be randomly selected, analyzed separately, and the results obtained with each sample compared.

To obtain accurate geometric parameters, digital images must be taken with a microscope. The images shown in this chapter have been taken with a digital microscope (Motic Spain SL, model DMWB1-223), which incorporates a digital camera and a 4x lens (resolution of 10.5 mm $pixel^{-1}$).

The little portions of the screens are placed on a microscope slide on which the measurement area is delimited (e.g., 1–2 cm^2). The previous calibration of the microscope for each of the objectives allows to know the equivalent in metric units of each pixel. In function of the objective used and the area covered by each image, a different number of images are needed to analyze the screen sample of 1–2 cm^2.

The image analysis process can be developed in four steps:

i. Capture of color digital images of the insect-proof screens (Figure 11.2a). These images must be converted to black and white.
ii. Black and white conversion. The software *Euclides* v1.4 requires that the pores be colored in gray (Figure 11.2b). Later, the software *Euclides* v1.4 performs a color conversion of the digital image, leaving the area corresponding to the pores in white and the area occupied by the threads in black (Figure 11.2c).

iii. Pore vertices recognition (Figure 11.2d). *Euclides* v1.4 automatically recognizes the vertices of each pore (Valera et al., 2006; Álvarez, 2010; Alvarez et al., 2012). In the event that the software does not identify a vertex correctly, a manual correction can be made.
iv. Geometric data calculation. When the user verifies that the vertices of all the pores have been correctly identified, the density of threads (number of weft and warp threads per cm^2), the porosity φ [m^2 m^{-2}], and several geometric parameters can be numerically calculated (Figure 11.3).

FIGURE 11.2 Images of insect-proof screens: original capture with the microscope (a), with black and white threads and gray pores (b), with black threads and white pores (c), and fully processed image, with identifying circles at the vertices of the pores (d).

Source: Adapted from A. López-Martínez, 2013. http://creativecommons.org/licenses/by/4.0/)

The number of threads per unit of length [threads cm^{-1}] in the directions of weft and warp can been calculated as (Álvarez et al., 2012):

$$\rho_x = \frac{1}{L_{py} + D_{hx}} \tag{11.1}$$

$$\rho_y = \frac{1}{L_{px} + D_{hy}} \tag{11.2}$$

where L_{px} is the length of the pore in the weft direction and L_{py} is the length of the pore in the warp direction [μm]; D_{hx} is the diameter of the weft threads and D_{hy} is the diameter of the warp threads [μm].

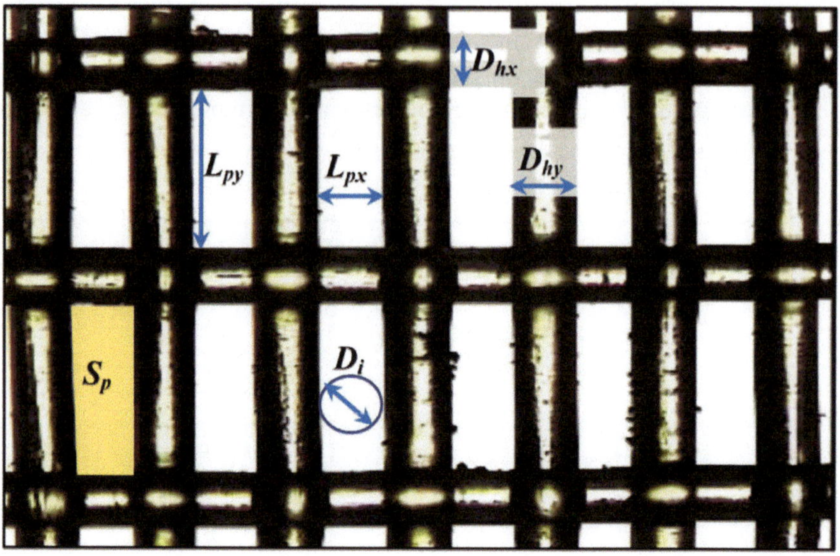

FIGURE 11.3 Parameters used for bidimensional geometric characterization of the insect-proof screens. L_{px}, length of the pore in the weft direction [μm]; L_{py}, length of the pore in the warp direction [μm]; D_{hx}, diameter of the weft threads [μm]; D_{hy}, diameter of the warp threads [μm]; D_i, inside circumference of the pore diameter [μm]; S_p, pore area [mm^2].

Source: Reprinted with permission from A. López-Martínez, et al., 2020. https://creativecommons.org/licenses/by/4.0/

The two-dimensional geometric characterization gives the screen porosity φ [m^2 m^{-2}], described as the area of the pores S_p in comparison to the total surface area S_t. The porosity can be calculated as (Valera et al., 2006; Álvarez et al., 2012; López et al., 2013):

$$\alpha = \frac{S_p}{S_t} = \frac{L_{px} L_{py}}{\left(L_{px} + D_{hy}\right)\left(L_{py} + D_{hx}\right)} \tag{11.3}$$

The denomination of weft and warp is its own of the textile industry for insect-proof screens that are made on looms. The threads that are fixed on the loom correspond with the weft; these threads are the ones that determine the width of the mesh. These fixed threads are divided into two groups (red and green threads in Figure 11.4) that are consecutively divided to make the threads of the warp pass between them (blue thread in Figure 11.4).

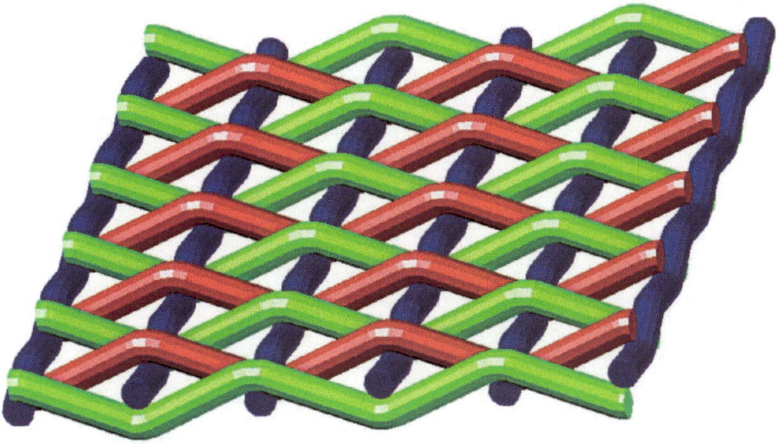

FIGURE 11.4 3D design of an insect-proof screen. Weft (red and green) and warp threads (blue).

Source: Reprinted from López, 2011

11.2.2 MEASURING THE THICKNESS

The thickness ***e*** of the insect-proof screens was essential to determine the aerodynamic parameters of the nets. The thickness can be determined accurately by using automatic video-based machines for noncontact measurement (Figure 11.5), with uncertainty lower than 10 μm.

To determine the net thickness, it can be placed between two plates of methacrylate (Figure 11.6).

Table 11.1 shows a resume of the geometric parameters of different types of insect-proof screens analyzed following the methodology previously explained.

11.3 AERODYNAMIC CHARACTERIZATION

Insect-proof screens affect natural ventilation capacity of the greenhouses increasing the pressure drop generated by the vent openings as a consequence of their geometric characteristics (height, width, and opening angle). Pressure loss produced by the screens can be correlated to the air velocity (Teitel, 2007; Molina-Aiz et al., 2009), with a quadratic term ($\Delta P = au^2$) or a second degree polynomial ($\Delta P = au^2 + bu$). With the first expression, drag generated by the screens can be described only with one parameter (F_φ, pressure drop coefficient). With the second expression, its necessary two parameters are: K_p, permeability and Y, inertial factor (Nield and Bejan, 1999).

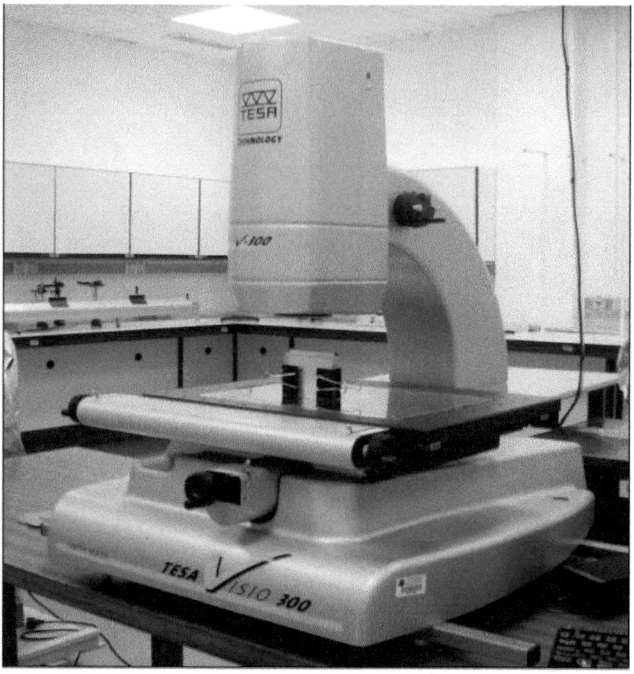

FIGURE 11.5 Automatic video-based machine for noncontact thickness measurement (TESA SA; model VISIO 300).

Source: Reprinted from López, 2011

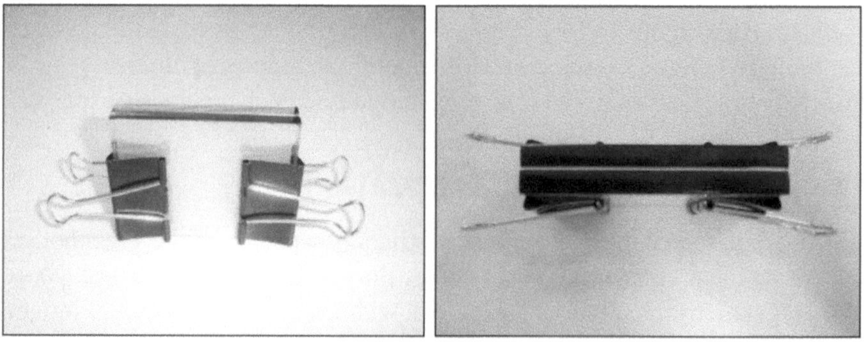

FIGURE 11.6 Support for the determination of the thickness of insect-proof screens.

Source: Reprinted from López, 2011

TABLE 11.1 Geometric Characteristics for Different Types of Insect-Proof screens (D_f, Thread Density According to the Manufacturer [Threads cm^{-2}]).

N	D_f	φ	L_{px}	L_{py}	D_{hx}	D_{hy}	D_h	D_i	S_p	e
1	13 × 30	39%	0.16	0.59	0.17	0.16	0.17	0.17	0.10	0.40
2	10 × 20	34%	0.23	0.73	0.28	0.27	0.27	0.24	0.17	0.56
3	14 × 27	39%	0.19	0.59	0.18	0.18	0.18	0.19	0.11	0.40
4	10 × 20	38%	0.23	0.84	0.25	0.25	0.25	0.24	0.20	0.53
5	10 × 20	38%	0.26	0.74	0.26	0.24	0.25	0.26	0.19	0.48
6	13 × 30	26%	0.11	0.61	0.19	0.21	0.20	0.11	0.07	0.46
7	10 × 20	35%	0.24	0.75	0.27	0.26	0.26	0.24	0.18	0.56
8	15 × 30	24%	0.11	0.46	0.20	0.21	0.20	0.11	0.05	0.51
9	10 × 20	36%	0.24	0.76	0.27	0.25	0.26	0.24	0.18	0.51
10	10 × 20	38%	0.25	0.83	0.26	0.25	0.25	0.26	0.21	0.50
11	10 × 20	38%	0.25	0.78	0.25	0.25	0.25	0.26	0.20	0.54
12	10 × 20	35%	0.24	0.76	0.26	0.26	0.26	0.24	0.18	0.54

φ, porosity [%]; L_{px}, length of the pore in the weft direction [mm]; L_{py}, length of the pore in the warp direction [mm]; D_{hx}, diameter of the weft threads [mm]; D_{hy}, diameter of the warp threads [mm]; D_h, threads diameter [mm]; D_i, inside circumference of the pore, diameter [mm]; S_p, pore area [mm^2]; e, thickness [mm].

Source: Reprinted from López et al., 2020.

Low-speed wind tunnel tests can be carried out to determine the aerodynamic characteristics of the screens (Bailey et al., 2003; Klose and Tantau, 2004; Molina-Aiz et al., 2006; Valera et al., 2006; Castellano et al., 2016; Pérez-Vega et al., 2016). Figures 11.7 and 11.8 show the wind tunnel of the research group AGR198 of the CIAIMBITAL Research Centre. The wind tunnel control system can be consulted with Espinoza et al. (2015).

For each insect-proof screen, it is necessary to carry out at least three tests; samples of the screens are randomly chosen for each test (Espinoza et al., 2015; López et al., 2018). With this wind tunnel, tests can be carried out up to a speed of 10 m s^{-1} or a pressure drop of 200 Pa.

From wind tunnel measurements, we can obtain the curve that relates the pressure drop ΔP [Pa] with air velocity u [m s^{-1}] for each screen (Figure 11.9). Insect-proof screens can be described aerodynamically as a porous media by the following parameters: the screen permeability K_p [m^2] that varies with the geometry of the net and is independent of the type of fluid (Nield and Bejan, 1999), and the inertial factor or drag constant Y [dimensionless]. Alternatively, drag effect of the insect-proof screen in the airflow can be characterized by only one parameter, the pressure drop coefficient F_φ.

FIGURE 11.7 Wind tunnel designed at the University of Almería.

Source: Reprinted with permission from A. López-Martínez, et al., 2020. https://creativecommons.org/licenses/by/4.0/

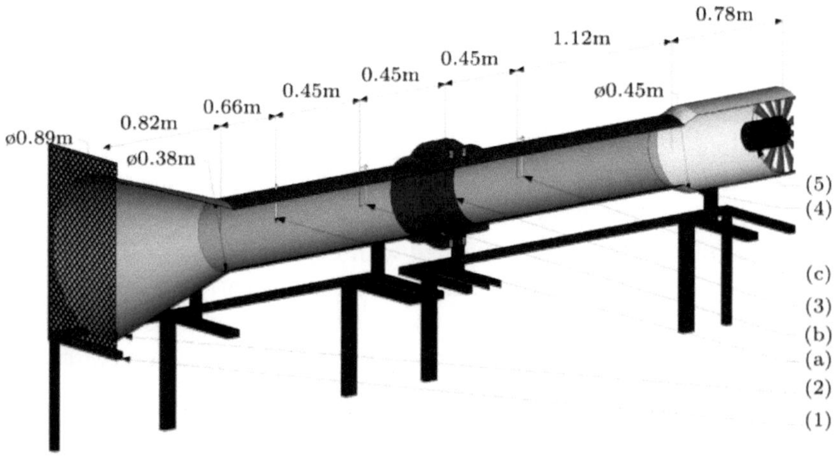

FIGURE 11.8 Principal components of the wind tunnel of the research group AGR198 of the CIAIMBITAL research center: (1) flow conditioner; (2) contraction; (3) test section; (4) diffuser; (5) elastic joint clamp and fan. (a) Hot-film anemometer and temperature probe; (b and c) Pitot tubes.

Source: Reprinted/Adapted from Espinoza et al., 2015. https://creativecommons.org/licenses/by/4.0/

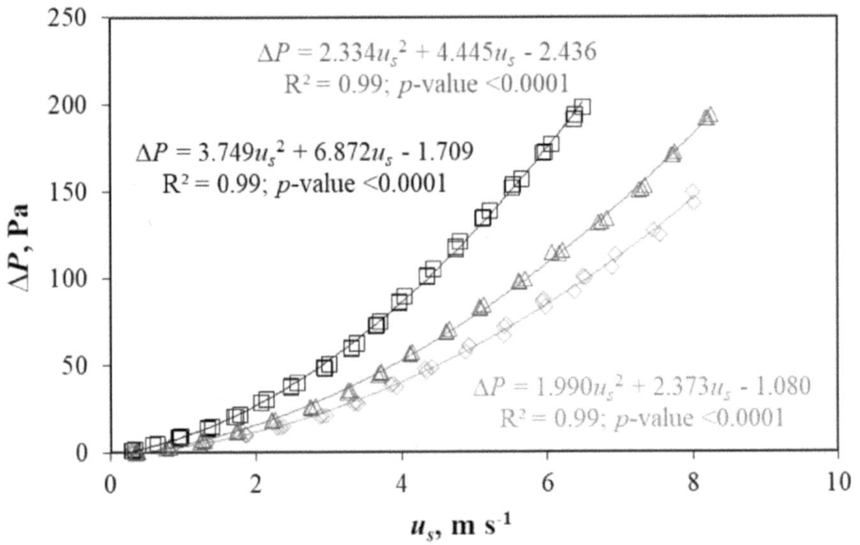

FIGURE 11.9 Curves of pressure drops ΔP [Pa] obtained in wind tunnel for three different insect-proof screens: 10 × 20 threads cm^{-2} and 0.350 of porosity (Δ); 13 × 30 threads cm^{-2} and 0.263 of porosity (□); 10 × 20 threads cm^{-2} and 0.360 of porosity (◊).

Pressure gradient originated by the insect-proof screens can be explained by a modification of Darcy's equation (Forchheimer, 1901):

$$\frac{\partial P}{\partial x} = -\left(\frac{\mu}{K_p}u + \rho_a\left(\frac{Y}{K_p^{1/2}}\right)|u|u\right) \tag{11.4}$$

where $\boldsymbol{P}$ is the pressure [Pa], $\boldsymbol{x}$ is the direction of the main flow, $\boldsymbol{u}$ is the velocity of air through the screen [m s^{-1}], $\boldsymbol{\mu}$ is the air dynamic viscosity [kg s^{-1} m^{-1}], and $\boldsymbol{\rho_a}$ is the density of air [kg m^{-3}]. To determine $\boldsymbol{\mu}$ and $\boldsymbol{\rho_a}$, it's necessary to know the temperature of air (Molina-Aiz et al., 2004). We can use a second-degree polynomial of $\boldsymbol{u}$ to fit the values of $\boldsymbol{\Delta P}$ measured experimentally (Miguel et al., 1997; Dierickx, 1998; Muñoz et al., 1999; Molina-Aiz et al., 2006; Valera et al., 2006). The zero-order term can be ignored (Miguel et al., 1997; Molina-Aiz et al., 2009; López et al., 2014), resulting the following expression:

$$\Delta P = au^2 + bu \tag{11.5}$$

Comparing the first- and second-order coefficients in eqs 11.5 and 11.4, parameters K_p and Y can be calculated (Molina-Aiz et al., 2009):

$$K_p = e\frac{\mu}{b} \tag{11.6}$$

$$Y = \frac{a\,K_p^{0.5}}{\rho_a\ \mathrm{e}} \tag{11.7}$$

The use of eqs 11.6 and 11.7 needs to measure the insect-proof screen thickness ***e***. Alternatively, the relation between $\boldsymbol{\Delta P}$ and $\boldsymbol{u}$ can be explained using the Bernoulli's equation (Kosmos et al., 1993; Pearson and Owen, 1994; Montero et al., 1997; Teitel and Shklyar, 1998; Bailey et al., 2003):

$$\Delta P = -\tfrac{1}{2} F_\varphi \rho_a u^2 \tag{11.8}$$

where F_φ is the pressure drop coefficient of the insect-proof screen that can be calculated with eqs 11.3 and 11.8, considering $\partial P/\partial x = \Delta P/e$ (Molina-Aiz et al., 2009):

$$F_\varphi = \frac{2\,e}{K_p^{0.5}}\left(\frac{1}{Re_p} + Y\right) \tag{11.9}$$

Teitel (2001) indicated that F_φ can be applied to estimate the pressure drop caused for a screen below the determined specific Reynolds number limit. For Reynolds number Re_p determined as a function of permeability, this specific limit can be established to $Re_p < 10^5$ (Molina-Aiz et al., 2009). The Reynolds number Re_p can be obtained from (Nield and Bejan, 1999):

$$Re_p = \frac{\sqrt{K_p}\,u\rho_a}{\mu} \tag{11.10}$$

Following the indicated methodology, we can determine the aerodynamic characteristics (Table 11.2) of the insect-proof screens indicated in Table 11.1. This characterization is essential to apply microclimatic simulation models in greenhouses (ventilation models, CFD simulations, etc.).

11.4 MODELS CHARACTERIZING INSECT-PROOF SCREENS' AERODYNAMICS

To accurately determine the parameters associated with the screens' aerodynamics, it is necessary to determine their geometric characteristics and carry out wind tunnel tests precisely. Different authors have obtained models that allow estimating the aerodynamic characteristics of the screens based on certain geometric parameters. There are models that allow estimating $\boldsymbol{F_\varphi}$ as a function of the porosity $\boldsymbol{\varphi}$ and a Reynolds number $\boldsymbol{Re_d}$ that is function of the thread diameter (Hayama et al., 2000; Pinker and Herbert, 1967; Linker et al., 2002; Bailey et al., 2003). To determine the thread diameter $\boldsymbol{D_h}$ and porosity $\boldsymbol{\varphi}$, digital image processing can be used, as described in Section 11.2. Also, the porosity can be estimated by counting black and white pixels from digital images of the insect-proof screens. Specific softwares can determine screen porosity by detecting the vertices of the pores; this method is more appropriate as it maintains a suitable ratio between the areas of the net occupied by threads and pores (López et al., 2020). The thread diameter can also be measured directly with a micrometer, but this task can be very laborious. Several models predicting the value of $\boldsymbol{F_\varphi}$ from the porosity $\boldsymbol{\varphi}$ [m^2 m^{-2}] and a Reynolds number $\boldsymbol{Re_d}$ are presented below.

TABLE 11.2 Aerodynamic Characteristics for Different Types of Insect-Proof Screens: Permeability K_p [m^2], Inertial Factor Y, and Pressure Drop Coefficient F_φ Depending on Re_p.

N	K_p	Y	F_φ
1	1.85×10^{-9}	0.16	$18.2 \times (Re_p^{-1} + Y)$
2	2.45×10^{-9}	0.19	$22.8 \times (Re_p^{-1} + Y)$
3	2.03×10^{-9}	0.19	$17.8 \times (Re_p^{-1} + Y)$
4	5.18×10^{-9}	0.22	$14.6 \times (Re_p^{-1} + Y)$
5	3.93×10^{-9}	0.23	$15.3 \times (Re_p^{-1} + Y)$
6	1.44×10^{-9}	0.27	$24.2 \times (Re_p^{-1} + Y)$
7	3.15×10^{-9}	0.20	$20.1 \times (Re_p^{-1} + Y)$
8	1.51×10^{-9}	0.38	$26.1 \times (Re_p^{-1} + Y)$
9	5.61×10^{-9}	0.25	$13.6 \times (Re_p^{-1} + Y)$
10	7.00×10^{-9}	0.27	$12.0 \times (Re_p^{-1} + Y)$
11	6.43×10^{-9}	0.26	$13.3 \times (Re_p^{-1} + Y)$
12	3.42×10^{-9}	0.20	$18.3 \times (Re_p^{-1} + Y)$

Source: Reprinted from López et al., 2020.

Pinker and Herbert (1967) carried out tests on a selection of eight single gauzes placed normal to the axis of a duct, providing a relation between the loss factor and the Reynolds number, with a correction factor depending on gauzes porosity:

$$F_\varphi = \left(\frac{13.0}{Re_d} + 0.82\right)\left(\frac{1-\varphi^2}{\varphi^2}\right) \quad (11.11)$$

Hayama et al. (2000) proposed a method based on the combination of expressions for the enthalpy balance and for the balance between aerodynamic resistance and buoyancy force in a ventilation model. They obtained a relationship of the flow resistance coefficient values for wire nets based on the Reynolds number and the net porosity coefficient:

$$F_{\varphi} = 28\left(\frac{Re_d\,\varphi^2}{1-\varphi}\right)^{-0.95} \tag{11.12}$$

Bailey et al. (2003) used a modified fan test facility to determine the pressure differences required to create airflows through five insect-proof screens. They proposed an empirical correlation of the pressure loss coefficient with a function of Reynolds number (based on fiber diameter) and screen porosity:

$$F_{\varphi} = \left[\frac{18}{Re_d} + \frac{0.75}{log\left(Re_d + 1.25\right)} + 0.055\,log\left(Re_d\right)\right]\left[\frac{1-\varphi^2}{\varphi^2}\right] \tag{11.13}$$

The Reynolds number Re_d can be calculated as (Nield and Bejan, 1999):

$$Re_d = \frac{D_h u \rho_a}{\mu} \tag{11.14}$$

An interesting alternative for estimating the pressure drop produced by insect-proof screens is to estimate the permeability $\boldsymbol{K_p}$ and the inertial factor $\boldsymbol{Y}$ from their geometric properties. We can relate $\boldsymbol{K_p}$ with the porosity $\boldsymbol{\varphi}$ [$m^2\ m^{-2}$] and fiber thread diameter $\boldsymbol{D_h}$ [m] using the semiempirical Kozeny equation (Kozeny, 1927; Nield and Bejan, 1999; Berg, 2014):

$$K_p = \frac{D_h^{\,2}\varphi^3}{\beta\left(1-\varphi\right)^2} \tag{11.15}$$

where $\beta = 1/\tau$ is the inverse of the tortuosity of the porous medium $\boldsymbol{\tau}$ expressing the relative difference between the microscopic (interstitial) head gradient along the streamline and the macroscopic head gradient (Berg, 2014).

On the other hand, we can use an expression that related $\boldsymbol{Y}$ with the fiber diameter and pore size (Beavers et al., 1973, Nield and Bejan, 1999):

$$Y = A + B\frac{D_h}{D_i} \tag{11.16}$$

in which $\boldsymbol{A}$ and $\boldsymbol{B}$ are constants depending on the nature of the porous medium. For the screens presented here, we can use the thread diameter $\boldsymbol{D_h}$ and the inner pore diameter $\boldsymbol{D_i}$ as characteristic lengths.

Finally, K_p and Y can be estimated using uniquely the porosity φ with the expressions proposed by Miguel (1998):

$$K_p = 3.44\times10^{-9}\varphi^{1.6} \tag{11.17}$$

$$Y = 4.30\times10^{-2}\varphi^{-2.13} \tag{11.18}$$

Recently, López et al. (2020) presented different expressions for estimating F_φ from the porosity φ [m^2 m^{-2}] and a Reynolds number Re_d that is function of the thread diameter D_h or a Reynolds number Re_t that is function of the thickness e (easy to measure with a micrometer):

$$F_\varphi = \left(0.481002+\frac{11.5331}{Re_d}\right)\left(\frac{1-\varphi^2}{\varphi^2}\right) \tag{11.19}$$

$$F_\varphi = \left(0.475502+\frac{26.2114}{Re_t}\right)\left(\frac{1-\varphi^2}{\varphi^2}\right) \tag{11.20}$$

The Reynolds number Re_t can be calculated as:

$$Re_t = \frac{eu\rho_a}{\mu} \tag{11.21}$$

Values of constants β, A, and B, in eqs 11.15 and 11.16 obtained from Nield and Bejan (1999) were calculated for several insect-proof screens by López et al. (2020). These authors proposed the following expression for estimating K_p (with a value β = 2.0679) from thread diameter D_h [m] and porosity φ [m^2 m^{-2}]:

$$K_p = \frac{D_h^2\varphi^3}{2.0679(1-\varphi)^2}+3.8362\times10^{-10} \tag{11.22}$$

A second expression can be applied to calculate the inertial factor of screens Y (with fixed values of A = 0.0571195 and B = 0.135966) from thread diameter [m] and from the diameter of the circumference that inscribes the pore D_i [m]:

$$Y = 0.0571195+0.135966\frac{D_h}{D_i} \tag{11.23}$$

Values of $F_{\varphi,sim}$ of the different types of insect-proof screens presented in Table 11.1 can be estimated using eqs 11.11–11.13, 11.19, and 11.20. Comparing with the values of $F_{\varphi,obs}$ determined from wind tunnel measurements (eq 11.9 and Table 11.1), a higher error is observed for low air velocities (Figure 11.10). The increasing error at low velocities produced

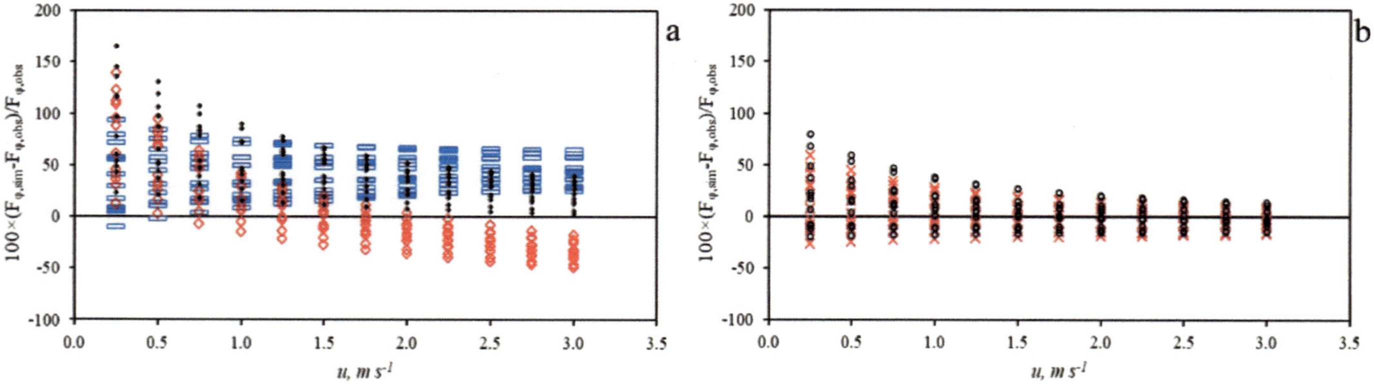

FIGURE 11.10 Difference between $F_{\varphi,obs}$ and $F_{\varphi,sim}$ (%) in function of the air velocity. (a) $F_{\varphi,sim}$ estimated using eqs 11.11 (═), 11.12 (◊), and 11.13. (b) $F_{\varphi,sim}$ assessed using eqs 11.19 (×) and 11.20 (○).

Source: Adapted from López-Martínez et al. 2020.

by these models is not suitable because in naturally ventilated greenhouses, air velocity through the screens installed in the vent openings generally is under the limit of 1.5 m s^{-1} (Molina-Aiz et al., 2009; López et al., 2012). López et al. (2020) observed that eq 11.11 produced an error greater than 20% for 28% of the screens analyzed. Using eqs 11.12 and 11.13, errors produced estimating $\boldsymbol{F}_{\varphi}$ were larger than 20% for 88 and 89% of screens analyzed, respectively. The best results were obtained using eqs 11.19 and 11.20, with errors in estimation of $\boldsymbol{F}_{\varphi}$ superior to 20% for only 6 and 8% of screens, respectively. Castellano et al. (2016) used an expression similar to eq 11.13 for estimating the aerodynamic parameters of different nets (11 flat woven nets with porosities between 34.4 and 71.6%; four knitted nets with porosities between 35.5 and 84.0%) with satisfactory results, but these authors only used the expression for air velocities above 4 m s^{-1}.

Pinker and Herbert (1967) obtained eq 11.11 performing tests with eight woven wire gauzes (porosities between 30 and 70%). These woven wire gauzes have an arrangement of the threads similar to the insect-proof screens. However, the threads of these woven wire gauzes were metallic, while the threads used for manufacture the insect-proof screens are normally made of polyethylene. The worst performance of the model presented in the eq 11.11 compared to the models presented in eqs 11.19 and 11.20 can be due to the differences existing between the woven wire gauzes used by Pinker and Herbert (1967) and the insect-proof screens used in greenhouses.

In the same way, Hayama et al. (2000) obtained the eq 11.12 performing tests with nine metallic wire nets (porosities between 37.1 and 77.8%). The differences between the material and porosities used to obtain eq 11.12 and the insect-proof screens used in greenhouses would explain their inaccuracy.

Bailey et al. (2003) obtained the eq 11.13 from tests with five insect-proof screens (porosity: 25, 45, 53, 66, and 68%). Four of these five screens have porosity values much higher than the screens presented in Table 11.1. This difference between porosities justifies the results offered by eq 11.13. Equations 11.19 and 11.20 proposed by López et al. (2020) improve the results obtained with eqs 11.11–11.13.

Figure 11.11 shows the values of $\boldsymbol{Y}$ and $\boldsymbol{K}_p$ estimated using eqs 11.17 and 11.18, deduced by Miguel (1998) using only the porosity $\boldsymbol{\varphi}$, and those estimated using eqs 11.22 and 11.23 from porosity $\boldsymbol{\varphi}$ and characteristic pore $\boldsymbol{D}_i$ and thread $\boldsymbol{D}_h$ diameters (Nield and Bejan, 1999; López et al., 2020). For these four equations, the estimated values have been represented with respect to the experimentally observed values (Figure 11.11).

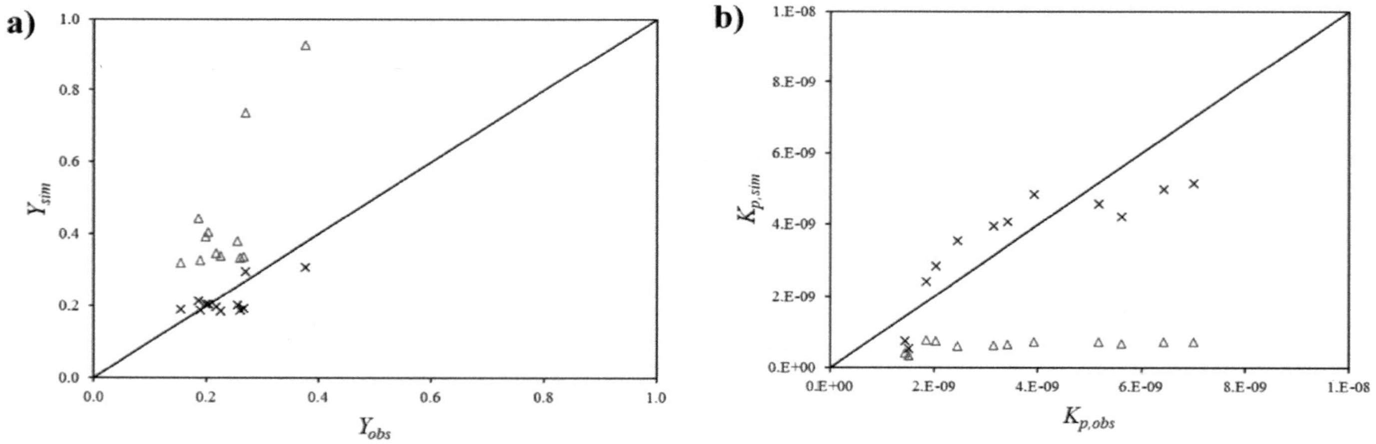

FIGURE 11.11 Simulated values of Y (a) and K_p (b) versus experimentally observed values. (Δ), eqs 11.17 and 11.18 (Miguel, 1998); (×), eqs 11.22 and 11.23 (López et al., 2020).

Source: Adapted from López-Martínez et al. 2020.

With the eqs 11.17 and 11.18, proposed by Miguel (1998), estimated values of $\boldsymbol{Y}$ and $\boldsymbol{K}_p$ are far from the values calculated from the tests in the wind tunnel. These equations overestimate the values of $\boldsymbol{Y}$ (Figure 11.11a) and underestimate the values of $\boldsymbol{K}_p$ (Figure 11.11b). As opposed to that, estimated values of $\boldsymbol{Y}$ and $\boldsymbol{K}_p$ obtained with eqs 11.22 and 11.23 are close to the values calculated from the tests in the wind tunnel (Figure 11.11).

Considering the results obtained with eqs 11.17 and 11.18, we recommend using eqs 11.22 and 11.23 to estimate the $\boldsymbol{Y}$ and $\boldsymbol{K}_p$ values of insect-proof screens for greenhouses. Miguel (1998) obtained the eqs 11.17 and 11.18 performing tests with 14 screens, including thermal screens (eight flat woven screens, insect-proof screens, with porosities between 4 and 90%; six knitted nets, shade nets, with porosities between 45 and 89%). This high variability of screens type may be the reason for the inaccuracy offered by these equations to estimate the $\boldsymbol{Y}$ and $\boldsymbol{K}_p$ values of the insect-proof screens presented in Table 11.1.

On the other hand, eqs 11.22 and 11.23 were obtained by wind tunnel tests of 35 insect-proof woven screens (rectangular pores, weft, and warp polyethylene threads), all of these screens with relatively similar porosity values (from 23.7 to 55.6%). To estimate the $\boldsymbol{Y}$ and $\boldsymbol{K}_p$ values of insect-proof screens similar to those indicated in Table 11.1, the use of eqs 11.22 and 11.23 is recommended.

Equations 11.22 and 11.23, however, require a more detailed geometric characterization of insect-proof screens. To correctly apply eqs 11.22 and 11.23, it is necessary to determine the value of the diameter of the threads, the diameter of maximum circle inscribed in pore, and the porosity of the net. Once the values of $\boldsymbol{Y}$ and $\boldsymbol{K}_p$ have been estimated, it is necessary to measure the thickness of the screen in order to obtain the pressure drop curve using eq 11.4. Also, to perform CFD simulations of the airflow through this type of screens, it is necessary to know these parameters ($\boldsymbol{Y}$ and $\boldsymbol{K}_p$) that characterize the mesh aerodynamically (Molina-Aiz et al., 2017).

11.5 CONCLUSIONS

In this chapter, a description of the models from the literature has been made, which can be used to estimate the parameters that aerodynamically characterize the insect-proof screens used in greenhouses. These screens have a weaving of polyethylene threads that are divided into two groups (weft and warp threads).

To model the pressure drop–velocity curve generated by the airflow resistance of insect-proof screens, there are two options:

1. With a quadratic term ($\Delta P = au^2$), needing the knowledge of the pressure drop coefficient $\boldsymbol{F_\varphi}$ of the screen.
2. With a quadratic polynomial ($\Delta P = au^2 + bu$), needing the knowledge of two parameters: permeability $\boldsymbol{K_p}$ and inertial factor $\boldsymbol{Y}$.

The existing models for estimating the pressure drop coefficient $\boldsymbol{F_\varphi}$ of insect-proof screens use the porosity of the screens $\boldsymbol{\varphi}$ and a Reynolds number to characterize the airflow through insect-proof screens. The Reynolds numbers used in different models are based on the diameter of the thread $\boldsymbol{Re_d}$ or the thickness of the screen $\boldsymbol{Re_t}$.

All existing models to estimate the permeability $\boldsymbol{K_p}$ and the inertial factor $\boldsymbol{Y}$ of insect-proof screens use the porosity of the screens $\boldsymbol{\varphi}$. More simple and inaccuracy models only use this parameter, and more complex and accuracy models include the diameter of the thread $\boldsymbol{D_h}$ and the diameter of the circumference that inscribes the pore $\boldsymbol{D_i}$.

KEYWORDS

- **insect-proof screens**
- **porous media**
- **pressure drop**
- **permeability**
- **airflow**

REFERENCES

Álvarez, A. J. Estudio de las características geométricas y del comportamiento aerodinámico de las mallas antiinsectos utilizadas en los invernaderos como medida de protección vegetal. PhD Dissertation, Universidad de Almería, Almería, Spain, 2010.

Álvarez, A. J.; Oliva, R. M.; Valera, D. L. Software for the Geometric Characterisation of Insect-Proof Screens. *Comput. Electron. Agric.* **2012,** *82*, 134–144.

Baeza, E. J.; Pérez-Parra, J. J.; Montero, J. I.; Bailey, B. J.; López, J. C.; Gázquez, J. C. Analysis of the Role of Sidewall Vents on Buoyancy-Driven Natural Ventilation in Parral-Type Greenhouses with and without Insect Screens Using Computational Fluid Dynamics. *Biosyst. Eng.* **2009,** *104*, 86–96.

Bailey, B. J.; Montero, J. I.; Pérez Parra, J.; Robertson, A. P.; Baeza, E.; Kamaruddin, R. Airflow Resistance of Greenhouse Ventilators with and without Insect Screens. *Biosyst. Eng.* **2003,** *86* (2), 217–229.

Baker, J. R.; Jones, R. K. Screening as Part of Insect and Disease Management in the Greenhouse. *NC Flower Growers' Bull.* **1989,** *34*, 1–9. https://hortscans.ces.ncsu.edu/uploads/s/c/screenin_53e4f3cf4132a.pdf

Bartzanas, T.; Boulard, T.; Kittas, C. Numerical Simulation of the Airflow and Temperature Distribution in a Tunnel Greenhouse Equipped with Insect-Proof Screen in the Openings. *Comput. Electron. Agric.* **2002,** *34*, 207–221.

Beavers, G. S.; Sparrow, E. M.; Rodenz, D. E. Influence of Bed Size on the Flow Characteristics and Porosity of Randomly Packed Beds of Spheres. *J. Appl. Mech.* **1973,** *40* (3), 655–660.

Berg, C. F. Permeability Description by Characteristic Length, Tortuosity, Constriction and Porosity. *Transp. Porous Med.* **2014,** *10*, 81–400.

Berlinger, M. J.; Leblush-Mordechl, S.; Fridja, D.; Mor, N. The Effect of Types of Greenhouse Screens on the Presence of Western Flower Thrips: A Preliminary Study. *OILB-SROP Bull.* **1992,** *16* (2), 13–19.

Berlinger, M. J.; Mordechl, S.; Leeper, A. In *Application of Screens to Prevent Whitefly Penetration into Greenhouses in the Mediterranean Basin*, Proceedings of the Working Group Integrated Control in Protecting Crops under Mediterranean Climate; Alassio, Italy, 1991; pp 105–110.

Berlinger, M. J.; Taylor, R. A. J.; Lebiush-Mordechi, S.; Shalhevet, S.; Spharim, I. Efficiency of Insect Exclusion Screens for Preventing Whitefly Transmission of Tomato Yellow Leaf Curl Virus of Tomatoes in Israel. *Bull. Entomol. Res.* **2002,** *92*, 367–373.

Bethke, J. A; Paine, T. D. Screen Hole Size and Barriers for Exclusion on Insect Pest of Glasshouse Crops. *J. Entomol Sci.* **1991,** 26, 169–177.

Castellano, S.; Scarascia-Mugnozza, G.; Russo, G.; Briassoulis, D.; Mistriotis, A.; Hemming, S.; Waaijenberg, D. Plastic Nets in Agriculture: A General Review of Types and Applications. *Appl. Eng. Agric.* **2008**, *24*, 799–808.

Castellano, S.; Starace, G.; De Pascalis, L.; Lippolis, M.; Mugnozza, G. S. Test Results and Empirical Correlations to Account for Air Permeability of Agricultural Nets. *Biosyst. Eng.* **2016,** *150*, 131–141.

Dierickx, I. E. Flow Reduction of Synthetic Screens Obtained with Both a Water and Airflow Apparatus. *J. Agric. Eng. Res.* **1998,** *71*, 67–73.

Espinoza, K.; Valera, D. L.; Torres, J. A.; López, A.; Molina-Aiz, F. D. An Auto-Tuning PI Control System for an Open Circuit Low-Speed Wind Tunnel Designed for Greenhouse Technology. *Sensors* **2015,** *15*, 19723–19749.

Fatnassi, H.; Boulard, T.; Bouirden, L. Simulation of Climatic Conditions in Full-Scale Greenhouse Fitted with Insect-Proof Screens. *Agric. For. Meteorol.* **2003,** *118*, 97–111.

Fatnassi, H.; Boulard, T.; Demrati, H.; Bouirden, L.; Sappe, G. Ventilation Performance of a Large Canarian-Type Greenhouse Equipped with Insect-Proof Nets. *Biosyst. Eng.* **2002,** *82* (1), 97–105.

Fatnassi, H.; Boulard, T.; Poncet, C.; Chave, M. Optimisation of Greenhouse Insect Screening with Computational Fluid Dynamics. *Biosyst. Eng.* **2006,** *93* (3), 301–312.

Forchheimer, P. Wasserbewegung durch Boden [Water movement from ground]. *Zeitschrift des Vereines Deutscher Ingenieure* **1901,** *45*, 1782–1788.

Harmanto; Tantau, H. J.; Salokhe, V. M.; Microclimate and Air Exchange Rates in Greenhouses Covered with Different Nets in the Humid Tropics. *Biosyst. Eng.* **2006,** *94* (2), 239–253.

Hayama, S.; Peng, G.; Ishizuka, M. In *Measurement of Flow Resistance Coefficients for Wire Nets in Natural Air Convection Flow*, Proceedings of Sixth Triennal International Symposium on Fluid Control, Measurement and Visualization (FLUCOME 2000); Sherbrooke, Canada, 2000; p 12. https://www.tib.eu/en/search/id/BLCP%3ACN071523984/Measurement-of-Flow-Resistance-Coefficients-for// (Accessed 8 April 2021).

Katsoulas, N.; Bartzanas, T.; Boulard, T.; Mermier, M.; Kittas, C. Effect of Vent Openings and Insect Screens on Greenhouse Ventilation. *Biosyst. Eng.* **2006,** *93* (4), 427–436.

Kittas, C.; Boulard, T.; Bartzanas, T.; Katsoulas, N.; Mermier, M. Influence of an Insect Screen on Greenhouse Ventilation. *Trans. ASAE* **2002,** *45* (4), 1083–1090.

Kittas, C.; Katsoulas, N.; Bartzanas, T.; Mermier, M.; Boulard, T. The Impact of Insect Screens and Ventilation Openings on the Greenhouse Microclimate. *Trans. ASABE* **2008,** *51* (6), 2151–2165.

Klose, F.; Tantau, H. J. Test of Insect Screens-Measurement and Evaluation of the Air Permeability and Light Transmission. *Europ. J. Hort. Sci.* **2004,** *69* (5), 235–243.

Kosmos, S. R.; Riskowski, G. L.; Christianson, L. L. Force and Static Pressure Resulting from Airflow Through Screens. *Trans. ASAE.* **1993,** *36* (5), 1467–1472.

Kozeny, J. Ueber Kapillare Leitung des wassers im boden [Capillary Conduction of the Water in the Ground]. *Wien Akad. Wiss.* **1927**, *136* (2a), 271–306.

Linker, R.; Tarnopolsky, M.; Seginer, I. Increased Resistance to Flow and Temperature-Rise Resulting from Dust Accumulation on Greenhouse Insect-Proof Screens, ASAE Annual International Meeting: Chicago, USA, 2002; p 9. https://doi.org/10.13031/2013.10475

López, A.; Molina-Aiz, F. D.; Valera, D. L.; Peña, A. Wind Tunnel Analysis of the Airflow Through Insect-Proof Screens and Comparison of their Effect when Installed in a Mediterranean Greenhouse. *Sensors* **2016,** *16*, 690.

López, A.; Molina-Aiz, F. D.; Valera, D. L.; Peña, A.; Espinoza, K. Effect of Material Ageing and Dirt on the Behaviour of Greenhouse Insect-Proof Screens. *Spanish J. Agric. Res.* **2018,** *16* (4), 0205. https://doi.org/10.5424/sjar/2018164-11711

López, A.; Valera, D. L.; Molina-Aiz, F. D.; Peña, A. Sonic Anemometry Measurements to Determine Airflow Patterns in Multi-Tunnel Greenhouse. *Spanish J. Agric. Res.* **2012,** *10* (3), 631–642.

López, A.; Valera, D. L.; Molina-Aiz, F. D.; Peña, A.; Marín, P. Field Analysis of the Deterioration After Some Years of Use of Four Insect-Proof Screens Utilized in Mediterranean Greenhouses. *Spanish J. Agric. Res.* **2013,** *11* (4), 958–967.

López, A.; Valera, D. L.; Molina-Aiz, F. D.; Peña, A.; Marín, P. Microclimate Evaluation of a New Design of Insect-Proof Screens in a Mediterranean Greenhouse. *Spanish J. Agric. Res.* **2014,** *12* (2), 338–352.

López-Martínez, A.; Molina-Aiz, F. D.; Valera, D. L.; Espinoza-Ramos, K. E. Models for Characterising the Aerodynamics of Insect-Proof Screens from their Geometric Parameters. *Biosyst. Eng.* **2020,** *192*, 42–55.

Miguel, A. F. Airflow Through Porous Screens: From Theory to Practical Considerations. *Ener. Build.* **1998,** *28*, 63–69.

Miguel, A. F.; Silva, A. M. Porous Materials to Control Climate Behaviour of Enclosures: An Application to the Study of Screened Greenhouses. *Ener. Build.* **2000,** *31*, 195–209.

Miguel, A. F.; Van de Braak, N. J.; Bot, G. P. A. Analysis of the Airflow Characteristics of Greenhouse Screening Materials. *J. Agric. Eng. Res.* **1997,** *67*, 105–112.

Molina-Aiz, F. D.; Norton, T.; López, A.; Reyes-Rosas, A.; Moreno, M. A.; Marín, P.; Espinoza K.; Valera D. L.; Using Computational Fluid Dynamics to Analyse the CO_2 Transfer in Naturally Ventilated Greenhouses. *Acta Hort.* **2017,** *1182*, 283–292.

Molina-Aiz, F. D.; Valera, D. L.; Álvarez, A. J. Measurement and Simulation of Climate Inside Almería-Type Greenhouses Using Computational Fluid Dynamics. *Agric. For. Meteorol.* **2004,** *125*, 33–51.

Molina-Aiz, F. D.; Valera, D. L.; Álvarez, A. J.; Madueño, A. A Wind Tunnel Study of Airflow Through Horticultural Crops: Determination of the Drag Coefficient. *Biosyst. Eng.* **2006,** *93* (4), 447–457.

Molina-Aiz, F. D.; Valera, D. L.; Peña, A. A.; Gil, J. A.; López, A. A Study of Natural Ventilation in an Almerı´a-Type Greenhouse with Insect Screens by Means of Tri-sonic Anemometry. *Biosyst. Eng.* **2009,** *104*, 224–242.

Möller, M.; Tanny, J.; Yan, L.; Cohen, S. Measuring and Predicting Evapotranspiration in an Insect-Proof Screenhouse. *Agric. For. Meteorol.* **2004,** *127*, 35–51.

Montero, J. I.; Muñoz, P.; Antón, A. Discharge Coefficients of Greenhouse Windows with Insect-Proof Screens. *Acta Hort.* **1997,** *443*, 71–77.

Muñoz, P.; Montero, J. I.; Antón, A.; Giuffrida, F. Effect of Insect-Proof Screens and Roof Openings on Greenhouse Ventilation. *J. Agric. Eng. Res.* **1999,** *73*, 171–178.

Nield, D. A.; Bejan, A. In *Convection in Porous Media,* 2nd ed.; Springer: New York, 1999.

Nucifora, S. Mallas antivirus para la defensa del cultivo de tomate protegido contra *Bemisia tabaci*, vector del TYLCV [Antivirus Meshes for the Defense of Tomato Crop Protected Against *Bemisia tabaci*, TYLCV Vector]. *Phytoma España.* **1994,** *57*, 40–48.

Pearson, C. C.; Owen, J. E. The Resistance to Air Flow of Farm Building Ventilation Components. *J. Agric. Engng Res.* **1994,** *57*, 53–65.

Pérez-Vega, C.; Ramírez Arias, J. A.; López-Cruz, I. L. Aerodynamic Characteristics of Anti-Insect Mesh Windows used in Greenhouses in Mexico. *Revista Mexicana de Ciencias Agrícolas* **2016,** *7* (3), 493–506.

Pinker, R. A.; Herbert, M. V. Pressure Loss Associated with Compressible Flow Through Square-Mesh Wire Gauzes. *J. Mech. Eng. Sci.* **1967,** *9* (1), 11–23.

Soni, P.; Salokhe, V. M.; Tantau, H. J. Effect of Screen Mesh Size on Vertical Temperature Distribution in Naturally Ventilated Tropical Greenhouses. *Biosyst. Eng.* **2005,** *92* (4), 469–482.

Taylor, R. A. J.; Shalhevet, S.; Spharim, I.; Berlinger, M. J.; Lebiush-Mordechi, S. Economic Evaluation of Insect-Proof Screens for Preventing Tomato Yellow Leaf Curl Virus of Tomatoes in Israel. *Crop Prot.* **2001,** *20*, 561–569.

Teitel, M. The Effect of Insect-Proof Screens in Roof Openings on Greenhouse Microclimate. *Agric. For. Meteorol.* **2001,** *110*, 13–25.

Teitel, M. The Effect of Screened Openings on Greenhouse Microclimate. *Agric. For. Meteorol.* **2007,** *143* (3-4), 159–175.

Teitel, M. Using Computational Fluid Dynamics Simulations to Determine Pressure Drops on Woven Screens. *Biosyst. Eng.* **2010,** *105*, 172–179.

Teitel, M.; Shklyar, A. Pressure Drop Across Insect-Proof Screens. *Trans. ASAE.* **1998,** *41* (6), 1829–1834.

Teitel, M.; Wenger, E. Air Exchange and Ventilation Efficiencies of a Monospan Greenhouse with one Inflow and One Outflow Through Longitudinal Side Openings. *Biosyst. Eng.* **2014**, *119*, 98–107. https://doi.org/10.1016/j.biosystemseng.2013.11.001.

Ticó, J.; Figuls, M.; Aparicio, V.; García, M. M.; Rodríguez, M. P. Eficacia de la malla fotoselectiva BioNet en la prevención de la mosca blanca y TYLCV en cultivos protegidos [Efficacy of the BioNet Photoselective Mesh in the Prevention of Whitefly and TYLCV in Protected Crops]. *Phytoma España.* **2002,** *135*, 87–92.

Valera, D. L.; Álvarez, A. J.; Molina-Aiz, F. D. Aerodynamic Analysis of Several Insect-Proof Screens used in Greenhouses. *Spanish J. Agric. Res.* **2006,** *4* (4), 273–279.
Valera, D. L.; Belmonte, L. J.; Molina-Aiz, F. D.; López, A. In *Greenhouse Agriculture in Almería. A Comprehensive Technoeconomic Analysis*, 1st ed.; Cajamar Caja Rural: Almería, Spain, 2016. https://www.publicacionescajamar.es/series-tematicas/economia/greenhouse-agriculture-in-almeria-a-comprehensive-techno-economic-analysis (accessed April 8, 2021).

CHAPTER 12

Ventilation Models in Greenhouses

F. D. MOLINA-AIZ[1], D. L. VALERA[1], A. LÓPEZ-MARTÍNEZ[1], and H. FATNASSI[2]

[1]*CIAMBITAL (Centro de Investigación en Agrosistemas Intensivos Mediterráneos y Biotecnología Agroalimentaria), University of Almería, Ctra. Sacramento s/n, Almería, Spain*

[2]*The International Center for Biosaline Agriculture (ICBA), Academic City, Dubai, United Arab Emirates*

ABSTRACT

Natural ventilation is the main method of climate control in greenhouses in warm climates. This technique allows acting on the main environmental parameters that affect crops: temperature, humidity, CO_2 concentration, and air velocity. Natural ventilation has the advantage of being a technique with minimal energy consumption (for motor drive when automated) as it is mainly based on a renewable energy such as wind. Natural ventilation is a complex physical phenomenon as it is the result of the interaction of wind and buoyancy forces. Several models allow determining approximately the airflow circulating in a greenhouse. However, most models do not allow to determine how air moves within the greenhouse. Only computational fluid dynamics (CFD) models can represent a spatial distribution of study variables and provide information on the airflow pattern generated by natural ventilation. More than 10 different forms of ventilation can be produced in a same greenhouse, depending on the combinations of roof and side openings, wind speed and direction, the inside–outside temperature difference, and the presence of a crop within the greenhouse. Energy balance and models based in the Bernouilli's equation determine the air change rate, with the

Protected Cultivation: Structural Design, Crop Management, Modeling and Automation.
Mahesh Chand Singh and K. K. Sharma (Eds.)

assumption of homogeneity in inside air. This chapter presents the different semiempirical models of analysis of natural ventilation in greenhouses.

12.1 INTRODUCTION

The first work about ventilation in greenhouses was carried out in the middle of the last century (Morris and Neale, 1954; Businger, 1963; Walker, 1965) and since then various experimental techniques and simulation methods have been used to study the influence of ventilation systems on the inner microclimate of the greenhouses. Given the importance of natural ventilation in climate control, many research works have been carried out with the aim of establishing a numerical function to directly link the ventilation rate with different environmental variables, such as wind direction and speed, solar radiation or outside and inside temperatures.

To make an accurate description of the flow of a fluid in a three-dimensional domain, it is necessary to determine the fields of velocity, pressure, and other variables such as enthalpía, concentration of chemical substances, etc., which can be transported by the flow throughout the domain. Fluid is considered to be a continuous medium in which conservation laws can be written as (Ferziger and Peric, 2002; Bournet et al., 2007):

$$\frac{\partial \phi}{\partial t}+\frac{\partial (u \cdot \phi)}{\partial x}+\frac{\partial (v \cdot \phi)}{\partial y}+\frac{\partial (w \cdot \phi)}{\partial z}=\tilde{\mathrm{A}}_{\phi} \cdot \nabla^2 \phi + S_{\phi} \tag{12.1}$$

where, ϕ represents the concentration of the transported quantity through the domain in non-dimensional form (mass, momentum, enthalpy, turbulence parameters, and concentration of substances as water vapor or CO_2). Symbols $\boldsymbol{u}$, $\boldsymbol{v}$, and $\boldsymbol{w}$ represent the components of the velocity vector, $\boldsymbol{\Gamma}_{\phi}$ is the diffusion coefficient and $\boldsymbol{S}_{\phi}$ is the source term (values were reported in Molina-Aiz et al. 2004; Boulard et al., 2017; Kato, 2018).

The conservation eq (12.1) for the fluid mass is the continuity equation, for the three momentum components is the Navier–Stokes equation and for enthalpy is the energy conservation equation. Due to the nonlinearity of the conservation equations, it is not possible to analytically determine the solutions of the system of equations given by (12.1) and the different values of ϕ except for rare situations in one-dimensional or two-dimensional flows where multiple simplifications of equations can be performed (Schlichting, 1955). Airflow patterns in greenhouses are characterized by three-dimensional geometries, with complex external flows around the structure, involving important pressure gradients, curvature of streamlines, eddies and vortex, flow separation, and reattachment with increased turbulence. In these conditions, it is impossible to obtain an analytical solution for the fields of velocity, pressure, and other amounts transported (Boulard et al., 2002).

Different methods have been developed (Figure 12.1) to determine microclimate and mass and energy transfers in greenhouses. The rapid development of numerical computation methods has helped to obtain approximate solutions to phenomenological micro-models (Figure 12.1) in which the space occupied by the greenhouse is discretized in geometric elements with an order of magnitude of decimeters or even less. These methods called computational fluid dynamics (CFD) are based on the deduction of the variables that characterize the flow (density $\boldsymbol{\rho}$, velocity $\boldsymbol{v}$, pressure $\boldsymbol{P}$, temperature $\boldsymbol{T}$, turbulent kinetic energy $\boldsymbol{\kappa}$, dissipation rate of turbulent kinetic energy ε, absolute humidity $\boldsymbol{x}$, and CO_2 concentration $\boldsymbol{C}$) from the conservation equation of each variable for all points defined by the geometric elements in which the calculation domain is divided.

The main advantages of the development of mathematical models for greenhouses ventilation are (Schauberger et al., 2000; Norton et al., 2007; Hemming et al., 2019):

- They quantify the needs of air change throughout the greenhouse openings, contributing to the development of optimization strategies aimed at improving productivity.
- They can be incorporated as part of climate controllers as well as used for the design of ventilation systems.
- They allow to combine space- and time-dependent parameters that affect the emission and consumption of substances (such as water vapor, CO_2, or fungal spores) to efficiently calculate the temporal variation of their concentration.
- Artificial intelligence (AI) based in climate control models could help to develop "autonomous greenhouses" in the future.

Models applied to greenhouses can be used for the design of new greenhouses or for climate control of existing structures. Models can be divided into empirical or black box, semi-empirical, and physical or phenomenological models (Figure 12.1).

12.1.1 EMPIRICAL MODELS

Empirical or black-box models (Figure 12.1) are based on experimental observations without the consideration of physical principles, being appropriate

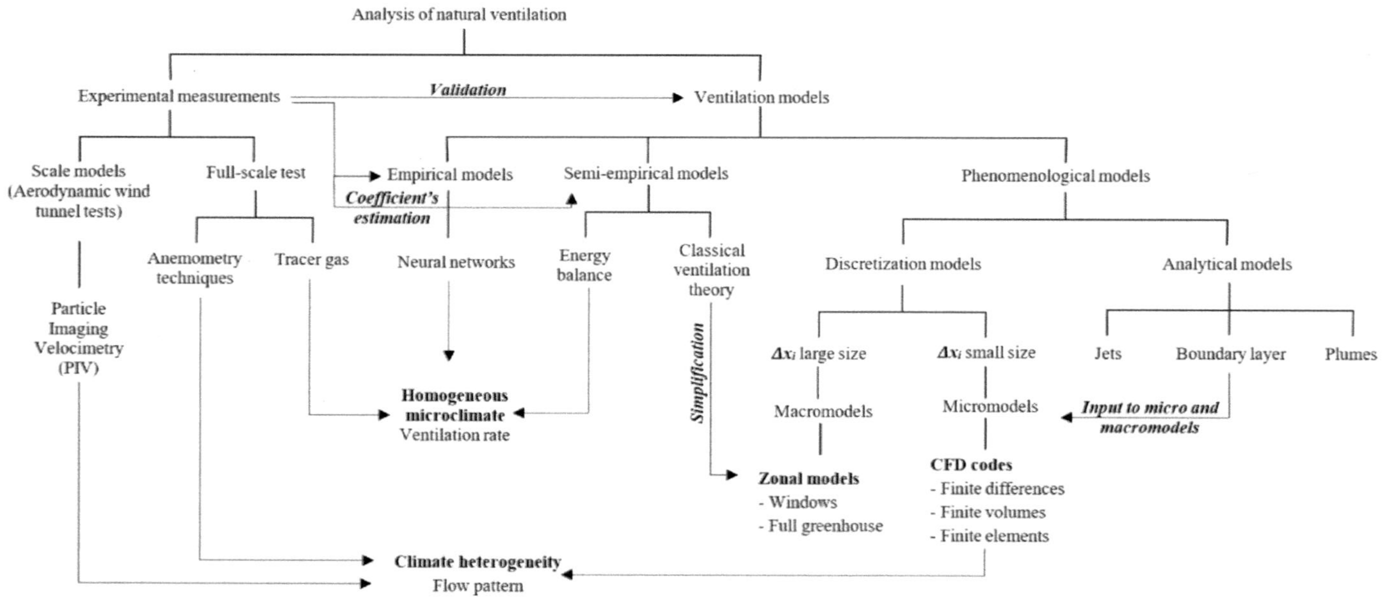

FIGURE 12.1 Various methods of study of natural ventilation in greenhouses

Source: Adapted from Krauss et al., 1997.

for use in computer climate control algorithms (Bot, 1980; Udink ten Cate and Van de Vooren, 1977) or as part of more complex climate, growth, and economic models (Krug and Liebig, 1980).

The empirical approach has also been used to calculate greenhouse heating needs (Schockert and Von Zabeltiz, 1980; Strom and Amsen, 1981; García et al., 1998), establishing mathematical relationships that depend on the variables of the outside environment of the greenhouse. These empirical functions present the disadvantage of being specific to the type of greenhouse, region, and weather conditions for which they are developed.

These empirical models have also been used for the study of ventilation rate in "Venlo" greenhouses (Bot, 1983), relating the ventilation flow to the type of windows, the opening angle, the wind speed and direction, and the temperature differences between inside and outside (De Jong, 1990). Black-box models (with no explicit physical meaning) are able to make accurate microclimate predictions within the greenhouse with a sufficient number of adjustable parameters, as long as all factors that have a significant influence on the results are present as input of the model (Seginer et al., 1994). These models use three types of variables: climatic variables (such as outside temperature or net radiation), control variables (ventilation airflow), and state variables, such as crop size (which varies slowly and affects indoor temperature, humidity, and CO_2 concentration), associating each state variable with a differential equation (Boaventura et al., 1997; Linker et al., 2000). Several neural network models have been developed to climate control in naturally ventilated greenhouses (Seginer et al., 1994; Kok et al., 1994; Patil et al., 2008; Taki et al., 2016; Atia and El-Madany, 2017; Castañeda-Miranda and Castaño, 2017; Jung et al., 2020; Moon and Son, 2021). In the last years, new types of climate model have been used in greenhouse as the single-input single-output (SISO) (Lafont et al., 2015) and multiple-input multiple-output (MIMO) models (Lammari et al., 2012).

12.1.2 SEMI-EMPIRICAL MODELS

Another alternative is to model the physical and physiological phenomena that occur within the greenhouse by formulating conservation equations based on mass and energy balances (Boulard and Baille, 1993; Boulard et al., 1996b). In this case, the model parameters have a physical meaning, being essential to have an accurate estimation of the solar radiation, mass and heat transfer coefficients (Choab et al., 2019).

Energy balance modeling and process-based greenhouse climate simulation, and its application to climate control in greenhouses were initiated in the early 1980s (Udink ten Cate, 1980; Bot, 1983). Several climate models have been developed to describe the energy exchanges inside greenhouses and validated at various locations, climates, and crops (Sethi et al., 2013). Over the past two decades, research in greenhouse modeling by various authors has contributed to the understanding of the physical aspects of the greenhouse such as light transmission (Kozai, 1977; Kozai and Kimura, 1977; Critten, 1983; Amsen, 1985; Nijskens et al., 1985; Critten, 1987, 1988; Kurata et al., 1991; Pieters and Deltour, 1997; Kittas and Baille, 1998; Papadakis et al., 1998; Abdel-Ghany and Kozai, 2006b), ventilation (Kozai and Sase, 1978; De Jong, 1990; Boulard and Baille, 1995; Boulard et al., 1997, 1998, 1999; Wang, 1998), and physiological processes in the greenhouse such as crop transpiration (Stanghellini, 1987; Boulard and Wang, 2000) and photosynthesis (Chalabi and Fernandez, 1994; Nederhoff and Vegter, 1994).

12.1.2.1 STATIONARY MODELS OF THE GREENHOUSE

Simulation methods provide a fast, cheaper, and more flexible and repeatable way compared to experimental trials. Some mathematical models were developed with the aim of forecasting the ambien conditions inside greenhouse as a function of the weather conditions and the characteristic parameters of the components of the greenhouse and were usually based on both static and dynamic energy balances. Static models are built into energy balances for the different components of the greenhouse (roof, indoor air, plants, and soil) under stationary conditions and were used to characterize the mean behavior of these components in time. However, these models distressed the heat storage capacity of these components (Kimball, 1973; Nijsken et al., 1984; Albright et al., 1985; Jolliet, 1994; Jolliet et al., 1999).

The first work in this category was done by Businger (1963), who described most transfer processes using a stationary model and by Walter (1965) who used an energy balance to predict air temperature in a ventilated greenhouse. Kimball (1973) later gave a more detailed description of convection heat transfer processes within the greenhouse. Kindelan (1980) and Avissar and Mahrer (1982) developed and adjusted a greenhouse energy balance model to determine the most favorable growth conditions for crop.

Studies on natural ventilation were also carried out based on estimates of an overall air exchange rate and the simulation of a homogeneous air temperature within the greenhouse using energy balance models that included the crop effect (Wang and Boulard, 2000a; Demrati et al., 2001). The energy balance allows the prediction of temperature within ventilated greenhouses when environmental conditions or ventilation rates vary (Walker, 1965; Walker et al., 1976; Kozi and Sase, 1978; Kindelan, 1980; Arinze et al., 1984; Fernandez and Bailey, 1992; Al-Jamal, 1994).

12.1.2.2 DYNAMIC MULTI-COMPONENT MODELS

Greenhouse microclimate modeling is necessary to predict instantaneous indoor climate values in response to external conditions and determine inputs to control systems (Bailey and Chalabi, 1995; Lafont et al., 2015). The dynamic models of greenhouse climate can be classified in mechanistic and black-box models. Mechanistic models are principally concentrated on the knowledge of the physical interaction inside greenhouses whereas black-box models are more used for applications, including: control, optimization, and design of the greenhouse system (López-Cruz et al., 2018). Mechanistic dynamic models based on mass and energy balances have been developed focusing on the different physical aspects of the greenhouse climate (Takakura et al., 1971; Kindelan, 1980; Bot, 1983; Zhang et al., 1997; Sharma et al., 1999; Sharma and Tiwari, 1999; Wang and Boulard, 2000a; Tavares et al., 2001; Ghosal et al., 2003; Chou et al., 2004; Luo et al., 2005a; Singh et al., 2006; Van Beveren et al., 2015; Singh et al., 2018). These models are useful tools for evaluating the energy transfer process in the greenhouse. Although most of them were developed and validated for northern European countries (Boulard et al.,1999; Boulard and Wang, 2000; Pieters and Deltour, 1997; Korner, 2003; Vanthoor et al., 2011) some models have been developed or adapted for warm climate areas (Fatnassi et al., 2004; Vanthoor et al., 2011; Shamshiri and Ismail, 2013; Mashonjowa et al., 2013; Fatnassi et al., 2013; Joudi and Farhan, 2015; Montoya et al., 2016; Mobtaker et al., 2016; Reyes-Rosas et al., 2017; Salazar-Moreno et al., 2018; Mohammadi et al., 2018).

For the design, analysis, and simulation of the greenhouse, it is essential that the models are based on knowledge of the static and dynamic behavior of the system (Bot, 2001). Dynamic models include a term for heat storage on at least some energy balance, while they usually assume the stationary state for the crop. This approach is similar to models in which the greenhouse is considered a stationary subsystem of a larger dynamic model (Price and Peart, 1973; Takami and Uchijima, 1977).

Abdel-Ghany and Kozai (2006a) developed a simulation model using a dynamic energy balance for heat transfer and water vapor in a greenhouse with natural ventilation and mist to predict air, plant, roof, soil surface, and relative humidity in the greenhouse. Impron et al. (2007, 2008) simulated the microclimate of the greenhouse using a simple dynamic model to optimize the properties of roofing material and ventilation, including the effect of crop transpiration. Van Ooteghem (2010) developed a system for optimizing the design of climate control strategies in greenhouses with respect to production and sustainability using a dynamic greenhouse and crop model.

12.1.2.3 SEMI-EMPIRICAL VENTILATION MODELS

Semi-empirical ventilation models are based on simplified physical equations and statistically adjusted parameters. Such models are often sufficient to describe the greenhouse climate from a global point of view. Okada and Takakura (1973) deduced an equation for the exchange of air in greenhouses, expressing the ventilation rate as a sum of two terms, the first proportional to the speed of the outside wind and the second to the square root of the temperature difference, which was correctly adjusted to the experimental data.

Kozai and Sase (1978) used Bernouilli's theorem and continuity theory to develop a computer program that estimated the rate of natural ventilation in multispan greenhouses with continuous side and roof windows under various climatic conditions. Another approach to semi-empirical modeling of greenhouse ventilation was presented by Kozai et al. (1980), who developed a model for assessing the natural ventilation rate in multispan greenhouses based on relationships between the flow rate and the pressure difference due to buoyancy or wind effect for an opening. They compared the data provided by this model with those obtained in tests carried out in a wind tunnel with satisfactory results (Sase et al., 1984), obtaining a new algorithm for natural ventilation control (Sase and Nara, 1985).

12.1.3 PHENOMENOLOGICAL MODELING

This type of modeling is the most academic, in the sense that it is based on the differential description of the physical system, using conservation equations. CFD models facilitate the analysis of climate heterogeneity by solving transport equations in closed (Nara, 1979; Lamrani et al., 2001) and ventilated greenhouses (Mistroitis et al., 1997a, 1997b). Given the large capacity of CFD simulations for the study of the environment within buildings, engineers have adopted this technique for the analysis of different types of systems including greenhouses (Norton et al., 2007).

Navier–Stokes' equations, presented below, represent the behavior of a fluid regimen, both temporally and spatially. Efficient quantification of ventilation in greenhouses can also be carried out using macromodels based on the Bernoulli equation (Roy et al., 2002) which is a simplification of Navier–Stokes's. The simplifications involved in these macromodels represent a significant loss of information such as temperature distribution or flow pattern that can lead to inaccuracies in the calculation of airflow variables (Boulard et al., 2002).

Macromodels give a description of the flow in which the fundamental principle of conservation of turbulent kinetic energy is not considered, leading to destruction or creation of it within the greenhouse (Norton et al., 2007). The turbulent kinetic energy available in the greenhouse is a function of the level of wind turbulence and the characteristics of the windows (Roy et al., 2002). Macromodels use a window load loss coefficient (Verlinde et al., 1988) or discharge coefficient C_d that is not a physical property of the flow, but whose value depends on the flow rate. So, it must be systematically determined by adjusting the model to experimental data (Kittas et al., 1997). The use of approximate values of this coefficient limits the accuracy of the solution provided by these macromodels (Axley and Chung, 2005).

In contrast to macromodels, micromodels do not need to adjust parameters experimentally. Micromodels establish changes in the momentum of fluid particles because of changes in pressure and viscous dissipation forces acting within the fluid. Micromodels are based on knowledge of boundary conditions and flow type, with very few physical assumptions needing to be made that affect their accuracy (Norton et al., 2007).

The energy balance models or tracer gas experimental measurements (Figure 12.1) do not allow clear identification of the energy components (sensible and latent) exchanged through the windows. In addition, they do not provide information on the airflow pattern at the crop level and through the windows. At the climate control level of a greenhouse, it may be sufficient to know the average temperature and humidity values within the greenhouse. However, homogeneous conditions do not adequately represent real conditions in greenhouses. In fact, the special heterogeneity of air velocity and environmental conditions (temperature, humidity, and CO_2 concentration) within large greenhouses interferes with plant activity. Since the local air velocity near the leaves affects mass and heat transfers from plants to greenhouse air and vice versa, it is interesting to determine airflow patterns within the canopy (Shilo et al., 2004), as enabled by CFD simulations (Boulard et al., 1998b).

The CFD technique allows easy access to conduct simulation studies from simple to complex greenhouse models (Lee et al., 2013). CFD programs have been used for the study of natural ventilation in greenhouses (Lee and Short, 2000; Fatnassi et al., 2003; Bartzanas et al., 2004; Molina-Aiz et al., 2004; Perez-Parra et al., 2004b; Teitel et al., 2008b; Tong et al., 2009; Teitel et al., 2009; Bournet and Boulard, 2010; Teitel et al., 2014; Benni et al., 2016; Aich et al., 2016; Lee et al., 2018; Pakari et al., 2019; Akrami et al., 2021), analyzing the effect of insect-proof screens (Teitel and Shklyar, 1998; Fatnassi et al., 2002; Bartzanas et al., 2002; Fatnassi et al., 2003; Fatnassi et al., 2006; Teitel, 2010; Madjoubi et al., 2016), mechanical ventilation (Flores-Velazquez et al., 2014), including other climate control systems such as heating systems (Kittas and Bartzanas, 2007; Chen et al., 2015), thermal screens (Montero et al., 2005), removable back walls in Chinese greenhouses (He et al., 2018; Tong et al., 2018), fogg (Kim et al., 2008), pad-fan cooling (Sapounas et al., 2008; Franco et al., 2011; Chen et al., 2014), or photovoltaic panels (Fatnassi et al., 2015; Amara et al., 2021; Akrami et al., 2021).

Other authors have studied with CFDs different design options of both greenhouses and windows (Campen and Bot, 2003; Lee et al., 2006; Baeza et al., 2006). Another innovation has been the study of the transport of substances such as spores of fungal diseases (Boulard et al., 2008) or pesticide drops (Barzanas et al., 2006). Currently, CFD simulations have reached a high level of complexity by including radiation exchanges between the atmosphere and greenhouse surfaces (Ould Khaoua et al., 2006; Bournet et al., 2007; Fidaros et al., 2010; Baxevanou et al., 2018), transfers of heat and air vapor with crop (Kuroyanagi et al., 2008; Roy et al., 2008; Majdoubi et al., 2009; Fatnassi et al., 2009), vapor condensation (Piscia et al., 2012), and plant photosynthesis (Roy et al., 2014; Molina-Aiz et al., 2017; Boulard et al., 2017).

In general, a good correlation has been observed between experimental and CFD-calculated data that underscore the convective origin of movement and air renewal in greenhouses along with the significant spatial heterogeneity of vector and scalar fields. CFD approximations have already been shown to provide realistic simulations for a wide range of geometric and boundary conditions and help to improve greenhouse control and design (Boulard et al., 2002).

The characteristics required in a CFD code for greenhouse ventilation analysis (Boulard and Wang, 2002) are the ability to model fluid properties, implement user-defined functions (to include heat and water vapor exchanges with plants), and model flow through porous media (to include the effects of insect-proof screens and plant aerodynamic resistance). The possibility of performing different types of numerical meshing is another feature of programs that facilitates the modeling of flow in and around complicated geometry structures (Norton et al., 2007).

12.2 THEORETICAL ANALYSIS OF NATURAL VENTILATION USING SEMI-EMPIRICAL MODELS

The classical theory of natural ventilation in greenhouses was developed from the works of Whittle and Lawrence (1960) in glasshouses and Bruce (1975) and Zhang et al. (1989) for livestock buildings. Bruce (1975, 1978, 1982) calculated the natural ventilation rate in agricultural structures considering the effects of wind and thermal buoyancy. Subsequently, many other works were carried out in which the same theory was applied to different types of agricultural constructions (Timmons and Baughman, 1981; Brockett and Albright, 1987; Foster and Down, 1987; Zhang et al., 1989; Down et al., 1990). This theory was also applied in Mediterranean greenhouses (Boulard and Baille, 1995; Papadakis et al., 1996; Baptista et al., 1999), with good results in the simulation of the inside temperature of the greenhouse depending on the windows opening.

Passive ventilation is a process that occurs naturally due to air convection due to density variations, which are generated by temperature differences between the areas inside the greenhouse and the outside (usually colder), and by the overpressure created by the wind blowing around the structure. The airflow in greenhouses is, therefore, the result of natural convection produced by the transfer of energy from the surface of the soil heated by solar radiation to the air and forced convection due to the action of the wind at high speed. The ventilation airflows are influenced both by greenhouse characteristics, the aerodynamic resistance to airflow of the openings and the insect-proof screens, the inertia of the air mass in the openings and the compressibility of inside air, and by frequency characteristics of wind pressures, their power spectra and the correlation among them (Haghighat et al., 1991).

The description of ventilation processes can be done by a simplified approximation in which the airflow through the windows is linked to the pressure differences on both sides produced by the forces of thermal and wind buoyancy. Bernoulli's equation with semi-empirical parameters has been used to express the mechanisms by which air circulates inside greenhouses. Coupling of thermal and wind effect mechanisms requires the use of two fundamental parameters, the pressure drop coefficient through the windows, C_d, and a wind effect coefficient, C_w, which integrates the stationary and turbulent aspects of wind action. The values of these parameters are usually obtained by on-site determinations, comparing experimentally measured air exchange rates with those obtained by ventilation models.

The static effect of the wind results in a vertical and horizontal pressure distribution, between the side and roof windows, and between the windward and leeward within the same window, respectively (Boulard et al., 1997a). In addition, it is necessary to consider the vertical distribution of static pressures between the inside and outside of the greenhouse resulting from the buoyancy of the air by effect of the temperature difference. This air movement called stack effect or chimney effect (Bruce, 1982) results in an airflow into the interior, at the bottom of the opening (or in the window at a lower height) and a flow outward in the upper part of the window (or in the window higher).

The complex mechanisms involved between thermal and wind action are not yet very clear, so Boulard and Baille (1995) studied several models in which different hypotheses of interaction between these two driving forces of natural ventilation were used. Boulard and Baille (1995) observed a linear dependence between the ventilation surface and wind speed on the one hand, and the rate of greenhouse renewal on the other.

There are currently several models for predicting ventilation in greenhouses based on previously obtained experimental results. Tests carried out by wind tunnels with scale models (Figure 12.1) in which side and roof windows were available, make it easy to study the effect of wind direction and speed (Kozai et al., 1980; Sase et al., 1984). However, most of the data comes from real-scale greenhouse tracer gas tests equipped with only roof windows (Bot, 1983; De Jong, 1990; Fernandez and Bailey, 1992; Boulard and Baille, 1995) or with roof and side opening combined (Kittas et al., 1997; Boulard et al., 1997a; Zhao et al., 2001).

This section presents some of the macro-models of natural ventilation in greenhouses based on the homogeneity of the interior air, which allow to estimate the ventilation rate. These theoretical and experimental approaches are compatible and can be associated with mass and energy global balances in the greenhouse (Van Meurs and Stanghellini, 1989; Joillet, 1994; Bailey et al., 1997).

Knowledge of the physical processes governing general ventilation has not yet reached a sufficient level to accurately predict the air exchange rate in the most common greenhouses (Bailey, 2000). Further, the study of ventilation is needed to confirm dependence on the overall pressure coefficient with respect to wind speed, to quantify airflow through individual windows, and to determine how flows through windward and leeward windows interact to produce the total air renewal rate (Bailey, 2000).

12.2.1 BERNOULLI EQUATION

General fluid flow equations are a micro-model that helps accurately describe dynamic behavior at each point in space within a calculation domain. In the case of natural ventilation in greenhouses, some considerations may be taken into account in order to determine a model representing the spatial domain of interest (Roy et al., 2002):

a) Steady flow: variations in flow (e.g., outside wind) are considered negligible for a sufficiently long period of time.
b) Non-viscous flow: the diffuse effects of viscosity are limited to restricted areas (limit layers, streems, etc.) and the viscous effects are, therefore, despised in this model.
c) Homogeneous and unrrotional flow: physical properties, and flow rate field are considered uniform in each region of the calculation domain.
d) Gravitational flow: the resulting of external forces is limited to vertical gravitational force.

When these considerations are applied to the general equations of fluid flow, a macro-model of the flow is obtained with a significant simplification of the equations. The equation of motion of fluids has the general form (Bird et al., 2009):

$$\frac{\partial}{\partial t}\rho\cdot\vec{v} = -\left[\nabla\left(\rho\cdot\vec{v}\cdot\vec{v}\right)\right] - \nabla P - \left[\nabla\cdot\vec{\tau}\right] + \vec{F} \tag{12.2}$$

Where ∇P is the vector called the gradient of the scalar $\boldsymbol{P}$, $[\nabla\cdot\tau]$ is a vector called the divergence of the stress tensor $\vec{\tau}$ and $\left[\nabla\left(\rho\cdot\vec{v}\cdot\vec{v}\right)\right]$ is the vector divergence of the product of the density $\boldsymbol{\rho}$ of the fluid [kg·m^{-3}] and the velocity vector $\vec{v}$. $\vec{F}$ is the volumetric external force on the fluid.

For constant values of fluid properties (density $\boldsymbol{\rho}$ and viscosity $\boldsymbol{\mu}$), and using the Newtonian expression for the stress tensor into the equation of motion (12.2) lead to the very famous Navier–Stokes equation (Navier, 1827; Stokes, 1845; Bird et al., 2009):

$$\rho\cdot\frac{D}{Dt}\cdot\vec{v} = -\nabla P + \mu\cdot\nabla^2\,\vec{v} + \vec{F} \tag{12.3}$$

Neglecting the viscous forces in the equation of motion, the Euler equation for inviscid fluid is obtained (Euler, 1755a, 1755b):

$$\rho\cdot\frac{D}{Dt}\cdot\vec{v} = -\nabla P + \vec{F} \tag{12.4}$$

Considering the assumptions of steady and non-viscous flow, the Navier–Stokes equation of momentum [eq (12.3)] of the fluid for each position within the flow can be expressed as:

$$\nabla(\rho \cdot \vec{v} \cdot \vec{v}) = -\nabla P + \rho \cdot \vec{g} \tag{12.5}$$

where, the external force on the fluid per unit volume in the case of airflows in greenhouses is reduced to gravitational attraction $\vec{F} = \rho \cdot \vec{g}$.

After some changes, eq (12.5) can be rewritten for a streamline where ***s*** is the distance along it as (Bird et al., 2009):

$$\frac{d}{ds}\left(\frac{1}{2} \cdot v^2\right) = -\frac{1}{\rho} \cdot \frac{d}{ds} P - g \cdot \frac{d}{ds} y \tag{12.6}$$

where ***y*** is the elevation in the gravitational field [m].

Integrating eq (12.6) along a streamline between two points ***A*** and ***B***, the following scalar equation is obtained:

$$\frac{1}{2} \cdot \rho \cdot v_A^2 + \rho \cdot g \cdot y_A + P_A = \frac{1}{2} \cdot \rho \cdot v_B^2 + \rho \cdot g \cdot y_B + P_B \tag{12.7}$$

This equation is known as Bernoulli's equation (Bernoulli, 1738) for non-viscous flows that expresses the conservation of mechanical energy consisting of three components: kinetic energy, potential energy, and pressure energy (all terms have units of work per unit volume). According to the unrrotational flow hypothesis, this relationship is also valid for expressing energy conservation between two sections perpendicular to the flow (Roy et al., 2002).

The assumption of an inviscid flow leads to the absence of diffuse effects on flow (unrrotional behavior) and energy dissipation due to shear stress on the flow. Energy dissipation is an irreversible process that prevents its absolute conservation, which is of great importance when this process appears in most practical situations where the viscosity of the fluid cannot be ignored, as is the case of a contraction (windows) or a porous medium (insect-proof screens and crops). The easiest way to take this term into account in eq (12.7) is to express the loss of energy between points ***A*** and ***B*** as an experimentally determined load loss by pressure measurements. Using an additional pressure drop in a singularity ΔP_{AB}, we obtain the following equation (Roy et al., 2002):

$$\frac{1}{2} \cdot \rho \cdot v_A^2 + \rho \cdot g \cdot y_A + P_A = \frac{1}{2} \cdot \rho \cdot v_B^2 + \rho \cdot g \cdot y_B + P_B + \Delta P_{AB} \tag{12.8}$$

This additional pressure drop can be related to the kinetic energy available by the following expression (Baptista et al., 1999; Roy et al., 2002):

$$\Delta P_{AB} = \frac{1}{2} \cdot F \cdot \rho \cdot v^2 \tag{12.9}$$

where F is the pressure drop coefficient (dimensionless) of the singularity (window, insect-proof screen, plant canopy, etc.), whose value depends on the flow conditions (flow regime, geometry, etc.).

12.2.2 NATURAL VENTILATION DRIVING FORCES IN GREENHOUSES

This section discusses the forces that cause air movement within a greenhouse. This allows to determine the parameters (some of them empirical) on which the airflow depends through the windows and to be able to quantify the airfñow of ventilation. Airflow through a window is caused by pressure differences that occur on either side of it. According to Bernouilli's law, admitting that the air is incompressible and that the viscous effects are not important, when opening a window, the pressure difference between the two sides of the window (potential energy) is converted into kinetic energy, resulting in a pressure drop (Pearson and Owen 1994):

$$\Delta P_{AB} = \frac{1}{2} \cdot \rho \cdot \frac{1}{C_d^2} \cdot v^2 \tag{12.10a}$$

being $C_d = F^{-1/2}$ the characteristic discharge coefficient of the geometry of the window.

If horizontal velocity is considered to vary with height, the vertical velocity profile through window *v(y)* can be used to determine the vertical distribution of the pressure drop ***ΔP****(y)* (Bruce, 1975; De Gids, 1978; Zhang et al., 1989):

$$\Delta P(y) = \frac{1}{2} \cdot \rho \cdot F \cdot v(y)^2 = \frac{1}{2} \cdot \rho \cdot \frac{1}{C_d^2} \cdot v(y)^2 \tag{12.10b}$$

Similarly, from eq (12.10b) can be inferred the vertical distribution of air velocity through the window knowing the pressure drop (Boulard et al., 1998a):

$$v(y) = \frac{|P(y)|}{P(y)} \cdot C_d \cdot \sqrt{\frac{2}{\rho} \cdot |P(y)|} \tag{12.11}$$

The sign of $|\Delta P|/\Delta P$ indicates the direction of airflow through the window.

Once the air velocity is known at each point of the window, the volumetric ventilation air flow rate can be calculated integrating it through the air inlet or outlet surface:

$$G = \int_0^{H_v} v(y) \cdot dS_v = L_v \cdot \int_0^{H_v} v(y) \cdot dy \qquad (12.12)$$

where, G is the airflow that enters or exits the greenhouse by ventilation through the window [$m^3 \cdot s^{-1}$], H_v is the height of the window opening through which the air enters or exits [m], S_v is the surface of the cross section of the window [m^2] and L_v is the length of the ventilation opening [m]. In the case of a greenhouse with a single ventilation opening, the net airflow G through the entire window would be around zero, with the same amount of air mass coming and leaving the greenhouse (the airflow is not equal as consequence of density differences between air inside and outside the greenhouse).

Considering in eq (12.11), a constant value of the mean pressure difference on both sides of the window, the average velocity value would also be obtained through the window. Replacing this constant mean value in eq (12.12), we can obtain the value of the airflow that enters or exits through the window (Baptista et al., 1999):

$$G = \frac{S_v}{2} \cdot C_d \cdot \sqrt{\frac{2}{\rho} \cdot \Delta P} \qquad (12.13)$$

The main forces governing the ventilation of greenhouses equipped with side and roof openings are (Kittas et al., 1997):

- The chimney effect, due to the thermal buoyancy forces (Bruce, 1975; Kittas et al., 1995), that produces a vertical distribution of pressures between the side and roof openings.
- The static effect of wind due to the mean component of wind speed, which induces a spatial distribution of pressures on the greenhouse cover (Bruce, 1975).
- The turbulent effect of wind, linked to pressure fluctuations caused by wind along all openings (Boulard and Baille, 1995; Boulard, 1996), which generates an input and output flow in the same window (Boulard et al., 1995).

12.2.3 VENTILATION BY THERMAL BUOYANCY—AIRFLOW DUE TO TEMPERATURE DIFFERENCE

A hot fluid like air inside the greenhouse tends to rise due to the force produced by the joint action of density differences and the Earth's gravitational field. Similarly, moist air rises mainly due to the lower density of the water vapor contained in the mixture (Mills, 1995), although this convective process is not usually considered in the study of ventilation in greenhouses. Natural convection flows in greenhouses are the result of density variations due to temperature gradients in the fluid, which are related to each other by the volumetric dilation coefficient $\boldsymbol{\beta_T}$. As a result, there is a dependency between the momentum and the energy conservation equations.

Temperature dependence on density in the Navier–Stokes equations [eq 12.3], introduces a problem in the resolution of this equation.

The simplest approximation for variations in density in conservation equations which admits buoyancy is the Boussinesq's approach. This approximation supposes that density is constant except when it directly causes buoyant forces [$\vec{F}$ in eq (12.3)], that all other fluid properties are constant and that the viscous dissipation is negligible (Gray and Giorgini, 1976).

Taking a reference thermodynamic state (density $\boldsymbol{\rho_0}$ and temperature $\boldsymbol{T_0}$) at the bottom of the window, density can be expressed using a Taylor expansion (Mills, 1995):

$$\rho = \rho_0 \cdot \left(1 - \frac{T - T_0}{T_0} + \frac{(T - T_0)^2}{T_0^2} - \ldots\right) \tag{12.14}$$

This approach can also be applied to other variables that influence density, such as pressure, but the influence of temperature is predominant for free convection problems in greenhouses.

Limiting eq (12.14) to a first-order development, the Boussinesq approximation provides the density value based on the temperature (Roy et al., 2002). The value of the drive volumetric force at eqs (12.3) to (12.5) is given by:

$$\vec{F}(y) = \vec{g} \cdot \rho\left[T(y)\right] = \vec{g} \cdot \rho(T_0) \cdot \left(1 - \frac{T(y) - T_0}{T_0}\right) \tag{12.15}$$

where $\vec{g}$ is the gravitational acceleration vector [m·s^{-2}].

From the pressure difference between the two sides of a window at its lower end $\boldsymbol{\Delta P_0}$ and the vertical distributions of the air temperature on both sides of the window, the vertical distribution of the pressure difference can be inferred, using the Boussinesq approximation.

The pressure difference between the two sides of the window for each height $\boldsymbol{y}$ can be calculated from the volumetric force as:

$$\Delta P_T(y) = \int_0^y \vec{F}(y) \cdot dy = \Delta P_0 - \int_0^y \rho(T_0) \cdot \left(\frac{T(y) - T_0}{T_0}\right) \cdot g \cdot dy \tag{12.16}$$

This equation can be simplified assuming that temperatures inside T_i [K] and outside T_o [K] the greenhouse are uniform (Roy et al., 2002). Considering that the outside air is colder (used as reference conditions), and therefore, located at the bottom of the window, the pressure difference can be calculated from the ratio of temperature to density, given by the Boussinesq approximation, such as (Bruce, 1982; Zhang et al., 1989; Walker and Wilson, 1993; Roy et al., 2002):

$$\Delta P_T(y) = \Delta P_0 - \rho_o \cdot g \cdot \frac{\Delta T_{i0}}{T_0} \cdot y \qquad (12.17)$$

where ΔT_{io} is the temperature difference between the inside and outside of the greenhouse [K] and ρ_o is the air density for the outside temperature [kg·m^{-3}].

By substituting this value in eq (12.11) is obtained:

$$v_T(y) = \frac{|P(y)|}{P(y)} \cdot C_d \cdot \sqrt{\frac{2}{\rho} \cdot \left(\Delta P_0 - \rho_o \cdot g \cdot \frac{\Delta T_{i0}}{T_0} \cdot y \right)} \qquad (12.18)$$

The integration of eq (12.12) using the value of $v_T(y)$ given by eq (12.18) allows to calculate the airflow in the greenhouse due to the thermal effect G_T. For the calculation of this value, three hypotheses with respect to airflow through ventilation openings can be considered (Boulard and Baille, 1995; Roy et al., 2002):

a) A first-order approach in which a simplified model is used considering constant the entrance and exit velocities.
b) Second-order approximation in which the velocity $v_T(y)$ integrated over the total height of the window, considering a variable interior temperature profile $T_i(y)$.
c) Third-order approximation in which temperature profiles are considered variables both outside $T_o(y)$ and inside the greenhouse $T_i(y)$.

12.2.3.1 FIRST-ORDER APPROACH

In the case of a single window, eq (12.18) can be simplified assuming that the air velocity has a constant value v over the entire input surface and $-v$ for the output surface. So, if h_I is the height above the ground of the geometric center of the input surface, the air velocity is (Boulard and Baille, 1995):

$$v_T(y) = \frac{|P(y)|}{P(y)} \cdot C_d \cdot \sqrt{\frac{2}{\rho} \cdot \left(\Delta P_0 - \rho_o \cdot g \cdot \frac{\Delta T_{i0}}{T_0} \cdot h_I \right)} \qquad (12.19)$$

If a neutral axis is defined (Figure 12.2), that is, a pressure reference level at a height ***h'*** for which there is no pressure difference between inside and outside $\Delta P(h') = 0$ and where the velocity is null $v(h') = 0$.

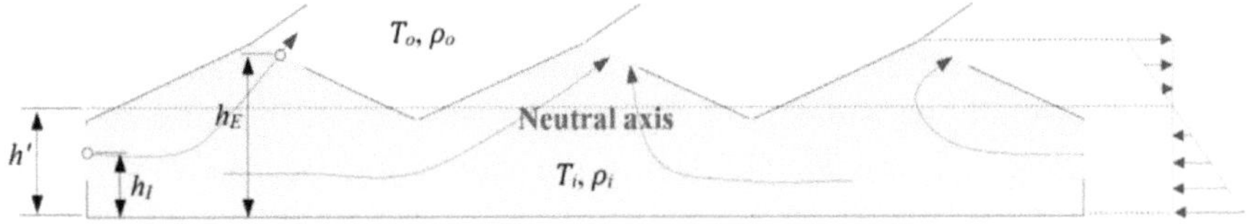

FIGURE 12.2 Airflow due to temperature difference

Source: Adapted from Brockett and Albright, 1987.

Replacing the null value of velocity in eq (12.12), the pressure difference at the neutral axis level is obtained:

$$\Delta P_0(h') = \rho_o \cdot g \cdot \frac{\Delta T_{i0}}{T_0} \cdot h' \tag{12.20}$$

Replacing this expression of ΔP_0 in (12.19), the velocity based on the difference in height with the neutral axis is obtained (Bruce, 1978; Zhang et al., 1989):

$$v_T = C_d \cdot \sqrt{2 \cdot g \cdot \frac{\Delta T_{i0}}{T_0} \cdot (h' - h_I) \cdot \frac{|h' - h_I|}{h' - h_I}} \tag{12.21}$$

where the sign indicating the air inlet or outlet is given by the value of $|h-h_I|/(h-h_I)$.

Replacing this expression of velocity in eq (12.12) and integrating it with respect to the height (y) from 0 to $H_V/2$ [air inlet surface (Figure 12.3)] or from $H_V/2$ to H_V (the exit surface of the window) yields a simple expression for the ventilation rate due to thermal buoyancy, G_T:

$$G_T = \frac{S_v}{2} \cdot C_d \cdot \sqrt{2 \cdot g \cdot \frac{\Delta T_{i0}}{T_0} \cdot \frac{H_V}{4}} \tag{1.22}$$

where S_V is the effective ventilation surface [m^2].

Figure 12.3 shows how the neutral axis is in the middle of the window, so the height difference with the centers of the input or exit sections $(h'-h_I)$ is equal to $H_V/4$.

Bruce (1978) proposed a similar equation to determine the ventilation flow rate produced by natural convention (thermal effect) in a building with roof and side opening (Figure 12.2) using the difference of density and the outside density as reference:

$$G_T = \frac{S_v}{2} \cdot C_d \cdot \sqrt{2 \cdot g \cdot \frac{\Delta\rho_{i0}}{\rho_0} \cdot h_{IE} \cdot \frac{1}{S_{vI}^2 + S_{vE}^2}} \tag{12.23}$$

where $\boldsymbol{h_{IE}}$ is the difference in height between the inlet ($\boldsymbol{h_I}$) and exit ($\boldsymbol{h_E}$) openings heights [m] and $\boldsymbol{S_{VI}}$ and $\boldsymbol{S_{VE}}$ are the surfaces of the inlet and exit openings [m^2], respectively. Replacing the values for the case of a single window $\boldsymbol{h_{IE}} = H_v/2$ and $\boldsymbol{S_{vI}} = \boldsymbol{S_{vE}} = S_v/2$ in eq (12.23) matches eq (12.22).

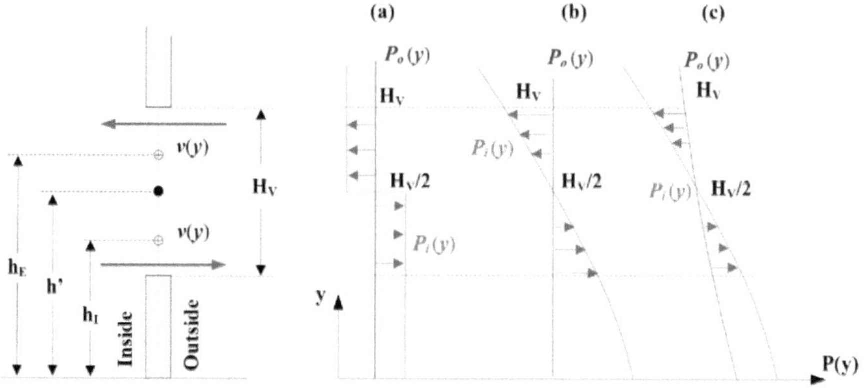

FIGURE 12.3 Vertical air velocity and pressure distributions in a window, according to a first- (a), second- (b), and third-order (c), scheme

Source: adapted from Brockett and Albright, 1987; Boulard and Baille, 1995; Roy et al., 2002.

12.2.3.2 SECOND-ORDER APPROACH

This approach considers a vertical velocity profile $v(y)$ given by eq (12.18) at the inlet surface at the bottom of the window or on the outlet surface at the top (Figure 12.3b). In eq (12.18), the pressure difference between the two sides of the window at the reference height is null [$\Delta P_0(H_v/2) = 0$]. Combining eqs (12.12) and (12.18) and integrating the airflow due to the temperature on the inner surface (between 0 and $H_v/2$ or between $H_v/2$ and H_v) yields (Roy et al., 2002):

$$G_T = \frac{S_v}{3} \cdot C_d \cdot \sqrt{2 \cdot g \cdot \frac{\Delta T_{i0}}{T_0} \cdot \frac{H_V}{2}} \tag{1.24}$$

12.2.3.3 THIRD-ORDER APPROACH

For mixed convection situations where the effects of the outside wind and the buoyancy are of the same order of magnitude, it is necessary to determine the temperature difference on each side of the window to obtain good accuracy of the model (Roy et al., 2002). The assumption of a linear temperature profile on each side of the openings results in a new approach where the two profiles are described by the following relationships (Roy et al., 2002):

- The inlet profile: $T_I(y)=T_{0,I}+b_I{\cdot}y$ (12.25)
- The exit profile: $T_E(y)=T_{0,E}+b_E{\cdot}y$ (12.26)

where $\boldsymbol{T_{0,I}}$ and $\boldsymbol{T_{0,E}}$ are the temperatures of air incoming and exiting from the greenhouse at ground level (y = 0), respectively. $\boldsymbol{b_I}$ and $\boldsymbol{b_E}$ are the slopes of the respective temperature profiles.

Considering again the approach of Boussinesq [eq (12.15)], eq (12.11) for air velocity is (Kirpatrick and Hill, 1988; Roy et al., 2002):

$$v(y)=\frac{|P(y)|}{P(y)}\cdot C_d\cdot\sqrt{\frac{2\cdot\Delta P_0}{\rho}+\frac{2\cdot g}{T_0}\cdot\left(\left(T_{0,E}-T_{0,I}\right)\cdot y+\left(b_E-b_I\right)\cdot\frac{y^2}{2}\right)} \quad (12.27)$$

The height of the neutral plane, ***h'***, can be determined by setting the above equation [eq (12.27)] at zero (Kirpatrick and Hill, 1988):

$$h'=\frac{-\left(T_{0,E}-T_{0,I}\right)}{b_E-b_I}\pm\cdot\sqrt{\left(\frac{T_{0,E}-T_{0,I}}{b_E-b_I}\right)^2-\frac{2\cdot T_0\cdot\Delta P_0}{g\cdot\rho\cdot\left(b_E-b_I\right)}} \quad (12.28)$$

The value of $\boldsymbol{G_T}$ would be obtained integrating eq (12.12) using the value of $v(y)$ given by eq (12.27) as was done in the previous two cases.

12.2.4 VENTILATION BY WIND EFFECT

The wind produces a distribution of pressures on the surface of the greenhouse creating areas of overpressure and suction (Hoxey and Richardson, 1983) that results in the movement of air in the greenhouse (Figure 12.4). The wind-driven ventilation is generally considered to consist of two components – a mean component driven by the mean pressure field at the ventilation openings, and a turbulent component driven by the fluctuating pressures and unsteady flow around the openings (Bot, 1983; Boulard and

Baille, 1995; Kittas et al., 1996; Boulard et al., 1998a; Straw et al., 2000; Roy et al., 2002; Teitel and Tanny, 2005; Molina-Aiz et al., 2009). Both medium and turbulent components of the ventilation flow can be estimated from measures of the pressure difference (Boulard et al., 1998a).

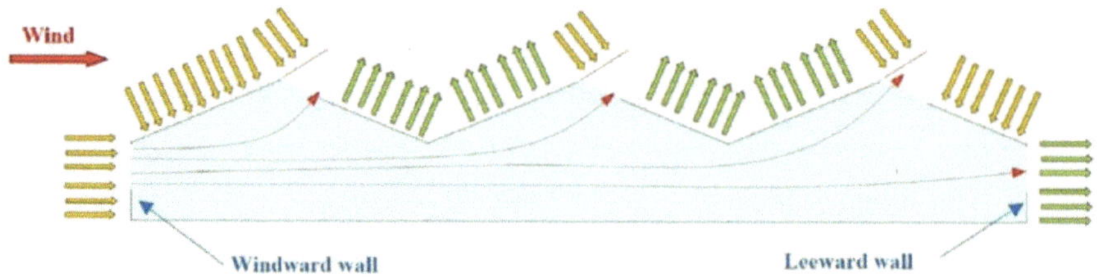

FIGURE 12.4 Airflow due to the outside wind, overpressure (■), and suction (■) zones.

12.2.4.1 STATIC EFFECT OF THE WIND—AVERAGE PRESSURE DIFFERENCES

The first component of airflow due to wind is a static or medium effect induced by an average pressure distribution, related to the average air velocity by a dimensional pressure coefficient (Boulard et al., 1996):

$$\overline{\Delta P_V} = \frac{1}{2} \cdot \rho \cdot \bar{C} \cdot v_R^2 \tag{12.29}$$

where $\overline{\Delta P_V}$ is the difference in average wind pressure [Pa], $\bar{C}$ is the dimensionless mean coefficient of pressure and v_R is the average wind speed measured at a reference level [$m \cdot s^{-1}$], usually corresponding to a height of 4 m (Kittas et al., 1996) or 4.5 m (Kittas et al., 1997).

Wind-induced airflow moves from overpressure zones (windward side) to areas where a suction effect occurs (leeward side). In the case of a simple structure (Figure 12.5), the effect that the obstacle produces on the airflow is the generation of turbulence (eddy formation) and the reversal of the flow in the side leeward (Wegley et al., 1980; Straw et al., 2000; Ohba et al., 2001). The zone of disturbance with turbulent flow, characterized by the formation of eddies and vortices, can be extended to a distance of 20 times the height of the building (Wegley et al., 1980). This phenomenon tends to produce the thrust of the air on the bottom of the leeward side and suction on the top (Figure 12.5a). In the case of greenhouses with simple structure, as a tunnel (vaulted-roof), strong wind can induce negative pressure values in the leeward part (Mistriotis and Briassoulis, 2002). In the case of multispan greenhouses with roof and side windows, the pressure distribution can be very different along the structure.

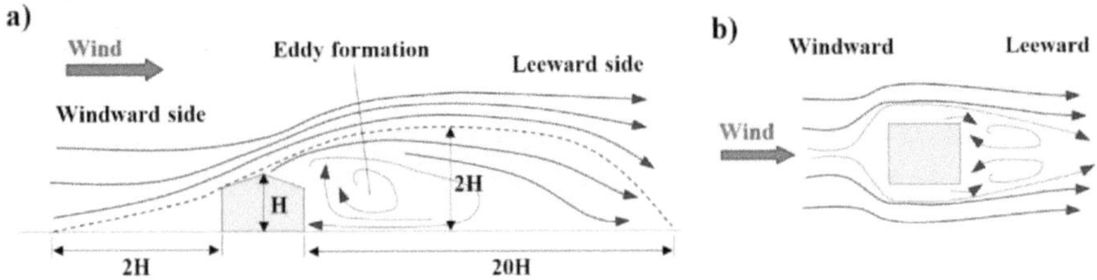

FIGURE 12.5 Schematic illustration of airflow around a small single building: vertical section (a) and horizontal plan (b) views, and zone of disturbed flow (■)

Source: Adapted from Wegley et al., 1980 and Ohba et al., 2001.

For greenhouses longer than 10–15 m, the air entering through the windward windows may become insufficient to reach the opposite side, producing the inversion of the flow inside the greenhouse (Lee et al., 2002, 2003; Molina-Aiz et al., 2006–2011). The reverse flow can produce the inlet of air through the leeward side openings. In multispan greenhouses with only roof ventilation, for perpendicular winds blowing from the back of the roof openings (leeward ventilation), the airflow at the mid-span opening can also have an opposite direction to that of the outside wind, resulting in an inflow at this region of the openings (Teitel et al., 2005).

In the case of a single window (and/or in similar conditions such as that of a wind parallel to the greenhouse axis), the average static flow through the window can be null (Liddament, 1986), and therefore, the corresponding average air velocity along the opening is also equal to zero. From the Bernoulli equation [eq (12.4)], which assumes that the pressure difference is converted to kinetic energy with a negligible loss (Bruce, 1978), the mean difference in static pressure at the window level can be considered equal to zero (Boulard et al., 1996), obtaining:

$$\overline{\Delta P} = \overline{\Delta P_T} + \overline{\Delta P_V} \tag{12.30}$$

where, $\overline{\boldsymbol{\Delta P}}$ is the static or mean value of the pressure differences [Pa] and the static or mean value of the pressure differences from $\overline{\boldsymbol{\Delta P}_T}$ the thermal effect [Pa]. As $\overline{\boldsymbol{\Delta P}} = 0$ we can deduce that (Kittas et al., 1996):

$$\overline{\Delta P_T} = -\overline{\Delta P_V} \tag{12.31}$$

From eq (12.17), the value of the static pressure difference due to the effect of temperature can be estimated as (Kittas et al., 1996):

$$\overline{\Delta P_T} = -\overline{\Delta P_0} + \rho_o \cdot g \cdot \frac{\Delta T_{io}}{T_o} \cdot h_v \tag{12.32}$$

where, $\overline{\Delta P_0}$ represents the static pressure difference between the inside and outside greenhouse, averaged in time at the soil level [Pa] and h_v is the average height of the window above the ground [m]. Considering, the mean pressure differences by the wind effect [eq. (12.29)], and the static pressure difference by thermal effect [eq (12.32)], and the relationship between the two values [eq (12.31)], yield:

$$\overline{\Delta P_0} = \frac{1}{2} \cdot \rho_o \cdot \overline{C} \cdot v_R^2 + \rho_o \cdot g \cdot \frac{\Delta T_{io}}{T_o} \cdot h_v \tag{12.33}$$

12.2.4.2 TURBULENT EFFECT OF WIND PRESSURE DIFFERENCE FLUCTUATIONS

The second component of airflow duc to wind is thc turbulent effect of wind due to fluctuation in pressure difference (Boulard and Baille, 1995; Kittas et al., 1996). The air exchange in a greenhouse is not only induced by the mean pressure differences averaged over time, but also depends on the pressure fluctuations linked to the turbulent characteristics of the wind that can be characterized by a turbulent pressure coefficient (Boulard, 1996; Boulard et al., 1996a):

$$\Delta P'_v = \frac{1}{2} \cdot \rho \cdot C' \cdot v_R^2 \tag{12.34}$$

where, $\Delta P'_v$ is the standard deviation of the fluctuations of the pressure difference through the window over a time interval [Pa] and C' is the dimensionless coefficient of turbulent pressure.

For pressure fluctuations, the Bernoulli equation is not applicable and an equation like eq (12.33) cannot be used for C'. However, as a first approximation, the standard deviations of pressure fluctuations over a certain period at ground level $\Delta P'_0$ can be considered to estimate the pressure drop fluctuations through the window (Papadakis et al., 1996). Soil-level fluctuations can be calculated as:

$$\Delta P'_0 = \sqrt{\left(\Delta P_0(t) - \overline{\Delta P_0}\right)^2} \tag{12.35}$$

where t is the time [s] and $\Delta P_0(t)$ is the instantaneous pressure difference at the soil level [Pa].

Replacing $\Delta P'_0$ in the eq (12.35) we get (Kittas et al., 1996; Boulard et al., 1998a):

$$\Delta P'_0 = \sigma_{\Delta P} = \frac{1}{2} \cdot \rho \cdot C' \cdot v_R^2 \tag{12.36}$$

Measurements of the total pressure difference at soil-level ΔP_0, of wind speed and air temperatures inside and outside of the greenhouse allow to calculate $\overline{\Delta P_0}$ and $\Delta P'_0$ (calculating mean and standard deviation) and estimate $\bar{C}$ and C' from eqs (1.29) and (1.36) (Kittas et al., 1996).

Despite the complexity of turbulence phenomena, the effect of fluctuations on the pressure difference generated by wind can be considered in a simple way from the standard deviation of pressure differences as (Boulard and Baille, 1995; Boulard et al., 1996):

$$\Delta P'_0 = \sigma_{\Delta P} = \sqrt{\overline{\left(\Delta P_0(t) - \overline{\Delta P_0}\right)^2}} \tag{12.37}$$

where $\overline{\Delta P}$ is the average value of pressure differences [Pa].

This second turbulent component of the wind effect is characterized by the coefficient C' that describes fluctuations in wind pressure. Although there are only a few references about the values of the C' coefficient, it has a lower spatial variability than the mean coefficient (Table 12.1).

The turbulent wind pressure coefficient C' increases when the wind is disturbed by the presence of obstacles (Kittas et al., 1996; Molina-Aiz et al., 2009). Like pressure due to wind effect, air velocity through a ventilation opening can be separated into mean and turbulent components (Cockroft and Robertson, 1976; Haghighat et al., 1991; Boulard et al., 1998a; Vassiliou et al., 1998):

$$\Delta P_v = \overline{\Delta P_v} + \Delta P'_v \tag{12.38}$$

$$v = \bar{v} + v' \tag{12.39}$$

where v' is the fluctuating component of air velocity [$m \cdot s^{-1}$] and $\bar{v}$ is the mean component [$m \cdot s^{-1}$].

The eq (12.11) of the vertical distribution of the velocity in the opening for steady-state (mean) relates the air velocity through opening to the pressure difference on both sides of it. In order to simplify turbulence modeling, changes in velocity v can be considered proportional to changes in pressure ΔP_v (Haghighat et al., 1991). Differentiation of eq (12.11) yields (Boluard et al., 1998a):

TABLE 12.1 Values of the Average $\bar{C}$ and Turbulent C' Wind Pressure Coefficients Measured in Different Greenhouse Types.

Greenhouse type	Measurement point	$\bar{C}$	C'	Source
Twin span greenhouse with a continuous roof vent	Roof (v_R = 0–2 m·s^{-1})	0.66	0.40	Boulard et al., 1996
	Roof (v_R = 2–4 m·s^{-1})	0.62	0.30	
	Roof (v_R = 4–6 m·s^{-1})	0.56	0.25	
	Roof (v_R = 6–9 m·s^{-1})	0.54	0.20	
Tunnel with two continuous side openings	Side openings	0.12	0.20	Kittas et al., 1996
Twin span greenhouse with side and roof openings	Roof vent		0.186	Papadakis et al., 1996
	Side vent		0.154	
	Roof and side		0.178	
Twin span greenhouse with a continuous roof vent	Windward end	−0.48	0.40	Boulard et al., 1998a
	Center	−0.28	0.43	
	Leeward end	−0.11	0.50	
Five spans Almería-type greenhouse with two roof and two side openings	Windward side vent	0.013–0.039	0.014–0.125	Molina-Aiz et al., 2009
	Windward roof vent		0.070–0.116	
	Leeward roof vent		0.014–0.099	
	Leeward side vent		0.054–0.127	

$$dv = \frac{C_d \cdot d\Delta P_v}{\sqrt{2 \cdot \rho \cdot \Delta P_v}} \tag{12.40}$$

If $\boldsymbol{\sigma_v}$ and $\boldsymbol{\sigma_{\Delta P}}$ represent, the values of the standard deviation of the turbulent air velocity and pressure fluctuations, respectively, then the following ratio can be obtained (Boulard et al., 1998a):

$$\sigma_v = \frac{C_d \cdot \sigma_{\Delta P}}{\sqrt{2 \cdot \rho \cdot \Delta P_v}} \tag{12.41}$$

As $\Delta \boldsymbol{P_v}$ approaches zero, this relation is no longer valid. The proportionality between the turbulent flow $\boldsymbol{\sigma_v}$ and the pressure $\boldsymbol{\sigma_{\Delta P}}$ was corroborated by Boulard et al. (1998a) using measurements with three-dimensional sonic anemometers. However, the proportionality coefficient differed from that expected according to the previous ratio [eq (12.41)] especially in positions near the window where $\Delta \boldsymbol{P_v}$ was very small or close to zero.

Since the separation of the two static and turbulent components from the pressure difference is excessively complex for the practical study of ventilation in greenhouses, their effect is usually included in a single pressure difference, and in a general way, the difference in wind pressure, $\Delta \boldsymbol{P_v}$, can be expressed according to an overall pressure coefficient $\boldsymbol{C_w}$ (Okada and Takakura, 1973; Swami and Chandra, 1987; Boulard and Baille, 1995; Kittas et al., 1995, 1996; Papadakis et al., 1996; Baptista et al., 1999; Bailey, 2000; Fatnassi et al., 2002):

$$\Delta P_V = \frac{1}{2} \cdot \rho \cdot C_w \cdot v_R^2 \tag{12.42}$$

Air moves from high pressure regions to low pressure ones (+ and − in Figure 12.6, respectively). When a vent opening is located in an area of the roof or walls where the overall wind effect coefficient $\boldsymbol{C_w}$ is negative, the air moves from inside to outside, and when $\boldsymbol{C_w}$ is positive the air exits from the greenhouse (Figure 12.6).

Combining eq (12.13), which establishes the relationship between the airflow circulating through a window and the mean pressure difference on both sides of it, and eq (12.42) representing the pressure difference caused by the wind speed, the rate of fluid flow can be calculated due to the effect of the wind as (Kittas et al., 1996):

$$G_v = \frac{S_v}{2} \cdot C_d \cdot \sqrt{C_w} \cdot v_R \tag{12.43}$$

Knowledge of the turbulent component of ventilation is not absolutely necessary to determine airflow because the overall wind effect coefficient

C_w deducted from pressure difference measurements can be used (Boulard et al., 1998a).

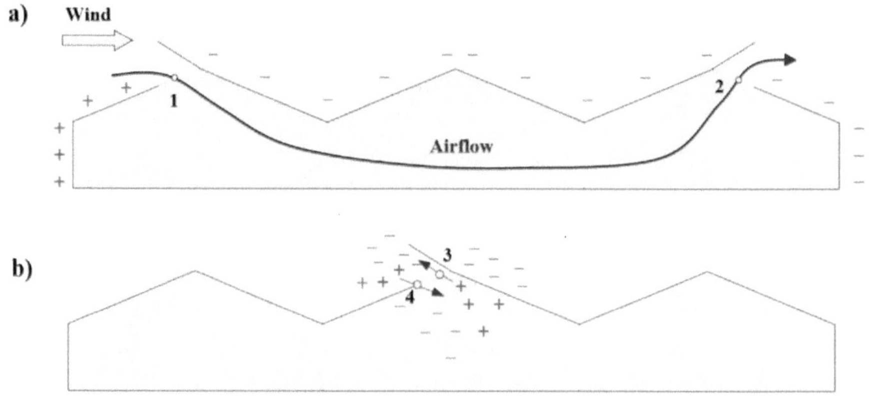

FIGURE 12.6 Scheme of the stationary and turbulent effects of the wind. Example of the distribution of the signs of the pressure $\bar{C}$ coefficient, on the side walls and cover of a greenhouse (positive signs indicate overpressure and negative suction): (a) Direction of the airflow generated by the pressure difference between the windward and leeward windows, and (b) airflow generated by the fluctuation of wind speed withing a single window.

Source: Adapated from Baille and Boulard, 1995.

12.2.4.3 ESTIMATE OF THE VENTILATION AIR EXCHANGE RATE IN GREENHOUSES BY EFFECT OF TURBULENCE

The effect of turbulence infiltration is especially significant when the mean pressure differences across openings are low while their turbulent components are large (Haghighat et al., 1991). This can happen in numerous situations, such as single-opening greenhouse (Boulard et al., 1998), two parallel openings (Potter, 1979), buildings with multiple openings (Etheridge, 2000), for wind parallel to the openings (Wang and Chen, 2012) and when buoyancy and wind effect have neutralized each other (Espinoza et al., 2017), etc. Several techniques have been proposed to calculate the turbulent infiltration resulting from wind fluctuations that produce two effects: pulsating flow and eddy penetration (Haghighat et al., 1991; Wang and Chen, 2012). The pulsating flow is caused by a variation in the pressure in the opening, which induces a fluctuation in the flow due to the compressibility of the air in the interior space (Haghighat et al., 1991). Thus, there is a pulsation of the flow through the openings that depends on the magnitude of the pressure fluctuation and the size of the greenhouse. The eddy penetration has a higher frequency of fluctuation that allows the air exchange in the openings. Eddy penetration is caused by fluid transfer due to eddy in unstable shear layers that exist when the external air flows across the opening (Straw et al., 2000). However, eddy penetration can be zero for wind perpendicular to the openings, due to zero parallel velocity (Wang and Chen, 2012).

Cockroft and Robertson (1976) studied the pulsation of airflow in a single opening building, subjected to a turbulent impacting airstream. They derived a simple theoretical model to help understand the physical phenomenon that causes airflow through the openings. In this model, the pulsation of the flow of a building with a single ventilation opening is caused by longitudinal wind fluctuations and the compressibility of air within the building.

Assuming an adiabatic process, the ventilation air exchange rate can be calculated considering the fluctuating component of airflow within a building with a single opening (Cockroft and Robertson, 1976):

$$G = \frac{\mathrm{dv}}{\mathrm{dt}} = \pm C_d \cdot S_v \cdot \left| v_x^2 - \left(\frac{2 \cdot \gamma \cdot P_a}{\rho \cdot V_{gre}} \right) \cdot V_{im} \right|^{1/2} \tag{12.44}$$

where γ is the polytropic exponent that is equal to 1.4 and 1.0 for adiabatic or isotherms flows, respectively, $\boldsymbol{P_a}$ is the pressure at the conditions in the free airstream [Pa], $\boldsymbol{V_{gre}}$ is the volume of the greenhouse [m^3] and $\boldsymbol{V_{im}}$ represents the decrease in volume of the original mass of air inside the greenhouse when the pressure inside is equal to the free airstream pressure ($\boldsymbol{P_i} = \boldsymbol{P_a}$) at time $\boldsymbol{t} = 0$.

Cockroft and Robertson (1976) argued that all the air flowing into a building, ***G***, does not necessarily contribute to effective ventilation. Some of the inflowing "pulsations" of air will be exhausted before mixing has a chance to occur. The net mean effective flow rate of ventilating air can then be calculated as (Cockroft and Robertson, 1976):

$$G_{efe} = f_M \cdot G = \frac{1}{2} \cdot f_p \cdot \left| \overline{G'} \right| \ [\mathrm{m^{3}s^{-1}}] \tag{12.45}$$

The proportion of mixed air $\boldsymbol{f_M}$ depends on the frequency spectrum of the turbulent airstream. Cockroft and Robertson (1976) got a value for this parameter of 0.38.

This pulsation model describes the effect of air compressibility in turbulent flows, but does not take into account air inertia (variations in air velocity in openings) or the characteristics of wind-induced pressure frequency.

12.2.4.4 DIRECT ESTIMATION OF AIRFLOW IN GREENHOUSE OPENINGS

Several methods can be used to estimate ventilation rates through wind driven ventilated buildings (Shen et al., 2012). Local estimations of total, medium and turbulent airflows and sensitive heat through greenhouse openings can be obtained using an eddy correlation system, comprising a sonic anemometer and a fine wire thermocouple (Boulard et al., 1996; Molina-Aiz et al., 2012). Considering the transport of heat with a velocity $\boldsymbol{v}$ normal to the opening, an elementary volume of air can transport heat at a rate $\rho \cdot v \cdot c_p \cdot T$ across the unit surface of the opening (Boulard et al., 1996). Using the Reynold's formulation, the temperature value can be expressed as:

$$T = \overline{T} + T' \tag{12.46}$$

where $\boldsymbol{T'}$ denotes instantaneous deviations from the average temperature, $\overline{T}$ [K]. Considering together, the average speed and temperature values can be written (Boulard et al., 1996a, 1997b, 1998a):

$$\rho \cdot c_p \cdot \overline{v \cdot T} = \rho \cdot c_p \cdot \overline{v} \cdot \overline{T} + \rho \cdot c_p \cdot \overline{v' \cdot T'} \tag{12.47}$$

where $\boldsymbol{c_p}$ is the specific heat of the air at constant pressure [J-·kg^{-1} K^{-1}]. This expression represents how the heat flux is composed of a mean normal airflow $\rho \cdot c_p \cdot \overline{v} \cdot \overline{T}$ and a portion due to eddy motion $\rho \cdot c_p \cdot \overline{v' \cdot T'}$, proportional to the covariance between $\boldsymbol{T}$ and $\boldsymbol{v}$ represented by the value $\overline{v' \cdot T'}$ (Boulard et al.,1998a). From this ratio can be calculated the average ventilation air exchange rate as:

$$G = \frac{S_v}{\Delta T_{io}} \cdot \left(\overline{v} \cdot \overline{T} + \overline{v' \cdot T'} \right) \tag{12.48}$$

Assuming the thermal effect is negligible, eq (12.48) allows to calculate the ventilation flow generated by the wind effect, $\boldsymbol{G_v}$.

Another way to determine the different components of airflow in openings is by simultaneously measurements of mean and turbulent pressure differences at ground level between inside and outside (Boulard et al., 1998a).

Applying eq (12.40) to measurements of the turbulent airflow and pressure differences at the opening level, the airflow can be expressed according to a natural wind coordinate system, in which the $\boldsymbol{x}$-axis is parallel to the mean flow, with $\boldsymbol{x}$ increasing in the direction of the flow (Boulard et al., 1998a). Considering the air velocity in the natural coordinate system $\boldsymbol{v_c}$, applying a coordinate rotation according to the Wesely formulas (Thurtell et al., 1970) we get the calculation of the standard deviation of the turbulent air velocity in the natural coordinate system σ_{vc} from $\overline{v_x}$, $\overline{v_y}$, $\boldsymbol{\sigma_{vx}}$, and $\boldsymbol{\sigma_{vy}}$:

$$\sigma_{vc} = \frac{\sigma_{vx} \cdot \overline{v_x} + \sigma_{vy} \cdot \overline{v_y}}{\sqrt{v_x^2 + v_y^2}} \tag{12.49}$$

Considering the pressure difference in the new ***x***-axis coordinate $\Delta P_c = \Delta P_x + \Delta P_c$ then eq (12.41) can be expressed as (Boulard et al., 1998a):

$$\sigma_{vc} = \frac{C_d \cdot \sigma_{\Delta P}}{\sqrt{2 \cdot \rho \cdot \Delta P_x + \Delta P_y}} \tag{12.50}$$

Combining eqs (12.49) and (12.50), and expressing $\boldsymbol{\Delta P}$ according to the wind speed and the mean pressure coefficient $\overline{C}$ given by eq (1.29) yields:

$$\frac{\sigma_{vx} \cdot \overline{v_x} + \sigma_{vy} \cdot \overline{v_y}}{\sqrt{v_x^2 + v_y^2}} = \frac{C_d \cdot \sigma_{\Delta P}}{\rho \cdot v_R \cdot \sqrt{\overline{\Delta C_x} + \overline{\Delta C_y}}} \tag{12.51}$$

where, $\overline{\Delta C_x}$ and $\overline{\Delta C_y}$ are the longitudinal and transverse variation of the mean pressure coefficient along the ventilation window, respectively (Boulard et al., 1998a).

From the above equation, the standard deviation of pressure fluctuations can be calculated as (Boulard et al., 1998a):

$$\sigma_{\Delta P} = \rho \cdot v_R \cdot \sqrt{\overline{\Delta C_x} + \overline{\Delta C_y}} \cdot \frac{\sigma_{vx} \cdot \overline{v_x} + \sigma_{vy} \cdot \overline{v_y}}{C_d \cdot \sqrt{v_x^2 + v_y^2}} \tag{12.52}$$

The terms $\overline{v_x}$, $\overline{v_y}$, σ_{vx}, and σ_{vy} can be measured with a three-dimensional sonic anemometer and $\overline{\Delta C_x}$ and $\overline{\Delta C_y}$ can be calculated using eq (12.29) from the measured pressure difference values along the length and height of the opening, respectively.

From the value of the standard deviation $\boldsymbol{\sigma_{\Delta P}}$ of the pressure difference fluctuations through the window $\Delta P'_V$ [eq (12.52)], and the mean pressure differences due to the wind effect $\overline{\Delta P_V}$ measured along an opening, can be calculated the total pressure difference due to wind $\boldsymbol{\Delta P_v}$ according to eq (12.38). By means of this value, the overall wind coefficient $\boldsymbol{C_w}$ can be obtained using eq (12.42), and finally the ventilation air exchange rate produced by the wind effect $\boldsymbol{G_v}$ can be obtained from eq (12.43).

Assuming the homogeneity of the air temperature inside and outside the greenhouse, the airflow exchanged can also be obtained from direct velocity measurements through the openings as (Swami and Chandra, 1987; Boulard et al., 1998a; Molina-Aiz et al., 2009):

$$G_v = \pm \sum_{j=1}^{n} S_{vj} \cdot \overline{v}_j = \frac{S_v}{2} \cdot \overline{v} \qquad (12.53)$$

where $\overline{v}_j$ is the mean components of the air velocity perpendicular to the plane of the opening to the elementary surface $\boldsymbol{S_{vj}}$ at the $\boldsymbol{n}$ locations of measurements (Molina-Aiz et al., 2009). The mean velocity of air can be related to pressure coefficients (Boulard et al., 1998a):

$$\left|\overline{v(x)}\right| = C_d \cdot \left|\overline{C_e(x) - C_i}\right|^{0.5} \cdot v_R \qquad (12.54)$$

where $\left|\overline{C_e(x) - C_i}\right|$ represents the abosulte value of the average difference (along the length of the opening) of the pressure coefficient differences between the interior (considered fixed) and the outside (depending on the area of the opening considered).

The Vickery model (Vickery and Surry, 1983) allows to calculate the airflow through each opening $\boldsymbol{j}$ in function of the pressure coefficient differences (Swami and Chandra, 1987):

$$G_{wj} = C_{dj} \cdot S_{vj} \cdot \frac{C_{ej} - C_{ij}}{\left|C_{ej} - C_{ij}\right|^{0.5}} \cdot v_R \qquad (12.55)$$

A similar expression can be obtained for the overall ventilated airflow in the greenhouse replacing eq (12.54) in eq (12.53), obtaining:

$$C_w = C_d \cdot \frac{S_v}{2} \cdot \left|\overline{C_e - C_i}\right|^{0.5} \cdot v_R \qquad (12.56)$$

By comparing the above equation [eq (1.56)] with eq (1.43), we can obtain the overall wind effect coefficient $\boldsymbol{C_w}$ as (Bruce, 1975; Brockett and Albright, 1987):

$$C_w = \left|\overline{C_e - C_i}\right| \qquad (12.57)$$

By replacing this ratio at eq (1.54), the average airflow velocity can be calculated through each ventilation opening produced by the pressure difference generated on both sides, considering only wind as driving force (Montero et al., 2001b; Campen and Bot, 2003; Molina-Aiz et al., 2009):

$$\overline{v} = C_d \cdot C_w^{0.5} \cdot v_R \qquad (12.58)$$

Similarly, the overall wind effect coefficient can be calculated from the mean velocity experimentally measured at each vent opening as (Molina-Aiz et al., 2009):

$$C_w = \left(\frac{\overline{v}}{C_d \cdot v_R} \right)^2 \tag{12.59}$$

Boulard et al. (1998a) compared the mean and turbulent airflows deducted from direct pressure measurements [eq (12.56)], with those obtained by sonic anemometers [eq (12.48) and (12.53)] and using the tracer gas technique [eq (12.43)]. The results obtained provided similar values of the ventilation efficiency coefficient [$\boldsymbol{E}_w = \boldsymbol{C}_d \cdot \boldsymbol{C}_w^{0.5}$].

12.3 COMBINED EFFECT OF THERMAL BUOYANCY AND WIND FORCE

The airflow through each window is essentially caused by the pressure difference between the outside and inside of the window, which is induced by the effects of buoyancy and wind forces (Whittle and Lawrence, 1960; Okada and Takakura, 1973). This pressure field is the result of a non-linear interaction between wind pressures around the greenhouse cover and buoyancy effect pressures caused by the air density difference between indoor and outdoor. These independent wind and thermal effects interact to set the indoor pressure that maintains a balance between the overall entering and leaving mass flow rates (Walker and Wilson, 1993).

Three main hypotheses can be considered when evaluating the thermal and wind effects together:

1. The total pressure difference can be obtained as the sum of the pressure differences due to each of the two effects (Bruce, 1978; Brockett and Albright, 1987; Zhang et al., 1989; Walker and Wilson, 1993; Boulard and Baille, 1995):

$$\Delta P(y) = \Delta P_T(y) + \Delta P_v \tag{12.60}$$

   This expression is valid when the effect of the wind is mainly caused by the mean distribution of pressures and not by the turbulent diffusion of air.

2. Another hypothesis considers the resulting flow as the vectorial sum of flows due to temperature differences and wind effect (Sherman and Grimsud, 1980; ASHRAE, 1985; Zhang et al., 1989; De Jong, 1990; Kittas et al., 1997):

$$G = \sqrt{G_T^2 + G_v^2} \tag{12.61}$$

In that hypothesis, the flow is considered driven by a pressure field equal to the sum of the two independent (stack and wind) pressure fields (Zhang et al., 1989). The resulting expression is obtained from the relation given by eq (12.13), which shows the proportionality between the pressure difference and the square of the flow rate [$\Delta P = 2 \cdot \rho \cdot G^2/(S_v \cdot C_d)^2$]. This expression can also be conceptually represented as the flow vector module, ***G***, composed of a vertical component, G_T (due to the upward movement of the air by thermal effect) and a horizontal component G_v, generated by the wind effect.

3. As the dynamics of the mean and turbulent flows are different, they can be considered independent and that the resulting flow as the algebraic sum of the two individual flows, one static due to the buoyancy effect, and the other turbulent due to the effect of the wind (Walker and Wilson, 1993; Boulard and Baille, 1995; Kittas et al., 1997):

$$G = G_T + G_v \quad (12.62)$$

In this case, it is assumed that the direction in which the air moves by the thermal and wind effects coincide. This method will produce large errors when magnitude of the thermal and wind effects are very similar (Walker and Wilson, 1993).

In fact, none of the three hypotheses is entirely true and depending on the geometry of the greenhouse and the windows will be closer or greater to the actual ventilation flow values. In function of building structure, characteristic of the wind (speed and direction) and configuration of vent openings (location, geometry, type, etc.), the wind force can either assist the buoyancy force or oppose the air flow (Li and Delsante, 2001). As consequence in a greenhouse with several side and roof openings, interaction between wind and thermal effect can derive in more than 10 airflow patterns (Espinoza et al., 2017).

12.3.1 VENTILATION MODELS IN GREENHOUSE WITH SINGLE OPENING

As a result of the three calculation hypothesis above and the possibility of determining the thermal effect according to a first- or second-order approximation, different models for the estimation of the ventilation air exchange rate ***G*** can be obtained:

- **Model 1.** Considering the first hypothesis of sum of *pressures,* $\Delta P = \Delta P_T + \Delta P_v$ and obtaining ΔP_T according to a second-order approximation and being ΔPv inferred from eq (12.42), yields the following expression (Boulard and Baille, 1995):

$$G = \frac{L_v}{3 \cdot g \cdot \left(\frac{\Delta T_{io}}{T_o}\right)} \cdot C_d \cdot \left[\left(2 \cdot g \cdot \frac{\Delta T_{io}}{T_o} \cdot \frac{H_v}{2} + C_w \cdot v_R^2 \right)^{3/2} - \left(C_w \cdot v_R^2\right)^{3/2} \right] \quad (12.63)$$

where L_v is the length of the opening [m].

- **Model 2.** The second hypothesis of vectorial sum of the flow rates is considered $G = \sqrt{G_T^2 + G_v^2}$, where the volume flow rate by thermal effect, G_T, is obtained as in the previous case according to the approximation of the second order [eq (12.24)] and the flow produced by the wind is calculated according to eq (12.43):

$$G = \frac{S_V}{3} \cdot C_d \cdot \sqrt{2 \cdot g \cdot \frac{\Delta T_{io}}{T_o} \cdot \frac{H_v}{2} + \frac{9}{4} \cdot C_w \cdot v_R^2} \quad (12.64)$$

If different values of the pressure coefficient are considered by wind effect on the outside and inside of the window (Bruce, 1978; Brockett and Albright, 1987), the following expression can be used (Kamaruddin, 1999; Kamaruddin et al., 2002a):

$$G = \frac{S_V}{3} \cdot C_d \cdot \sqrt{2 \cdot g \cdot \frac{\Delta T_{io}}{T_o} \cdot \frac{H_v}{2} + \frac{9}{4} \cdot \frac{(C_o - C_i)^2}{|C_o - C_i|} \cdot v_R^2} \quad (12.65)$$

The relationship between the two expressions [eqs (12.60) and (12.61)] is given by the eq (1.57).

- **Model 3.** It is considered the second hypothesis of vectorial sum of rate of fluid flow, $G = \sqrt{G_T^2 + G_v^2}$ obtaining the ventilation volumetric flow rate by thermal effect, G_T, according to the approximation of the first order [eq (12.22)] and that of the wind is calculated from eq (12.43), obtaining (Boulard and Baille, 1995; Boulard, 1996):

$$G = \frac{S_V}{2} \cdot C_d \cdot \sqrt{2 \cdot g \cdot \frac{\Delta T_{io}}{T_o} \cdot \frac{H_v}{4} + C_w \cdot v_R^2} \quad (12.66)$$

- **Model 4.** The third hypothesis of sum of flow rates is considered, $G = G_T + G_v$, calculating the flow rate due to the temperature differences G_T, according to a second-order approximation by (eq 12.24) and inferring G_v from eq (12.43), resulting (Boulard and Baille, 1995; Boulard, 1996):

$$G = \frac{S_V}{3} \cdot C_d \cdot \left[\sqrt{g \cdot \frac{\Delta T_{io}}{T_o} \cdot H_v} + \frac{3}{2} \cdot \sqrt{C_w} \cdot v_R \right] \quad (12.67)$$

- **Model 5.** It is considered the third hypothesis of sum of flow rates, $G = G_T + G_v$, with the flow rate due to temperature differences, $\boldsymbol{G_T}$, calculated according to a first-order approximation by the eq (12.22) and $\boldsymbol{G_v}$ is obtained from eq (12.43), resulting (Okada and Takakura, 1973; Boulard and Baille, 1995; Boulard, 1996):

$$G = \frac{S_V}{2} \cdot C_d \cdot \left[\sqrt{2 \cdot g \cdot \frac{\Delta T_{io}}{T_o} \cdot \frac{H_v}{4}} + \sqrt{C_w} \cdot v_R \right] \tag{12.68}$$

- **Model 6.** In this model, the thermal effect is neglected, considering only the effect of the wind becoming eq (12.63) the same that eq (12.43) used to analyze the wind effect ventilation:

$$G = \frac{S_V}{2} \cdot C_d \cdot \sqrt{C_w} \cdot v_R \tag{12.69}$$

Considering different values of the pressure coefficient, the following expression can be used for this model (Kamaruddin et al., 2002b):

$$G = \frac{S_V}{2} \cdot C_d \cdot \frac{(C_o - C_i)}{\sqrt{|C_o - C_i|}} \cdot v_R \tag{12.70}$$

Of the six models, the first seems to predict more accurate values (Boulard and Baille, 1995). However, the last expression [eq (12.69)] has been widely used in ventilation studies in greenhouses (Boulard and Baille, 1995; Papadakis et al., 1996; Kittas et al., 1996; Montero et al., 1997; Boulard et al., 1998a; Muñoz et al., 1999; Bailey, 2000; Lee and Short, 2000; Demrati et al., 2001; Kittas et al., 2002; Perez-Parra, 2002; Kamaruddin et al., 2002a; Campen and Bot, 2003; Brugger et al., 2005; Fatnassi et al., 2004; Pérez-Parra et al., 2004b; Abreu et al., 2005) due to its simplicity.

Bucklin et al. (1991) and Nääs et al. (1998) suggested an equivalent empirical equation to determine the rate of wind ventilation in single openings:

$$G = \frac{S_V}{2} \cdot E_w \cdot v_R \tag{12.71}$$

where $\boldsymbol{E_w}$ is the effectiveness of the opening that indicates the capacity of a window surface $\boldsymbol{S_v}$ to generate a flow $\boldsymbol{G}$ when the outside wind speed is $\boldsymbol{v_R}$. Comparing eqs (12.69) and (12.71), the effectiveness of the window corresponds to the overall coefficient of the wind effect:

$$E_w = C_d \cdot \sqrt{C_w} \tag{12.72}$$

Ventilation flow can also be obtained from the average air velocity at the opening, assuming that half of the ventilation opening acts as an inlet and the other half acts as an outlet, using an equation similar to eq (12.71) in which the opening angle of the windows is considered (Campen and Bot, 2003):

$$G = \frac{S_V}{2} \cdot \sqrt{\left(\frac{C_w}{F \cdot f_a}\right)} \cdot v_R \tag{12.73}$$

In this expression, the factor f_a contains all the effects related to the window aperture angle, which determines the surface available for air exchange and hydraulic resistance of the opening (Bot, 1983).

The result of using an effectiveness coefficient, E_w of the windows is that the ventilation flow for the greenhouse is proportional according to a linear relationship to the wind speed at the reference height and the effective area of the ventilation openings. This correlation has been confirmed through many trials in various greenhouse types. In eq (12.68), the coefficient C_w represents the relationship between the wind speed at the reference height and the effective pressure difference over all greenhouse openings (Kittas et al., 1997). However, C_w depends on the situation of the openings (obstacles, constructions, or neighboring greenhouses), its geometry and size and wind direction. The existence of obstacles close to the greenhouse side openings can reduce significantly the mean airflow and increase turbulence (López et al., 2011). Thus, the wind coefficient can change significantly in function of the wind direction (Reyes-Rosas et al., 2017; López-Martínez et al., 2019). Calibration of C_w can consist of a simple fit of the ventilation eq (1.67), which can be achieved by experimental measurements (tracer gas, energy balance, and anemometry) of the ventilation rate G. Continuous measurements along the time of the airflow rate in an Almería-type greenhouse showed a variation in the coefficient C_w along the day (Molina-Aiz et al., 2017).

Two different expressions can be proposed to calculate the total aerodynamically effective opening surface (Boulard and Draoui, 1995). One is the area, S_v (Figure 12.7) occupied by an insect-proof screen installed in the opening and calculated using the cord joining the extremities of the vent when it is open $H_v = 2 \cdot a_v \cdot \sin(\alpha_v/2)$, and the second is the vertical projection of this surface S_{v0} obtained from the vertical opening of the window $H_{v0} = a_v \cdot [\sin \beta_c + \sin(\alpha_v - \beta_c)]$

when $\beta_c > \alpha_v$ or $H_{v0} = a_v \cdot [\sin \beta_c - \sin (\beta_c - \alpha_v)]$ when $\beta_c > \alpha_v$ [being equivalent both equations because $+ \sin(\alpha_v - \beta_c) = - \sin(\beta_c - \alpha_v)$].

Morris and Neale (1954) identified the lineal relationship between the air exchange rate and the wind speed for a given greenhouse window. Bot (1983) proposed a function $G_d(\alpha_V)$ that allowed to obtain dimensionless ventilation rate in order to compare the results obtained in different types of greenhouses, called normalized volumetric flow rate (Fernandez and Bailey, 1992; Boulard and Draoui, 1995; Kittas et al., 1995; Baptista et al., 1999):

$$G_d(\alpha_v) = \frac{G}{A_v \cdot v_R} \tag{12.74}$$

being α_V the opening angle of the rotary-flap window (Figure 12.7) and $A_V = a_v \cdot L_v$ the surface of the flap [m^2], from which the effective surface of the window $S_V = H_v \cdot L_v$ can be calculated as (Boulard and Draoui, 1995):

$$S_v = L_v \cdot \left[2 \cdot a_v \cdot sin(\alpha_v/2)\right] = 2 \cdot A_v \cdot sin(\alpha_v/2) \tag{12.75}$$

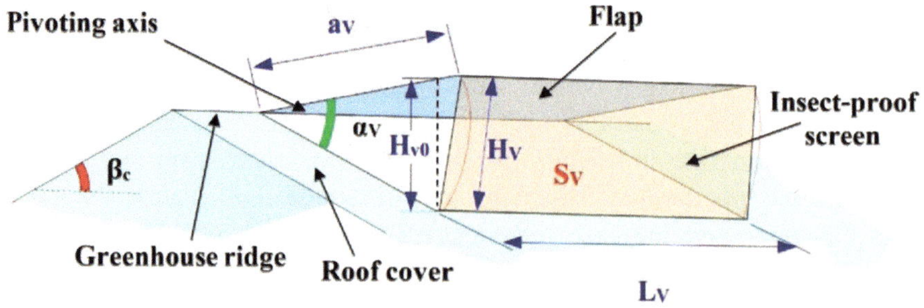

FIGURE 12.7 Scheme of the aerodynamically effective opening surface S_v in a continuous roof light top-hung window equipped with insect-proof screen.

Boulard and Draoui (1995) found a linear relationship between the dimensional airflow $G_d(\alpha_v)$ and the opening angle of the folding windows, α_v:

$$G_d(\alpha_v) = a_G \cdot \alpha_v + G_d(0) \tag{12.76}$$

where a_G is a constant and $G_d(0)$ is the offset of ventilation or leakeage when $\alpha_v = 0$.

The values of the dimensionless flow obtained by field tests for different types of greenhouses are presented in Table 12.2.

From eqs (12.68) and (12.70), you can obtain the value of the ventilation flow (Boulard and Baille, 1995):

TABLE 12.2 Dimensionless Ventilation Rate $G_\phi(\alpha_V)$ Determined by Several Authors for Different Types of Greenhouses.

Authors	$G_\phi(\alpha_V)$	Greenhouse type	S_c [m^2]	Number of spans	Opening types
Bot, 1983	$1.07 \times 10^{-3}\,\alpha_{V}.\exp(-\alpha_V/50)$	Venlo	1382	12	RO
De Jong, 1990	$2.29 \times 10^{-2}[1 - \exp(-\alpha_V/21.1)]$	Venlo	3210	24	RO
Fernandez and Bailey, 1992	$1.71 \times 10^{-3} + 8.31 \times 10^{-2}\,\alpha_V$	Venlo	422	4	RO
Kittas et al., 1995	$4.0 \times 10^{-3} + 0.27 \cdot \sin(\alpha_V/2)$	Single gable	179	2	RO
Boulard and Draoui, 1995	$1.8 \times 10^{-2} + 1.8 \times 10^{-3}\,\alpha_V$	Multispan	416	2	RO
Wang and Deltour, 1996	$1.0 \times 10^{-3} + 1.73 \times 10^{-3}\,\alpha_V$	Multispan	2150	18	RO
Deltour and Wang, 1999	$1.2 \times 10^{-3} + 1.82 \times 10^{-3}\,\alpha_V$	Multispan	1700	12	RO
Wang and Boulard, 2000a	$1.758 \times 10^{-3}\,\alpha_V$	Multispan	416	2	RO

Note: $\boldsymbol{S_c}$ area of the greenhouse soil covered by the greenhouse and ***RO*** roof opening.

$$G = E_w \cdot A_v \cdot v_R \cdot sin(\alpha_v / 2) \quad (12.77)$$

The air velocity in naturally ventilated buildings also depends on the angle of incidence of the wind θ_w penetrating the building (Nääs et al., 1998). Pressure coefficient C_w, together with the discharge coefficient, C_d can be used to estimate wind-produced ventilation (Zhang et al., 1989); however, it is difficult to estimate its value depending on the angle of wind incidence θ_w. Bucklin et al. (1991) and Nääs et al. (1998) estimated the effectiveness of side openings according to the angle of incidence of the wind θ_w (Table 12.3).

TABLE 12.3 Values of the Effectiveness Coefficient of the Window Depending on the Angle of Incidence of the Wind (Bucklin et al., 1991; Nääs et al., 1998).

Window effectiveness, E_w	Wind direction, θ_w [°]
0.60	85°–90°
0.50	70°–85°
0.35	50°–70°
0.30	30°–50°
0.25	10°–30°

Source: Adapted from Bucklin et al., 1991; Nääs et al., 1998

Nääs et al. (1998) developed an algorithm from scale tests in a wind tunnel to predict the effectiveness of the openings of naturally ventilated buildings:

$$E_w = 16.33 \times \left[(0.2 \cdot \rho \cdot v_R) / \mu \right]^{-0.3515} \cdot (sin\theta_w)^{1.201} \cdot \left[4 \cdot \left(\frac{H_v}{L_v} \right) \right]^{-0.1213} \cdot [sin\beta_c]^{-0.1531} \quad (12.78)$$

where, μ is the dynamic viscosity of the air [kg·s^{-1}·m^{-1}], θ_w the angle of incidence of the wind (20°–90°), β_c is the roof slope of the greenhouse (10°–30°) and H_V/L_V is the ratio of opening height to opening length (1/4–1/8). The discharge coefficient is affected by this ratio of openings (Table 12.4).

Besides of the angle of incidence of the wind, the effect of the crop on ventilation can be particularly important in greenhouses with side openings where the airflow can be partly obstructed by the crop rows (Fatnassi et al., 2009). The dependence of the ventilation efficiency coefficient of a greenhouse tunnel on crop height h_c [m] and wind direction θ_w [°] can be estimated as (Fatnassi et al., 2009):

$$E_w = 0.15 \cdot \left(cos\theta_w \right)^2 - 0.1 \cdot h_c + 0.36 \tag{12.79}$$

The characteristic of airflow across a greenhouse opening, relating the ventilation volumetric flow rate to the pressure difference, is given by the geometry and dimensions of the opening (de Jong and Bot, 1992). Two similar types of openings frequently used in naturally ventilated greenhouses, are the one-side-mounted casement windows used in Venlo greenhouse, and the continuous roof light top-hung windows used in Mediterranean greenhouses. In both cases, the window opening area can be modified changing the angle $\boldsymbol{\alpha_v}$ between the flap and the plane of the greenhouse cover (Figure 12.7).

TABLE 12.4 Values of the Dimensionless Discharge Coefficient of the Opening Determined by Several Authors.

Authors	C_d	Type of window
Brown and Solvason, 1962	0.6–0.8	Vertical rectangular
Kozai et al., 1980	0.6–0.7	Greenhouse roof opening
Bruce, 1982	0.6	Rectangular opening
Bot, 1983	$0.64 + 0.001 \cdot \alpha_V$	Venlo greenhouse roof opening
Bot, 1983	$\{1.75 + 0.7 \cdot \exp [-(L_V/(A_V \cdot \sin\alpha_V)/32.5]\}^{-0.5}$	Venlo greenhouse roof opening
Kirkpatrick and Hill, 1988	0.73	Vertical rectangular
Zhang et al., 1989	0.61	Vertical rectangular
De Jong, 1990	0.74	Vertical rectangular
Montero et al., 1997	$0.22 \ln[L_V/H_V] + 0.11$	Multispan roof opening [a]
Muñoz, 1998	$1.206 + 3{,}755\text{-}\exp[-0.1 \cdot L_V/H_V]\text{-}^{-0.5}$	Multispan roof opening [a]
Muñoz, 1998	$\{1.466 + 4.137 \cdot \exp [-0.048 \cdot L_V/H_V]\text{-}^{-0.5}$	Multispan roof opening [b]
Bailey et al., 2003	$\{1.9 + 0.7 \cdot \exp[-(L_V/(32.5) H_V \cdot \sin \alpha_V)]'^{-0.5}$	Greenhouse roof opening
Bailey et al., 2003	$-0.198 + 0.157 \cdot \alpha_V + 0.00108 \cdot L_V/H_V$	Large openings ($L_v/H_v \geq 5$)

Notes: [a]openings in side spans; [b] openings in internal spans.

In the case of a single window, the ventilation volumetric flow rate can be calculated in function of the roof slope $\boldsymbol{\beta_c}$ where the window is located (Bot, 1983):

$$G = \frac{L_v \cdot H_v^{3/2}}{3} \cdot C_d \cdot \left[sin\beta_c - sin(\beta_c - \alpha_v) \right]^{3/2}$$

$$\cdot \sqrt{g \cdot \frac{\Delta T_{io}}{T_o}} \cdot \left[1 + 0.6198 \cdot \left(\frac{H_v \cdot sin(\beta_c - \alpha_v)}{sin\beta_c - sin(\beta_c - \alpha_v)} \right)^{3/2} \right] \quad (12.80)$$

Abreu et al. (2005) considered a first modified version (denoted as b) of models 1, 3, and 6 including a fixed infiltration flow $\boldsymbol{G_{v0}}$ [m^3s^{-1}], and a second (denoted as c) with a variable ventilation flow depending on an equivalent vertical vent opening for leakage $\boldsymbol{H_{V0}}$ [m] or an equivalent cross-section $\boldsymbol{S_{v0}}$ [m^2]:

- **Model 1b.** This model included the fixed infiltration flow $\boldsymbol{G_{v0}}$ in eq (12.63):

$$G = \frac{L_v}{3 \cdot g \cdot \left(\frac{\Delta T_{io}}{T_o} \right)} \cdot C_d \cdot \left[\left(2 \cdot g \cdot \frac{\Delta T_{io}}{T_o} \cdot \frac{H_v}{2} + C_w \cdot v_R^2 \right)^{3/2} - \left(C_w \cdot v_R^2 \right)^{3/2} \right] + G_{v0} \quad (12.81)$$

- **Model 3b.** This model adds a fixed infiltration flow $\boldsymbol{G_{v0}}$ to eq (12.66):

$$G = \frac{S_V}{2} \cdot C_d \cdot \sqrt{2 \cdot g \cdot \frac{\Delta T_{io}}{T_o} \cdot \frac{H_v}{4} + C_w \cdot v_R^2} + G_0 \quad (12.82)$$

- **Model 6b.** This model complements eq (12.69) with a fixed infiltration flow $\boldsymbol{G_{v0}}$:

$$G = \frac{S_V}{2} \cdot C_d \cdot \sqrt{C_w} \cdot v_R + G_0 \quad (1.83)$$

- **Model 1c.** This model adds equivalent vertical vent opening for leakage $\boldsymbol{H_{V0}}$ to eq (12.63):

$$G = \frac{L_v}{3 \cdot g \cdot \left(\frac{\Delta T_{io}}{T_o} \right)} \cdot C_d \cdot \left\{ \left[2 \cdot g \cdot \frac{\Delta T_{io}}{T_o} \cdot \left(\frac{H_v + H_{v0}}{2} \right) + C_w \cdot v_R^2 \right]^{3/2} - \left(C_w \cdot v_R^2 \right)^{3/2} \right\} \quad (12.84)$$

- **Model 3c.** This model incorporates the equivalent cross-section $\boldsymbol{S_{v0}}$ in eq (1.63):

$$G = \left(S_{v0} + \frac{S_v}{2} \right) \cdot C_d \cdot \sqrt{\left(2 \cdot g \cdot \frac{\Delta T_{io}}{T_o} \cdot \frac{H_v}{4} \right) + C_w \cdot v_R^2} \quad (1.85)$$

- **Model 6c.** This model includes the equivalent cross-section S_{v0} in the airflow given by eq (12.63):

$$G = \left(S_{v0} + \frac{S_v}{2} \right) \cdot C_d \cdot \sqrt{C_w} \cdot v_R \tag{12.86}$$

- **Model 7.** From the analysis of experimental data of the ventilation airflow, G obtained with the technique of tracer gas (Boulard and Draoui, 1995; Boulard and Baille, 1995) or using continuous velocity measurements with sonic anemometers (Molina-Aiz et al., 2017), we can suggest an influence of wind velocity in the value of C_w. To try to take into account the influence of wind speed on the value of C_w, Boulard and Baille (1995) inferred an empirical model with less dependence on wind:

$$G = \frac{S_v}{2} \cdot C_d \cdot \sqrt{\left(2 \cdot g \cdot \frac{\Delta T_{io}}{T_o} \cdot \frac{H_v}{4} \right) + C_w \cdot v_R^{1.45}} \tag{12.87}$$

12.3.2 VENTILATION IN GREENHOUSES WITH SIDE AND ROOF OPENINGS

When a greenhouse has windows located at different heights, the static effect of the wind produces a vertical distribution of pressures between the side and roof openings (Bruce, 1975) and a horizontal distribution of pressures between the windward and the leeward part of the greenhouse (Hoxey and Richardson, 1983) and the effect of the permeability of the side wall that modified the pressure distribution, the "side-wall effect" or "surface-edge effect" analyzed by several authors (De Jong, 1990; Fernandez and Bailey, 1992). These wind effects together with the buoyancy effect produce two main airflows:

- A vertical ventilation airflow due to the "chimney" effect and the vertical distribution of pressures due to wind.
- A horizontal flow of ventilation due to the effect of the "side-wall effect" and the turbulent effect.

12.3.2.1 THERMAL EFFECT IN GREENHOUSES WITH SIDE AND ROOF VENTILATION

Ventilation in greenhouses with side and roof openings due to the effect of the temperature difference between inside and outside (thermal buoyancy force) can be considered as a result of the pressure difference that is generated between the two types of openings. Considering the ventilation system as two resistances to airflow in series, the total pressure difference produced by thermal effect can be calculated as the sum of the individual pressure drop at each of the windows (Pearson and Owen, 1994):

$$\Delta P_T = \Delta P_{TI} + \Delta P_{TE} \quad (12.88)$$

where ΔP_{TI} is the pressure difference in the air inlet windows due to the temperature difference [Pa] and ΔP_{TE} is the pressure variation in the outlet windows [Pa].

From eqs (12.9) and (12.78), it yields:

$$\Delta P_T = \frac{1}{2} \cdot \rho_I \cdot F_I \cdot v_I^2 + \frac{1}{2} \cdot \rho_E \cdot F_E \cdot v_E^2 \quad (12.89)$$

where ρ_I and ρ_E are the fluid densities [kg·m^{-3}], F_I and F_E are the dimensionless pressure loss coefficients, and v_E and v_S are the air velocity [m·s^{-1}] of incoming and exiting air streams in the side and roof openings, respectively.

Assuming stationary conditions and considering the conservation of the mass:

$$\rho_I \cdot S_I \cdot v_I = \rho_E \cdot S_E \cdot v_E \quad (12.90)$$

where S_I and S_E are the surfaces of the inlet and outlet openings, respectively [m^2].

From the above equation, an expression can be obtained to the air velocity of the incoming air streams through the side opening:

$$v_I = \frac{\rho_E \cdot S_E \cdot v_E}{\rho_I \cdot S_I} \quad (12.91)$$

Substituting eq (12.91) in eq (1.89) yields:

$$\Delta P_T = \frac{1}{2} \cdot \rho_I \cdot F_I \cdot \left(\frac{\rho_E \cdot S_E \cdot v_E}{\rho_I \cdot S_I} \right)^2 + \frac{1}{2} \cdot \rho_E \cdot F_E \cdot v_E^2 \quad (12.92)$$

which can take the form:

$$\Delta P_T = \frac{1}{2} \cdot \rho_E \cdot v_E^2 \cdot \left(\frac{\rho_E \cdot S_E^2}{\rho_I \cdot S_I^2} \cdot F_I + F_E \right) \quad (12.93)$$

For a column of air that rises due to the thermal buoyancy force through the greenhouse (Figure 12.8), the pressure difference that occurs can be expressed based on the difference in density between the inlet and outlet windows (Pearson and Owen, 1994):

$$\Delta P_T = g \cdot h_{IE} \cdot (\rho_I - \rho_E) \tag{12.94}$$

where h_{IE} is the difference in height between air inlet and outlet (side and roof openings) [m].

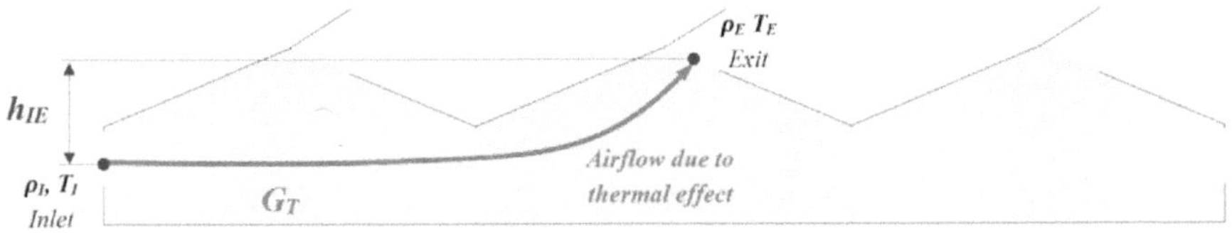

FIGURE 12.8 Airflow through two consecutive openings due to the buoyancy force

Source: Adapted from Pearson and Owen, 1994).

Equating the overall pressure losses, gives by eq (12.93), and the buoyancy force moving the air, gives by eq (12.94) yields:

$$g \cdot h_{IE} \cdot (\rho_I - \rho_E) = \frac{1}{2} \cdot \rho_E \cdot v_E^2 \cdot \left(\frac{\rho_E \cdot S_E^2}{\rho_I \cdot S_I^2} \cdot F_I + F_E \right) \tag{12.95}$$

or:

$$v_E = \sqrt{\frac{2 \cdot g \cdot h_{IE} \cdot \Delta\rho_{IE}}{\rho_E \cdot \left(\frac{\rho_E \cdot S_E^2}{\rho_I \cdot S_I^2} \cdot F_I + F_E \right)}} \tag{12.96}$$

where $\Delta\rho_{IE}$ is the density difference between the air incoming and exiting of the greenhouse [kg·m^{-3}].

Multiplying by the surface of the opening acting as exit, S_E, we can obtain the volumetric flow rate of air exiting of the greenhouse by ventilation (Pearson and Owen, 1994) [m^3·s^{-1}]:

$$G_E = v_E \cdot S_E = \sqrt{\frac{2 \cdot g \cdot h_{IE} \cdot \Delta\rho_{IE}}{\rho_E \cdot \left(\frac{\rho_E}{\rho_I} \cdot \frac{F_I}{S_I^2} + \frac{F_E}{S_E^2} \right)}} \tag{12.97}$$

Considering the air as an ideal gas and taking into account that in the case of air movement in greenhouses, the absolute pressure at the inlet and outlet is very similar ($P_I/P_E \approx 1$), densities can be related to temperatures at the two windows:

$$\frac{\rho_E}{\rho_I} = \frac{T_I}{T_E} \tag{1.98}$$

where, T_E and T_I are the temperatures of incoming and exiting air streams [K].

- **Model 8.** When a greenhouse is ventilated by both side and roof windows, and buoyancy force prevails (the wind effect is negligible), the ventilation rate can be calculated by replacing the density values in eq (12.97) with temperatures [ρ_E/ρ_I according to eq (1.88) and $\Delta\rho_{IE}/\rho_E$ replaces by $\Delta T_{IE}/T_E$] (Pearson and Owen, 1994):

$$G = \sqrt{2 \cdot g \cdot \frac{h_{IE} \cdot \Delta T_{IE}}{T_E \cdot \left(\frac{T_E}{T_I} \cdot \frac{F_I}{S_I^2} + \frac{F_E}{S_E^2} \right)}} \tag{12.99}$$

The ventilation flow in the case of input and output windows with the same geometry (F_E=F_S) and considering that the ratio between the input and output temperatures is next to one (T_E/T_S~1) (Bruce, 1978; Bot, 1983):

$$G = C_d \cdot \sqrt{2 \cdot g \cdot \frac{h_{IE} \cdot \Delta T_{IE}}{T_E \cdot \left(\frac{1}{S_I^2} + \frac{1}{S_E^2} \right)}} \tag{12.100}$$

Being $\boldsymbol{C_d} = \boldsymbol{F_I}^{-1/2} = F_E^{-1/2}$.

- **Model 9.** When a greenhouse is ventilated by side and roof windows, temperature and humidity differences can occur between the ridge and the bottom of the greenhouse, which may be in the same order as the differences between indoor and outdoor air. In this case, if the temperature and humidity are distributed linearly with the height, the ventilation rate can be calculated as (Zhao et al., 2001):

$$G = \sqrt{2 \cdot g \cdot \frac{h_{IE} \cdot (2 \cdot \rho_o - \rho_I - \rho_E)}{\rho_o \cdot \left(\frac{F_I}{S_I^2} + \frac{F_E}{S_E^2} \right)}} \tag{12.101}$$

where $\boldsymbol{\rho_o}$ is the density of the air outside the greenhouse [kg·m^{-3}].

12.3.2.2 COMBINED EFFECT ON GREENHOUSES WITH SIDE AND ROOF VENTILATION

The airflow produced by the combination of wind effects and air buoyancy in greenhouses equipped with roof and side openings can be represented by an expression similar to those of greenhouses with a single type of windows, where the "chimney" effect depends on the size of the side and roof openings and the distance separating the two types of windows and the effect of the wind depends on the wind coefficient and the outside speed as on the models applied to a single type of openings (Section 12.3.1).

- **Model 10.** This model considers the second hypothesis combining wind and thermal effects, calculating the ventilation rate as vectorial sum ($G = \sqrt{G_T^2 + G_v^2}$) of vertical (thermal effect) and horizontal (wind effect) flows. The effect of temperature can be quantified by (eq 1.100) and the effect of wind by eq (12.43) obtaining (Zhang et al.,1989; Kittas et al., 1997; Boulard et al., 1997a):

$$G = C_d \cdot \sqrt{\left(\frac{S_{RV} \cdot S_{SV}}{\sqrt{S_{RV}^2 + S_{SV}^2}}\right)^2 \cdot \left(2 \cdot g \cdot \frac{\Delta T_{io}}{T_o} \cdot h_{SR}\right) + \left(\frac{S_{RV} + S_{SV}}{2}\right)^2 \cdot C_w \cdot v_R^2} \quad (12.102)$$

where, $\boldsymbol{h_{SR}}$ is the vertical distance between the midpoint of side and roof openings [m] and $\boldsymbol{S_{RV}}$ and $\boldsymbol{S_{SV}}$ are the roof and side ventilation areas, respectively [m^2].

The use of expressions combining wind action and the "chimney" effect improves the predictability of models for greenhouses equipped with side and roof windows (Boulard et al., 1997a).

If different coefficients of static wind pressure of the roof and side windows are considered, a new expression is obtained (Kittas et al., 1997a):

$$G = C_d \cdot \sqrt{\left(\frac{S_{RV} \cdot S_{SV}}{\sqrt{S_{RV}^2 + S_{SV}^2}}\right)^2 \cdot \left(2 \cdot g \cdot \frac{\Delta T_{io}}{T_o} \cdot h_{SR} + \left|C_{wRV} - C_{wSV}\right| \cdot v_R^2\right) + \left(\frac{S_{RV} + S_{SV}}{2}\right)^2 \cdot C_w \cdot v_R^2} \quad (12.103)$$

where $\boldsymbol{C_{wRV}}$ and $\boldsymbol{C_{wSV}}$ are static wind pressure coefficient at the level of roof and side openings (dimensionless), respectively.

The value of the difference in these coefficients obtained by Kittas (1980) for a multispan greenhouse with roof and side windows on both sides was $(C_{wRV} - C_{wSV}) = 0.0007$. This positive value indicates that the vertical airflow due to the effect of static wind pressure has the same direction as the airflow induced by the "chimney" effect, that is, the air enters through the side openings and exits through the roof windows. However, its value is very small and statistically non-significant. This means that the vertical flow generated by the vertical pressure distribution is negligible or, that a large part of the static effect of the wind is already explained by the coefficient $\boldsymbol{C_w}$

(Kittas et al., 1997a). However, CFD simulations (Molina-Aiz et al., 2011) and measurements with anemometers (Molina-Aiz, 2010; López-Martinez, 2010; Espinoza et al., 2017) have showed that in certain conditions the wind and thermal effect can oppose. This opposition of the two driven forces of the natural ventilation occurs when a strong wind produces that air entry inside the greenhouse through the windward roof openings where the warm inside air tries to exit as consequence of the buoyancy force.

- **Model 11.** The third hypothesis of algebraic sum of ventilation flow rates due to the thermal and wind effects, $G = G_T + G_v$, can also be considered The thermal ventilation flow can be quantified by eq (12.100) and the wind effect by eq (12.43), obtaining the following expression (Kittas et al., 1997a):

$$G = C_d \cdot \frac{S_{RV} \cdot S_{SV}}{\sqrt{S_{RV}^2 + S_{SV}^2}} \sqrt{2 \cdot g \cdot \frac{\Delta T_{io}}{T_o} \cdot h_{SR}} + \frac{S_{RV} + S_{SV}}{2} \cdot C_d \cdot \sqrt{C_w} \cdot v_R \quad (12.104)$$

- **Model 12.** Considering different values of the pressure coefficients by wind effect on the outside and inside of the windows, a new expression is obtained (Kamaruddin, 1999; Kamaruddin et al., 2002a):

$$G = C_d \cdot \sqrt{\left(\frac{S_{RV} \cdot S_{SV}}{\sqrt{S_{RV}^2 + S_{SV}^2}}\right)^2 \cdot \left(2 \cdot g \cdot \frac{\Delta T_{io}}{T_o} \cdot h_{SR}\right) + \left(\frac{S_{RV} + S_{SV}}{2}\right)^2 \cdot \frac{(C_o - C_i)^2}{\left|C_o - C_i\right|} \cdot v_R^2} \quad (12.105)$$

In greenhouses with side and roof windows, the hypothesis of sum of the pressure fields, corresponding to the vectorial sum of the flows, eq (12.102) improves statistical adjustments (Molina-Aiz, 2010; López-Martínez, 2011) and provides values of the model parameters (**C_d** and **C_w** listed in Table 12.5) closer to those indicated in the literature (Kittas et al., 1997a).

We can obtain three additional models using a different discharge coefficient for the side (**C_{dSV}**) and roof openings (**C_{dRF}**) (López-Martinez et al., 2019) and considering that the wind can either assist the buoyancy force or oppose the airflow (Li and Delsante, 2001; Molina-Aiz, 2010; Molina-Aiz et al., 2017; López-Martinez et al., 2019).

- **Model 13.** It is considered that flow is driven by the sum of the two independent pressure fields and the vectorial sum of the airflows generated by thermal and wind $G = \sqrt{G_T^2 \pm G_v^2}$:

$$G=\sqrt{\left(\frac{(C_{dRV}\cdot S_{RV})\cdot(C_{dSV}\cdot S_{SV})}{\sqrt{(C_{dRV}\cdot S_{RV})^2+(C_{dSV}\cdot S_{SV})^2}}\right)^2\cdot\left(2\cdot g\cdot\frac{\Delta T_{io}}{T_o}\cdot h_{SR}\right)\pm\left(\frac{C_{dRV}\cdot S_{RV}+C_{dSV}\cdot S_{SV}}{2}\right)^2\cdot C_w\cdot v_R^2} \quad (12.106)$$

– **Model 14.** Considering the algebraic sum of flows by thermal and wind effects, $G = G_T \pm G_v$:

$$G=\left[\frac{(C_{dRV}\cdot S_{RV})\cdot(C_{dSV}\cdot S_{SV})}{\sqrt{(C_{dRV}\cdot S_{RV})^2+(C_{dSV}\cdot S_{SV})^2}}\right]\cdot\sqrt{2\cdot g\cdot\frac{\Delta T_{io}}{T_o}\cdot h_{SR}}\pm\frac{C_{dRV}\cdot S_{RV}+C_{dSV}\cdot S_{SV}}{2}\cdot\sqrt{C_w}\cdot v_R \quad (12.107)$$

– **Model 15.** This model only considerate the wind effect, neglecting contribution of stack effect:

$$G=\frac{C_{dRV}\cdot S_{RV}+C_{dSV}\cdot S_{SV}}{2}\cdot\sqrt{C_w}\cdot v_R \quad (12.108)$$

Depending on the models considered, different values of the discharge coefficients and wind effect are obtained as reflected in Table 12.5. The values obtained by the different authors are quite similar which seems to indicate the independence of their value with respect to the surface of the greenhouse. Recently, has been published a detailed review of existing literature on the wind pressure coefficients on greenhouse structures (Maraveas, 2020).

The mean and standard deviation of the values listed in Table 12.5 for the different parameters are: $\boldsymbol{C_d} = 0.644 \pm 0.117$ (for unscreened greenhouses), $\boldsymbol{C_w} = 0.123 \pm 0.079$ and $\boldsymbol{E_w} = C_d{\cdot}C_w^{\ 0.5} = 0.200 \pm 0.079$. Despite the good predictability of these models in some types of greenhouses, many unresolved theoretical problems remain. Thus, the most important question is the nature (mean or turbulent) of the forces that generate the effect of the wind and the mechanisms by which the combination with the thermal effect occurs. In addition, the estimation of an air renewal rate, based on an overall flow measurement, does not identify each component which makes it necessary to take complementary measures for estimating local flows in each of the ventilation openings.

12.4 INFLUENCE OF INSECT-PROOF SCREENS ON VENTILATION

Insect-proof screens placed in the greenhouse openings produce an additional resistance to the ventilation airflow (Teitel et al., 1999; Teitel, 2001;

TABLE 12.5 Values of the Discharge C_d and Wind Effect C_w Coefficients and Effectiveness of the Windows $E_w = C_d\,Cw_v^{0.5}$ Obtained by Various Authors and in Different Types of Greenhouses for the Different Ventilation Models (M).

Authors	M	C_d	C_w	$C_d \cdot C_w^{0.5}$	G_0 [$m^3 \cdot s^{-1}$]	S_{V0} [m^2]	Type of greenhouse	S_c [m^2]	Number of spans	Type of windows
Boulard and Baille, 1995	1	0.75	0.09	0.205	–	–	Multispan	416	2	RV
Abreu et al., 2005	1	0.646	0.251	0.324	–	–	Multispan	179	2	SV
Abreu et al., 2005	1b	0.454	0.271	0.236	0.585	–	Multispan	179	2	SV
Boulard and Baille, 1995	3	0.848	0.07	0.202	–	–	Multispan	416	2	RV
Teitel et al., 1999	3	0.530	0.051	0.043	–	–	Multispan	960	4	RV[a]
Papadakis et al., 1996	3	0.745	0.079	0.210	–	–	Multispan	416	2	RV+SV
Sase et al., 2002	3	0.56	0.098	0.175	–	–	Multi-layer	242	4	RV+SV
Sase et al., 2002	3	0.89	0.116	0.303	–	–	Multi-layer	242	4	RV+SV
Pérez-Parra et al., 2004a	3	0.649	0.100	0.205	–	–	Almería	881	5	RV[b]
Pérez-Parra et al., 2004a	3	0.649	0.037	0.125	–	–	Almería	881	5	RV[c]
Pérez-Parra et al., 2004a	3	0.645	0.006	0.046	–	–	Almería	881	5	RV[d]
Pérez-Parra et al., 2004a	3	0.666	0.006	0.047	–	–	Almería	881	5	SV[d]
Abreu et al., 2005	3	0.685	0.129	0.246	–	–	Multispan	179	2	RV
Abreu et al., 2005	3	0.758	0.035	0.142	–	–	Multispan	179	2	SV
Abreu et al., 2005	3	0.69	0.093	0.210	–	–	Multispan	179	2	RV+SV
Abreu et al., 2005	3b	0.457	0.268	0.237	0.585	–	Multispan	179	2	SV
Abreu et al., 2005	3c	0.713	0.099	0.224	–	1.234	Multispan	179	2	SV
Majdoubi et al., 2007	3	0.25	0.14	0.09	–	–	Canary	11,250	9	RV+SV
Boulard and Baille, 1995	4	0.5	0.14	0.185	–	–	Multispan	416	2	RV
Boulard and Baille, 1995	5	0.467	0.15	0.184	–	–	Multispan	416	2	RV

TABLE 12.5 *(Continued)*

Authors	M	C_d	C_w	$C_d \cdot C_w^{0.5}$	G_0 [$m^3 \cdot s^{-1}$]	S_{V0} [m^2]	Type of greenhouse	S_c [m^2]	Number of spans	Type of windows
Boulard and Baille, 1995	6	–	–	0.207	–	–	Multispan	416	2	RV
Kittas et al., 1996	6	–	0.12	0.23	–	–	Tunnel	900	1	RV
Montero et al., 1997	6	0.769	0.32	0.435	–	–	Multispan	230 (E1/3)	3	RV[e]
Montero et al., 1997	6	0.579	0.32	0.328	–	–	Multispan	230 (E1/3)	3	RV[f]
Boulard et al., 1998a	6	–	–	0.22	–	–	Multispan	416	2	RV
Muñoz et al., 1999	6	0.504	0.400	0.319	–	–	Multispan	230	3	RV
Baptista et al., 1999	6	0.65	0.10	0.205	–	–	Venlo	204.8	4	RV
Bailey, 2000	6	0.65	0.11	0.215	–	–	Venlo	38,700	60	RV
Lee and Short, 2000	6	–	–	0.28	–	–	Multispan	1153	4+½	RV+SV
Demrati et al., 2001	6	–	–	0.27	–	–	Multispan	10,000	32	RV+SV
Kittas et al., 2002	6	0.7	0.169	0.288	–	–	Venlo	416	2	RV
Authors	**M**	C_d	C_w	$C_d \cdot C_w^{0.5}$	G_0 [$m^3 \cdot s^{-1}$]	S_{V0} [m^2]	**Greenhouse**	S_c [m^2]	**spans**	**Openings**
Kamaruddin et al., 2002[a]	6	–	–	0.13	–	–	Multispan	600	1	RV+SV
Kamaruddin et al., 2002[b]	6	–	–	0.221	–	–	Tunnel	18	1	RV
Campen and Bot, 2003	6	–	–	0.05	–	–	Almería	881	5	RV[d]
Brugger et al., 2005	6	–	–	0.053	–	–	Almería	881	5	RV[d]
Fatnassi et al., 2004	6	–	–	0.264	–	–	Tunnel	368	1	SV
Fatnassi et al., 2004	6	–	–	0.131	–	–	Tunnel	368	1	SV+RV
Pérez-Parra et al., 2004a	6	0.649	0.104	0.205	–	–	Almería	881	5	RV[b]

TABLE 12.5 *(Continued)*

Authors	M	C_d	C_w	$C_d \cdot C_w^{0.5}$	G_0 [$m^3 \cdot s^{-1}$]	S_{V0} [m^2]	Type of greenhouse	S_c [m^2]	Number of spans	Type of windows
Pérez-Parra et al., 2004a	6	0.649	0.039	0.125	–	–	Almería	881	5	RV[c]
Pérez-Parra et al., 2004a	6	0.645	0.007	0.046	–	–	Almería	881	5	RV[d]
Pérez-Parra et al., 2004a	6	0.406	0.007	0.029	–	–	Almería	881	5	RV[d]
Pérez-Parra et al., 2004a	6	0.666	0.008	0.047	–	–	Almería	881	5	SV[d]
Pérez-Parra et al., 2004b	6	0.64	0.4	0.405	–	–	Almería	881	5	RV[g]
Perez-Parra, 2002	6	0.656	0.074	0.025	–	–	Almería	881	5	SV+RV[d]
Abreu et al., 2005	6	–	–	0.336	–	–	Multispan	179	2	SV
Boulard et al., 1997a	6b	–	–	0.08–0.17	4.9–9.3	–	Tunnel	240	1	RV
Boulard et al., 1997a	6b	–	–	0.14–0.51	0–16	–	Tunnel	240	1	SV
Boulard et al., 1997a	6b	–	–	0.07–0.18	8.9–21	–	Tunnel	240	1	RV+SV
Boulard et al., 1997a	6b	–	–	0.18	6.5	–	Multispan	416	2	RV
Boulard et al., 1997a	6b	–	–	0.26	8.5	–	Multispan	416	2	SV
Boulard et al., 1997a	6b	–	–	0.19	0.5	–	Multispan	416	2	RV+SV
Boulard et al., 1997a	6b	–	–	0.21	13.0	–	Multi-layer	177	3	RV
Baptista et al., 1999	6b	0.64	0.09	0.192	0.33	–	Venlo	204	4	RV
Pérez-Parra et al., 2003	6b	–	–	0.03	5.26	–	Almería	861	5	RV[b]
Pérez-Parra et al., 2003	6b	–	–	0.03	4.98	–	Almería	861	5	RV[c]
Pérez-Parra et al., 2003	6b	–	–	0.162	2.71	–	Almería	861	5	RV[d]
Pérez-Parra et al., 2003	6b	–	–	0.09	0.305	–	Almería	861	5	RV[d]
Abreu et al., 2005	6b	–	–	0.243	0.594	–	Multispan	179	2	SV
Katsoulas et al., 2006	6b	–	–	0.078	0.135	–	Multispan	160	1	SV

TABLE 12.5 *(Continued)*

Authors	M	C_d	C_w	$C_d \cdot C_w^{0.5}$	G_0 [$m^3 \cdot s^{-1}$]	S_{V0} [m^2]	Type of greenhouse	S_c [m^2]	Number of spans	Type of windows
Abreu et al., 2005	6c	–	–	0.261	–	0.953	Multispan	179	2	SV
Kittas et al., 1995	6d	0.7	0.14	0.26	0.5	1.4	Multispan	177	1	SV+RV
Boulard and Baille, 1995	7	0.664	0.24	0.325	–	–	Multispan	416	2	RV
Kittas et al., 1997b	9	–	–	0.100	–	–	Multispan	416	2	SV+RV
Kittas, 1980	10	0.75	0.07	0.198	–	–	Multispan	416	2	SV+RV
Perez-Parra, 2002	10	0.656	0.002	0.025	–	–	Almería	881	5	SV+ RV[d]
Kittas et al., 1996	10b	0.6	0.12	0.208	1.7	–	Multispan	416	2	SV+RV
Kittas et al., 1997a	10d	0.7	0.13	0.252	1.4	0.5	Multispan	416	2	SV+RV
Kittas et al., 1997a	10	0.75	0.07	0.198	–	–	Multispan	416	2	SV+RV
Kittas et al., 1997a	11	0.42	0.18	0.178	–	–	Multispan	416	2	SV+RV
Greenhouses Equipped with Insect-Proof Screens on Windows										
Authors	**M**	**C_d**	**C_w**	**$C_d \cdot C_w^{0.5}$**	**Porosity, φ**		**Greenhouse**	**S_c [m^2]**	**Spans**	**Openings**
Teitel et al., 1999	3	0.199	0.043	0.110	0.35		Multispan	960	4	RV[b,c]
Montero et al., 1997	6	0.509	0.32	0.288	0.45		Multispan	230 (E1/3)	3	RV
Montero et al., 1997	6	0.369	0.32	0.209	0.25		Multispan	230 (E1/3)	3	RV
Montero et al., 1997	6	0.379	0.32	0.214	0.45		Multispan	230 (E1/3)	3	RV[f]
Muñoz et al., 1999	6	0.555	0.075	0.152	0.45		Multispan	230	3	RV[b,c]
Kittas et al., 2002	6	0.329	0.171	0.136	0.60		Venlo	416	2	RV[b,c]

TABLE 12.5 *(Continued)*

Authors	M	C_d	C_w	$C_d \cdot C_w^{0.5}$	G_0 [$m^3 \cdot s^{-1}$]	S_{V0} [m^2]	Type of greenhouse	S_c [m^2]	Number of spans	Type of windows
Fatnassi et al., 2002a	6	0.42	0.11	0.14	0.69		Canary	5600	1	RV+SV[h]
Fatnassi et al., 2002a	6	0.51	0.04	0.10	0.69		Canary	5600	1	RV+SV[i]
Brugger et al., 2005	6	–	–	0.156	0.34		Almería	881	5	RV[b,c]
Pérez-Parra et al., 2003	6b	–	–	0.018	0.39		Almería	861	5	RV[d]
Katsoulas et al., 2006	6b	–	–	0.052	0.313		Multispan	160	1	SV
Katsoulas et al., 2006	6b	–	–	0.028	0.220		Multispan	160	1	RV
Katsoulas et al., 2006	6b	–	–	0.096	0.340		Multispan	160	1	SV+RV
Teitel et al., 2008a	10	0.253	0.075	0.069	0.62		Single span	74.4	4.5	SV
Molina-Aiz et al., 2009	10	0.159	0.035	0.081	0.34		Almería	1588	5	RV+SV[j]
Molina-Aiz et al., 2009	10	0.229	0.061	0.056	0.34		Almería	1588	5	RV+SV[k]

Notes: S_c soil surface covered by the greenhouse, ***SV*** side ventilation, and ***RV*** roof ventilation. [a] with 20% shading mesh; [b] windward and [c] leeward roof flap ventilators; [d] roof rolling ventilators; [e] Roof windows in side spans; [f] Roof windows in central spans; [g] First roof flap window open to windward. [h] Wind perpendicular and [i] parallel to the openings; [j] *Levante* wind from northeast (NE) [k] *Poniente* wind from southwest (SW)

Fatnassi et al., 2002a; Bartzanas et al., 2002; Kittas et al., 2002; Fatnassi et al., 2003; Teitel et al., 2005; Molina-Aiz et al., 2012; Fatnassi et al., 2013; López et al., 2017). The use of insect-proof screens can decrease around 50% the air velocity through the greenhouse openings, and therefore, the ventilation rate (Molina-Aiz, 2010). This reduction in ventilation rates produces higher temperature and humidity, increasing gradients in both the vertical and the horizontal directions (Teitel, 2007).

12.4.1 CONSTANT PRESSURE-LOSS COEFFICIENT OF INSECT-PROOF SCREENS

The discharge coefficient method based on Bernoulli's equation [eq (12.7)] can be used to calculate the pressure drop through woven wire insect-proof screens over a large range of Reynolds numbers (Teitel et al., 2001). The aerodynamic resistance of unscreened openings is function of the shape, generally rectangular and the relation between height and length, $\boldsymbol{H_v/L_v}$ and the opening angle $\boldsymbol{\alpha_v}$ (Table 12.4). All these factors are characterized by the discharge coefficient $\boldsymbol{C_{dLH}}$ or its equivalent pressure drop coefficient $\boldsymbol{F_{LH}} = 1/\boldsymbol{C_{dLH}}^2$. For unscreened openings, the discharge coefficient shows low variability, with values around $\boldsymbol{C_{dLH}}$ = 0.56–0.89 (Tables 12.4 and 12.5). However, when insect-proof screens are installed in the openings, the discharge coefficient can be reduced drastically ($\boldsymbol{C_{dLH}}$ = 0.199–0.555 in Table 12.5) as consequence of the additional resistance to airflow produced by the nets.

The overall discharge coefficient of a screened opening can be calculated considering a system with two resistances in series (Arbel et al., 2000). The total pressure difference on both sides of a screened opening can be considered as the sum of the pressure drops across the inlet opening ΔP_{HL} and across the screen ΔP_{φ} (Arbel et al., 2000; Kittas et al., 2002):

$$\Delta P = \Delta P_{LH} + \Delta P_{\varphi} = \frac{1}{2}\cdot\rho\cdot F_{HL}\cdot v^2 + \frac{1}{2}\cdot\rho\cdot F_{\varphi}\cdot v^2 \qquad (12.109)$$

Considering that the air velocity is the same across the opening and the screen in eq (12.109), a dimensionless pressure drop coefficient of a screened openings can be calculated as (Arbel et al., 2000; Kittas et al., 2002):

$$F = F_{HL} + F_{\varphi} \qquad (12.\ 110)$$

The equivalent discharge coefficient of the opening with a net $C_d = 1/F^2$ can be calculated as (Arbel et al., 2000; Teitel et al., 2001; Kittas et al., 2002):

$$C_d = \sqrt{\frac{1}{\frac{1}{C_{d,LH^2}} + \frac{1}{C_{d,\varphi^2}}}} \tag{12.111}$$

where, $C_{d,LH}$ is the discharge coefficient due to the shape of the opening and $C_{d,\varphi}$ is the discharge coefficient of the insect-proof screen given by eq (12.112).

Several authors have used a coefficient of discharge to characterize the airflows resistance of insect-proof screens installed in the greenhouse openings (Brundrett, 1993; Kosmos et al., 1993; Pearson and Owen, 1994; Montero et al., 1996; Muñoz et al., 1999; Teitel et al., 1999; Fatnassi et al., 2002; Kittas et al., 2002; Bailey et al., 2003).

12.4.2 PERMEABILITY OF INSECT-PROOF SCREENS AND VARIABLE DISCHARGE COEFFICIENT

The first method does not allow prediction of pressure drop when the texture of the screen is complex and a characteristic diameter of the threads cannot be defined or at low Reynolds numbers, lower than 150 (Teitel et al., 2001). Taking into account that air velocities through screened openings in naturally ventilated greenhouses ranged between 0.1 and 1.5 m·s^{-1} (Teitel, 2001; Kitttas et al., 2002; Kittas and Bartzanas, 2007; Molina-Aiz et al., 2009; López et al., 2012), a constant discharge coefficient seems not adequate.

The pressure drop in an insect-proof screen can be calculated more accurately using the equation applied to the porous media known under the names of Dupuit (1863) and Forchheimer (1901), modifying Darcy's equation adding a term depending on the square of velocity (Joseph et al., 1976; Dullien, 1979; Nield and Bejan, 2017):

$$\frac{\partial P}{\partial x} = -\left(\frac{\mu}{K_p}u + \rho_a\left(\frac{Y}{K_p^{1/2}}\right)|u|\,u\right) \tag{12.112}$$

where $\boldsymbol{\mu}$ is the dynamic viscosity of air [kg·s^{-1}·m^{-1}], $\boldsymbol{K_p}$ is the porous medium permeability depending on the geometry of the medium [m^2], and $\boldsymbol{Y}$ is a dimensionless form-drag constant dependent on the nature of the porous medium, called inertial factor.

Equation (1.112) is not a purely empirical expression, as it can be derived from the Navier–Stokes equation [eq (12.3)] for a one-dimensional, stationary and incompressible flow of a Newtonian fluid into a rigid porous media (Dullien, 1979). The Forchheimer equation can be used to calculate the pressure drop through insect-proof screens with a complex texture and in woven threads screens with regular shaped holes (Teitel et al., 2001).

Comparing eqs (12.109) and (12.112), considering $\partial P/\partial x = \Delta P/e$, Molina-Aiz et al. (2009) deduced a new expression for the pressure drop coefficient due to the presence of an insect-proof screen F_φ variable with the Reynold number $\boldsymbol{Re_p}$:

$$F_\varphi = \frac{2e}{K_p^{0.5}}\left(\frac{1}{Re_p} + Y\right) \quad (12.113)$$

Various authors have used this second approach to study the airflow resistance of several insect-proof screens (Miguel et al., 1997; Bartzanas et al., 2002; Fatnassi et al., 2003, 2004; Molina-Aiz et al., 2004; Valera et al., 2006; Molina-Aiz et al., 2009; López-Martínez et al., 2017).

12.5 VENTILATION REQUIREMENTS IN GREENHOUSES

The cooling capacity of a greenhouse by ventilation is usually quantified by the ventilation volumetric flow rate circulating across the greenhouse openings $\boldsymbol{G}$ [$m^3 \cdot s^{-1}$]. A second parameter used to quantify ventilation capacity is the air change rate $\boldsymbol{R}$ [h^{-1}] that is the ratio of the volume of changed air in an hour to the greenhouse volume and represents the number of times over 1h period that all of the air within the internal volume of the greenhouse is replaced (Okada and Takakura, 1973). It is also common to express the ventilation rate as the volume of air exchanged per unit of time and per covered floor surface $\boldsymbol{Q_V}$ [$m^3 \cdot s^{-1} \cdot m^{-2}$] (ASABE, 2008).

Several experimental techniques can be used to measure and predict ventilation and infiltration rates such as tracer gas techniques, energy balances, velocity sensing with sonic anemometers, measurements of pressure differences between inside and outside, and CFD simulations (Figure 12.1). One of the most widely used is the tracer-gas-decay method (Bot, 1983; De Jong, 1990; Boulard and Baille, 1995; Fernandez and Bailey, 1992; Boulard et al. 1997; Baptista et al., 1999; Fatnassi et al., 2002; Pérez-Parra et al., 2004a; Abreu et al., 2005). This technique is based on a mass balance of a tracer gas [as carbon dioxide (CO_2), nitrous oxide (N_2O) and less commonly sulfur hexafluoride (SF_6), methane (CH_4), hydrogen (H_2)] inside the greenhouse (Baptista et al., 1999). This technique allows us to obtain an overall air change rate, in the same way that energy balances. However, anemometry allows to measure the volumetric airflow at each greenhouse openings individually and to distinguish which acts as air inlet or outlet (Molina-Aiz et al., 2009). The tracer-gas technique is based on the hypothesis of a perfectly mixed space. However, this assumption may not be valid in naturally ventilated greenhouses, and consequently ventilation rates deduced from tracer-gas measurements can yield incorrect values of the air change rate (Teitel et al., 2014). The volumetric airflow can be related to the air change rate using the greenhouse volume:

$$R = 3600\,[s \cdot h^{-1}] \cdot \frac{G}{V_{gre}}[h^{-1}] \quad (12.114)$$

12.5.1 ADEQUATE VENTILATION LEVEL

Natural ventilation is considered one of the important factors of greenhouse environment because it is a driving force of heat loss in heating period, greenhouse cooling in hot days, and the supplement of CO_2 which is absorbed by crops photosynthesis (Okada and Takakura, 1973). The ventilation air exchange rate for a given greenhouse determines the capacity to cooling air and to control inside temperature and humidity (Walker and Duncan, 1973). The ventilation airflow is related to the thermal gradient between the inside and outside of the greenhouse (Buffington et al., 1987), as well as to moisture variation. The main factors affecting the ventilation requirement for temperature control are the volume of the greenhouse, the maximum inside temperature that plants can tolerate in function of their stage of development, and the outside environment conditions (Walker and Duncan, 1973).

Ventilation requirements are variable and must be adjusted usually to the environmental needs of crop development and mainly to outdoor climatic conditions. Maximum rates of ventilation are required during the warm spring and hot summer seasons and lower rates will be necessary on cool or cloudy days (Walker and Duncan, 1973). Thus, the necessary air change rates in a greenhouse can range from 2 air changes per hour in winter to 60 h^{-1} in summer (Buffington et al.,1987; Gates and Duncan, 1999). According to Businger (1963), during the summer a minimum air change rate of 10 h^{-1} under favorable light conditions is required, with recommended values of 50–100 h^{-1}. Similarly, the air change rate values required to obtain a certain air temperature rise ΔT_{io} varies depending mainly on the level of evapotranspiration of the crop, between 35 and 90 h^{-1} (ASAE, 2008), with an optimal value of 45–75 h^{-1} (Walker and Duncan, 1973).

12.5.2 INFLUENCE OF WIND AND THERMAL EFFECTS ON VENTILATION RATE

As seen above, the main driving forces of natural ventilation are first, the pressure field created by the interaction of the structure and wind, and secondly, the airflow caused by the temperature differences between inside and outside (Okada and Takakura, 1973; Buffington et al., 1987; Boulard and Draoui, 1995; Boulard et al., 1997a). Wind forces alone largely explain ventilation flows in the case of Venlo-type greenhouses (Bot, 1983), while the combination of wind and temperature influence are considered important in Mediterranean greenhouses (Boulard et al., 1997a; Molina-Aiz et al., 2009).

The effect of wind is considered the most important, both in greenhouses equipped only with roof windows (Bot, 1983; de Jong, 1990; Fernandez and Bailey, 1992; Boulard and Draoui, 1995; Kittas et al., 1996) as in greenhouses with roof and side openings (Boulard and Baille,1995; Papadakis et al., 1996). The ventilation opening surface with the wind speed can explain the major variation in the air exchange rate (Boulard and Draoui, 1995; Kittas et al., 1996; Boulard et al., 1997a). The thermal effect increases with the square root of the temperature gradient $\Delta \boldsymbol{T}_{ie}$, while the wind effect increases linearly with the wind speed, and therefore, its influence grows rapidly compared to that of the thermal effect. In addition, when the wind speed decreases, the thermal gradient can increase, and therefore, ventilation due to the temperature difference reduces. In the case of a typical Dutch greenhouse (Venlo) with only roof ventilation, Bot (1983) calculated that wind effects dominate over thermal ones when:

$$v_R > \frac{\Delta T_{io}^{1/2}}{3} \tag{12.115}$$

In Venlo-type greenhouses equipped only with roof windows, limit speed values have been obtained, from which the wind effect is predominant and the thermal buoyancy effect of the air can be negligible, from 1.2 m·s^{-1} (Bot, 1983). Thus, temperature-induced ventilation is generally only significant for low speeds, which in the case of greenhouses equipped only with roof windows are less than 1 m·s^{-1} (Baptista et al., 1999) or 2 m·s^{-1} (Bot, 1983; Papadakis et al., 1996).

In the case of multispan (Kittas et al., 1997) and Almería-type (Molina-Aiz, 2010) greenhouses with both roof and side openings, the thermal effect seems increase, becoming significant when:

$$V_R < \Delta T_{io}^{1/2} \quad (12.116)$$

In general, the thermal or buoyancy effect can be neglected both in multispan greenhouses with only side windows, when the air velocity is greater than 1.5 $m \cdot s^{-1}$ (Kittas et al., 1996), as in greenhouses with roof and side openings, at 1.8 $m \cdot s^{-1}$ (Papadakis et al., 1996). This value of wind speed (1.8 $m \cdot s^{-1}$) the limit above which the wind effect predominates was also reported by other authors (Bot, 1983; Boulard, 1993).

However, Boulard et al. (1997a) noted that the "chimney" effect is important when side and roof openings are open simultaneously and for winds less than 2–2.5 $m \cdot s^{-1}$. For Almería-type greenhouses with roof and side openings equipped with insect-proof screens, this limit is similar, 2 $m \cdot s^{-1}$ (Molina, 2010). Naturally ventilated greenhouses should be provided with vent openings on both sides of the ridge and on both sidewalls (ASABE, 2008), allowing the greenhouse to be ventilated by both wind pressure and thermal gradients (Buffington et al., 1987). The use of both types of openings can multiply the air change rate by up to 10 with respect to a single opening (for wind speed less than 1 $m \cdot s^{-1}$). The combined sidewall vent area should equal the combined ridge vent area, and each should be at least 15–20% of the floor area (ASABE, 2008). Thus, a ventilation system with side and roof openings is more efficient than ventilation only with roof vents for low wind speeds ($\boldsymbol{v_R}$ < 2.5 $m \cdot s^{-1}$) for normal thermal conditions ($\boldsymbol{\Delta T_{ie}}$ = 5°C) (Kittas et al., 1997). In Almería-type greenhouses with screened roof and side openings, a minimum ventilation area of 35% is necessary to achieve an air change rate of 40 h^{-1} with a wind speed of $\boldsymbol{v_R}$ < 1 $m \cdot s^{-1}$, limiting the temperature rise to $\boldsymbol{\Delta T_{ie}}$ = 5°C (Molina-Aiz, 2010).

12.6 CONCLUSIONS

Natural ventilation is the main climate control technique in greenhouses that allows to reducing indoor temperature and humidity and increasing the concentration of CO_2. The airflow through the greenhouse openings is generated by two main forces: the buoyancy of the hot air and the outside wind, which produces pressure differences between the inside and the outside $\boldsymbol{\Delta P}$. The air moves across the openings from areas with more to zones with less pressure, being the airflow difficulties by the geometry of the windows (height/length ratio and opening angle of the flaps), the presence of the plants inside the greenhouse and the placement of insect-proof screens in the openings.

The thermal effect produces a movement by natural convection. The air ascends by effect of the gravitational field as consequence of reduction of the density when is heated by the soil and plants absorbing solar radiation. In general, this effect generates a vertical upward movement of the interior air toward the top of the individual windows or toward the roof openings. The wind effect can be divided in a mean component (that generates overpressure in the windward area and suction in leeward zones), and a turbulent component (that produces the air inlet and outlet in each individual window). Usually, the air movement produced by the wind effect is horizontal from the windward to the leeward zones. The use of flap in roof vents allows to regulate the air inlet or outlet depending on the opening angle and the windward or leeward orientation. In general, roof opening to windward tend to favor the entry of air through the roof of the greenhouse, and when open to leeward facilitates the air outlet.

To obtain an accurate description of ventilation in greenhouses, it is necessary to consider the differential equations of Navier–Stokes, which must be solved by numerical methods used in CFD models. Semi-empirical models can be used to design ventilation systems or to calculate airflow rate in existing greenhouses in order to quantify heat exchange in energy balances. These models are based on the Bernoilli equation which is a simplification of the differential equations. This equation relates velocity across the openings with the inside–outside pressure difference $\boldsymbol{\Delta P_{io}}$, and a discharge coefficient $\boldsymbol{C_d}$ characterizing the aerodynamic resistance of the openings.

The airflow through the windows is directly proportional to their surface $\boldsymbol{S_v}$, the discharge coefficient, the temperature gradient between the inside and outside $\boldsymbol{\Delta T_{io}}$ (thermal effect), the wind speed $\boldsymbol{v_R}$ and a wind coefficient $\boldsymbol{C_w}$ characteristic of the windows (wind effect). The presence of insect-proof screens modifies the discharge coefficient, reducing about half when the porosity of the meshes $\boldsymbol{\varphi}$ is around 30%. Although the discharge coefficient related to the opening geometry is quite constant (around 0.70), the presence of the nets makes their value dependent on the flow rate. To calculate $\boldsymbol{C_d}$ coefficients in function of air velocity, it is necessary to carry out wind tunnel tests determining the parameters that characterize the aerodynamic behavior of the screens (permeability $\boldsymbol{K_p}$ and inertial factor $\boldsymbol{Y}$) or the use of numerical models that relate them to their geometry (porosity, size of the porous, and thickness of the threads).

Most greenhouse ventilation studies used constant values of $\boldsymbol{C_w}$ characteristic of each greenhouse and its ventilation system. However, recent research has demonstrated the need of different values of wind coefficients for windward or leeward openings, to consider the position of the windows and the greenhouse with respect to the surrounding obstacles. To obtain different coefficients is necessary to perform measurements with anemometers or CFD simulations to determine the airflow in each greenhouse openings.

Three main hypotheses of interaction between thermal and wind effect can be considered to quantify ventilation flow rate ***G***: the sum of the pressures produced by each of the two effects ($\Delta P = \Delta P_T + \Delta P_v$), the vectorial sum of the flow rates ($G = \sqrt{(G_T^2 + G_v^2)}$) or the algebraic sum of the flow rates ($G = G_T + G_v$). Considering different options to characterize the thermal and wind effects and the possibility of having a single window or roof and side opening 15 different ventilation models can be used. Recent experimental studies with anemometry and numerical CFD simulations have shown the inaccuracy of these traditional models in certain greenhouses. This error is the result of a negative interaction of the upward air motion (generated by a high thermal gradient) that can counteract the effect of a moderate wind in the windward roof openings (where the outside air enters with a downward movement).

To maintain an adequate thermal rise inside the greenhouses equipped with insect-proof screens ($\boldsymbol{\Delta T_{io}}$ = 5°C), it is necessary to install side and roof openings with ventilation surfaces of 15–20% in each case. In this way, the maximum recommended ventilation rates of 45 h^{-1} can be reached for summer days with the highest ventilation requirements.

KEYWORDS

- **airflow**
- **wind effect**
- **buoyancy**
- **air change rate**
- **computational fluid dynamics**

REFERENCES

Abdel-Ghany, A. M.; Kozai, T. Dynamic Modeling of the Environment in a Naturally Ventilated, Fog-Cooled Greenhouse. *Renew. Energy* **2006a,** *31*, 1521–1539.

Abdel-Ghany, A. M.; Kozai, T. On the Determination of the Overall Heat Transmission Coefficient and Soil Heat Flux for a Fog Cooled, Naturally Ventilated Greenhouse, Analysis of Radiation and Convection Heat Transfer. *Energy Conv. Manage.* **2006b,** *47*, 2612–2628.

Abreu, P. E.; Boulard, T.; Mermier, M.; Meneses, J. F. Parameter Estimation and Selection of a Greenhouse Natural Ventilation Model and its use on an Energy Balance Model to Estimate the Greenhouse Air Temperature. *Acta Hort.* **2005,** *691*, 611–618.

Aich, W.; Kolsi, L.; Borjini, M. N.; Aissia, H. B.; Öztop, H.; Abu-Hamdeh, N. Three-Dimensional CFD Analysis of Buoyancy-Driven Natural Ventilation and Entropy Generation in a Prismatic Greenhouse. *Therm. Sci.* **2016,** *52*, 1–12.

Akrami, M.; Mutlum, C. D.; Javadi, A. A.; Salah, A. H.; Fath, H. E. S.; Dibaj, M.; Farmani, R.; Mohammed, R. H.; Negm, A. Analysis of Inlet Configurations on the Microclimate Conditions of a Novel Standalone Agricultural Greenhouse for Egypt using Computational Fluid Dynamics. *Sustainability* **2021**, *13*, 1446. https://doi.org/10.3390/su13031446.

Albright, L. D.; Seginer, I.; Marsh, L. S.; Oko, A. In Situ Thermal Calibration of Unventilated Greenhouses. *J. Agric. Eng. Res.* **1985,** *31* (3), 265–281.

Ali, H. B.; Bournet, P. E.; Cannavo, P.; Chantoiseau, E. Using CFD to Improve the Irrigation Strategy for Growing Ornamental Plants Inside a Greenhouse. *Biosyst. Eng.* **2019,** *186*, 130–145.

Al-Jamal, K. Greenhouse Cooling in Hot Countries. *Energy* **1994,** *19,* 1187–1192.

Amara, H. B.; Bouadila, S.; Fatnassi, H.; Arici, M.; Guizani, A. A. Climate Assessment of Greenhouse Equipped with South-Oriented PV Roofs: An Experimental and Computational Fluid Dynamics Study. *Sustain. Energy Technol. Assess.* **2021,** *45*, 101100. https://doi.org/10.1016/j.seta.2021.101100.

Amsen M. G. A Simple Method to Calculate Improvements of Diffuse Light Distribution in Detached Greenhouses. *Acta Hort.* **1985,** *174*, 105–109.

Arbel, A.; Shklyar, A.; Barak, M. Buoyancy-Driven Ventilation in a Greenhouse Cooled by a Fogging System. *Acta Hort.* **2000,** *534*, 327–334.

Arinze, E. A.; Schoenau, G. J.; Besant, R. W. A Dynamic Thermal Performance Simulation Model of an Energy Conserving Greenhouse with Thermal Storage. *Trans. ASAE* **1984,** *27*, 508–519.

ASABE. Heating, Ventilating and Cooling Greenhouses. Standard ANSI/ASAE EP406.4 JAN2003 (R2008), St. Joseph (US), 2008. https://elibrary.asabe.org/abstract.asp?aid=45682&t=2.

Atia, D. M.; El-madany, H. T. Analysis and Design of Greenhouse Temperature Control using Adaptive Neuro-Fuzzy Inference System. *J. Electr. Syst. Inf. Technol.* **2017,** *4* (1), 34–48.

Avissar, R.; Mahrer, Y. Verification Study of a Numerical Greenhouse Microclimate Model. *Trans. ASAE.* **1982,** 25 (6), 1711–1720.

Axley J. W.; Chung D. H. Mechanical Power Balances for Multizone Building Airflow Analysis. *Int. J. Vent.* **2005,** *4* (2), 95–112.

Baeza E. J. Optimización del diseño de los sistemas de ventilación en invernadero de tipo parral multicapilla mediante CFD. [Optimizing the Design of Multispan *Parral* Greenhouse Ventilation Systems using CFDs]. Ph Dissertation, Universidad de Almería (Spain), 2007.

Baeza, E. J.; Pérez-Parra, J.; Montero, J. I. Effect of Ventilator Size on Natural Ventilation in Parral Greenhouse by Means of CFD Simulations. *Acta Hort.* **2005,** *691,* 465–472.

Baeza, E. J.; Pérez-Parra, J. J.; López, J. C.; Montero, J. I. CFD Simulation of the Natural Ventilation of a Parral Greenhouse with a Baffle Device Below the Greenhouse Vents. *Acta Hort.* **2008,** *801*, 885–892.

Baeza, E. J.; Pérez-Parra, J. J.; López, J. C.; Montero. J. I. CFD Study of the Natural Ventilation Performance of a Parral Type Greenhouse with Different Numbers of Spans and Roof Vent Configurations. *Acta Hort.* **2006,** *719*, 333–340.

Baeza, E. J.; Pérez-Parra, J. J.; Montero, J. I.; Bailey, B. J.; López, J. C.; Gázquez. J. C. Analysis of the Role of Sidewall Vents on Buoyancy-Driven Natural Ventilation in Parral-Type Greenhouses with and Without Insect Screens using Computational Fluid Dynamics. *Biosyst. Eng.* **2009,** *104*, 86–96.

Bailey, B. J. Constraints, Limitations and Achievements in Greenhouse Natural Ventilation. *Acta Hort.* **2000,** 534, 21–30.

Bailey, B. J.; Chalabi, Z. S. Improving the Cost Effectiveness of Greenhouse Climate Control. *Comput. Electron. Agric.* **1995,** *10*, 203–214.

Bailey, B. J.; Chalabi, Z. S.; Aikman, K. E.; Cockshull, K. E. Improved Strategies for Controlling CO_2 Enrichment in Tomatoes Greenhouses. *Acta Hort.* **1997,** *443*, 155–162.

Bailey, B. J.; Montero, J. I.; Pérez-Parra, J.; Robertson, A. P.; Baeza, E.; Kamaruddin, R. Airflow Resistance of Greenhouse Ventilators with and Without Insect Screens. *Biosyst. Eng.* **2003,** *86* (2), 217–229.

Baptista, F. J.; Bailey, B. J.; Randall, J. M.; Meneses, J. F. Greenhouse Ventilation Rate, Theory and Measurement with Tracer Gas Techniques. *J. Agric. Eng. Res.* **1999,** *72*, 363–374.

Bartzanas, T.; Boulard, T.; Kittas, C. Effect of Vent Arrangement on Windward Ventilation of a Tunnel Greenhouse. *Biosyst. Eng.* **2004,** *88* (4), 479–490.

Bartzanas, T.; Boulard, T.; Kittas, C. Numerical Simulation of the Airflow and Temperature Distribution in a Tunnel Greenhouse Equipped with Insect-Proof Screen in the Openings. *Comput. Electron. Agric.* **2002,** *34* (1-3), 207–221.

Bartzanas, T.; Katsoulas, N.; Kittas, C.; Sapounas, A. A. Dispersion of Pesticides from a Naturally Ventilated Greenhouse: A CFD Approach. *Acta Hort.* **2006,** *718*, 307–314.

Baxevanou, C.; Fidaros, D.; Bartzanas, T.; Kittas, C. Yearly Numerical Evaluation of Greenhouse Cover Materials. *Comput. Electron. Agric.* **2018,** *149*, 54–70.

Benni, S.; Tassinari, P.; Bonora, F.; Barbaresi, A.; Torreggiani, D. Efficacy of Greenhouse Natural Ventilation: Environmental Monitoring and CFD Simulations of a Study Case. *Energy Build.* **2016,** *125*, 276–286.

Bernoulli, D. *Hydrodynarnica sive de viribus et motibus fluidorum commentarii* ; Johann Reinhold Dulsseker, Argentorati; Strasbourg (France), 1738. https://archive.org/details/bub_gb_3yRVAAAAcAAJ/page/n6/mode/1up

Bird, R. B.; Steward, W. E.; Lightfoot, E. N. In *Transport Phenomena*, 2nd ed.; John Wiley & Sons: New York (US), 2009.

Boaventura, C. J.; Couto, C.; Ruano, A. E. Real-Time Parameter Estimation of Dynamic Temperature Models for Greenhouse Environmental Control. *Cont. Eng. Prac.* **1997,** *5* (10), 1473–1481.

Bot, G. P. A. Validation of a Dynamical Model of Greenhouse Climate. *Acta Hort.* **1980,** *106*, 149–158.

Bot, G. P. A. Developments in Indoor Sustainable Plant Production with Emphasis on Energy Saving. *Comput. Electron. Agric.* **2001,** 30, 151–165.

Bot, G. P. A. Greenhouse Climate: From Physical Processes to a Dynamic Model. PhD Dissertation, Agricultural University of Wageningen, Wageningen (The Netherland), 1983. https://library.wur.nl/WebQuery/wurpubs/fulltext/188427.

Boulard T.; Draoui, B.; Neirac F. Calibration and Validation of a Greenhouse Climate Control Model. *Acta Hort.* **1996b,** *406*, 49–61.

Boulard, T., Caracterisation et modelisation du climat des serres: application a la climatisation estivale. [Characterization and Modelling of the Greenhouse Climate: Application to Summer Climatic Control]. PhD Dissertation, Ecole Nationales Suérierure Agronomique de Montpellier, Motpellier (France), 1996.

Boulard, T.; Baille, A. A Simple Greenhouse Climate Control Incorporating Effects of Ventilation and Evaporative Cooling. *Agric. For. Meteorol.* **1993,** *65*, 145–157.

Boulard, T.; Baille, A. Modelling of Air Exchange Rate in a Greenhouse Equipped with Continuous Roof Vents. *J. Agric. Eng. Res.* **1995,** *61*, 37–48.

Boulard, T.; Chave, M.; Fatnassi, H.; Poncet, C.; Roy, J. C. Botrytis Cinerea Spore Balance of a Greenhouse Rose Crop. *Agric. For. Meteorol.* **2008,** *148* (3), 504–511.

Boulard, T.; Draoui, B. Natural Ventilation of a Greenhouse with Continuous Roof Vents: Measurements and Data Analysis. *J. Agric. Engng Res.*, **1995,** *61* (1), 27–35.

Boulard, T.; Feuilloley, P.; Kittas C. Natural Centilation Performance of Six Greenhouse and Tunnel Types. *J. Agric. Eng. Res.* **1997a,** *67*, 249–266.

Boulard, T.; Haxaire, R.; Lamrani, M. A.; Roy, J. C.; Jaffrin, A. Characterization and Modelling of the Air Fluxes Induced by Natural Ventilation in a Greenhouse. *J. Agric. Eng. Res.* **1999,** *74* (2), 135–144.

Boulard, T.; Kittas, C.; Papadakis, G.; Mermier, M. Pressure Field and Airflow at the Opening of a Naturally Ventilated Greenhouse. *J. Agric. Eng. Res.* **1998a,** *71* (1), 93–102.

Boulard, T.; Kittas, C.; Roy, J. C.; Wang, S. SE-Structures and Environment: Convective and Ventilation Transfers in Greenhouses, Part 2: Determination of the Distribued Greenhouse Climate. *Biosyst. Eng.* **2002b,** *83* (2), 129–147.

Boulard, T.; Lamrani, M. A.; Roy, J. C.; Jaffrin, A.; Bouirden, L. Natural Ventilation by Thermal Effect in a One-Half Scale Model Mono-Span Greenhouse. *Trans. ASAE* **1998b,** *41* (3), 773–781.

Boulard, T.; Meneses, J. F.; Mermier, M.; Papadakis, G. The Mechanisms Involved in the Natural Ventilation of Greenhouses. *Agric. For. Meteorol.* **1996a,** *79* (1-2), 61–77.

Boulard, T.; Papadakis, G.; Kittas, C.; Mermier, M. Air Flow and Associated Sensible Heat Exchanges in a Naturally Ventilated Greenhouse. *Agric. For. Meteorol.* **1997b,** *88* (1-4), 111–119.

Boulard, T.; Roy, J. C.; Pouillard, J. B.; Fatnassi, H.; Grisey, A. Modelling of Micrometeorology, Canopy Transpiration and Photosynthesis in a Closed Greenhouse Using Computational Fluid Dynamics. *Biosyst. Eng.* **2017,** *158,* 110–133.

Boulard, T.; Wang, S. Greenhouse Crop Transpiration Simulation from External Climate Conditions. *Agric. For. Meteorol.* **2000,** *100* (1), 25–34.

Boulard, T.; Wang, S.; Haxaire, R. Mean and Turbulent Air Flows and Microclimatic Patterns in an Empty Greenhouse Tunnel. *Agric. For. Meteorol.* **2000,** *100,* (2-3), 169–181.

Bournet, P. E.; Ould Khaoua, S. A.; Boulard, T. Numerical Prediction of the Effect of Vent Arrangements on the Ventilation and Energy Transfer in a Multi-Span Glasshouse using a Bi-Band Radiation Model. *Biosyst. Eng.* **2007,** *98* (2), 224–234.

Bournet, P. E.; Ould Khaoua, S. A.; Boulard, T.; Migeon, C.; Chassériaux, G. Effect of Roof and Side Opening Combinations on the Ventilation of a Greenhouse using Computer Simulation. *Trans. ASABE* **2007,** *50 (1)*, 201–212.

Bournet, P. E.; Boulard, T. Effect of Ventilator Configuration on the Distributed Climate of Greenhouses: A Review of Experimental and CFD Studies. *Comput. Electron. Agric.* **2010,** *74*, 195–217.

Brockett, B. L.; Albright, L. D. Natural Ventilation in Single Airspace Building. *J. Agric. Eng. Res.* **1987,** *37* (3-4), 141–154.

Brown, W. G.; Solvason, K. R. Natural Convection Through Rectangular Openings in Partitions-1 Vertical Partitions. *Int. J. Heat Mass Trasnsf.* **1962,** *5* (9), 859–862.

Bruce, J. M. A Computer Program for the Calculation of Natural Ventilation due to Wind. *Farm Build. Res. Dev. Stud.* **1975,** *7*, 1–3.

Bruce, J. M. Natural Convection Through Openings and its Applications to Cattle Building Ventilation. *J. Agric. Eng. Res.* **1978,** *23* (2), 151–167.

Bruce, J. M. Ventilation of a Model Livestock Building by Thermal Buoyancy. *Trans. ASAE* **1982,** *25* (6), 1724–1726.

Brugger, M.; Montero, J. I.; Baeza, E. J.; Perez-Parra, J. Computational Fluid Dynamic Modeling to Improve the Design of the Spanish Parral Style Greenhouse. *Acta Hort.* **2005,** *691*, 425–432.

Brundrett, E. Prediction of Pressure Drop for Incompressible Flow Through Screens. *J. Fluids Eng.* **1993,** *115* (2), 239–242.

Bucklin, R. A.; Nääs, I. A.; Zazueta, F. S.; Walker, W. R. Natural Ventilation in Swine Houseing. IFAS Bulletin 270, Florifa Cooperaitve Extension Service/Institute of Food and Agric. Sciences, University of Florida, Ginesville (US), 1991. https://ufdc.ufl.edu/UF00008531/00001/1j

Buffington, D. E.; Bucklin, R. A.; Henley, R. W.; McConnell, D. B. Greenhouse Ventilation. Florida Cooperative Extension Service, Institute of Food and Agricultural Sciences, University of Florida, Fact Sheet AE-10, Gainesville (USA), 1987. https://edis.ifas.ufl.edu/pdf%5CAE%5CAE03000.pdf.

Businger, J. A. The Glasshouse Climate. In *Physics of Plant Environment*; van Wijk, Ed.; North-Holland Publishing Co.: Amsterdam (The Netherland), 1963.

Campen, J. B.; Bot, G. P. A. Determination of Greenhouse-Specific Aspect of Ventilation Using Three-Dimensional Computational Fluid Dynamics. *Biosyst. Eng.* **2003,** *84* (1), 69–77.

Castañeda-Miranda, A.; Castaño, V. M. Smart Frost Control in Greenhouses by Neural Networks Models. *Comput. Electron. Agric.* **2017,** *137*, 102–114.

Chalabi, Z. S.; Fernández, J. E. Estimation of Net Photosynthesis of a Greenhouse Canopy Using a Mass Balance Method and Mechanistic Models. *Agric. For. Meteorol.* **1994,** *71*, 165–182.

Chaplin, G. C.; Randall, J. R.; Baker, C. J. The Turbulent Ventilation of a Single Opening Enclosure. *J. Wind. Eng. Ind. Aerodyn.* **2000,** *85* (2), 145–161.

Chen, J.; Xu, F.; Tan, D.; Shen, Z.; Zhang, L.; Ai, Q. A Control Method for Agricultural Greenhouses Heating Based on Computational Fluid Dynamics and Energy Prediction Model. *Appl. Ener.* **2015,** *141*, 106–118.

Chen, J.; Cai, Y.; Xu, F.; Hu, H.; Ai, Q. Analysis and Optimization of the Fan-Pad Evaporative Cooling System for Greenhouse Based on CFD. *Adv. Mech. Eng.* **2014,** *6*, 712–740.

Choab, N.; Allouhi, A.; El Maakoul, A.; Kousksou, T.; Saadeddine, S.; Jamil, A. Review on Greenhouse Microclimate and Application: Design Parameters, Thermal Modeling and Simulation, Climate Controlling Technologies. *Sol. Ener.* **2019,** *191*, 109–137.

Chou, S. K.; Chau, K. J.; Ho, J. C.; Ooi, C. L. On the Study of an Energy-Efficient Greenhouse for Heating and Dehumidification Application. *Appl. Ener.* **2004,** *77*, 355–373.

Cockroft, J. P.; Robertson, P. Ventilation of an Enclosure Through a Single Opening. *Build. Environ.* **1976,** *11*, 29–35.

Critten, D. L. Direct Sunlight Losses in North-South Aligned Multi-Span Greenhouses with Symmetric Roofs at UK Latitudes. *J. Agric. Eng. Res.* **1988,** *40*, 71–79.

Critten, D. L. The Evaluation of a Computer Model to Calculate the Daily Light Integral and Transmissivity of a Greenhouse. *J. Agric. Eng. Res.* **1983,** *28*, 545–563.

Critten, D. L. The Prediction of Multi-Span Greenhouse Light Transmittance with Particular Reference to Tunnels under Direct Winter Light Conditions. *J. Agric. Eng. Res.* **1987,** *38* (1), 57–64.

De Gids, W. F. Calculation Method for the Natural Ventilation of Buildings. *Verwarm. Vent.* **1978,** *35* (7), 551–564.

De Jong, T. Natural Ventilation of Large Multispan Greenhouses. PhD Dissertation, Wageningen Agricultural University, Wageningen (The Netherlands), 1990. https://library.wur.nl/WebQuery/wurpubs/fulltext/206452.

De Jong, T.; Bot, G. P. A. Flow Characteristics of One-Side-Mounted Windows. *Ener. Build.* **1992,** *19* (2), 105–112.

Deltour, J.; Wang, S. Ventilation Induced Interior Air Movement in Large Scale Venlo-type Greenhouses. *Acta Hort.* **1999,** *486*, 119–124.

Demrati, H.; Boulard, T.; Bekkaoui, A.; Bouirden L. Natural Ventilation and Microclimkatic Performance of a Large-Scale Banana Greenhouse. *J. Agric. Eng. Res.* **2001,** *80* (3), 261–271.

Down, M. J.; Foster, M. P.; McMahon, T. A. Experimental Verification of a Theory for Ventilation of Livestock Building by Natural Convection. *J. Agric. Eng. Res.* **1990,** *45*, 269–279.

Dullien, F. A. L. Porous Media-Fluid Transport and Pore Structure. Academic Press: New York (US.), 1979.

Dupuit, J. Études théoriques et pratiques sur le mouvement des eaux dans les canaux découverts et à travers les terrains perméables [Theoretical and Practical Studies on the Movement of Water in Exposed Channels and Through Permeable terrain], 2nd Ed., Dunod, Ed.; Paris (Francia), 1863. https://gallica.bnf.fr/ark:/12148/bpt6k62096061/f284.item.texteImage#.

Espinoza, K.; López, A.; Valera, D. L.; Molina-Aiz, F. D.; Torres, J. A.; Peña, A. Effects of Ventilator Configuration on the Flow Pattern of a Naturally-Ventilated Three-Span Mediterranean Greenhouse. *Biosyst. Eng.* **2017,** 164, 13–30.

Etheridge, D. W. Unsteady Flow Effects due to Fluctuating Wind Pressures in Natural Ventilation Design-Mean Flow Rates. *Build. Environ.* **2000,** *35* (2), 111—133.

Euler, L. Principes généraux de l'état d'équilibre d'un fluide. *Académie Roayle des Sciences et des Belles-Letres de Berlin, Mémoires* **1755a,** *11*, 217–273. https://archive.org/details/euler-e225/mode/2up.

Euler, L. Principes généraux du mouvement des fluides. *Académie Roayle des Sciences et des Belles-Letres de Berlin, Mémoires* **1755b,** 11, 274–315. https://archive.org/details/euler-e226/mode/2up

Fatnassi, H.; Boulard, T.; Bouirden, L. Development, Validation and use of a Dynamic Model for Simulate the Climate Conditions in a Large Scale Greenhouse Equipped with Insect-Proof Nets. *Comput. Electron. Agric.* **2013,** *98*, 54–61.

Fatnassi, H.; Boulard, T.; Bouirden, L. Simulation of Air Flux and Temperature Patterns in a Large Scale Greenhouse Equipped with Insect Proof Nets. *Acta Hort.* **2002a,** *578*, 329–338.

Fatnassi, H.; Boulard, T.; Bouirden, L. Simulation of Climatic Conditions in Full-Scale Greenhouse Fitted with Insect-Proof Screens. *Agric. For. Meteorol.* **2003,** *118* (1-2), 97–111.

Fatnassi, H.; Boulard, T.; Demrati, H.; Bouirden, L.; Sappe, G. Ventilation Performance of a Large Canarian-Type Greenhouse Equipe with Insect-Proof Nets. *Biosyst. Eng.* **2002b,** *82* (1), 97–105.

Fatnassi, H.; Boulard, T.; Lagier, J. Simple Indirect Estimation of Ventilation and Crop Transpiration Rates in a Greenhouse. *Biosyst. Eng.* **2004,** *88* (4), 467–478.

Fatnassi, H.; Boulard, T.; Poncet, C.; Chave M. Optimisation of Greenhouse Insect Screening with Computational Fluid Dynamics. *Biosyst. Eng.* **2006,** *93* (3), 301–312.

Fatnassi, H.; Leyronas, C.; Boulard, T.; Bardin, M.; Nicot, P. Dependence of Greenhouse Tunnel Ventilation on Wind Direction and Crop Height. *Biosyst. Eng.* **2009,** *103*, 338–343.

Fatnassi, H.; Poncet, C.; Bazzano, M. M.; Brun, R.; Bertin, N. A Numerical Simulation of the Photovoltaic Greenhouse Microclimate. *Sol. Ener.* **2015,** *120*, 575–584.

Fernandez, J. E.; Bailey, B. J. Measurement and Prediction of Greenhouse Ventilation Rates. *Agric. For. Meteorol.* **1992,** *58*, 229–245.

Ferziger, J. H.; Peric, M. In *Computational Methods for Fluid Dynamics*, Springer: Berlin (Germany), 2002. https://doi.org/10.1007/978-3-642-56026-2.

Fidaros, D.; Baxevanou, C. A.; Bartzanas, T.; Kittas, C. Numerical Simulation of Thermal Behavior of a Ventilated arc Greenhouse During a Solar Day. *Renew. Ener.* **2010,** *35*, 1380–1386.

Flores-Velazquez, J.; Montero, J. I.; Baeza, E. J.; Lopez, J. C. Mechanical and Natural Ventilation Systems in a Greenhouse Designed Using Computational Fluid Dynamics. *Int. J. Agric. Biol. Eng.* **2014,** *7* (1), 1–16.

Forchheimer, P. Wasserbewegung durch boden [Water Movement Through the Ground]. *Ver. Deut. Ing.* **1901,** 45, 1782–1788.

Foster, M. P.; Down, M. J. Ventilation of Livestock Buildings by Natural Convection. *J. Agric. Eng. Res.* **1987,** *37*, 1–13.

Franco, A.; Valera, D. L.; Peña, A.; Pérez, A. M. Aerodynamic Analysis and CFD Simulation of Several Cellulose Evaporative Cooling Pads used in Mediterranean Greenhouses. *Comput. Electron. Agric.* **2011,** *76*, 218–230.

García, J. L.; de la Plaza, S.; Navas, L. M.; Benavente, R. M.; Luna, L. Evaluation of the Feasibility of Alternative Energy Sources for Greenhouse Heating. *J. Agric. Eng. Res.* **1998,** *69*, 107–114.

Gates, R. S.; Duncan, G. A. Ventilation System Design Parameters for Greenhouses for Tobacco Production. Agricultural Engineering Extension Updates. University of Kentucky, Lexington (US), 1999. https://uknowledge.uky.edu/aeu_reports/2

Ghosal, M. K.; Tiwari, G. N.; Srivastava, N. S. L. Modeling and Experimental Validation of a Greenhouse with Evaporative Cooling by Moving Water Film Over External Shade Cloth. *Ener. Build.* **2003,** *35*, 843–850.

Gray, D. D.; Giorgini, A. The Validity of the Boussinesq Approximation for Liquids and Gases. *Int. J. Heat Mass Transf.* **1976,** *19*, 545–551.

Haghighat, F.; Rao, J.; Fazio, P. The Influence of Turbulent Wind on Air Change Rates-a Modelling Approach. *Build. Environ.* **1991,** *26* (2), 95–109.

Haghighat, F.; Rao, J.; Fazio, P. The Influence of Turbulent Wind on Air Change Rates - A Modelling Approach. *Build. Environ.* **1991,** *26* (2), 95–109.

He, X.; Wang, J.; Guo, S.; Zhang, J.; Wei, B.; Sun, J.; Shu, S. Ventilation Optimization of Solar Greenhouse with Removable Back Walls Based on CFD. *Comput. Electron. Agric.* **2018,** *149*, 16–25.

Hemming, S.; de Zwart, F.; Elings, A.; Righini, I.; Petropoulou, A. Remote Control of Greenhouse Vegetable Production with Artificial Intelligence-Greenhouse Climate, Irrigation, and Crop Production. *Sensors* **2019,** *19*, 1807.

Hoxey, R. P.; Richardson, G. M. Wind Loads on Film Plastic Greenhouses. *J. Wind Eng. Ind. Aerodyn.* **1983,** *11* (1-3), 225–237.

Impron, I.; Hemming, S.; Bot, G. P. A. Effects of Cover Properties, Ventilation Rate, and Crop Leaf Area on Tropical Greenhouse Climate. *Biosyst. Eng.* **2008,** *99*, 553–564.

Impron, I.; Hemming, S.; Bot, G. P. A. Simple Greenhouse Climate Model as a Design Tool for Greenhouses in Tropical Lowland. *Biosyst. Eng.* **2007,** *98* (1), 79–89.

Jolliet, O. HORTITRANS, A Model for Predicting and Optimizing Humidity and Transpiration in Greenhouses. *J. Agric. Eng. Res.* **1994,** *57*, 23–37.

Jolliet, O.; Danloy, L.; Gay, J. B.; Munday, G. L.; Reist, A. HORTICERN, An Improved Static Model for Predicting the Energy Consumption of a Greenhouse. *Agric. For. Meteorol.* **1999,** *55*, 265–294.

Joseph, D. D. In *Stability of Fluid Motions II*; Springer-Verlag: Berlin (Germany), 1976.

Joudi, K. A.; Farhan, A. A. A Dynamic Model and an Experimental Study for the Internal Air and Soil Temperatures in an Innovative Greenhouse. *Energy Convers. Manage.* **2015**, *91*, 76–82.

Jung, D. H.; Kim, H. S.; Jhin, C.; Kim, H. J.; Park, S. H. Time-Serial Analysis of Deep Neural Network Models for Prediction of Climatic Conditions Inside a Greenhouse. *Comput. Electron. Agric.* **2020,** *173*, 105402. https://doi.org/10.1016/j.compag.2020.105402.

Kamaruddin, R. A Naturally Ventilated Crop Protection Structure for Tropical Conditions. PhD Dissertation, Cranfield University, Cranfield (UK), 1999. http://dspace.lib.cranfield.ac.uk/handle/1826/11975.

Kamaruddin, R.; Montero, J. I.; Bailey B. J. A Naturally Ventilated Greenhouse for Temperate Vegetable Production in the Tropics. *Acta Hort.* **2002a,** *578*, 97–103.

Kamaruddin, R.; Montero, J. I.; Bailey, B. J. Effect of Ventilator Configuration on Ventilation in a Tropical Crop Protection Structure. *Acta Hort.* **2002b,** *578*, 105–111.

Kato, S. Review of Airflow and Transport Analysis in Building using CFD and Network Model. *Japan Architec. Rev.* **2018,** *1* (3), 299–309.

Katsoulas, N.; Bartzanas, T.; Boulard, T.; Mermier, M.; Kittas, C. Effect of Vent Openings and Insect Screens on Greenhouse Ventilation. *Biosyst. Eng.* **2006,** *93* (4), 427–436.

Kim, K.; Yoon, J. Y.; Kwon, H. J.; Han, J. H.; Son, J. E.; Nam, S. W.; Giacomelli, G. A.; Lee, I. B. 3-D CFD Analysis of Relative Humidity Distribution in Greenhouse with a Fog Cooling System and Refrigerative Dehumidifiers. *Biosyst. Eng.* **2008,** *100*, 245–255.

Kimball, B. A. Simulation of the Energy Balance of a Greenhouse. *Agric. Meteorol.* **1973,** *11*, 243–260.

Kindelan, M. Dynamic Modeling of Greenhouse Environment. *Trans. ASAE.* **1980,** *23* (5), 1232–1239.

Kirkpatrick, A. T.; Hill, D. D. Mixed Convection Heat Transfer in a Passive Solar Building. *Sol. Ener.* **1988,** *40 (1)*, 25–34.

Kittas, C. Etude théorique et expérimentale du bilan d'énergie des serres [Theoretical and Experimental Study of Greenhouse Energy Balance]. Ph Dissertation, Université de Perpignan, Perpignan (France), 1980.

Kittas, C.; Draoui, B.; Boulard, T. Quantification du taux d'aération d'une serre á ouvrant continu en toiture. *Agric. For. Meteorol.* **1995,** *77*, 95–111.

Kittas, C.; Baille, A. Determination of the Spectral Properties of Several Greenhouse Cover Materials and Evaluation of Specific Parameters Related to Plant Response. *J. Agric. Eng. Res.* **1998,** *71*, 193–202.

Kittas, C.; Bartzanas, T. Greenhouse Microclimate and Dehumidification Effectiveness under Different Ventilator Configurations. *Build. Environ.* **2007,** *42* (10), 3774–3784.

Kittas, C.; Boulard, T.; Bartzanas, T.; Katsoulas, N.; Mermier, M. Influence of an Insect Screen on Greenhouse Ventilation. *Trans. ASAE.* **2002,** *45* (4), 1083–1090.

Kittas, C.; Boulard, T.; Mermier, M.; Papadakis, G. Wind Induced Air Exchange Rates in a Greenhouse Tunnel with Continuous Side Openings. *J. Agric. Eng. Res.* **1996,** *65*, 37–49.

Kittas, C.; Boulard, T.; Papadakis, G. Natural Ventilation of a Greenhouse with Ridge and Side Openings: Sensitivity to Temperature and Wind Effects. *Trans. ASAE.* **1997a,** *40* (2), 415–425.

Kittas, C.; Boulard, T.; Papadakis, G. Greenhouse Ventilation Rates Throgh Combined Roof and Side Openings: An Experimental Study. *Acta Hort.* **1997b,** *443*, 31–38.

Kok R.; Lacroix R.; Clark G.; Taillefer E. Imitation of A Procedural Greenhouse Model with an Artificial Neural Network. *Can. Agric. Eng.* **1994,** *36* (2), 117–126.

Kosmos, S. R.; Riskowski, G. L.; Christianson, L. L. Force and Static Pressure Resulting from Airflow Through Screens. *Trans. ASAE.* **1993,** *36*, 1467–1472.

Kozai, T. Direct Solar Light Transmission into Single-Span Greenhouses. *Agric. Meteorol.* **1977,** *18*, 327–338.

Kozai, T.; Kimura, M. Direct Solar Light Transmission into Multispan Greenhouses. *Agric. Meteorol.* **1977,** *18*, 339–349.

Kozai, T.; Sase, S. A Simulated of Natural Ventilation for a Multi-Span Greenhouse. *Acta Hort.* **1978,** *87*, 39–49.

Kozai, T.; Sase, S.; Nara, M. A Modelling Approach to Greenhouse Ventilation Control. *Acta Hort.* **1980,** *106*, 125–138.

Krauss, G.; Depecker, P.; Kindagen, J. I. Méthode des résaux de neurones pour le calcul des coefficients de vitesse. Aplication aux écoulements de l'air dans les batiments. *Rev. Gen. Therm.* **1997,** *36 (3)*, 180–191.

Krug, H.; Liebig, H. P. A Plant Production Model to Optimize Temperature in Greenhouses on the Base of Ecological and Economical Data. *Acta Hort.* **1980,** *106*, 179–190.

Kurata, K.; Quan, Z.; Nunomura, O. Optimal Shapes of Parallel East-West Oriented Single Span Tunnels with Respect to Direct Light Transmissivity. *J. Agric. Eng. Res.* **1991,** *48*, 89–100.

Kuroyanagi, T.; Nagasaki, Y.; Nakamoto, Y.; Tanaka, H.; Ibuki, T. Simulation of Microclimate in a Sloping Greenhouse using CFD. *Acta Hort.* **2008,** *801*, 909–916.

Lafont, F.; Balmat, J. F.; Pessel, N.; Fliess, M. N.; Fliess, M. A Model-Free Control Strategy for an Experimental Greenhouse with an Application to Fault Accommodation. *Comput. Electron. Agric.* **2015,** *110*, 139–149.

Lammari, K.; Bounaama, F.; Draoui, B.; Mrah, B.; Haidas, M. GA Optimization of the Coupled Climate Model of an Order Two of a Greenhouse. *Ener. Proc.* **2012,** *18*, 416–425.

Lamrani, M. A.; Boulard, T.; Roy, J. C.; Jaffrin, A. Airflows and Temperature Patterns Induced in a Confined Greenhouse. *J. Agric. Eng. Res.* **2001,** *78*, 75–88.

Lee, I. B.; Hong, S. W.; Hwang, H. S.; Seo, I. H. Study on Ventilation Efficiencies of Naturally Ventilated Multi-Span Greenhouses in Korea. *Acta Hort.* **2006,** *719*, 341–347.

Lee, I. B.; Sase, S.; Okushima, L.; Ikeguchi, A.; Choi, K.; Yun, J. A Wind Tunnel Study of Natural Ventilation for Multi–Span Greenhouse Scale Models Using Two–Dimensional Particle Image Velocimetry (PIV). *Trans. ASAE.* **2003,** *46* (3), 763–772.

Lee, IG. In 2002 ASAE Annual International, CIGR XVth World Congress; Chicago, USA, 2002. https://doi.org/10.13031/2013.10468.

Lee, I. B.; Short, T. H. Two-Dimensional Numerical Simulation of Natural Ventilation in a Multi-Span Greenhouse. *Trans. ASAE.* **2000,** *43* (3), 745–753.

Lee, I. B.; Bitog, J. P. P.; Hong, S. W.; Seo, I. H.; Kwon, K. S.; Bartzanas, T.; Kacira, M. The Past, Present and Future of CFD for Agro-Environmental Applications. *Comput. Electron. Agric.* **2013,** *93*, 168–183.

Lee, S. Y.; Lee, I. B.; Kim, R. W. Evaluation of Wind-Driven Natural Ventilation of Single-Span Greenhouses Built on Reclaimed Coastal Land. *Biosyst. Eng.* **2018,** *171*, 120–142.
Li, Y.; Delsante, A. Natural Ventilation Induced by Combined Wind and Thermal Forces. *Build. Environ.* **2001,** *36* (1), 59–71.
Liddament, M. W. Air Infiltration Calculation Techniques–An Appications Guide. Air Infiltration and Ventilation Centre, Bracknell, (UK). 1986. https://www.aivc.org/sites/default/files/members_area/medias/pdf/Guides/GU02%20AIR%20INFILTRATION%20CALCULATION%20TECHNIQUES.PDF
Linker, R.; Gutman, P.; Seginer, I. Robust Model-Based Failure Detection and Identification in Greenhouses. *Comput. Electron. Agric.* **2000,** *26*, 255–270.
López, A.; Valera, D. L.; Molina-Aiz, F. D.; Moreno, M. A.; Marín, P. Analysis of the Microclimate of a Greenhouse with Two Anti-Insect Screens of Different Thread Density. *Acta Hort.* **2017,** *1170*, 227–234.
López, A.; Valera, D. L.; Molina-Aiz, F. D.; Peña, A. A. Effects of Surrounding Buildings on Air Patterns and Turbulence in Two Naturally Ventilated Mediterranean Greenhouses Using Tri-sonic Anemometry. *Trans. ASABE* **2011,** *54* (5), 1941–1950.
López-Cruz, I. L.; Fitz-Rodríguez, E.; Salazar-Moreno, R.; Rojano-Aguilar, A.; Kacira, M. Development and Analysis of Dynamical Mathematical Models of Greenhouse Climate: A Review. *Eur. J. Hort. Sci.* **2018,** *83* (5), 269–280.
López-Martínez, A. Contribución al conocimiento del microclima de los invernaderos mediterráneos mediante anemometrıía sónica y termografía [Contribution to the knowledge of Mediterranean Greenhouse Microclimate Through Sonic Anemometry and Thermography]. PhD Dissertation, University of Almería, Almería (Spain), 2011.
López-Martínez, A.; Molina-Aiz, F. D.; Valera-Martínez, D. L.; López-Martínez, J.; Peña-Fernández, A.; Espinoza-Ramos, K. E. Application of Semi-Empirical Ventilation Models in a Mediterranean Greenhouse with Opposing Thermal and Wind Effects. Use of Non-Constant C_d (Pressure Drop Coefficient Through the Vents) and C_w (Wind Effect Coefficient). *Agron.* **2019,** *9*, 736. https://doi.org/10.3390/agronomy9110736.
Luo, W.; De Zwart, H. F.; Dai, J.; Wang, X.; Stanghellini, C.; Bu, C. Simulation of Greenhouse Management in the Subtropics, Part I, Model Validation and Scenario Study for the Winter Season. *Biosyst. Eng.* **2005a,** *90* (3), 307–318.
Majdoubi, H.; Boulard, T.; Hanafi, A.; Bekkaoui, A.; Fatnassi, H.; Demrati, H.; Nya, M.; Bouirden L. Natural Ventilation Performance of a Large Greenhouse Equipped with Insect Screens. *Trans. ASABE* **2007**, *50* (2), 641–650.
Majdoubi, H.; Boulard, T.; Fatnassi, H.; Bouirden, L. Airflow and Microclimate Patterns in a One-Hectare Canary Type Greenhouse, An Experimental and CFD Assisted Study. *Agric. For. Meteorol.* **2009,** *149,* 1050–1062.
Madjoubi, H.; Boulard, T.; Fatnassi, H.; Senhaji, A.; Elbahi, S.; Demrati, H.; Mouqallid, M.; Bouirden, L. Canary Greenhouse CFD Nocturnal Climate Simulation. *Open J. Fluid Dyn.* **2016,** *6*, 88–100.
Maraveas, C. Wind Pressure Coefficients on Greenhouse Structures. *Agriculture.* **2020,** *10*, 149. https://doi.org/10.3390/agriculture10050149.
Mashonjowa, E.; Ronsse, F.; Milford, J. R.; Pieters, J. G. Modelling the Thermal Performance of a Naturally Ventilated Greenhouse in Zimbabwe Using a Dynamic Greenhouse Climate Model. *Sol. Ener*. **2013**, *91*, 381–393.
Miguel, A. F.; Van de Braak, N. J.; Bot, G. P. A. Analysis of the Airflow Characteristics of Greenhouse Screening Materials. *J. Agric. Eng. Res.* **1997,** *67,* 105–112.

Mills, A. F. In *Heat and Mass Transfer*; 1st ed.; Taylor & Francis Group: Boca Raton (US), 1995. https://doi.org/10.4324/9780203752173.

Mistriosis, A.; Bot, G. P. A.; Picuno, P.; Scarascia-Mugnozza, G. Analysis of the Efficiency of Greenhouse Ventilation Using Computational Fluid Dynamics. *Agric. For. Meteorol.* **1997a,** *85*, 217–228.

Mistriotis, A.; Arcidiacono, C.; Picuno, P.; Bot, G. P. A.; Scarascia-Mugnozza, G. Computacional Analysis of Ventilation in Greenhouses at Zero- and Low-Wind-Speeds. *Agric. For. Meteorol.* **1997b,** *88*, 121–135.

Mistriotis, A.; Briassoulis, D. Numerical Estimation of Internal and External Aerodynamic Coefficients of a Tunnel Greenhouse Structure with Openings. *Comput. Electron. Agric.* **2002,** *34*, 191–205.

Mobtaker, H. G.; Ajabshirchi, Y.; Ranjbar, S. F.; Matloobi, M. Solar Energy Conservation in Greenhouse: Thermal Analysis and Experimental Validation. *Renew. Ener.* **2016,** *96* (A), 509–519.

Mohammadi, B.; Ranjbar, S. F.; Ajabshirchi, Y. Application of Dynamic Model to Predict Some Inside Environment Variables in a Semi-Solar Greenhouse. *Inf. Process. Agric.* **2018,** *5*, 279–288.

Molina-Aiz, F. D. Simulación y modelación de la ventilación en invernaderos de Almería mediante la utilización de Dinámica Computacional de Fluidos [Simulation and Modelling of Ventilation in Almería Greenhouses Using Computational Fluid Dynamics]. PhD Dissertation, Universidad de Almería, Almería (Spain), 2010.

Molina-Aiz, F. D.; Norton, T.; López, A.; Reyes-Rosa, A.; Moreno, M. A.; Marín, P.; Espinoz, K.; Valera, D. L. Using Computational Fluid Dynamics to Analyse the CO_2 Transfer in Naturally Ventilated Greenhouses. *Acta Hort.* **2017,** *1182*, 283–292.

Molina-Aiz, F. D.; Valera, D. L. Configuration by Evaluating Ventilation Efficiency Based on Computational Fluid Dynamics. *Acta Hort.* **2011,** *893*, 669–677.

Molina-Aiz, F. D.; Valera, D. L.; Álvarez, A. J. Measurement and Simulation of Climate Inside Almería-Type Greenhouses Using Computational Fluid Dynamics. *Agric. For. Meteorol.* **2004,** *125*, 33–51.

Molina-Aiz, F. D.; Valera, D. L.; López A. Numerical and Experimental Study of Heat and Mass Transfers in an Almería-Type Greenhouse. *Acta Hort.* **2017,** *1170*, 209–218.

Molina-Aiz, F. D.; Valera, D. L.; López, A.; Álvarez, A. J. Analysis of Cooling Ventilation Efficiency in a Naturally Ventilated Almería-Type Greenhouse with Insect Screens. *Acta Hort.* **2012,** *927*, 551–558.

Molina-Aiz, F. D.; Valera, D. L.; Peña, A. A.; Alvarez, A. J.; Gil, J. A. Analysis of the Effect of Rollup Vent Arrangement and Wind Speed on Almería-Type Greenhouse Ventilation Performance using Computational Fluid Dynamics. *Acta Hort.* **2006,** *719*, 173–180.

Molina-Aiz, F. D.; Valera, D. L.; Peña, A. A.; Gil, J. A.; López, A. A Study of Natural Ventilation in an Almería-Type Greenhouse with Insect Screens by Means of Tri-Sonic Anemometry. *Biosyst. Eng.* **2009,** *104*, 224–242.

Montero, J. I.; Muñoz, P.; Anton, A. Discharge Coefficients of Greenhouse Windows with Insect-Proof Screens. *Acta Hort.* **1997,** *443*, 71–77.

Montero, J. I.; Muñoz, P.; Antón, A.; Iglesias, N. Computational Fluid Dynamic Modelling of Night-Time Energy Fluxes in Unheated Greenhouses. *Acta Hort.* **2005,** *691*, 403–409.

Montoya, A. P.; Guzmán, J. L.; Rodríguez, F.; Sánchez-Molina, J. A. A Hybrid-Controlled Approach for Maintaining Nocturnal Greenhouse Temperature: Simulation Study. *Comput. Electron. Agric.* **2016,** *123*, 116–124.

Moon, T.; Son, J. E. Knowledge Transfer for Adapting Pre-trained Deep Neural Models to Predict Different Greenhouse Environments Based on a Low Quantity of Data. *Comput. Electron. Agric.* **2021,** *185*, 106136.
Morris, L. G.; Neale, F. E. The Infra-Red Carbon Dioxide Gas Analyser and its use in Glasshouse Research. National Institute of Agricultural Engineering, Tech. Memo. No. 99, Silsoe (UK), 1954.
Muñoz, P. Ventilación Natural de invernaderos multitúnel. Ph Dissertation, Universitat de Lleida, Lleida (Spain), 1998.
Muñoz, P.; Montero, J. I.; Antón, A.; Giuffrida, F. Effect of Insect-Proof Screens and Roof Openings on Greenhouse Ventilation. *J. Agric. Eng, Res.* **1999,** *73*, 171–178.
Nääs, I. A.; Moura, D. J.; Bucklin, R. A.; Fialho, F. B. An Algorithm for Determing Opening Effectiveness in Natural Ventilation by Wind. *Trans. ASAE.* **1998,** *41* (3), 767–771.
Nara, M. Studies on Air Distribution in Farm Buildings-Two Dimensional Numerical and Experiment. *J. Soc. Agric. Struct.* **1979,** *9* (2), 18–25.
Navier, M. Mémoire sur les lois du mouvement des fluides. *Mém. Acad. Roy. Sci.* **1827,** *6*, 389–440.
Nederhoff, E. M.; Vegter, J. G. Canopy Photosynthesis of Tomatoes, Cucumber, and Sweet Pepper in Greenhouses, Measurements Compared to Models. *Ann. Bot.* **1994,** *73*, 421–427.
Nield, D. A.; Bejan, A. In *Convection in Porous Media*; Springer International Publishing: Berlin (Germany), 2017. https://doi.org/10.1007/978-3-319-49562-0.
Nijskens, J.; Deltour, J.; Coutisse, S.; Nisen, A. Heat Transfer Through Covering Materials of Greenhouses. *Agric. For. Meteorol.* **1984,** *33*, 193–214.
Nijskens, J.; Deltour, J.; Coutisse, S.; Nisen, A. Radiation Transfer Through Covering Materials, Solar and Thermal Screens of Greenhouses. *Agric. For. Meteorol.* **1985,** *35*, 229–242.
Norton, T.; Sun, D. W.; Grant, J.; Fallon, R.; Dodd, V. Applications of Computational Fluid Dynamics (CFD) in the Modelling and Design of Ventilation Systems in the Agricultural Industry, A review. *Bioresour. Technol.* **2007,** *98* (12), 2386–2414.
Ohba, M.; Irie, K.; Kurabuchi, T. Study on Airflow Characteristics Inside and Outside a Cross-Ventilation Model, and Ventilation Flow Rates Using Wind Tunnel Experiments. *J. Wind Eng. Indus. Aerodyn.* **2001,** 89 (14–15), 1513–1524.
Okada, M.; Takakura, T. Guide and Data for Greenhouse Air Conditioning. 3. Heat Loss Due to Air Infiltration of Heated Greenhouse. *J. Agric. Meteorol.* **1973,** *28* (4), 223–230.
Ould Khaoua, S. A.; Bournet, P. E.; Migeon, C.; Boulard, T.; Chassériaux, G. Analysis of Greenhouse Ventilation Efficiency Based on Computational Fluid Dynamics. *Biosyst. Eng.* **2006,** *95* (1), 83–98.
Pakari, A.; Ghani, S. Airflow Assessment in a Naturally Ventilated Greenhouse Equipped with Wind Towers: Numerical Simulation and Wind Tunnel Experiments. *Ener. Build.* **2019,** *199*, 1–11.
Papadakis, G.; Manolakos, D.; Kyritsis, S. Solar Radiation Transmissivity of a Single-Span Greenhouse Through Measurements on Scale Models. *J. Agric. Eng. Res.* **1998,** *71*, 331–338.
Papadakis, G.; Mermier, M.; Meneses, J. F.; Boulard, T. Measurement and Analysis of Air Exchange Rates in a Greenhouse with Continuous Roof and Side Openings. *J. Agric. Eng. Res.* **1996,** *63*, 219–228.
Patil, S. L.; Tantau, H. J.; Salokhe, V. M. Modelling of Tropical Greenhouse Temperature by Auto Regressive and Neural Network Models. *Biosys. Eng.* **2008,** *99*, 423–431.
Pearson, C. C.; Owen, J. E. The Resistance to Air Flow of Farm Building Ventilation Components. *J. Agric. Eng. Res.* **1994,** *57*, 53–65.

Pérez Parra, J. J. Natural Ventilation of "Parral" Type Greenhouses. PhD Dissertation, University of Córdoba, Córdoba (Spain), 2002.

Pérez-Parra, J. J.; Baeza, E. J.; López, J. C.; Pérez, C. Effects of Vent Types and Insect Screens on Ventilation of "Parral" Greenhouses. *Acta Hort.* **2003,** *614*, 393–400.

Pérez-Parra, J. J.; Baeza, E. J.; Montero, J. I.; Bailey, B. J. Natural Ventilation of Parral Greenhouses. *Biosyst. Eng.* **2004a,** *87* (3), 89–100.

Pérez-Parra, J. J.; Baeza, E. J.; Zaragoza, G.; López, J. C.; Montero. J. I. Modelling Natural Ventilation in a "Parral" Greenhouse with Continuous Flap Vents using Computational Fluid Dynamics and Wind Tunnel Measurements. *Acta Hort.* **2004b,** *659*, 135–142.

Pieters, J. C.; Deltour, J. M. Performance of Greenhouses with the Presence of Condensation on Cladding Materials. *J. Agric. Eng. Res.* **1997,** *68*, 125–137.

Piscia, D.; Montero, J. I.; Baeza, E.; Bailey, B. J. A CFD Greenhouse Night-Time Condensation Model. *Bios. Eng.* **2012,** *111*, 141–154.

Potter, I. N. Effect of Fluctuation Wind Pressures on Natural Ventilation Rates. *ASHRAE Trans.* **1979,** *85* (2), 445–457.

Price, D. R.; Peart, R. M. Simulation Model to Study the Utilization of Waste Heat Using a Combination Multiple Reservoir and Greenhouse Complex. *J. Environ. Qual.* **1973,** *2*, 216–224.

Reyes-Rosas, A.; Molina-Aiz, F. D.; Valera, D. L.; López, A.; Khamkure, S. Development of a Single Energy Balance Model for Prediction of Temperatures Inside a Naturally Ventilated Greenhouse with Polypropylene Soil Mulch. *Comput. Electron. Agric.* **2017,** *142*, 9–28.

Roy, J. C.; Boulard, T.; Kittas, C.; Wang S. Convective and Ventilation Transfers in Greenhouses, Part 1: The Greenhouse Considered as Perfectly Stirred Tank. *Biosyst. Eng.* **2002,** *83* (1), 1–20.

Roy, J. C.; Pouillard, J. B.; Boulard, T.; Fatnassi, H.; Grisey, A. Experimental and CFD Results on the CO_2 Distribution in a Semi Closed Greenhouse. *Acta Hort.* **2014,** *1037,* 993–1000.

Roy, J. C.; Vidal, C.; Fargues, J.; Boulard, T. CFD Based Determination of Temperature and Humidity at Leaf Surface. *Comput. Electron. Agric.* **2008,** *61*, 201–212.

Salazar-Moreno, R.; López-Cruz, I. L.; Sánchez Cruz, A. C. Dynamic Energy Balance Model in a Greenhouse with Tomato Cultivation: Simulation, Calibration an Evaluation. *Rev. Chapingo Serie Hort.* **2018,** *25* (1), 45–60.

Sapounas, A. A.; Nikita-Martzopoulou, Ch.; Bartzanas, T.; Kittas, C. Fan and Pad Evaporative Cooling System for Greenhouses, Evaluation of a Numerical and Analytical Model. *Acta Hort.* **2008,** *797*, 131–138.

Sase, S.; Nara, M. A Control Algorithm for Natural Ventilation Based on Wind Tunnel Testing. *Acta Hort.* **1985,** *174*, 75–79.

Sase, S.; Takakura, T.; Nara, N. Wind Tunnel Testing on Airflow and Temperature Distribution of a Naturally Ventilated Greenhouse. *Acta Hort.* **1984,** *148*, 329–336.

Sase, S.; Reiss, E.; Both, A. J.; Roberts, W. J. Developing a Natural Ventilation Model for Open-Roof Greenhouses. Center for Controlled Environment Agriculture, Cook College, Rutgers University, (US), 2002. http://horteng.envsci.rutgers.edu/ppt/papers/openroofmodel.pdf

Schauberger, G.; Piringer, M.; Petz, E. Steady State Balance Model to Calculate the Indoor Climate of Livestock Buildings, Demonstrated for Finishing Pigs. *Int. J. Biometeorol.* **2000,** *43*, 154–162.

Schlichting, H.; Gesten, K. In *Boundary-Layer Theory*; Springer: Berlin, (Germany), 2017. https://doi.org/10.1007/978-3-662-52919-5.

Schockert, K.; Von Zabeltiz, C. Energy Consumption of Greenhouses. *Acta Hort.* **1980,** *106*, 21–26.

Seginer, I.; Boulard, T.; Bailey, B. J. Neural Network Models of the Greenhouse Climate. *J. Agric. Engng. Res.* **1994,** *59* (3), 203–216.

Sethi, V. P.; Sumathy, K.; Lee, C.; Pal, D. S. Thermal Modeling Aspects of Solar Greenhouse Microclimate Control: A Review on Heating Technologies. *Sol. Ener*. **2013,** *96*, 56–82.

Shamshiri, R.; Ismail, W. I. W. A Review of Greenhouse Climate Control and Automation Systems in Tropical Regions. *J. Agric. Sci. Appl*. **2013,** *2*, 176–183.

Sharma, P. K.; Tiwari, G. N. Parametric Study of a Greenhouse by Using Runge-Kutta Methods. *Ener. Convers. Manage.* **1999,** *40*, 901–912.

Sharma, P. K.; Tiwari, G. N.; Sorayan, V. P. S. Temperature Distribution in Different Zones of the Microclimate of a Greenhouse, A Dynamic Model. *Ener. Convers. Manage.* **1999,** *40*, 335–348.

Shen, X.; Zhang, G.; Bjerg, B. Comparison of Different Methods for Estimating Ventilation Rates Through Wind Driven Ventilated Buildings. *Ener. Build.* **2012,** *54*, 297–306.

Sherman, M. H.; Grimsrud, D. T. Infiltration-Pressurization Correlation Simplified Physical Modeling. *Trans. ASHRAE* **1980,** *86* (2), 778–807.

Shilo, E.; Teitel, M. Mahrer, Y.; Boulard, T. Air-Flow Patterns and Heat Fluxes in Roof-Ventilated Multi-Span Greenhouse With Insect-Proof Screens. *Agric. For. Meteorol.* **2004,** *122*, 3–20.

Singh, G.; Singh, P. P.; Singh-Lubana, P. P.; Singh, K. G. Formulation and Validation of a Mathematical Model of the Microclimate of a Greenhouse. *Renew. Energy*. **2006**, *31* (10), 1541–1560.

Stanghellini, C. Transpiration of Greenhouse Crops. An Aid to Climate Management. Ph. Dissertaion, Agricultural University Wageningen, Wageningen (The Netherland), 1987. https://edepot.wur.nl/202121.

Stokes, G.G. On the Theories of the Internal Friction of Fluids in Motion and of the Equilibrium and Motion of Elastic Solids. *Trans. Cambridge Philos. Soc.* **1845,** *8,* 287–319.

Straw, M. P.; Baker, C. J.; Robertson, A. P. Experimental Measurements and Computations of the Wind-Induced Ventilation of a Cubic Structure. *J. Wind Eng. Indust. Aerodyn.* **2000,** *88* (2-3), 213–230.

Strom, J. S.; Amsen, M. G. Heat Comsumption Model for Greenhouse Nurseries. *Acta Hort.* **1981,** *115*, 503–515.

Swami, M. V.; Chandra, S. Procedure for Calculating Natural Ventilation Airflow Rates in Buildings. American Socity of Heating, Refrigerating and Air-Conditioning Engineers (ASHRAE), Final Report, FSEC-CR-163-86, 1987. http://www.fsec.ucf.edu/en/publications/pdf/fsec-cr-163-86.pdf.

Takami, S.; Uchijima, A. A Model for the Greenhouse Environment as Affected by the Mass and Energy Exchange of a Crop. *J. Agr. Met.* **1977,** *33*, 117–127.

Taki, M.; Ajabshirchi, Y.; Ranjbar, S. F.; Rohani, A.; Matloobi, M. Heat Transfer and MLP Neural Network Models to Predict Inside Environment Variables and Energy Lost in a Semi-Solar Greenhouse. *Ener. Build.* **2016,** *110* (1), 314–329.

Tavares, C.; Goncalves, A.; Castro, P.; Loureiro, D.; Joyce, A. Modeling an Agriculture Production Greenhouse. *Renew. Ener.* **2001,** *22*, 15–20.

Teitel, M. The Effect of Insect-Proof Screens in Roof Openings on Greenhouse Microclimate. *Agric. For. Meteorol.* **2001,** *110* (1), 13–25.

Teitel, M. The Effect of Screened Openings on Greenhouse Microclimate. *Agric. For. Meteorol.* **2007,** *143* (3-4), 159–175.

Teitel, M. Using Computational Fluid Dynamics Simulations to Determine Pressure Drops on Woven Screens. *Biosyst. Eng.* **2010,** *105*, 172–179.

Teitel, M.; Barak, M.; Berlinger, M. J.; Lebiush-Mordechai, S. Insect-Proof Screens in Greenhouses: Their Effect on Roof Ventilation and Insect Penetration. *Acta Hort.* **1999,** *507*, 25–34.

Teitel, M.; Dvorkin, D.; Haim, Y.; Tanny, J.; Seginer, I. Comparison of Measured and Simulated Flow Through Screens: Effects of Screen Inclination and Porosity. *Biosyst. Eng.* **2009,** *104* (3), 404–416.

Teitel, M.; Liran, O.; Tanny, J.; Barak, M. Wind Driven Ventilation of a Mono-Span Greenhouse with a Rose Crop and Continuous Screened Side Vents and its Effect on Flow Patterns and Microclimate. *Biosyst. Eng.* **2008a,** *101* (1), 111–122.

Teitel, M.; Shklyar, A. Pressure Drop Across Insect-Proof Screens. *Trans. ASAE.* **1998,** *41* (6), 1829–1834.

Teitel, M.; Tanny, J. Heat Fluxes and Airflow Patterns Through Roof Windows in a Naturally Ventilated Enclosure. *Flow Turbul. Combust.* **2005,** *74*, 21–47.

Teitel, M.; Tanny, J.; Ben-Yakir, D.; Barak, M. Airflow Patterns Through Roof Openings of a Naturally Ventilated Greenhouse and Their Effect on Insect Penetration. *Biosyst. Eng.* **2005,** *92* (3), 341-353.

Teitel, M.; Wenger, E. Air Exchange and Ventilation Efficiencies of a Monospan Greenhouse with One Inflow and One Outflow Through Longitudinal Side Openings. *Biosyst. Eng.* **2014,** *119*, 98–107.

Teitel, M.; Ziskind, G.; Lirana, O.; Dubovsky, V.; Letan, R. Effect of Wind Direction on Greenhouse Ventilation Rate, Airflow Patterns and Temperature Distributions. *Biosyst. Eng.* **2008b,** *101* (3), 351–369.

Thurtell, G. W.; Tanner, C. B.; Wesely, M. L. Three-Dimensional Pressure-Sphere Anemometer System. *J. Appl. Meteorol.* **1970,** *9* (3), 379–385.

Tong, G.; Christopher, D. M.; Li, B. Numerical Modelling of Temperature Variations in a Chinese Solar Greenhouse. *Comput. Electron. Agric.* **2009,** 68, 129–139.

Tong, G.; Christopher, D. M.; Zhang, G. New Insights on Span Selection for Chinese Solar Greenhouses using CFD Analyses. *Comput. Electron. Agric.* **2018,** *149*, 3–15.

Udink ten Cate, A. J. Remarks on Greenhouse Climate Control. *Acta Hort.* **1980,** *106*, 43–48.

Udink ten Cate, A. J.; van de Vooren, J. Digital Adaptive Cntrol of a Glasshouse Heating System. *IFAC Proc. Vol.* **1977,** *10* (16), 505–512.

Valera, D. L.; Álvarez, A. J.; Molina-Aiz, F. D. Aerodynamic Analysis of Several Insect-Proof Screens used in Greenhouses. *Spanish J. Agric. Res.* **2006,** *4* (4), 273–279.

Van Meurs, W. T.; Stanghellini, C. A Transpiration Based Control Algorithm. *Acta Hort.* **1989,** *245*, 191–204.

Van Beveren, P.; Bontsema, J.; Van Straten, G.; Van Henten, E. Minimal Heating and Cooling in a Modern Rose Greenhouse. *Appl. Ener.* **2015,** *137*, 97–109.

Van Ooteghem, R. J. C. Optimal Control Design for a Solar Greenhouse. *IFAC Proc. Vol.* **2010,** *43* (26), 304–309.

Vanthoor, B. H. E.; Stanghellini, C.; Van Henten, E. J.; De Visser, P. H.B. A Methodology for Model-Based Greenhouse Design: Part 1, A Greenhouse Climate Model for a Broad Range of Designs and Climates. *Biosyst. Eng.* **2011,** *110* (4), 363–377.

Vassiliou, N. N.; Martzopoulos, G. G.; Nikita-Martzopoulou, C. Air Flow Pattern Outside and Inside a Double Span Arch Type Greenhouse. *IFAC Proc. Vol.* **1998,** *31* (12), 231–237.

Verlinde, W.; Gabriels, D.; Christiaens, J. P. A. Ventilation Coefficient for Wind-Induced Natural Ventilation in Cattle Buildings. *Trans. ASAE.* **1998,** *41* (3), 783–788.

Vickery, P. J.; Surry, D. The Aylesbury Experiments Revisited-Further Wind Tunnel Tests and Comparisons. *J. Wind Eng. Indust. Aerodyn.* **1983,** 11 (1-3), 39–62.

Walker J. N.; Duncan G. A. Estimating Greenhouse Ventilation Requirements. University of Kentaky, College of Agriculture, Cooperative Extension Service, 1973. https://www.uky.edu/bae/sites/www.uky.edu.bae/files/AEN-09.pdf.

Walker, I. S.; Wilson, D. J. Evaluating Models for Superposition of Wind and Stack Effect in Air Infiltration. *Build. Environ.* **1993,** *28* (2), 201–210.

Walker, J. N. Predicting Temperatures in Ventilated Greenhouses. *Trans. ASAE* **1965,** *8* (3), 445–448.

Walker, J. N.; Peterson, W. O.; Duncan, G. A.; Anastasi, D. T. Temperature and Humidity in a Greenhouse Ventilated with Coal Mine Air. *Trans. ASAE* **1976,** *19* (2), 311–317.

Wang, H.; Chen, Q. A New Empirical Model for Predicting Single-Sided, Wind-Driven Natural Ventilation in Buildings. *Ener. Build.* **2012,** *54*, 386–394.

Wang, S. Measurement and Modelling of Natural Ventilation in a Large Scale Venlo-type Greenhouse. Ph. Dissertation, Faculté Universitaire des Sciences Agronomiques de Gembloux, Gembloux (Belgium), 1998.

Wang, S.; Boulard, T. Measurement and Prediction of Solar Radiation Distribution in Full-Scale Greenhouse Tunnels. *Agron.* **2000b,** *20* (1), 41–50.

Wang, S.; Boulard, T. Predicting the Microclimate in a Naturally Ventilated Plastic House in a Mediterranean Climate *J. Agric. Eng. Res*. **2000a,** *75* (1), 27–38.

Wang, S.; Deltour, J. An Experimental Ventilation Function for Large Greenhouses Based on a Dynamic Energy Balance Model. *Agric. Eng. J.* **1996,** *5* (3-4), 103–112.

Wegley, H. L.; Ramsdell, J. V.; Orgill, M. M.; Drake, R. L. In *Siting Handbook for Small Wind Energy Conversion Systems*. U.S. Department of Energy. Pacific Northwest Laboratory, Richland (US), 1980. https://doi.org/10.2172/5490541.

Whittle, R.; Lawrence, W. C. The Climatology of Glasshouses. II. Ventilation. *J. Agric. Eng. Res.* **1960,** *5*, 36–41

Zhang, J. S.; Janni, K. A.; Jacobson, L. D. Modeling Natural Ventilation Induced by Combined Thermal Buoyancy and Wind. *Trans. ASAE.* **1989,** *32* (6), 2165–2174.

Zhang, Y.; Mahrer, Y.; Margolin, M. Prediction the Microclimate Inside a Greenhouse, An Application of a Onedimensional Numerical Model in an Unheated Greenhouse. *Agric. For. Meteorol.* **1997,** *86* (3-4), 291–297.

Zhao, Y.; Teitel, M.; Barak, M. Vertical Temperature and Humidity Gradients in a Naturally Ventilated Greenhouse. *J. Agric. Eng. Res.* **2001,** *78* (4), 431–436.

CHAPTER 13

Greenhouse Climate Control for Flower Regulation in Ornamental Crops

RISHU SHARMA[1], POONAM KUMARI[2], and HOMRAJ SAHARE[3]

[1]*Department of Floriculture and Landscaping, Punjab Agricultural University, Ludhiana, Punjab, India*

[2]*Division of Agrotechnology, CSIR-Institute of Himalayan Bioresource Technology, Palampur, Himachal Pradesh, India*

[3]*Department of Horticulture, School of Agriculture, Lovely Professional University, Phagwara, Punjab, India*

ABSTRACT

Ornamental plants form the integral part in human life, be it birth or death, happiness or grief; flowers have become a way to channel your feelings into expressions. With increasing consumption, growers are now devising different ways and means to meet the demand for every occasion. Greenhouse technology with precise environmental control is the key in ensuring the supply of fresh ornamentals even in off-season. Flower regulation is highly complex and controlled by different factors, such as environmental factors, genetic factors, and their interactions. With the understanding of physiology and flowering responses to different environmental cues, we can manipulate the plant to flower at desired time. Ornamental crops sensitive to photoperiod can be regulated by altering their photoperiodic conditions, similarly by altering greenhouse temperature; we can regulate the flowering of thermo-sensitive crops. Supply of fresh produce throughout the year is not the only significance of flower regulation, but it also reduces the gluts in the market through regulated supply, avoids spoilage or wastage of produce,

Protected Cultivation: Structural Design, Crop Management, Modeling and Automation.
Mahesh Chand Singh and K. K. Sharma (Eds.)

increases growers' income with better returns in off season, and creates employment round the year. Use of sensors technology equipped with automation software tools in greenhouse can transform the flower production and strengthen this sector further.

13.1 INTRODUCTION

Ornamental plants including major cut flowers, loose flowers, and other plants of ornamental value are giving a forward push to the country's economy as with increase in urban settlements, industrial development, and modern lifestyle, demand of ornamental plants and floral products is also seeing a rise. Floral products are used everywhere nowadays, be it a wedding, welcoming ceremony, festival, conference, funeral, or any other occasions. Trend in the production of flower crops indicates that growing period of flower is mostly fixed where a particular flower is produced during its normal season following the routine cultivation practices. This results in bulk production of the crop during its normal growing season which makes it less remunerative in terms of returns as it fetches low prices. But, if we compare this with off-season or production during specific occasion, growers can fetch better price for their produce due to limited produce in the market. Therefore, schedule flower production through applying the techniques of flower regulation is very much critical for flower growers. Flower regulation basically is a technique to manipulate the flowering time through specific treatment given to the plant at ripeness-to-flower stage, so that it comes in flowering on a specific day (e.g. on Mother's Day), or during off-season which means flowering could be earlier or later than the normal period of blooming. To successfully achieve the flower regulation, one must understand the factors affecting the flowering and their control. As most of the ornamental crops are grown inside the protected environment, it is possible to control their growth manipulating factors, that is light, temperature, humidity, and other factors that affect the flowering directly or indirectly. When greenhouse systems are supplemented with automated control system, each and every input is regulated with intrinsic network of sensors that monitor and control different parameters like temperature, humidity, carbon dioxide, pH, electrical conductivity, shade and also provide external-climatic fluctuations via a weather station. Manipulation of crop growth parameters and programming of greenhouse control system according to flowering schedule, physiological basis of flowering need to be understood for each crop so that crop-wise schedule can be prepared for regulation.

13.2 PROCESS OF FLOWERING

Flowering signals the transition of plant from vegetative to reproductive stage is the most fundamental thing that determines the survival of plant over many generations. It is completed in different stages starting from floral initiation (floral primordia differentiation), floral organization (the distinction of the different floral parts), floral maturation (growth of flower parts, differentiation of the sporogeneous tissues, meiosis, pollen, and embryo sac development), and anthesis (opening of the complete flower). To initiate flowering, plants sense all the environmental cues viz., day length and temperature mainly, some endogenous factors *viz*., juvenile phase, available reserves, metabolic contents etc. and complete understanding is required to know the interaction of these factors in the initiation of flowering. Basic physiological and genetic studies related to flowering have been done in *Arabidopsis thaliana*. Several studies indicate that for the formation of flower, a special flower forming substance exists in the leaves which is hypothetically named as florigen, "a flower forming stimulus." With the revelation of genetic studies many transcription factors and regulatory proteins were identified playing role in flower regulation (Milec et al., 2014). Five genetically regulated pathways have been investigated to play role in flowering, *viz*., vernalization pathway related to exposure to cold stimulus for flowering (Sheldon et al., 1999); photoperiod pathway related to flowering response toward day length and light quality (Reeves and Coupland, 2000); gibberellins pathway referring to the gibberellic acid mediated flowering; autonomous pathway which works through endogenous plant regulators and is not dependent on the other pathways, for example, plant age. Seasonal influences also affect flowering, e.g., roses, marigold, chrysanthemum, and heliconia are among the plants which have little influence of seasonal changes on their flowering whereas jasmine, dendrobium orchids are greatly influenced by seasonal changes.

13.3 ENVIRONMENTAL FACTORS AFFECTING FLOWERING

13.3.1 EFFECT OF TEMPERATURE

Temperature functions as a major regulator of flowering time. Plants can be exposed to wide temperature ranges from near freezing to over 50°C depending upon their habitat. They are also subjected to diurnal and seasonal changes in temperature even in a single location. Ambient temperature, under which a plant grows throughout its vegetative growth, also affects flowering (Samach and Wigge, 2005). Any change in temperature alters the morphological and biochemical processes of the plants. Exposure to the temperature beyond the tolerable limits may cause the irreversible disruption of plant system. It has been believed that ambient temperature responses are closely associated with the metabolic fluctuations in the plant system. Slight increase in temperature in polyhouse helps in efficient and increased photosynthesis and results in early flowering for e.g., in easter lilies, whereas low temperature in polyhouse results in poor flowering and flower malformation in certain crops like carnation, roses etc. Basically, photosynthetically active radiations are important for performing the photosynthesis process where temperature further affects the rate of sugar synthesis by photosynthesis; therefore, there is need to maintain a proportional balance between radiation and temperature.

13.3.2 EFFECT OF LIGHT

Photoperiod or day length as well as light intensity or light quality has a huge role to play in control of flowering mechanisms in plants. The way plants respond to the variations in day length in a year affects the flowering time and reproduction. As per photoperiodic requirement to begin flowering, plants are categorized as long-day, short-day, and day-neutral. Several genes involved in photoperiodic signaling of flowering have been extensively studied in *Arabidopsis thaliana.* Photoperiod stimulus is perceived to lead to the initiation of floral organs from the shoot apical meristem. This suggests that a moving signal, florigen, moves from leaves to shoot apical meristem to signal the initiation of flowering (Chailakhyan, 1936). Flowering in long-day plants is initiated, when day length is longer than their critical photoperiod and in case of short-day plants, for flower initiation day length should be shorter than critical photoperiod, whereas in day-neutral plants flowering occurs irrespective of day length. Further photoperiodic response of short-day and long-day plants can be facultative or obligate, where flowering is progressed by photoperiod in facultative plants or a critical photoperiod is desired in obligate plants (Thomas et al., 1997). Apart from photoperiod, light intensity also acts on flowering. During the short days, light intensities are usually low and to compensate that, plants are supplemented with high-intensity lighting to increase their growth, and with low-intensity lighting to regulate flowering in photoperiod-sensitive plants. Plant responses are influenced under the interaction of light intensity, duration, and quality of light. Often the effect of light quality is under looked by many workers and is among the least considered environmental parameters influencing flowering. However, it is well understood now that with the use of different light sources emitting different light qualities for e.g., LEDs, different desired plant responses can be attained. Photosynthetically active radiations (PAR) fall under 400–700 nm wavelength. In general, orange-red light is considered most effective and cyan-green light as least effective in plant growth and flowering. However, it is ascertained that abnornalities can be seen in plants if grown under a single, narrow light waveband (Massa et al., 2008). Blue light is required for chlorophyll formation, stomatal activities, and for normal plant growth, whereas red light helps in flowering and reproductive development of plants. Supplementing the ambient day light enhances the total light uptake by the plant per day which is the daily light integral (DLI) and this increased DLI can promote plant growth through increased photosynthesis.

13.3.3 EFFECT OF HUMIDITY

Relative humidity (RH) is interpreted as the water vapor accommodated by air relative to the maximum amount of water vapor that the air can accommodate at a certain temperature. Relative humidity fluctuations can affect the plant growth by affecting stomatal conductance. During warm weather, plant may close its stomata to minimize water loss. The stomata also act as a cooling mechanism, as the temperature is too high plants close stomata for an extended period of time to save water; it has no means of moving carbon dioxide and oxygen molecules, slowly allowing the plant to suffocate on water vapor and its own transpired gases. A plant cannot effectively do transpiration process or uptake nutrients from the medium in the conditions of excessive humidity and with such prolonged condition, plant inevitably rots. When a plant is exposed to high temperature and low relative humidity, transpiration rate increases. Many researchers have elucidated that how changes in relative humidity control the growth and flowering process of ornamental plants. Gislerod et al. (1987) conducted an experiment on nine greenhouse species planted in growth chambers with relative air humidity (RH) of 55–60, 70–75, and 90–95% for 24 to 100 days. Twice a week, they were given a full nutrient solution. The increase of RH from lowest to highest significantly reduced transpiration rate in *Euphorbia pulcherrima* (60%), *Begonia hiemalis* (54%), *Saintpaulia ionantha* (54%), *Nephrolepis exaltata* (48%), and *Lycopersicon esculentum* (44%). By enhancing RH, the content of macronutrient elements in plant leaves decreased. When the plants were growing at high RH, macronutrient elements in the growing medium at the culmination of the experiment were lowest. Begonia plant diameter and dry weight increased with a rise in RH from 70 to 81%, but these decreased with a further increase to 93%. The air humidity for best plant quality was found to be 70% (Mortensen, 2000). According to Mortensen and Gislerd (2000), rising relative air humidity from 67 to 94% results in increased shoot length, leaf size, and shoot fresh weight in cut roses. Furthermore, roses grown at high RH (>85%) had a slightly shorter vase life when measured at low air humidity (1170–1710 Pa vpd), while the difference was less when observed at high humidity (440 Pa vpd). The ability of leaves to control water loss is closely related to the vase life of cut roses and the keeping quality of pot roses. Roses grown in high humidity have a lesser keeping quality and show signs of water stress such as bent neck, foliage wilting, and nonuniform flower opening and wilting (Mortensen and Fjeld, 1995; Torre and Fjeld, 2001). Kesumawati (2013) reported that intermittent mist improves flowering time of *Anthurium andraeanum* in rock wool medium (12 weeks) earlier than rock wool medium without mist (32 weeks). This method resulted in the production of higher quality flowers.

13.3.4 EFFECT OF CARBON DIOXIDE (CO_2)

Carbon dioxide is an essential component that plays role in photosynthesis (also known as carbon assimilation). In green plants, photosynthesis is a known chemical process that converts CO_2 and water into sugars using light energy. Via respiration, these sugars are then utilized for plant growth. The difference between photosynthesis and respiration rate enables plants to accumulate dry matter. Diffusion allows carbon dioxide to enter into the plant through stomatal openings. Stomata are specialized cells in the epidermal layer of the leaves that are mostly found on the underside of the leaves. Gas exchange takes place as the cells open and close. The rate of CO_2 absorption by the plant is highly influenced by the concentration of CO_2 outside the leaf. The greater the CO_2 absorption by the plant, the higher will be the CO_2 concentration outside the bud. Several factors regulate the opening and closing of stomata, including leaf temperature, ambient air temperature, light levels, relative humidity, water tension, and CO_2 and oxygen (O_2) concentrations in the air and on the leaf. The CO_2 enrichment reduces photosynthesis inhibition by oxygen and boosts net photosynthesis in plants. This is the basis for CO_2-induced enhanced growth rates at low and high light levels. Elevated CO_2 levels also elevate the optimal temperature for development (Mortensen, 1987). The CO_2 concentration is required to be optimized for better plant growth and quality as per specific plant species and genotypes. In *Arabidopsis*, the sugar signaling metabolite, trehalose-6-phosphate (T6P), depicted a clear association (r 2 = 0.94) with sucrose levels in the vegetative and shoot-apical meristem tissues. The T6P has been suggested as a prime indicator for floral induction, imparting details of tissue carbohydrate supply (Wahl et al., 2013). Dry weight increment, better plant height, leaf number, and lateral branching are all beneficial effects of CO_2 enrichment in pot plants, cut flowers, vegetables, and forest plants. The CO_2 enrichment also brings improvement in plant quality as computed by their growth habit and flowering. Rooting is often aided by greater CO_2 levels. In greenhouses, the optimum CO_2 concentration required for better growth and yield appears to be between 700 and 900 $\mu l\ l^{-1}$, and this CO_2 level is usually recommended. The CO_2 concentrations greater than 1000 $\mu l\ l^{-1}$ can cause growth retardation and leaf injury, and they almost certainly increase CO_2 leakage from the greenhouse (Mortensen, 1987). Long-term yield increases in roses grown in enriched environments are thought to be due to the maintenance of Rubisco and CA activities after extended exposure to CO_2-enriched atmospheres. Optimal concentration of CO_2 in Alstroemeria cv. "Fiona" for plant growth and flower production was 900 ppm; however in Alstroemeria cv. "Helios" it has been observed that at this concentration of CO_2 flower quality decreased significantly (Van Labeke and Dambre, 1998). At varying light levels (45, 130, 270, and 390 $\mu mol\ m^{-2}\ s^{-1}$), raising the CO_2 concentration from 330 to 900 $\mu l\ l^{-1}$ markedly increased the dry weight, leaf, and flower number and decreased the time before flowering. The CO_2 enrichment increased the mean relative growth rate by 16%. At the lowest light level, flowering initiated earlier by seven days in CO_2-enriched weather, but not at the maximum light level (Mortensen and Ulsaker, 1985). Increases in dry weight, plant height, leaf number, and lateral branching can all be related to CO_2 enrichment in greenhouses. Roses in a greenhouse need between 700 and 900 $\mu l\ l^{-1}$ of CO_2. A concentration of more than 900 $\mu l\ l^{-1}$can inhibit growth and cause leaf injury (Zeiger et al., 2002). Kromwijk et al. (2012) observed that up to 1000 ppm of CO_2 would improve photosynthesis and flowering in Phalaenopsis. Supplementation of CO_2 (700 and 1000 ppm) in phalaenopsis cut flower resulted in increase in spike number and flowers along with increased spike fresh weight (Endo and Ikushima, 1997). Similarly, Hwang and Jeong (2009) advocated that for better vegetative growth of Phalaenopsis cultivars it should be supplemented with 2000 ppm CO_2 rather than 1000 ppm of CO_2.

13.4 TECHNIQUES OF FLOWER REGULATION IN COMMERCIAL ORNAMENTAL CROPS

Flower regulation is aimed at production during the off-season specifically during the occasion on specific date. Blooms that are produced during normal season fetch low price due to overflowing produce in the market and owing to that growers have to face losses many times. Due to bulk production in normal season, unsold flowers are either left un-harvested or are spoiled after being harvested. Therefore, there is a need to encourage farmers to schedule their production of cut flowers during the off-season to fetch better price. Apart from that the flowers are required specifically to mark certain occasions or festivities like Christmas Day, Thanks giving Day, New Year, Mother's Day, Valentine's Day, Easter Sunday, and many others. Therefore, if growers regulate their production to harvest during these specific dates, for example, Red roses during valentine's day and Easter lilies during Easter then it will turn maximum benefits to them (Dhiman and Laishram, 2010). Flower regulation is carried out by growers to accomplish certain objectives *viz.*, to avoid bulk of normal season cut flowers, to avoid excessive spoilage of surplus cut flowers, to avoid diseases arising from flower waste, to generate employment round the year, to boost growers' income, to decrease flower import and balance trade, and to maintain the supply chain of flowers at the time of need. It is well established that the onset of flowering is regulated by various environmental as well as endogenous plant factors. So, by manipulating those factors we can regulate the flowering time in ornamental crops.

13.4.1 TEMPERATURE MANIPULATION

Flowering is affected by air temperature as well as temperature of the growing media. There is a need to maintain high temperature for efficient photosynthesis inside greenhouse under diurnal conditions, whereas under nocturnal conditions it is not desired as plant is not in active stage (Pawlowski, 2009). Desired temperature inside the greenhouse can be attained either by cooling or heating systems in the greenhouse (Singh et al., 2016, 2018a,b, 2021). Generally, we employ greenhouse cooling for flower regulation in bulbous ornamentals. Cooling is done by natural ventilation where ambient air replaces the enclosed air mass via natural way by providing sufficient open area in the greenhouse structure (about 20% of the floor area). For efficient cooling mechanical ventilation is used where the air from the enclosure is drawn out with auxiliary power using equipment like exhaust fans, vents, and provision for inlet of ambient air. Evaporative cooling is another mode of cooling used in the summer months of the year (Singh et al., 2018c, 2019a,b). Warm air is expelled through exhaust fans in one side of the wall which is replaced by air entering through wet pads fixed on the opposite side of the wall. The entering air is cooled by evaporation of water in pads (Gupta et al., 2007).

13.4.1.1 FLOWER REGULATION IN EASTER LILIES (LILIUM LONGIFLORUM)

Regulation of flowering in Easter lilies is achieved by controlling greenhouse temperature and using leaf counting technique. In this, rate of leaf unfolding is monitored during the period between flower bud initiation and visible bud formation stage which is usually 5–6 weeks before flowering or when plants are 3–4 inches tall. This technique includes several steps for execution. First, index plants are selected in different sections of the greenhouse to sample growing conditions and on the selected plants number of leaves that the plant will bear will be determined. It is done by counting number of leaves and leaf initials during flower bud formation in the greenhouse. Flower buds can first be observed under microscope on plants from precooled bulbs (six weeks at 4.44°C) approximately six weeks after potting in a greenhouse maintained at 21°C day and 15.55°C night minimum temperature. Then on the index plants leaf unfolding rate (daily or weekly) is determined.

After that number of leaves yet to unfold before the buds in visible stage are calculated. It will give the number of leaves yet to be unfold which will be equal to total leaf count minus leaves unfolded. Calculate days or weeks remaining for flower buds to become visible. It will be equal to leaves yet to unfold divided by leaf unfolding rate. Based upon this, determine whether the plants are behind or ahead of schedule and accordingly raise or reduce forcing temperatures to hasten or delay the rate of leaf unfolding. Bulb precooling has an important role to play here as inadequately precooled bulbs; the rate of leaf unfolding will increase with elevated temperatures. But, in deficiently precooled bulbs, the rate of leaf unfolding will not respond readily to the temperature regulation. Wait for about six weeks from the time the buds are visible among the terminal cluster of unfolding leaves until they open. The greenhouse grower should follow his usual growing guidelines after buds become visible. The length of this period is easily determined.

Time is foreseen when greenhouse grower gets bulbs with predictable leaf counts. For this sake, bulb industry will need to make two critical determinations, viz. leaf number in the bulbs at the time of harvest and the number of leaves the plants will initiate between harvest and flower bud initiation in the greenhouse. Easter lilies, unlike tulips and hyacinths, form no flower buds in storage (Blaney and Roberts, 1967).

13.4.1.2 FLOWER REGULATION IN FREESIA (FREESIA HYBRIDA)

Temperature governs the growth as well as flowering time in Freesia plant. Basic annual thermo periodic cycle of warm (30°C)–cool (10°C)–warm (30°C) is required for the induction of flowering. Congenial temperature for the induction and formation of flower in freesia is 12−15°C. Year-round production in freesia has been achieved when night temperature is maintained at 15°C. Flower induction happened after the emergence of 7 or more leaves by decreasing the temperature below 15°C (Singh and Singh, 2006).

13.4.1.3 FLOWER REGULATION IN ALSTROEMERIA (ALSTROEMERIA SPP.)

Temperature of growing medium is the regulator of flowering in Alstroemeria. Flowering can be achieved for longer duration if temperature of growing medium is kept at 16°C regardless of the air temperature. Plant sends flowering shoots at a temperature from 9−13°C up to a continuous 17°C, whereas at 21°C devernalization of rhizomes will occur and plant will produce only vegetative shoots (Talukdar, 2006).

13.4.1.4 FLOWER REGULATION IN TULIPS (TULIPA SPP.)

Like any other bulbous crops, tulips are also forced to flower by cold temperature treatment and there are numerous year-round production schedules used by growers round the world. After harvesting in June–July, bulbs are left uncooled and planted outdoors or under protected conditions for flowering around April for natural period blooming. To achieve early flowering, after harvesting bulbs are stored at moderate (20°C) or high (30−35°C) temperatures for 1−2 weeks under dry conditions followed by low temperatures (2−9°C) treatment either dry or in moist conditions for a certain period. Then after that the bulbs are transferred to a temperature of 17−20°C in a rooting room or directly to the greenhouse. By this technique, first flower can be observed in November. Under higher storage temperature, or delayed duration of transfer from high to low temperatures, flowering will be delayed. To schedule delayed flowering (summer or autumn), bulbs are stored in frozen conditions stored (known as ice-tulips). For ice-tulips, bulbs are stored at moderate temperatures (17–20°C) until November or December. Then they are rooted in trays at 9 or 5°C for three to six weeks (Table 13.1). After that bulbs are kept at −2 to −4°C for freezing. After summer, the trays are kept in the greenhouse or outside and flowering can be achieved in autumn (Jhon and Neelofar, 2006).

13.4.1.5 FLOWER FORCING IN DAFFODILS (NARCISSUS PSEUDONARCISSUS)

For inducing early flowering that is for Christmas, the bulbs should be exposed to 34°C for one week. Bulbs are the held at 13°C till shipping or till precooled. For precooling bulbs are stored at 9°C for 15 weeks and this duration varies with the cultivars. Bulbs are induced to rooting at 9°C in the last phase of precooling treatment. When flowering is to be delayed, temperature in rooting room is decreased to 5°C and later to 1–2°C in the dark. Rooted bulbs are forced in the greenhouse at 13–15°C and 1000–2500 foot candles light intensity (Table 13.2). Temperature above 17°C should be avoided as it tends to produce short stems (Singh and Singh, 2006).

13.4.1.6 FLOWER FORCING IN HYACINTHS (HYCINTHUS ORIENTALIS)

Hyacinths flowering can be obtained for Christmas with specially "prepared" bulbs. It requires warm–cool–warm temperature sequence for growth and flowering. The bulbs require high temperature (25.5°C) for initiation and development of floral parts, low temperature (9–13°C) for mobilization of floral stalk, and warm temperature (17–23°C) for floral stalk elongation. For forcing, only big-sized bulbs (15 cm or above) are used. For getting early flowering, the bulbs need to be exposed to 34°C for 5 days and then at 25.5°C for rapid floral development. When the uppermost floret reached to P_2 stage (inner 3 perianth primordia distinguishable), the bulbs should be exposed to 17°C for 3 weeks. For late forcing, the initial 34°C treatment is not given. In such cases, the bulbs are placed at 20°C for 10 days immediately after harvest and then at 25.5°C. Thereafter, the bulbs are planted in pots at 9–13°C for development of roots and mobilization of flower stalk and leaves. Minimum approximate cold treatment requirements for prepared and regular hyacinths are 10 and 13 weeks, respectively. Rooted bulbs are forced in greenhouses at 23°C for early forcing (Table 13.3). Most growers use however 16–18°C which is more economical (Singh and Singh, 2006).

13.4.1.7 FLOWER REGULATION IN GLADIOLUS (GLADIOLUS GRANDIFLORUS)

In case of gladiolus, after drying and curing of corms, low temperature at 4–5°C is used to break dormancy in corms for a duration of 4–8 weeks. Then treatment under high temperature of 23–30°C for 4–6 weeks is helpful in promoting development of shoot and root during storage and subsequent rapid growth of the daughter plants. Chilling of gladiolus corms either wet or dry at 6 and 10°C took less number of days to induce flowering as compared to low-temperature treatment at 1°C (Imanishi et al., 2000).

13.4.2 LIGHT MANIPULATION

13.4.2.1 PHOTOPERIOD MANIPULATION

Many flower crops are scheduled or regulated for their flowering time through the phenomenon of photoperiodic regulation. Plants can be categorized into different classes knowing how they respond to day length for their flower initiation (Table 13.4).

In photoperiodic regulation, duration of dark period is more imperative than length of light period as short-day plant can be made to flower if grown under short days, but if the long-night period is interrupted with a light flash without any change in day length, flowering will not occur. Such photosensitive plants can be made to flower at any desired time of year either by providing supplementary light to achieve artificial LDs or by keeping the plant in the dark to achieve artificial SDs. Short days can be created by covering the plants with dark covers such as black plastic or tarpaulin cloth to inhibit the entry of light. Dense black polythene sheeting (150 gauge thick) is the most suitable material for complete black outs. Light levels should be <0.5 foot candles in a blackout situation. Blacking out at either end of the day is considered to be the best. In chrysanthemum plants are covered at 5 pm and removed at 9 am. Long days can be obtained by shortening the dark period either by night interruption or by day-extension. Day length can be extended by night interruption or mum lighting which is done by turning on the greenhouse lights for continuous four hours. Day extension lighting can also be used where lights are tuned on at sunset and then turned off after the completion of critical photoperiod. Incandescent or high-pressure sodium lamps (HPS) are used for photoperiodic lighting. But nowadays, cyclic lighting is more popular with the growers and it is done by using timer-regulated incandescent lamps that are set to turn on in between the dark period for a specific cycle.

13.4.2.1.1 Flower Regulation in Chrysanthemum (Denranthema grandiflora)

On the basis of its photoperiodic response, chrysanthemum is a qualitative short-day plant with temperature modifying the photoperiodic response. Flowering initiates when it gets shorter day-length than the critical day-length and for vegetative growth, longer day-length than the critical day-length is required. However, two critical photoperiods are there in chrysanthemum, one for initiation of flowers and the other for flower development. Moreover, critical photoperiod varies depending on the cultivar and temperature. Year-round flowering in chrysanthemum was first suggested by Post in 1947. To schedule crop for year-round production or for single crop, certain principles must be kept in mind to grow high-quality flowers. Essentials for year-round production include planting of rooted vegetative cuttings followed by pinching to induce branching. When single stems or branches of pinched plants attain a given height, they are provided with short days to induce flowering. Up to this point, the plants are kept under long-day conditions. Plants are given short-day conditions (minimum of 12-hour day length) until induced flower buds are developed to a stage no longer affected by day length. On an appropriate stage, the flowers are harvested with sufficient stem length. After the harvesting the plant stubbles are removed and land is prepared for next planting (Datta and Gupta, 2012). All the cultivars are not suitable for year-round schedule as they differ in their short-day requirement and belong to different response groups (Table 13.5).

Under controlled photoperiod different chrysanthemum selections had taken less number of days (116.53) to initiate flowering, whereas in case of natural photoperiod more number of days (192.15) were taken to initiate flowering (Rani, 2010).

13.4.2.1.2 Flower Regulation in Carnation (*Dianthus caryophyllus*)

Modern carnation cultivars belong to the quantitatively long-day plants as per their photoperiodic response, i.e., flower initiation is not inhibited by short days but promoted by long days. Extended lightening durations can produce significant differences for number of days taken to flowering. If provided with long days, plants take 117.43 days to full bloom whereas under normal photoperiod, it takes 140.01 days (Dalal et al., 2006).

13.4.2.1.3 Flower Regulation in Poinsettia (*Poinsettia pulcherrima*)

On the basis of photoperiodic response, poinsettia can be categorized as a short-day plant where critical day length required for flowering is nearly 12 h 40 min. To obtain off-season flowering, during natural long-day conditions, plants can be covered with black cloth at around 17:00 and opened at 8:00 to give plants a 14-hour dark period and thus stimulate flowering. Similarly, to delay flowering during occurrence of natural short days, artificial lighting can be provided to plant continuously or intermittently in the night between 22:00 and 2:00.

13.4.2.2 MANIPULATION OF LIGHT INTENSITY

Inside greenhouse, light intensity can be manipulated by using different light sources *viz.*, incandescent bulbs, metal halide lamps, high-pressure sodium (HPS) lamps which works on with the automatic control. To obtain long-day lighting, incandescent lamps have been commonly used. Nowadays, the light-emitting diodes (LEDs) are also finding their way in plant applications. In greenhouse study of ornamental crops, supplemental lighting with high-intensity discharge (HID) shortened the production time in geranium (*Pelargonium* spp.); time to anthesis in cut rose (Quatchak et al., 1986; Tsujita, 1987). Shade management by erecting shade nets inside the greenhouse helps in manipulating light levels inside the greenhouse. Shade-grown cut flowers result in longer stem length in comparison to grown in full-sun conditions. Flowering can be achieved early or it can be delayed by altering the irradiance, but this response is highly variable in flower crops (Craig and Walker, 1963; Armitage and Wetzstein, 1984). Upon giving 100, 200, or 300 $\mu mol \cdot m^{-2} \cdot s^{-1}$ irradiance, *Achillea millefolium* flowers after 57, 45, and 37 days respectively (Zhang et al., 1996), whereas herbaceous ornamental species flowered early under the ambient day light along with high-pressure sodium lamps emitting 25–50 $\mu mol \cdot m^{-2} \cdot s^{-1}$ light intensity (Erwin and Warner, 2005). HPS lamp also advanced the flowering time in some geranium cultivars in comparison to the plants provided with natural daylight conditions (Armitage and Tsujita, 1979).

13.4.3 HUMIDITY MANIPULATION

There are many techniques to control relative humidity around plants. To increase humidity, misting and fogging are used commonly but can increase the risk of diseases, so if used, should be early in the day so that foliage gets drier before the onset of cooler night-time temperatures. A humidifier can also be used attached to the heating or ventilating system in the greenhouse. Trays containing gravel with a constant moisture level can be kept under pots or containers and with evaporation of moisture around the pebbles, relative humidity in the plants vicinity is increased. Close planting or close grouping of pots in greenhouse is another way to increase humidity. Humidity can be reduced by following several practices like maintaining properly drained floors by regular watering, providing them warmness, reducing moisture content, having properly spaced plant arrangement, keeping greenhouse dry overnight, weed management, using bottom heats, wetting agents, incorporation of proper ventilation and heating in combination to allow air exchange and for bringing outdoor temperature to optimum temperature respectively, fans for cooling and after it heater for bringing temperature up during sunrise and sunset three times in an hour. Insect nets depending on mesh size also increase humidity by altering the air flow in the house (Rane, 1989; Bartok, 1990; Ling, 2002). Based on air cooling and water movement evaporative cooling systems can be: Fogging (misting), fan-pad evaporative cooling, and roof evaporative cooling. Water spray producing fog particles prevents leaf dehydration and maintains foliar water profile (Ishii et al., 2016). It is found that using fogging in rose greenhouse 40–45% of fogging water is efficient in cooling rose plants and also leads to less stressed plants (Katsoulas et al., 2001). Humidity control by high-pressure fogging systems leads to increased production of roses during summers, especially when water stress on plants is more (Darlington et al., 1992). Fan-Pad Evaporative Cooling system, comprising of evaporative cooling pad at one side and exhaust fan or blower at the other side, is used to raise the relative humidity levels. Roof Evaporative Cooling uses thin water film at the roof to reduce air temperature inside house. Passive cooling is obtained through shading and reflection. Due to high solar irradiation levels various regions require shading by cloths, blinds, or reflective materials to lower the coming radiation (Castellano et al., 2014). Photovoltaic covered greenhouse has shown good results in growing various ornamental plants (Ganguly et al., 2010).

13.4.4 CO_2 MANIPULATION

CO_2 can be enriched in greenhouse using several methods like by using different polyethylene tubes below the crop canopy in flower and vegetable production greenhouse which contains small holes and allows uniform CO_2 distribution at greenhouse. Burners or generators can also be used for CO_2 enrichment, in its burning of natural gas and propane it leads to production of heat, moisture, and CO_2. Burners are distributed above the crops throughout the house. A single burner is estimated to cover the area of 4,800 sp. feet. The CO_2 enrichment is mainly required during day when its level is low as most plants undergo photosynthesis during day. It is recommended to do enrichment after one hour of sunrise and before three hours of sunset (Lopez et al., 2018). CO_2 can be introduced using dry ice containers in greenhouse and by using pure carbon dioxide from pressurized container (Yong et al., 1999). For achieving better supplementation of $CO_{2,}$ automated monitoring and control is used through CO_2 meter placed close to crops.

13.5 USE OF CLIMATE SENSORS AND ARTIFICIAL INTELLIGENCE

In the regulation of environmental inputs inside the greenhouse, manual control becomes laborious and lacks precision. To facilitate these operations, use of sensor technology and artificial intelligence has immense to contribute. In smart greenhouses, different sensors that sense soil parameters, temperature, and humidity along with actuators include shade screens and LED lights are connected and linked to the automation software which controls the whole climate inside the greenhouse. Modern flower growers prefer to integrate the automated climate control system using real-time sensors which will help flower crops to grow round the year. An automated climate control system automates the greenhouse to attain crop-specific desired temperature along with monitoring humidity, shading, fogging, and others. In a fully automated greenhouse system that employs wireless communication, real-time control and remote monitoring makes the farming completely effortless as everything can be operated with smartphones sitting at any remote location. Whole process is accomplished by the network of real-time sensors with wireless communication in the greenhouse through mesh WiFi. Temperature sensors (thermostats/thermocouples) are used to measure air, substrate, and plant temperature. Humidity sensors are the most expensive sensors and are difficult to maintain which makes its use less practical in greenhouse industry. Basically, three types of humidity sensors are used, *viz.*, capacitive, resistive, and wet/dry bulb. Pyrometers and PAR sensors are used as light sensors and for measuring $CO_{2,}$ an infrared gas analyzer (IRGA) is used as a sensor. Computer-controlled sensors and controllers monitor and maintain maximum and minimum set points for CO_2 levels. Therefore, the controller can inject CO_2 from compressed or liquid sources, or be used to switch CO_2 burners on or off. Additionally, CO_2 levels can be integrated with other environmental cues like light intensity, temperature, and humidity. All sensors are periodically calibrated to ensure accuracy. Nowadays, Wireless Sensor Network (WSN) has become important to control the greenhouse climate. In this system, control actions will be calculated and executed as per system requirements, for example, when foggers must be switched on or when ventilation must be turned off as per changes in climatic conditions outside the greenhouse (Pawlowski et al., 2009). Different control algorithms have been designed which are based on linear models for each operating point controlling each climatic factor inside the greenhouse. Such greenhouses equipped with artificial intelligence of software applications performing on their own based upon inputs received from sensors and other control system have made the life of growers easier as they do not have to devote their entire time on monitoring their crop and management, but now they can just sit and regulate each and everything while sitting at a comfortable location. Several greenhouses professionals and researchers have developed different sensor technologies and software solutions to automate greenhouse control which can augment the modern growing techniques with more accuracy and control.

13.6 CONCLUSION AND FUTURE PROSPECTS

Ornamental crops have opened up more avenues for the farmers to enhance their returns from per unit area. Quality production of ornamental crops inside greenhouse is garnering much attention as with manipulation of climatic factors inside the greenhouse; crops can be made to grow round the year. With flower regulation growers can schedule their crop for bloom during specific dates which can fetch better prices and better management of the produce. Although several advances have already been made in this field through research but still in developing countries like India, the growers are still using the old traditional practices to grow flowers which lack international quality standards and could not export their produce. So, to reap the benefits of flower regulation and increase the export potential of flower crops, proper technology dissemination and infrastructure provision at subsidized rates should be made feasible for the growers. Apart from that research work is also needed in the field of flower regulation with automated systems, i.e., with the application of sensors and artificial intelligence to make precise schedule for regulation. Regulation schedule, including ideal varieties, time, specific dates for scheduling, needs to be standardized for all the potential flower crops. In a country like India where agriculture is the prime profession but farmer's income is still low, flower crops can be a way to earn more per unit area and with the precise knowledge of flower regulation farmers can schedule its production as per market demand.

TABLE 13.1 Different Varieties of Tulip (*Tulipa spp.*) with Respect to Their Forcing Requirements

Suitable varieties for forcing	Earliest date to bring indoors
Apricot Beauty, Kansas, Bellona, Christmas Gold, Christmas Marvel, First Lady, Paul Richter	Early January
Merry Widow, Attilla, Orange Wonder	Mid-January
Garden Party, Gudoshnik, Peach Blossom, Jewel of Spring, Blizzaard, Apeldoorn, Make Up	Early February

Source: Adapted from Trinklein (2012).

TABLE 13.2 Different Forcing Time for Daffodil (*Narcissus Pseudonarcissus*) Along with Suitable Varieties

	Forcing time	
January to February	**January to April**	**March to April**
Suitable varieties		
February Gold, Flower Record, Carlton	Fortune, Ice Follies, Las Vegas, Mt. Hood, Dutch Master, Cassata, Barrett Browning	Accent, Bridal Crown

Source: Adapted from Trinklein (2012).

TABLE 13.3 Forcing Specifications for Hyacinth (*Hycinthus Orientalis*) with Respect to Suitable Varieties

Forcing seasons		
Early	**Mid**	**Late**
(Can be potted till 1st October and need to bring indoors in late December)	(Can be potted till 1st October and need to bring indoors in early January)	(Can be potted till 15th October and need to bring indoors in mid-February)
Suitable varieties		
Anne Marie, Delft Blue, Jan Bos, L'Innocence	Carnegie, Myosotis, Ostara, Pink Pearl	City of Harlem, King of the Blues, Lady Derby

Source: Adapted from Trinklein (2012).

TABLE 13.4 Photoperiodic Classification of Different Ornamental Plants

Category		**Plant species**	**Photoperiod requirement**
Short day	Obligate	Chrysanthmum, Poinsettia, *Tagetes erecta* (African marigold)	Daylight is less than a critical length 8 hrs light and 16 hrs dark (i.e. light period less than critical day length)
	Facultative	Celosia, Cosmos, Zinnia, *Tagetes tenuifolia (Signet marigold)*	
Long day	Obligate	Tuberous begonia, Rudbeckia, Sweet pea, Gazania, China aster, *Fuchsia x hybrid, Mimulus, Oenothera*	16 hrs light and 8 hrs dark (when light period is more than critical day length)
	Facultative	Pansy, Statice, Calendula, Dianthus, Tithonia, snapdragon	
Day neutral		Rose, Balsam, Gomphrena, Cleome, Stock, *Dianthus barbatus, Tagetes patula (French marigold)*	Blooms in all photoperiods ranging from 5 hrs to 24 hrs continuous exposure

Source: Adapted from Sharma et al. (2020) and Warner (2006).

TABLE 13.5 Different Response Groups and Their Specifications in Chrysanthemum (*Denranthema Grandiflora*) to be Referred for Scheduling or Forcing

Response group (as per no. of short days required for initiation of flowers)	**Natural blooming time with respect to planting and pinching dates**
7 week	22 September–9 October (planting on June 25 and pinching on July 9)
8 week	12 October–21 October (planting on July 2 and pinching on July 16)
9 week	22 October–1 November (planting on July 9 and pinching on July 23)

TABLE 13.5 *(Continued)*

Response group (as per no. of short days required for initiation of flowers)	Natural blooming time with respect to planting and pinching dates
10 week	2 November–11 November (planting on July 16 and pinching on July 30)
11 week	12 November–21 November (planting on July 23 and pinching on August 6)
12 week	22 November–4 December (planting on July 30 and pinching on August 13)
13 week	5 December–11 December (planting on August 6 and pinching on August 20)
14 week	12 December–26 December (planting on August 13 and pinching on August 27)
15 week	27 December–5 January (planting on August 20 and pinching on September 3)

Source: Adapted from Sharma et al. (2020).

KEYWORDS

- **flower regulation**
- **temperature**
- **light**
- **humidity**
- **carbon dioxide**
- **sensors**

REFERENCES

Armitage, A. M.; Tsujita, M. J. The Effect of Supplemental Light Source, Illumination and Quantum Flux Density on the Flowering of Seed-propagated Geraniums. *J. Hortic. Sci.* **1979,** *54* (3), 195–198.

Armitage, A. M.; Wetzstein, H. Y. Influence of Light Intensity on Flower Initiation and Differentiation in Hybrid Geranium. *Hort. Sci.* **1984,** *19* (1), 114–116.

Beeson, R. C.; Graham, M. E. D. CO2 Enrichment of Greenhouse Roses Affects Neither Rubisco Nor Carbonic Anhydrase Activities. *J. Am. Soc. Hortic. Sci.* **1991,** *116* (6), 1040–1045.

Blaney, L. T.; Roberts, A. N. Flower Count Compared to Total Bulb Weight. *Oregon Orn. Nursery Dig.* **1965,** *9*, 1–2.

Castellano, S. Photovoltaic Greenhouses: Evaluation of Shading Effect and its Influence on Agricultural Performances. *J. Agric. Eng.* **2014,** *45* (4), 168–175.

Chailakhyan, M. K. New Facts Supporting the Hormonal Theory of Plant Development. *Dokl. Akad. Nauk SSSR*, **1936,** 4, 77–81.

Craig, R.; Walker, D. E. The Flowering of Pelargonium Hortorum Bailey Seedlings as Affected by Cumulative Solar Energy. *Proc. Am. Soc Hort Sci.*, **1963,** 83, 772–6.

Dalal, S. R.; Nandre, D. R.; Bharad, S. G.; Utgikar, S.; Shinde, R. D. Effect of Pinching on Carnation cv. Yellow Solar under Polyhouse Condition. *Int. J. Agri. Sci.* **2006,** *2* (2), 356–357.

Datta, S. K.; Gupta, V. N. Year Round Cultivation of Garden Chrysanthemum (Chrysanthemum Morifolium Ramat.) Through Photoperiodic Response. *Sci. Cult.* **2012,** *78* (1/2), 71–77.

Dhiman, S. R.; Laishram N. In *Flower Regulation*, Proceedings of National Symposium on Lifestyle Floriculture: Challenges and Opportunities, Solan, India, Mar 19–21, 2010; pp 20–24.

Endo, M.; Ikushima, I. Effects of CO2 Enrichment on Yields and Preservability of Cut Flowers in Phalaenopsis. *J. Jpn. Soc. Hortic. Sci.* **1997,** *66* (1), 169–174.

Erwin, J. E.; Warner, R. M. In *Determination of Photoperiodic Response Group and Effect of Supplemental Irradiance on Flowering of Several Bedding Plant Species*. IV International ISHS Symposium on Artificial Lighting 580, Nov 2000; pp 95–99.

Ganguly, A.; Misra, D.; Ghosh, S. Modeling and Analysis of Solar Photovoltaic-Electrolyzer-fuel Cell Hybrid Power System Integrated with a Floriculture Greenhouse. *Energy Build.* **2010,** *42* (11), 2036–2043.

Gisleröd, H. R.; Selmer-Olsen, A. R.; Mortensen, L. M. The Effect of Air Humidity on Nutrient Uptake of Some Greenhouse Plants. *Plant Soil* **1987,** *102* (2), 193–196.

Gupta Y. C.; Verma V. K.; Sharma P. Greenhouse Enviornment. Mannual on Greenhouse Cultivation of Flower Crops. 2007; pp 5–7. https://www.greenhousegrower.com/production/monitoring-carbon-dioxide-in-the-greenhouse/

Hwang, S. J.; Jeong, B. R. Growth of Phalaenopsis Plants as affected by Nutrient Solution Composition. *Korean J. Hort. Sci. Technol.* **2009,** *27* (7), 73–80.

Imanishi, H.; Imae, Y.; Kaneko, E.; Sonoda, S. In *Effect of Temperature and Daylength on Flowering of Early Flowering Gladiolus*. VIII International Symposium on Flower bulbs 570, Aug 2000; pp 437–446.

Ishii, M.; Sase, S.; Moriyama, H.; Okushima, L.; Ikeguchi, A.; Hayashi, M.; Giacomelli, G. A. Controlled Environment Agriculture for Effective Plant Production Systems in a Semiarid Greenhouse. *Jpn. Agric. Res. Q.* **2016,** *50* (2), 101–113.

Jhon, A. Q.; Neelofar. Tulip. In *Bulbous Ornamental and Aquatic Plants*, Pointer Publishers, Jaipur, India, 2006; pp 133–142.

Katsoulas, N.; Baille, A.; Kittas, C. Effect of Misting on Transpiration and Conductances of a Greenhouse Rose Canopy. *Agric. For. Meteorol.* **2001,** *106* (3), 233–247.

Kesumawati, E. In *The Effect of Inclination Bench and Intermittent Mist on the Growth and Production of Anthurium Plant (*Anthurium Andraeanum*)*. Proceedings of The Annual International Conference, Syiah Kuala University-Life Sciences & Engineering Chapter, Vol. 3, No. 1, Nov 2013; pp 80–86.

Kromwijk, J. A. M.; Meinen, E.; Dueck, T. A. In *The Effect of Elevated CO2 on the Vegetative and Generative Growth of Phalaenopsis*. International Symposium on Orchids and Ornamental Plants 1025, Jan, 2012; pp 155–161.

Massa, G. D.; Kim, H. H.; Wheeler, R. M.; Mitchell, C. A. Plant Productivity in Response to LED Lighting. *Hort. Sci.* **2008,** *43* (7), 1951–1956.

Milec, Z.; Valárik, M.; Bartoš, J.; Šafář, J. Can a Late Bloomer Become an Early Bird? Tools for Flowering Time Adjustment. *Biotechnol. Adv.* **2014,** *32* (1), 200–214.

Mortensen, L. M.; Gislerod, H. R. Effect of Air Humidity on Growth, Keeping Quality, Water Relations, and Nutrient Content of Cut Roses. *Gartenbauwissenschaft* **2000,** *65* (1), 40–44.

Mortensen, L. M. Effects of Air Humidity on Growth, Flowering, Keeping Quality and Water Relations of Four Short-day Greenhouse Species. *Sci. Hortic.* **2000,** *86* (4), 299–310.

Mortensen, L. M.; Fjeld, T. In *High Air Humidity Reduces the Keeping Quality of Cut Roses*. VI International Symposium on Postharvest Physiology of Ornamental Plants 405, Jun 1995; pp 148–155.

Mortensen, L. M.; Ulsaker, R. Effect of CO2 Concentration and Light Levels on Growth, Flowering and Photosynthesis of Begonia× Hiemalis Fotsch. *Sci. Hortic.* **1985,** *27* (1–2), 133–141.

Mortensen, L. M. CO2 Enrichment in Greenhouses. Crop Responses. *Sci. Hortic.* **1987,** *33* (1–2), 1–25.

Pawlowski, A.; Guzman, J. L.; Rodríguez, F.; Berenguel, M.; Sánchez, J.; Dormido, S. Simulation of Greenhouse Climate Monitoring and Control with Wireless Sensor Network and Event-based Control. *Sensors* **2009,** *9* (1), 232–252.

Quatchak, D. J.; White, J. W.; Holcomb, E. J. Temperature, Supplemental Lighting, and Chloromequat Chloride Effects on Flowering of Geranium Seedlings. *J. Am. Soc. Hortic. Sci.* **1986,** *111* (3), 376–379.

Rani, U. Screening of Newly Developed Selections of Chrysanthemum (*Dendranthema Grandiflora* Tzvelev) for Off-season Flower Production. M.Sc. Thesis. 2010, Dr. Y. S. Parmar University of Horticulture and Forestry, Nauni, Solan.

Reeves, P. H.; Coupland, G. Response of Plant Development to Environment: Control of Flowering by Day Length and Temperature. *Curr. Opin. Plant Biol.* **2000,** *3* (1), 37–42.

Sharma, R.; Thakur, N.; Thaneshwari Kumari, P.; Sahare, H. A. Photoperiod: A Key Pathway for Manipulation of Flowering Time in Commercial Ornamental Crops. *Think India Journal* **2020,** *22* (30), 1529–1543.

Sheldon, C. C.; Burn, J. E.; Perez, P. P.; Metzger, J.; Edwards, J. A.; Peacock, W. J.; Dennis, E. S. The FLF MADS Box Gene: A Repressor of Flowering in Arabidopsis Regulated by Vernalization and Methylation. *Plant Cell* **1999,** *11* (3), 445–458.

Singh, K.; Singh, P. Physiology of Ornamental Bulbous Crops. In *Bulbous Ornamental and Aquatic Plants*, Pointer Publishers, Jaipur, India, 2006; pp. 290–310.

Singh, M. C.; Singh, J. P.; Singh, K. G. Mathematical Modeling of Greenhouse Microclimate under Vertically Trained Soilless Cropped Conditions. *Agric. Res.* **2021a,** *10.* https://doi.org/10.1007/s40003-021-00609-0

Singh, M. C.; Singh, J. P.; Pandey, S. K. Cutting, N. G.; Sharma, P.; Shrivastav, V.; Sharma, P. A Review of Three Commonly used Techniques of Controlling Greenhouse Microclimate. *Int. J. Curr. Microbiol. App. Sci.* **2018a,** *7* (1), 3491–3505.

Singh, M. C.; Singh, J. P.; Singh, K. G. Development of a Microclimate Model for Prediction of Temperatures Inside a Naturally Ventilated Greenhouse under Cucumber Crop in Soilless Media. *Comput. Electron. Agric.* **2018b,** *154*, 227–238.

Singh, M. C.; Singh, K. G.; Singh, J. P. Nutrient and Water Use Efficiency of Cucumbers Grown in Soilless Media under a Naturally Ventilated Greenhouse. *J. Agric. Sci. Technol.* **2019b,** *21* (3), 193–207.

Singh, M. C.; Singh, K. G.; Singh, J. P. Yield of Soilless Cucumbers Planted under Partially Controlled Greenhouse Environment in Relation to Deficit Fertigation. *Ind. J. Hort.* **2018c,** *75* (2), 259–264.

Singh, M. C.; Singh, K. G.; Singh, J. P.; Mahal, A. K. Performance of Soilless Cucumbers in Relation to Differential Fertigation under Naturally Ventilated Greenhouse Conditions. *J. Plant. Nutr.* **2019a,** *42* (11–12), 1316–1332.

Singh, M. C.; Yousuf, A.; Singh, J. P. Greenhouse Microclimate Modeling under Cropped Conditions-a Review. *Res. Environ. Life* **2016,** *9* (12), 1552–1557.

Talukdar M. Alstroemeria. In *Bulbous Ornamental and Aquatic Plants*; Pointer Publishers: Jaipur, India, 2006; pp 133–142.

Thomas, B.; Vince-Prue, D. *Photoperiodism in Plants*, 2nd ed.; Academic Press: New York, NY, USA, 1997; pp 1–26.

Torre, S.; Fjeld, T. Water Loss and Postharvest Characteristics of Cut Roses Grown at High or Moderate Relative Air Humidity. *Sci. Hortic.* **2001,** *89* (3), 217–226.

Trinklein, D. H. Forcing Bulbs for Indoor Bloom. *Lawn and Garden.* **2012**.

Tsujita, M. J. High intensity Supplementary Radiation of Roses. In *Roses*; Langhans, R.W., Ed.; Roses Inc.: Haslett, MI, USA, 1987; pp 171–186

Van Labeke, M. C.; Dambre, P. Effect of Supplementary Lighting and CO2 Enrichment on Yield and Flower Stem Quality of Alstroemeria Cultivars. *Sci. Hortic.* **1998,** *74* (4), 269–278.

Wahl, V.; Ponnu, J.; Schlereth, A.; Arrivault, S.; Langenecker, T.; Franke, A.; Schmid, M. Regulation of Flowering by Trehalose-6-phosphate Signaling in *Arabidopsis Thaliana. Science* **2013,** *339* (6120), 704–707.

Warrner, R. M. Supplemental Lighting on Bedding Plants - Making it Work for You. *OFA Bull. No.* **2006,** 899.

Yong, J. W. H.; Lim, E. Y. C.; Hew, C. S. Can We Use Elevated Carbon Dioxide to Increase Productivity in the Orchid Industry? *Malayan Orchid Rev.* **1999,** *36*, 75–81.

Zeiger, E.; Talbott, L. D.; Frechilla, S.; Srivastava, A.; Zhu, J. The Guard Cell Chloroplast: A Perspective for the Twenty-First Century. *New Phytol.* **2002,** *153* (3), 415–424.

Zhang, D.; Armitage, A. M.; Affolter, J. M.; Dirr, M. A. Environmental Control of Flowering and Growth of Achillea millefolium L. Summer Pastels'. *Hort. Sci.* **1996,** *31* (3), 364–365.

CHAPTER 14

Smart Farming Technologies for Protected Cultivation

VISHNU PRASAD[1], ABRAR YOUSUF[2], HASAN MIRZAKHANINAFCHI[3], ANURAG SATPATHI[4], and AEKESH KUMAR[5]

[1]*Water Technology Center, ICAR-Indian Agricultural Research Institute, New Delhi, India*

[2]*Regional Research Station of Punjab Agricultural University, Ballowal Saunkhri, Balachaur, SBS Nagar, Punjab, India*

[3]*Department of Farm Machinery and Power Engineering, Punjab Agricultural University, Ludhiana, Punjab, India*

[4]*College of Agriculture, G. B. Pant University of Agriculture and Technology, Pantnagar, Uttarakhand, India*

[5]*College of Technology, G. B. Pant University of Agriculture and Technology, Pantnagar, Uttarakhand, India*

ABSTRACT

The food safety has become an increasingly important challenge worldwide, particularly as the world population continues to increase and arable lands are diminishing due to raising problems like urban development, land degradation, soil loss/erosion, water shortages, etc. The smart farming is indeed a new idea that refers to managing farms with the use of cutting-edge technologies such as sensors-based systems, Internet of Things (IoT), robotics, drones, artificial intelligence (AI), cloud computing, and big data to enhance the quality and quantity of products while optimizing the human labor needed for production. The disciplines and expertise now required

Protected Cultivation: Structural Design, Crop Management, Modeling and Automation.
Mahesh Chand Singh and K. K. Sharma (Eds.)

for agricultural sector contain Global Positioning System (GPS), robotics/ cybernetics, image processing (computer-based), climate/micro-climate prediction etc. The protected cultivation environment is a relatively easy system for automated machinery due to its structured nature. The development of systems is easier under the protected cultivation environment. The automation systems for protected cultivation deal with micro-climate control, seedling production, spraying, harvesting, irrigation etc. Here, some of the currently used smart farming technologies have been discussed highlighting their necessity, applications, and the automation under protected cultivation.

14.1 INTRODUCTION

Agriculture and horticulture, along with hunting, fishing, and gathering, have been widely practiced since rudimentary times and were the economic backbones of all primordial civilizations. Farming, as a long-term industry, is undergoing rapid and deep changes due to a variety of socioeconomic variables and technological developments. The continuous rise in global population causes an increase in food demand, which leads to higher farm produce prices, changing trade regulations, supply chains, and reshaping conventional food delivery systems. Despite that global population is increasing continuously, the number of farm employees is steadily declining, signaling, for decades, there has been a manpower deficit in the sector. Furthermore, a widespread desire for organic and sustainably produced goods, particularly in western countries, means that more attention is devoted to growing and harvesting, adding new dimensions to the modern farming concept. Modern farming eventually finds itself on the verge of a conundrum. Growers want to cultivate more high-quality crops in large quantities, but the quantity of available hands to do so is constantly reducing. The influx of labor into the farming is declining, so the only option left is to use more efficient technologies for various farming operations.

The food safety has become an increasingly important challenge worldwide, particularly as the world population continues to increase and farming lands are diminishing due to raising problems like urbanization, desertification, soil loss, water scarcity, etc. As the world population continues to grow, the rising demand for crop production and food is significant. According to a study with an exponential rise in the world's population from 1 billion in 1820 to 7 billion in 2012 and a foreseen 10 billion in 2056, the requisition for food, energy, and natural resources (water, air, soil) is expected to rise significantly. The present world population is estimated by the United Nations to be 7.7 billion and it is expected to hit 8.6 and 9.8 billion in 2030 and 2050, respectively. It is projected that by 2050, about 42% more land for farming and 120% more water will be required to feed the whole world population (Godfray et al., 2010; Roser et al., 2020). According to a latest information by the United Nations's Intergovernmental Panel on Climate Change (IPCC), there are numerous effects of climate change on farming. Extreme weather occurrences, such as heavy rain, more violent storms, and heat waves, are becoming more common, all of which have the potential to reduce agricultural productivity. Heavy precipitation can cause flooding and soil water logging, while water shortages in arid areas of the world may become more acute. Climate change may result in more frequent environmental effects, such as changes in seasonal occurrences in plant and animal life cycles.

To combat with issues related to climate change, the United Nations' Food and Agriculture Organization (FAO) recommended that farming sectors be provided with novel tools and technologies, primarily digital technologies. Precision farming strives to maximize yield per unit of farmland by employing cutting-edge technology in a long-term sustainable manner to obtain the quantitative and qualitative improvements in production, and profits. It makes use of a variety of technologies to enhance crop productivity, including sensors GPS and big-data. Information and communication technologies (ICT)-based decision support system (DSS), supported up by means of real-time data, can provide information about all elements of agriculture. This promotes improved decision-making, resulting in less waste and optimal operational efficacy.

The smart farming technologies are based on the integration of ICT into farm machinery, equipment, and sensors, which allows for the generation of a significant volume of data and information with the gradual introduction of automation into the process. It depends on data transmission and the concentration of data in remote storage systems to assist the combination and analysis of various farm data for decision-making. It is a new descriptor that refers to managing farms using technology such as sensing devices, IoT, robotic systems, unmanned aerial vehicles, and AI to increase the quantity and quality of products generated while minimizing the costs of human work required. The disciplines and expertise now required for agricultural sector contain robotics/cybernetics, Global Positioning System (GPS), image processing, climate/micro-climate prediction, ecological controls, etc. To make the most use of all of these knowledge and technology, farmers and farm managers must be trained in their usage. Figure 14.1 shows the various components of smart farming.

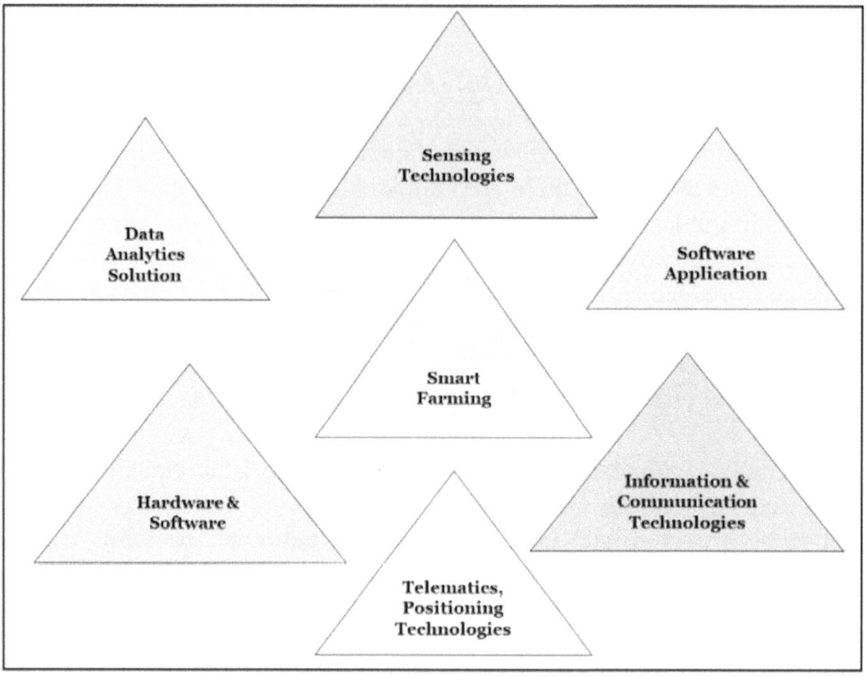

FIGURE 14.1 Components of smart farming.

Automation reigns supreme in today's realm of modern science. The crop production cycles can be streamlined with modern agricultural techniques and technologies, making them more efficient and labor-intensive. The state-of-the-art technologies with the application of automation makes the agriculture innovative. Protected cultivation system is a relatively easy environment for the introduction of automated machinery due to its structured nature. As a result, the automated system must simply cope with the farming product's changeability. Therefore, the development of systems is easier and simpler. Climate control, seedling production, spraying, harvesting, and irrigation etc. can be handled by automated systems under protected cultivation. This chapter gives a synoptic view of available various smart faming technologies, its necessity, and an application under protected cultivation.

14.2 NECESSITY OF SMART FARMING TECHNOLOGIES

There are many constraints to obtain the target of our food demands and feed humankind in near future. Many other problems including climate change, freshwater shortage, and decrease in area under arable land necessitate an alternate solution of it. Some of the reasons that why we need smart farming technologies are:

1. Increasing population
2. Decreasing area under arable lands
3. Increasing demand for fresh water
4. Impact of urbanization on rural labor supply.
5. Limited availability of energy especially fossil fuels.
6. Higher demand of agriculture-related products.
7. For optimum utilization of available natural resources.
8. For real-time monitoring of farming systems.
9. For advancement in farming for sustainable development.
10. To deal with climate change-related issues etc.

14.3 TECHNOLOGICAL EVOLUTION FOR SMART FARMING

Understanding the advancement of technology and its use in the farming sector is essential to comprehend the current state of modern farming. Figure 14.2 shows how the different technologies are used for the betterment of farming under protected cultivation. The present day concern is the application of Artificial Intelligence (AI), Internet of Things (IoT), and Big data in this sector. The cutting-edge technologies such as IoT, AI, Big-data, and cloud computing have great scope in upcoming days to accomplish the task related to farming and food security which will play a great role in the development of humankind upon the earth.

14.4 WIRELESS SENSOR NETWORKS (WSN$_S$) FOR SMART FARMING

The introduction of WSNs prompted a shift in research in the agricultural sector. In farming, the majority of WSNs-based solutions are used for a variety of purposes. It has recently become very popular in a variety of farming purposes. They are used to predict irrigation schedules by monitoring soil moisture and weather variables. It can also be used to predict crop growth, production, and quality. Because WSNs-based applications are accessible, the working efficiency of the present one could be upgraded to monitoring multiple parameters by merely adding more sensor nodes to the same framework (Ojha et al., 2015).

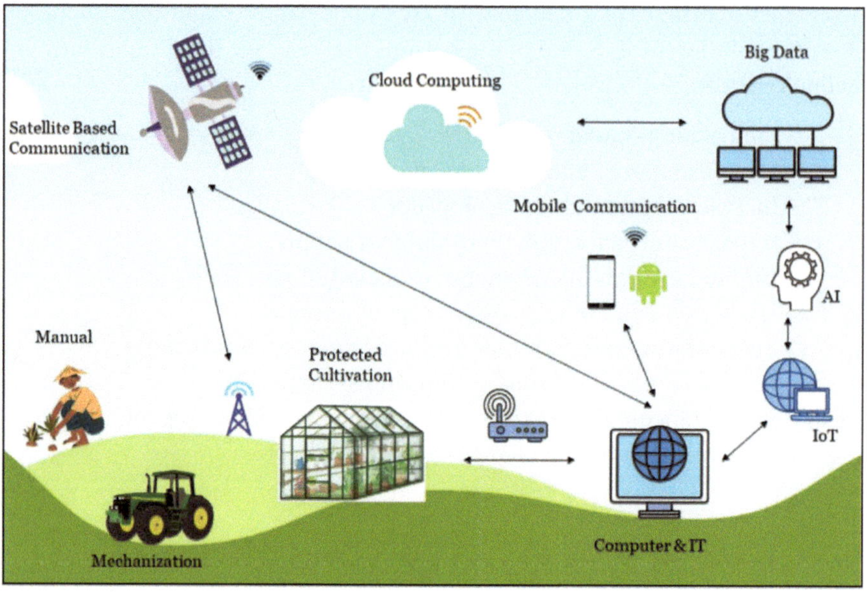

FIGURE 14.2 Evolution of different technologies used for protected cultivation.

WSNs are a network of wirelessly connected battery-powered sensors that are often placed to meet a specific operational use (Akyildiz and Kasimoglu, 2004; Akyildiz et al., 2002a,b). There are two types of WSNs, which are often specifically utilized in farming applications.

1. Terrestrial Wireless Sensor Networks (TWSNs)
2. Underground Wireless Sensor Networks (UWSNs)

14.4.1 TWSNs

The nodes/sensors in TWSNs are placed above the ground level. Figure 14.3(b) displays a basic sensor network for various applications of TWSNs in the field. The micro electromechanical systems (MEMS) technology developments have facilitated the creation of good, smart, small, and low cost sensors for farming applications. Such advanced sensors allow a sensor node to correctly gather information out of its surroundings. The nodes subsequently communicate among themselves to accomplish the application requirements based on the sensed data. For example, let us consider a scenario under precision farming in which WSNs are used to automate the irrigation system throughout the field. Each of these sensors works together to detect the soil moisture to decide time and length of irrigation on that area. The decision is then communicated to a sensor node attached to a water pump through the same network.

14.4.2 UWSNs

In UWSNs, the wireless sensors are buried inside the soil (Vuran and Akyildiz, 2008; Akyildiz and Stuntebeck, 2006). An application of underground sensor networks in farming is shown in Figure 14.3(c). Higher frequencies are severely attenuated in this environment, but lower frequencies can permeate the soil (Yu et al., 2013; Silva and Vuran, 2010). As a result, the network's communication radius is limited, and it requires a larger number of nodes to cover a large area. The use of wired sensors expands the network's reach by requiring a lesser number of sensors. However, the sensors and wiring in this system may be subject to agricultural practices. Unlike the TWSN-based applications, a sensor node is buried in the soil. A gateway node is also installed to send data from the subsurface sensor nodes to a surface sink located above the earth. Following that, the data can be transferred over the internet and stored in remote databases, as well as used to notify a mobile phone user. UWSNs require a larger number of nodes due to the comparably shorter communication expanse.

14.4.3 NETWORK ARCHITECTURE FOR SMART FARMING APPLICATIONS

14.4.3.1 SINGLE-TIER

This architectural style is most common for small-scale farming operations. In this kind of architecture, the data from on-field devices and sensor nodes are sent directly to a sink node close to the area of interest. It is also known as the single clustered architecture.

14.4.3.2 MULTI-TIER

In this type of structural design, several levels are there in the application hierarchy. The fundamental clusters are formed by the on-field sensor nodes, which remain at the bottom of the hierarchy. After that numerous clusters are

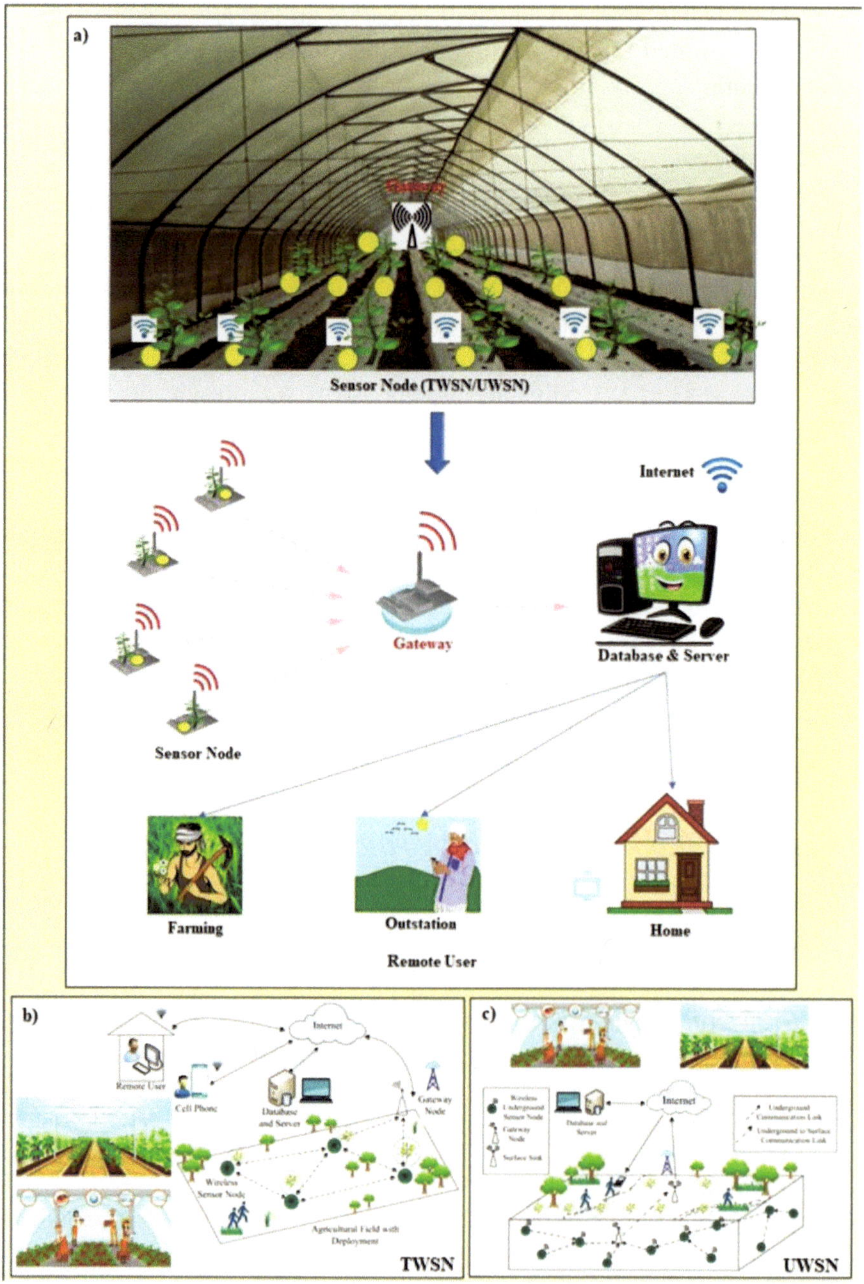

FIGURE 14.3 **a)** A typical multi-tier architecture for farming application under protected cultivation. **b)** Terrestrial wireless sensor network. **c)** Underground wireless sensor network.
Source: Adapted from Ojha et al. (2015) and Kumar and Reddy (2020).

used to reach the gateway nodes at the next tiers of hierarchy. It is typically made up of heterogeneous nodes. Figure 14.3(a) shows a multi-tier architecture for farming applications under protected cultivation with different levels of gateways.

14.4.4 APPLICATIONS OF WSNs IN SMART FARMING

This section discusses the various agricultural/farming applications that can be done using WSNs. It has a wide range of applications and is a very effective technology for improving farming sector. The applications are as follows:

- **System for irrigation-water management:** To optimize water utilization in farming, modern farming demands a higher-quality irrigation management system (Greenwood et al., 2010; Adamala et al., 2014). Another source of inspiration for the need of an innovative approach is the alarming drop in ground water levels. Micro-irrigation systems are cost-effective and water-efficient in this environment (von Westarp et al., 2004; Raina et al., 1998). The WSNs are used as the coordinating technology in this case (Moghaddam et al., 2010; Lorite et al., 2013; Gutiérrez et al., 2013; Lichtenberg et al., 2015).
- **Monitoring of agricultural systems:** Farming now employs a variety of upgraded systems and devices. In this aspect, a better system for managing these devices makes the overall operation easier and allows for automation in the faming process (Kim et al., 2014).
- **Controlling pests and diseases:** Pesticides and fertilizers are used in a controlled manner to improve crop quality while lowering farming costs. Therefore, to keep pesticide consumption under control, we must keep track of the likelihood and presence of pests in crops. We also need climate data including air temperature, air humidity, and wind speed of the surrounding area to anticipate the incidence of pests (Bhave et al., 2013; Matese et al., 2009). A WSN can monitor and forecast these occurrences autonomously throughout the area under consideration (Bhargava et al., 2014).
- **Controlling fertilizer applications:** The usage of fertilizers has a direct impact on plant growth and crop quality. Monitoring variations in soil nutrients such as N, P, K, and pH can help farmers control the use of fertilizers. As a result, soil nutrient balance can be attained, and crop production and quality can be enhanced with limited fertilizer applications.

- **Monitoring of greenhouse gases:** Farming and greenhouse emissions are inextricably linked. Greenhouse gases contribute to rising global temperatures, which has a direct effect on agriculture. Greenhouse gas emissions are produced by a variety of farming sources. Malaver et al. (2015) presented the development of a solar-operated unmanned aerial vehicle (UAV) and WSNs to monitor greenhouse gases (CH_4 and CO_2).
- **Remote controlling and diagnosis:** Remote control and monitoring of farm equipment such as pumps, lights, heaters, and valves in machineries are now possible due to the IoT (Coates et al., 2013; Fukatsu et al., 2011).

14.5 APPLICATION OF SENSORS IN PROTECTED CULTIVATION

Any instrument/device that measures certain chemical or physical aspects of the environment and provides the information through signals (electrical) to be collected and analyzed by the main automation computer for decision-making and operational reasons is referred to as a sensor probe e.g., if the pH of irrigation water is very high or low, a pH sensor will trigger an alarm so as to maintain the appropriate pH level. The microclimatic factors (i.e., temperature (air, plant, growing media/soil), humidity, light/solar radiation, vapor pressure deficit (VPD), etc.) are important characteristics to be monitored under protected cultivation system (Singh et al., 2016; 2017a,b; 2018a,b,c,d; 2021a,b; Shamshiri et al., 2018; Bai et al., 2018; Ma et al., 2019). The presence of mildew illness or condensation can be detected using sensors. Figure 14.4 depicts some of the most often used sensors under protected cultivation. For example, BlueDot BME280+TSL2591 is one of the compact cost-effective digital sensors having exceptional accuracy and provides a customizable solution for micro-climatic parameters. The BME280 is an instrument which has digital humidity, pressure, and temperature sensors that use well-established sensing principles to measure a variety of factors. Its tiny size and low power consumption make it ideal for use in battery-powered sensor nodes inside greenhouses. It also has a quick reply time with a great overall accuracy across an extensive range of temperature. The absolute barometric-pressure sensor offers a high level of accuracy, resolution, and sensitivity. Low noise and good resolution have been enhanced in the integrated temperature sensors. Its result is utilized to compensate the pressure and humidity sensors for temperature and can also be used to estimate the ambient temperature. The soil moisture sensors, such as the 10HS, use capacitance technology to determine the soil dielectric

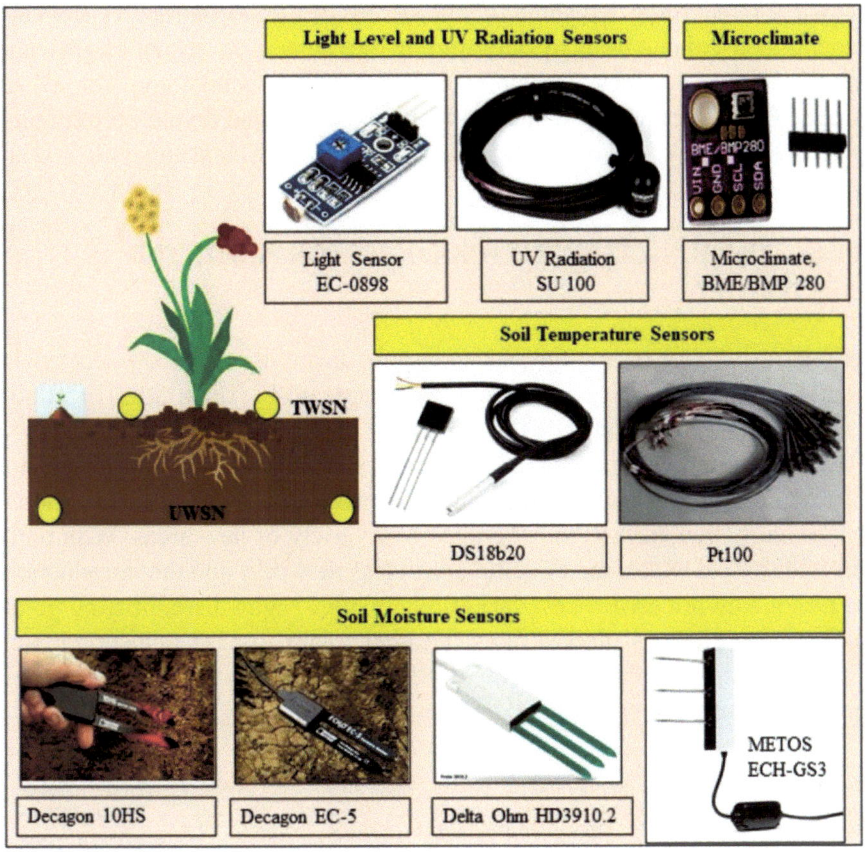

FIGURE 14.4 Sensor probes commonly used for measuring environmental variables under protected cultivation.
Source: Adapted from Shamshiri et al. (2021).

constant to determine the moisture content (volume basis) for technical studies and greenhouse applications. Such sensors are less sensitive to temperature and salt with little power requirements. Light level sensor, also well known as LDR (Light Dependent Resistor), offers great precision, quick reply, and higher resistance. The SQ-110 sensor, which was particularly calibrated to detect solar radiation, produces a voltage that is proportionate to the visible spectrum light intensity, and this is a crucial factor in photosynthesis processes. Other devices such as DS18b20 which is a waterproof, strong, and free from corrosion may be applied for recording soil-based temperatures. Another temperature sensor, Pt-1000, with its working based on the resistance (920 Ω and 1200 Ω) is considered useful in greenhouse applications. One more sensor for recording moisture in soil i.e. ECHO EC-5, computes the soil moisture on volume basis by determining the dielectric constant of growing media using frequency domain/capacitance-based technology. The EC-5 probe 70 MHz frequency reduces effects of salinity and texture, proving its perfectness in any kind of growing substrates/media (Shamshiri et al., 2021).

14.6 COMMUNICATION TECHNOLOGIES FOR PROTECTED CULTIVATION

Wireless (mobile) technology, in its various forms is a field of electronics that is emerging and expanding at a fast speed. Wi-Fi (IEEE 802.11), Bluetooth, Ultra-Wide Band (UWB), WiMax, Zigbee, and other wireless LAN (WLAN) technologies are all evolving and establishing their own opportunities in the marketplace. Consequently, wireless technologies are becoming increasingly popular and may be used in a variety of new areas. With both the widespread expansion of cellular wireless networks and the introduction of reasonably priced devices, various companies began disseminating point information such as market rates, weather, plant disease advisories, etc. Table 14.1 shows some commonly used communication technologies under protected cultivation.

TABLE 14.1 Some Commonly Used Communication Technologies under Protected Cultivation

Parameter	Standard/Protocol	Frequency bands	Transmission range
Wi-Fi	802.11a, b, g, n	2.4 GHz	20–100 m
Bluetooth	802.15.1	2.4 GHz	8–10 m
WiMAX	802.16a,e	2–66 GHz	<50 km
XBee-PRO 802.15.4 EU	802.15.4	2.4 GHz	750 m
XBee-PRO 802.15.4	802.15.4	2.4 GHz	1600 m
LoRaWAN EU	LoRaWAN	868 MHz	>15 km
LoRaWAN US	LoRaWAN	902–928 MHz	>15 km

Source: Adapted from Ojha et al. (2015) and Shamshiri et al. (2021).

14.7 ADVANCE TECHNOLOGIES FOR SMART FARMING

In current times, with the introduction of the novel technological tools/ appliances like IoT, big-data analytics, cloud computing, and Artificial

Intelligence (AI) new applications are envisioned. The brief description of such concepts and their futuristic applications are given here.

14.7.1 THE SENSOR-CLOUD COMPUTING TECHNIQUE

The term "cloud computing" has become a catchword in the world of information technology. The incorporation of cloud computing and IoT is often having great importance. Cloud computing that uses sophisticated storage, processing, and service capabilities with the IoT's ability to collect information creates a true network between people and things. WSNs have numerous obstacles and limitation in terms of processing, storage, memory, portability, safety and availability etc. To address such difficulties, sensor cloud computing has become part of current research. It is the integration of WSNs with cloud computing. Virtualization allows many applications to operate simultaneously on each sensor node. Henceforth, sensor resources are shared to allowing numerous applications to run at the same time. Traditional WSNs need us to examine the positions of all nodes if we want to know the status of a particular node's position, which is a cumbersome process. In this scenario, virtualization aids in the reduction of time. On-field WSNs applications that use cloud computing are referred to as sensor-cloud computing (Alamri et al., 2013; Ojha et al., 2014; Zhu et al., 2014). When using a unified framework, WSNs will have more processing power and storage capacity available to them. Sensor clouds also improve the management of data and control access while enhancing resource utilization. The agricultural domain can be used for a variety of goals, containing the followings:

- Establish a cloud-based storage system for spatial variations in soil and environmental profile over time.
- Application of mobile sensor-cloud services for crop health monitoring and yield prediction.
- Designing a large-scale sensor-cloud-controlled smart irrigation system.
- Making a sensor-cloud-controlled environment control system for off-season vegetable and flower production under protected cultivation etc.

14.7.2 THE INTERNET OF THINGS (IoT)

The IoT combines heterogeneous smart devices or "things" with interoperable communication technologies to extend ubiquitous computing notions (Morais et al., 2008; Al-Fuqaha et al., 2015; Atzori et al., 2010; Gubbi et al., 2013; Ye et al., 2013). It can be defined as the "things" which can identify, communicate, and interact with the neighboring environment. These pillars enable IoT to provide a flexible real-time control mechanism for on-field parameters, e.g., excessive fog, rain, and temperature lead to the formation of fungi in the leaves in hot and humid tropical climates (Shamshiri et al., 2017). Growers can have a prognosis of the situation and take the appropriate actions before an epidemic using IoT-based sensor data fusion combined with mathematical models. The IoT is an innovatory tool that bodes well for the futuristic computing and communication. Its growth is dependent on developing innovative solutions in a variety of sectors ranging from wireless sensor nodes to nanotechnology. First, a simple, inconspicuous, and cost-effective method of item identification is required to connect commonplace things and gadgets to massive databases and networks. The data about things can only be collected and processed after that. The radio-frequency identification allows the operations of such functions. Second, the capability to identify variations in status (physical) of the items utilizing sensor technologies will enhance data collection. By decentralizing data-handling abilities to the system's boundaries, embedded brain in a thing itself can boost the network's power. Finally, improvement in nanotechnology means that ever-smaller objects will be able to communicate and link with one another. All of these advancements will combine to form the IoTs, which will connect the world's items in both a sensory and intellectual manner. The key parts of an IoT-based data collecting and sharing system with several sensor nodes and repeaters for protected cultivation are depicted in Figure 14.5. Prior to the actual production, this framework allows growers to evaluate microclimate factors in relation to designs of different protected structures and claddings. As a result, the IoT could be a viable option for a variety of farming applications. The cost-effective agricultural supply chain management utilizing Radio-frequency identification (RFID) tags is one of the potential IoT-based agricultural applications.

- Animal movement in open pastures/agricultural field may be tracked remotely.
- Farms with automation for counting pests and remote recording.
- Pesticide sprays can be controlled and scheduled remotely at a user-defined rate and time.
- Remote water flow control and leak detection in a large-scale water supply in agriculture.
- Automation under protected cultivation, etc.

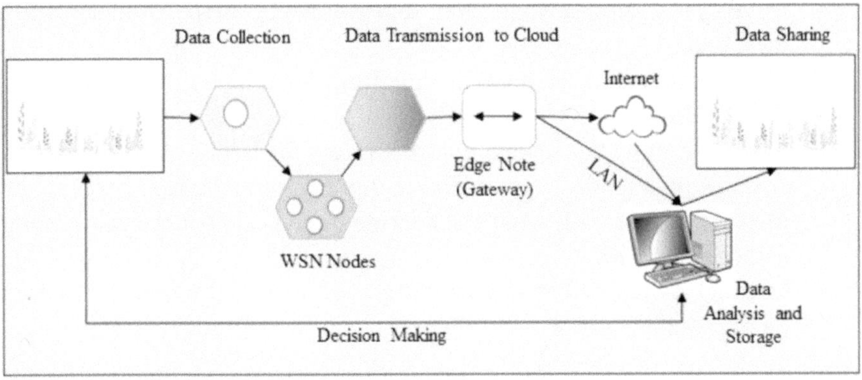

FIGURE 14.5 An IoT-based data acquisition and monitoring system for protected cultivation.
Source: Adopted from Shamshiri et al. (2020).

14.7.3 THE BIG-DATA ANALYTICS

Big data present technological transformation in the farming sector. Big data collection and analysis can help farmers to boost their productivity even while preventing a global food crisis. The best approach is to explore data in various fields such as banking, insurance, and other businesses. Although farming has been reluctant to implement big data analysis in recent years, it is now doing so. Farming operations generate a plethora of data, including economic models, agricultural yield, and crop diseases etc. (Shah et al., 2016). Real-time data on wind, weather, soil, crop maturity, wage costs, and even machinery can help make smart decisions (Kumari et al., 2016). The use of big data in agriculture reduces farmer failures (Arjun, 2016). When the analysis is done on a regular basis, the service will be better (Yadav et al., 2015). Big-data analytics approach can be utilized to extract valuable data from a large dataset of various forms (Kim et al., 2014; Nabrzyski et al., 2014) and can also be used to discover hidden connections, unknown patterns, corporate trends, client preferences, detect crimes, disasters, etc. Handling and regulating large amounts of data with conventional tools and procedures is extremely difficult. As a result, the big-data analytics framework has discovered several tools and approaches for managing data and making judgments. The effectiveness of these strategies is critical since it facilitates decision-making. The approaches currently employed in big data analytics are clustering, classification, and association (Kumari et al., 2016). A list of a few big-data applications for the farming sector is given as under:

- Developing crop-growth and disease-management models using farm data.
- Creating a web-based analytics platform for farmers to deliver better farming info.
- Simple farming equipment control system for large-scale farmlands.
- A decision-support service for increasing crop productivity at the lowest possible cost while considering large-scale contextual farming and climatic data.
- Optimal policy determination for government and enterprises based on data analytics, etc.

14.7.4 THE ARTIFICIAL INTELLIGENCE

Artificial intelligence (AI) is the simulation of human intelligence in machines that are programmed to think and act like humans. In the realm of agriculture, AI is a new technology. Today's agriculture system has been elevated to a new level, thanks to AI-based equipment and technologies. Crop production has improved because of this technology, as has real-time monitoring, harvesting, processing, and marketing (Yang et al., 2007). Automation and control of greenhouse conditions must deal with a variety of variables and disruptions that mathematical formulae cannot fully model (Shamshiri et al., 2017; Romantchik et al., 2017; Shamshiri et al., 2018; Pérez-González et al., 2018; Chen et al., 2020). When AI is used in greenhouse automation, it must work in harmony with all other components of the system, including various sensors, physical systems that control equipment, and actuators. The primary motivation for developing AI-based automation systems in greenhouse environment is to move towards more reliable, predictive, and adaptive control command techniques that minimize production costs and increase output. AI is a computer system that has been programmed to perceive the environment, make judgments, and act intelligently. AI can help with sustainable greenhouse production in a variety of ways, such as lowering climate control system, electrical energy consumption or lowering fertigation system water and chemical needs. AI algorithms, for example, can be used to update micro-climate set points focusing on a predetermined crop growth phases, light situations, and other external factors.

A comprehensive environmental control system should take advantage of not only wireless and IoT-based computerized tools, but likewise cutting-edge cooling and heating techniques based on intelligent algorithms. In this technique, data collected from many wireless sensors installed in various regions of the greenhouse is being used to train machine-learning algorithms based on experience and understanding systems and mathematical crop development models. The process's output command and decision signals are then employed to manage specific factors inside the crop growth micro-environment while also being optimized to reduce energy, chemical, and water needs. AI is:

- Assisting in the analysis of farm data.
- Management of labor shortages.
- Assisting in improving harvest quality and precision.

14.8 AUTOMATION UNDER PROTECTED CULTIVATION

Controlling and automating microclimate and fertigation under protected cultivation can help to improve the viability of confined environment farming by saving water, fertilizer, and energy use as well as maximizing production and profitability (Shamshiri et al., 2018). The direction in modern farming for monitoring systems is to move away from offline methods and move toward the wireless and cloud-based data collecting technology (Shamshiri et al., 2020). The greatest quality greenhouse production with yield prediction is now achievable because of advances in sensor technology. Parallel approaches for automation experts are available through digital technology such as the IoT, which can be designed especially for uses in protected cultivation. The sensors (wireless) and IoT-enabled devices are employed to monitor and regulate the greenhouse environment in real time via secure internet connections on mobile devices (Rezvani et al., 2020). Growers may observe all interior and exterior data and make any necessary changes to the environment in real time using various sensors that communicate data to a central computer that runs knowledge-based automation software. Figure 14.6 illustrates the input and output parameters in a protected structure.

14.8.1 A CASE STUDY

The given study outlines the development and implementation of an automated irrigation system in a greenhouse setup in Mexico, which included WSN (distributed), a gate-way, and a remote server (Gutiérrez et al. 2014). The objective of this study was to use a WSN system that could reduce water consumption. The system has basically three major components that are explained in the following steps:

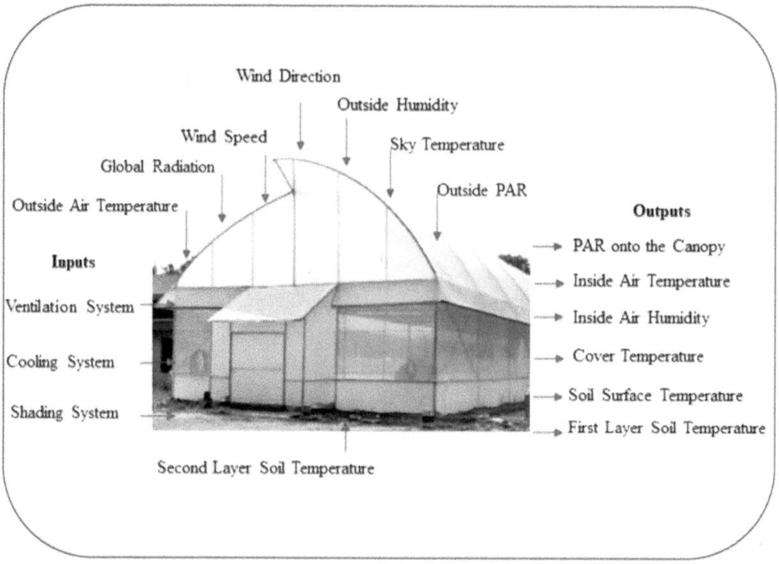

FIGURE 14.6 Inputs and outputs of controlled environment farming.
Source: Adapted from Shamshiri et al. (2018).

- **Wireless Sensor Units** (WSUs): It has soil moisture and temperature sensors buried under the ground for collecting information in various depths. The WSU, deployed on-field basically has four components, i.e., application-specific sensors, processing-unit, radio-transceiver, and battery-power. For saving of energy, there is a mechanism that the micro-controller rests in sleep when not in use. Each of the WSUs has a solar panel connected to recharge their batteries.
- **Wireless Info Unit** (WIU): It collects all the information from the WSUs using the ZigBee and turns as a leading node. It involves services supportive to both ZigBee and GPRS communications. It also has smart decision-making capabilities, such as automatic watering activation when soil moisture and temperature data exceed a predetermined threshold. All of the data about soil moisture and temperature is compared to a predefined threshold value, and the pumps are turned on for an estimated period of time. The received info and irrigation-related data are kept in the associated solid-state memory and then sent GPRS using the Hypertext Transfer Protocol to a distant server (HTTP). The pumping motors are driven via 2 electronic relays (40-A and 12 V DC). The WIU can be instructed to change the irrigation schedule from the distant server and is also fitted out with a key to accomplish irrigation manually.

- **Remote web server**: It shows a specific graphical user interface (GUI), which imagines the information from individual WSU, water intake, and irrigation action types. The web/network applications also enable the users with nonstop program designing/programming facilities of the planned irrigation arrangements/schedules and altering the threshold values on the basis of crop types and growing seasons. The remote server stores all the data.

14.9 CONCLUSION

This chapter provides an outline of the uses of sensors, IoT, AI, and big data for protected cultivation/smart farming, which will be helpful for sustainable production under protected cultivation. Further investigations are required to be undertaken for developing low-cost, low-power WSN solutions for agriculture applications or, for example, for environmental protection, so that they are cheap to farmers, and they may boost crop productivity and quality. As a result, designing, developing, and deploying an affordable wireless sensor network for agriculture applications necessitates a thorough understanding of sensor surveys, sensor node design, wireless technology, sensor node operating systems, routing protocols, data packet format, data aggregation techniques, and deployment techniques. IoT, AI, big data, and cloud computing are some of the most advanced technologies that will undoubtedly enhance agriculture productivity, but they are still in the research phase. In near future such technologies will play key role in accomplishing tasks related to food demand through automation under protected cultivation and among other sectors of farming.

KEYWORDS

- **GPS technology**
- **cloud computing**
- **Artificial intelligence**
- **Internet of Things**
- **protected cultivation**
- **smart farming**

REFERENCES

Adamala, S.; Raghuwanshi, N. S.; Mishra, A. Development of Surface Irrigation Systems Design and Evaluation Software (SIDES). *Comput. Electron. Agric.* **2014,** *100*, 100–109.

Akyildiz, I. F.; Kasimoglu, I. H. Wireless Sensor and Actor Networks: Research Challenges. *Ad Hoc Networks* **2004,** *2* (4), 351–367.

Akyildiz, I. F.; Stuntebeck, E. P. Wireless Underground Sensor Networks: Research Challenges. *Ad Hoc Networks* **2006,** *4* (6), 669–686.

Akyildiz, I. F.; Su, W.; Sankarasubramaniam, Y.; Cayirci, E. A Survey on Sensor Networks. *IEEE Commun. Mag.* **2002a,** *40* (8), 102–114.

Akyildiz, I. F.; SU, W.; Sankarasubramaniam, Y.; Cayirci, E. Wireless Sensor Networks: A Survey. *Comput. Netw.* **2002b,** *38* (4), 393–422.

Alamri, A.; Ansari, W. S.; Hassan, M. M.; Hossain, M. S.; Alelaiwi, A.; Hossain, M. A. A Survey on Sensor-cloud: Architecture, Applications, and Approaches. *Int. J. Distrib. Sens. Netw.* **2013,** *9* (2), 917–923.

Al-Fuqaha, A.; Guizani, M.; Mohammadi, M.; Aledhari, M.; Ayyash, M. Internet of Things: A Survey on Enabling Technologies, Protocols, and Applications. *IEEE Communications Surveys & Tutorials* **2015,** *17* (4), 2347–2376.

Arjun, S.; Joshi, A.; Das, H. P.; Amutha, R. Big Data Analytics for Agricultural Development in India. *Int. J. Eng. Res. (IJERT)*, **2016,** *4*, 22.

Atzori, L.; Iera, A.; Morabito, G. The Internet of Things: A Survey. *Computer Networks* **2010,** *54* (15), 2787–2805.

Bai, X.; Wang, Z.; Sheng, L.; Wang, Z. Reliable Data Fusion of Hierarchical Wireless Sensor Networks with Asynchronous Measurement for Greenhouse Monitoring. *IEEE Trans. Control Syst. Technol.* **2018,** *99*, 1–11.

Bhargava, K.; Kashyap, A.; Gonsalves, T. A. Wireless Sensor Network based Advisory System for Apple Scab Prevention. *IEEE Twentieth National Conference on Communications (NCC)* **2014,** 1–6.

Bhave, A. G.; Mishra, A.; Raghuwanshi, N. S. A Combined Bottom-up and Top-down Approach for Assessment of Climate Change Adaptation Options. *J. Hydrol.* **2014,** *518*, 150–161.

Chen, J.; Ma, Y.; Pang, Z. A Mathematical Model of Global Solar Radiation to Select the Optimal Shape and Orientation of the Greenhouses in Southern China. *Sol. Energy* **2020,** *205*, 380–389.

Coates, R. W.; Delwiche, M. J.; Broad, A.; Holler, M. Wireless Sensor Network with Irrigation Valve Control. *Comput. Electron. Agric.* **2013,** *96*, 13–22.

Fukatsu, T.; Kiura, T.; Hirafuji, M. A Web-based Sensor Network System with Distributed Data Processing Approach Via Web Application. *Comput. Stand. Interfaces* **2011,** *33* (6), 565–573.

Godfray, H. C. J.; Beddington, J. R.; Crute, I. R.; Haddad, L.; Lawrence, D.; Muir, J. F.; Toulmin, C. Food Security: The Challenge of Feeding 9 billion People. *Science* **2010,** *327* (5967), 812–818.

Greenwood, D. J.; Zhang, K.; Hilton, H. W.; Thompson, A. J. Opportunities for Improving Irrigation Efficiency with Quantitative Models, Soil Water Sensors and Wireless Technology. *J. Agric. Sci.* **2010,** *148* (1), 1–16.

Gubbi, J.; Buyya, R.; Marusic, S.; Palaniswami, M. Internet of Things (IoT): A Vision, Architectural Elements, and Future Directions. *Future Gener. Comput Syst.* **2013,** *29* (7), 1645–1660.

Gutiérrez, J.; Villa-Medina, J. F.; Nieto-Garibay, A.; Porta-Gándara, M. Á. Automated Irrigation System using a Wireless Sensor Network and GPRS Module. *IEEE Transactions on Instrumentation and Measurement* **2013,** *63* (1), 166–176.

Kim, J. K.; Tam, S. M. Data Integration by Combining Big Data and Survey Sample Data for Finite Population Inference. *Int. Stat. Rev.* **2020.**

Kim, Y. D.; Yang, Y. M.; Kang, W. S.; Kim, D. K. On the Design of Beacon based Wireless Sensor Network for Agricultural Emergency Monitoring Systems. *Comput. Stand. Interfaces* **2014,** *36* (2), 288–299.

Kumar, P.; Reddy, S. R. N. Wireless Sensor Networks: A Review of Motes, Wireless Technologies, Routing Algorithms and Static Deployment Strategies for Agriculture Applications. *CSI Transactions on ICT* **2020,** *8*, 331–345.

Kumari, S. V.; Bargavi, P.; Subhashini, U. Role of Big Data Analytics in Agriculture. *Int. J. Comput. Sci. Math. Eng.* **2016,** *3*, 3.

Lichtenberg, E.; Majsztrik, J.; Saavoss, M. Grower Demand for Sensor-controlled Irrigation. *Water Resour. Res.* **2015,** *51* (1), 341–358.

Lorite, I. J.; Santos, C.; García-Vila, M.; Carmona, M. A.; Fereres, E. Assessing Irrigation Scheme Water Use and Farmers' Performance using Wireless Telemetry Systems. *Comput. Electron. Agric.* **2013,** *98*, 193–204.

Ma, D.; Carpenter, N.; Maki, H.; Rehman, T. U.; Tuinstra, M. R.; Jin, J. Greenhouse Environment Modeling and Simulation for Microclimate Control. *Comput. Electron. Agric.* **2019,** *162*, 134–142.

Malaver, A.; Motta, N.; Corke, P.; Gonzalez, F. Development and Integration of a Solar Powered Unmanned Aerial Vehicle and a Wireless Sensor Network to Monitor Greenhouse Gases. *Sensors* **2015,** *15* (2), 4072–4096.

Matese, A. D. G. S. F.; Di Gennaro, S. F.; Zaldei, A.; Genesio, L.; Vaccari, F. P. A Wireless Sensor Network for Precision Viticulture: The NAV System. *Comput. Electron. Agric.* **2009,** *69* (1), 51–58.

Moghaddam, M.; Entekhabi, D.; Goykhman, Y.; Li, K.; Liu, M.; Mahajan, A.; Nayyar, A.; Shuman, D.; Teneketzis, D. A Wireless Soil Moisture Smart Sensor Web using Physics-based Optimal Control: Concept and Initial Demonstrations. *IEEE J. Sel. Top. Appl. Earth Observ. Rem. Sens.* **2010,** *3* (4), 522–535.

Morais, R.; Fernandes, M. A.; Matos, S. G.; Serôdio, C.; Ferreira, P. J. S. G.; Reis, M. J. C. S. A ZigBee Multi-powered Wireless Acquisition Device for Remote Sensing Applications in Precision Viticulture. *Comput. Electron. Agric.* **2008,** *62* (2), 94–106.

Nabrzyski, J.; Liu, C.; Vardeman, C.; Gesing, S.; Budhatoki, M. Agriculture Data for All-integrated Tools for Agriculture Data Integration, Analytics, and Sharing. *IEEE International Congress on Big Data* **2014,** 774–775.

Ojha, T.; Bera, S.; Misra, S.; Raghuwanshi, N. S. In *Dynamic Duty Scheduling for Green Sensor-cloud Applications*. IEEE 6th International Conference on Cloud Computing Technology and Science, 2014, 841–846.

Ojha, T.; Bera, S.; Misra, S.; Raghuwanshi, N. S. In *Dynamic Duty Scheduling for Green Sensor-cloud Applications*. Proceedings of IEEE Cloud Com, Singapore, **2014.**

Ojha, T.; Misra, S.; Raghuwanshi, N. S. Wireless Sensor Networks for Agriculture: The State-of-the-Art in Practice and Future Challenges. *Comput. Electron. Agric.* **2015,** *118*, 66–84.

Pérez-González, A.; Begovich-Mendoza, O.; Ruiz-León, J. Modeling of a Greenhouse Prototype using PSO and Differential Evolution Algorithms based on a Real-time LabView™ Application. *Applied Soft Computing*, **2018,** *62*, 86–100.

Raina, J. N.; Thakur, B. C.; Bhandari, A. R. Effect of Drip Irrigation and Plastic Mulch on Yield, Water use Efficiency and Benefit-cost Ratio of Pea Cultivation. *J. Indian Soc. Soil Sci. (India)* **1998**.

Rezvani, S. M. E.; Abyaneh, H. Z.; Shamshiri, R. R.; Balasundram, S. K.; Dworak, V.; Goodarzi, M.; Muhammad, S.; Mahns, B. IoT-Based Sensor Data Fusion for Determining Optimality Degrees of Microclimate Parameters in Commercial Greenhouse Production of Tomato. *Sensors* **2020,** *20* (22), 6474.

Romantchik, E.; Ríos, E.; Sánchez, E.; López, I.; Sánchez, J. R. Determination of Energy to be Supplied by Photovoltaic Systems for Fan-pad Systems in Cooling Process of Greenhouses. *Appl. Therm. Eng.* **2017,** *114*, 1161–1168.

Roser, M.; Ritchie, H.; Ortiz-Ospina, E. *World Population Growth.* https://ourworldindata.org/world-population-growth?source=post_page——d904819ea029
(accessed on 30 November 2020).

Shah, P.; Hiremath, D.; Chaudhary, S. In *Big Data Analytics Architecture for Agro Advisory System.* IEEE 23rd International Conference on High Performance Computing Workshops (HiPCW), 2016, 43–49.

Shamshiri R. R., Measuring Optimality Degrees of Microclimate with Asynchronous Measurement for Greenhouse Monitoring. *IEEE Trans. Control Syst. Technol.* **2018,** *99*, 1–11.

Shamshiri, R. R.; Bojic, I.; van Henten, E.; Balasundram, S. K.; Dworak, V.; Sultan, M.; Weltzien, C. Model-based Evaluation of Greenhouse Microclimate using IoT-Sensor Data Fusion for Energy Efficient Crop Production. *J. Clean. Prod.* **2020,** *263*, 121303.

Shamshiri, R. R.; Hameed, I. A.; Thorp, K. R.; Balasundram, S. K.; Shafian, S.; Fatemieh, M.; Muhammad, S.; Benjamin, M.; Samiei, S. Greenhouse Automation Using Wireless Sensors and IoT Instruments Integrated with Artificial Intelligence. 2021.

Shamshiri, R. R.; Kalantari, F.; Ting, K. C.; Thorp, K. R.; Hameed, I. A.; Weltzien, C.; Shad, Z. M. Advances in Greenhouse Automation and Controlled Environment Agriculture: A Transition to Plant Factories and Urban Agriculture. 2018.

Shamshiri, R. R.; van Beveren, P.; Che Man, H. Dynamic Assessment of Air Temperature for Tomato (Lycopersicon esculentum Mill) Cultivation in a Naturally Ventilated Net-Screen Greenhouse under Tropical Lowlands Climate. *J. Agric. Sci. Technol.* **2017,** *19*, 59–72.

Silva, A. R.; Vuran, M. C. Communication with Aboveground Devices in Wireless Underground Sensor Networks: An Empirical Study. *IEEE International Conference on Communications* **2010,** 1–6.

Singh, M. C.; Singh, J. P.; Singh, K. G. Mathematical Modeling of Greenhouse Microclimate under Vertically Trained Soilless Cropped Conditions. *Agric. Res.* **2021a,** *10.* https://doi.org/10.1007/s40003-021-00609-0

Singh, M. C.; Singh, J. P.; Singh, K. G.; Gupta, O. P.; Kumar, G. Computation of Vapour Pressure Deficit and Crop Transpiration Via Development of a Web based Computer Module. *J. Agrometeorol.* **2021b,** *23* (4), 243–248.

Singh, M. C.; Kachwaya, D. C., Kalsi, K. Soilless Cucumber Cultivation under Protective Structures in Relation to Irrigation Coupled Fertigation Management, Economic Viability and Potential Benefits–A Review. *Int. J. Curr. Microbiol. App. Sci.* **2018b,** *7* (03), 2451–2468.

Singh, M. C.; Singh, J. P.; Pandey, S. K. Cutting, N. G.; Sharma, P.; Shrivastav, V.; Sharma, P. A Review of Three Commonly used Techniques of Controlling Greenhouse Microclimate. *Int. J. Curr. Microbiol. App. Sci.* **2018a,** *7* (1), 3491–3505.

Singh, M. C.; Singh, J. P.; Pandey, S. K.; Mahay, D.; Srivastava, V. Factors Affecting the Performance of Greenhouse Cucumber Cultivation-A Review. *Int. J. Curr. Microbiol. App. Sci.* **2017a,** *6* (10), 2304–2323.

Singh, M. C.; Singh, J. P.; Singh, K. G. Development of a Microclimate Model for Prediction of Temperatures Inside a Naturally Ventilated Greenhouse under Cucumber Crop in Soilless Media. *Comput. Electron. Agric.* **2018f,** *154*, 227–238.

Singh, M. C.; Singh, J. P.; Singh, K. G. Development of Mathematical Models for Predicting Vapour Pressure (Deficit) Inside a Greenhouse Independently from Internal and External Climate. *J. Agrometeorol.* **2018e,** *20* (3), 238–241.

Singh, M. C.; Singh, J. P.; Singh, K. G. Optimal Operating Microclimatic Conditions for Drip Fertigated Cucumbers in Soilless Media under a Naturally Ventilated Greenhouse. *Ind. J. Ecol.* **2017b,** *44*, 821–826.

Singh, M. C.; Yousuf, A.; Singh, J. P. Greenhouse Microclimate Modeling under Cropped Conditions-a Review. *Res. Environ. Life* **2016,** *9* (12), 1552–1557.

von Westarp, S.; Chieng, S.; Schreier, H. A Comparison Between Low-cost Drip Irrigation, Conventional Drip Irrigation, and Hand Watering in Nepal. *Agric. Water Manag.* **2004,** *64* (2), 143–160.

Vuran, M. C.; Akyildiz, I. F. In *Cross-layer Packet Size Optimization for Wireless Terrestrial, Underwater, and Underground Sensor Networks*. In IEEE INFOCOM 2008-The 27th Conference on Computer Communications 2008, 226–230.

Yadav, R.; Rathod, J.; Nair, V. Big Data Meets Small Sensors in Precision Agriculture. *Int. J. Comput. Appl.* **2015,** *975*, 88–87.

Yang, H.; Liusheng, W.; Hongli Junmin X. In *Wireless Sensor Networks for Intensive Irrigated Agriculture*. 4th IEEE Consumer Communications and Networking Conference, 2007. CCNC, Las Vegas, Nevad, 2007; pp 197–201.

Ye, K.; Jiang, X.; He, Y.; Li, X.; Yan, H.; Huang, P. vHadoop: A Scalable Hadoop Virtual Cluster Platform for MapReduce-based Parallel Machine Learning with Performance Consideration. *IEEE International Conference on Cluster Computing Workshops* **2012,** 152–160.

Yu, X.; Wu, P.; Han, W.; Zhang, Z. A Survey on Wireless Sensor Network Infrastructure for Agriculture. *Comput. Stand. Interfaces* **2013,** *35* (1), 59–64.

Zhu, C.; Shu, L.; Hara, T.; Wang, L.; Nishio, S.; Yang, L. T. A Survey on Communication and Data Management Issues in Mobile Sensor Networks. *Wirel. Commun. Mob. Comput.* **2014,** *14* (1), 19–36.

CHAPTER 15

Economic Feasibility, Status, and Future Prospects of Protected Cultivation

S. M. VANITHA[1] and RAVI S. C.[2]

[1]ICAR-Indian Institute of Soil and Water Conservation (IISWC), Research Centre, Udhagamandalam, Tamil Nadu

[2]ICAR-Central Institute for Sub-tropical Horticulture (CISH), Lucknow, Uttar Pradesh

ABSTRACT

Growing the high value crops under controlled climatic conditions even in off-season is usually termed as protected farming. Protected cultivation creates employment opportunities, increases quality and quantity of the produce thus leading to increase in income. It provides assured yield and returns from small and marginal land with effective and efficient utilization of resources mainly land, water, and nutrients. A fundamental requisite for adoption of protected cultivation technology (PCT) is that, it should assure more returns which is competitive in comparison with other crops. Many studies on PCT have shown the highest cost of initial investment as one of the major economic drawback for the adoption of this technology, despite its several advantages. There should be compromise between the huge investment and equipment costs in greenhouse with its performance to generate high quality commodities competitively. Several methodologies are been followed for assessing the techno-economic feasibility analysis of PCT, the important and commonly used are cost and returns analysis, measuring increased productivity per unit area, net returns, cash flows and profitability analysis with comparison between open cultivated and protected cultivated

Protected Cultivation: Structural Design, Crop Management, Modeling and Automation.
Mahesh Chand Singh and K. K. Sharma (Eds.)

crops, measurement of both tangible and imperceptible profits added to the produce quality fetching premium market price, amortization of preliminary cost of investment, internal rate of return (IRR), net present value (NPV), benefit cost ratio (BCR), payback period, partial budgeting technique, input/resource use efficiency. Sensitivity of analysis and many socio-economic determinant variables influencing the adoption of PCT and economic risk are discussed in detail. To promote PCT at a larger scale various schemes, programmes, and policies including subsidies should be made available to the needy without incurring much of the transaction costs. Awareness regarding such schemes and programs implemented at state and national level is also important to sensitize the farmers regarding the benefits of PCT and its socio-economic impact.

15.1 INTRODUCTION

Protected cultivation is a technology in which vegetables, flowers, and others high-value crops (Table 15.1) are grown in protected structures under controlled environmental conditions. It is evident from the past studies that PCT has prospective for producing more and increasing the per unit of yield along with increased quality of produce compared to open field cultivation of crops (Singh et al., 2017a, b; 2018a, b). Ever increasing population, climate change, land fragmentation and marginalization, huge demand for nutritive, healthy and quality food have resulted in shifting of farmers toward modern technologies of crop production like protected cultivation. Protected cultivation could be an alternative for maximizing crop productivity and increasing the produce quality of vegetables round the year (Chakraborty and Sethi, 2015; Singh et al., 2019a, b; 2020a, b).

TABLE 15.1 Crops Grown under Protected Structures.

Type of the crop	Name of the crop
Fruits	Strawberry
Vegetables	Tomato, colored capsicum (yellow, red, and green), cucumber, broccoli, red cabbage, radish, leafy vegetables, etc.
Flowers	Chrysanthemum, carnation, gerbera, rose, lilium, orchid, gladiolus, etc.
Seedlings and nurseries	Vegetables, flowers, seedlings of agroforestry plants, nurseries via tissue culture, grafting of fruits plants such as guava, citrus, mango, etc.

There are ample opportunities in the protected cultivation technology (PCT) for young entrepreneurs, attracting more youth toward agriculture through providing livelihood, creating job opportunities, and giving better dividends. At present, India has just 50,000 ha area under protected cultivation while, in China it is 2 million ha. There exists a vast potential to increase the area of protected farming covering various protected structures during forthcoming years. Enhanced productivity by 3 to 5% as compared to cultivate in open fields, in addition high nutrient and water use efficiency can be noticed under protected cultivation (Singh, 2014). Important technological interventions like use of plastic mulch, low tunnels (crop cover), walk-in-tunnels, naturally ventilated polyhouse, net houses, green houses with controlled environment, vertical farming, soil-less farming (hydroponics, aeroponics, and aquaponics), and vegetable grafts are needed to be scaled up for wide adoption.

15.2 ECONOMIC BENEFITS OF PCT

15.2.1 REDUCTION IN COSTS OF PLANT PROTECTION CHEMICALS

The PCT provides protection to crops from abiotic stress, from physical or mechanical damages such as very high/low temperatures, excess/deficit rainfall/water, hot and cold waves, and also from the damages caused by biotic factors like pest and disease incidences, etc. Thus, PCT minimizes the pesticide consumption to cultivate the crops and is reported that greenhouses provide high protection against pests and diseases for high-value crops (FAO, 2018).

15.2.2 EFFICIENT ENERGY UTILIZATION THROUGH GREENHOUSE EFFECT

The short wave radiations which enter into the greenhouse refract from the surfaces and get changed into long wave radiations. The green house traps these long wave radiations not allowing them to escape resulting in increased temperature. This is called as greenhouse effect. Hence inside the greenhouse builds higher temperature than the ambient/outside environment. However, after sunset it starts losing stored heat through conduction, convection, and radiation.

15.2.3 EFFICIENT USE OF RESOURCES

In Protected cultivation, there is intensive use of resources like land, water, energy, agrochemicals including fertilizers and pesticides (Nair and Barche, 2014; Nimbrayan et al., 2018). More than 75% water usage can be saved through micro irrigation in PCT when it is compared with open field conditions. There is also minimum weed infestation. In protected cultivation, resource-use efficiency is improved multifold in comparison to open field for the production of various crops. The farmers of protected cultivation can be benefited by reducing the wastage of resources with better and precise scheduling of irrigation and climate control (Stanghellin and Montero, 2012). Greenhouse leads five and seven times increase in the land and water productivity than open cultivated lands (FAO, 2018).

15.2.4 SCALE EFFICIENCY

Increased productivity per unit area (increase scale of production) results in increased returns to scale. Pests and diseases are found to be responsible to cause approximately one-third of the crop loss under open field cultivation whereas 2–3 fold higher yield can be harvested in greenhouse by protecting crop from considerable damage caused by aforesaid factors. Polyhouse farming helps the farmers to make earnings throughout the year by cultivating several crops and get premium prices for off-season vegetables as an extra income (Kumar et al., 2017). The gross and net returns over total variable cost were greater for tomato production under polyhouse than open field conditions (Kumar et al., 2016b). Similarly, the net returns in capsicum cultivation under polyhouses were higher than open field conditions despite higher cost of cultivation in polyhouses (Kumar et al., 2016a).

15.2.5 PREMIUM/REMUNERATIVE MARKET PRICE

The PCT provides high value, improved quality of horticultural produce, propagation of improved planting materials to increase germination percentage, disease-free planting material, and better hardening of grafted plants. Production of disease-free and genetically improved transplants, year-round production of flower, vegetable, or fruit crops helps in fetching a remunerative/premium price due to early/off-season market demand for the better quality produce of the farmers having less chemical pesticides (Figure 15.1).

FIGURE 15.1 Benefits of PCT.

15.3 ECONOMICS OF CROP PRODUCTION UNDER PROTECTED CULTIVATION

In order to popularize any technology, it should be economically feasible. As stated earlier, protected cultivation is used in nursery for raising seedlings of many fruits and vegetables. Singh et al. (2010) investigated the economics of raising nursery seedlings of cucurbits in plug tray (Table 15.2) and it was found that, in an area of about 500 m^2, 3 lakh quality seedlings were raised in each batch totaling to 12 lakh seedlings in 4 batches fetching a profit of Rs. 1.5 lakh besides generating an employment of 600 man days.

TABLE 15.2 Economics and Employment Generation in Plug-Tray Nursery Raising of Cucurbits.

Size of nursery greenhouse	No. of possible seedling batches	Total number of seedlings (lakh)	Net profit (lakh)	Employment generation
500 m^2	4 (three for winter and one for rainy season)	12.0 (in four batches)	1.5	150 days × 4 persons = 600 man days

Source: Adapted from Singh et al., 2010.

An economic feasibility study on tomato cultivation under protected structures was conducted by Laxmi and Chakraborty (2017). The study revealed that, greenhouse cultivation of tomato fetched highest net returns of Rs. 8,80,007 per ha compared to shade net house (Rs. 5,88,434 per ha) and open field cultivation (Rs. 81,344 per ha). Cultivation of tomato in greenhouse and shade net house resulted in higher productivity of 438 and 346 quintals per ha as against only 179 quintals per ha under open field cultivation. The remuneration time of investment done for tomato cultivation under greenhouse and shade net house was 4 and 2 years, respectively (Table 15.3). Mehta et al. (2020) studied the comparative productivity of different crops under open as well as protected cultivation. It was reported that, there was an increase in productivity under protected cultivation. The productivity increased by 415, 162, and 145% in protected cultivation as against open field cultivation of cucumber, tomato, and capsicum (Table 15.4). This shows the potential of protected cultivation to meet the food demands of the increasing population.

TABLE 15.3 Economic Viability of Cultivating Tomato under Protected Structures.

Particular	Cost of cultivation (Rs./ha)	Yield (q/ha)	Net returns (Rs/ha)	Payback period in years
Greenhouse	4,34,250	438.09	8,80,007.41	3.76
Shade net house	4,50,900	346.44	5,88,434.2	2.28
Open field	4,56,750	179.36	81,344.41	1.48

Source: Adapted from Laxmi and Chakraborty, 2017.

TABLE 15.4 Comparison of Productivity of High-Value Crops under Protected and Open Field Conditions (q/400 m^2).

Crop	Capsicum	Tomato	Cucumber
Protected conditions	21.92	31.17	38.07
Open-field conditions	8.95	11.88	7.39
% difference	144.92	162.37	415.16

Source: Adapted from Mehta et al., 2020.

Protected cultivation technology resulted in enhanced income, employment, socio-economic status, and provided decent standard of living to the resource poor small and marginal farmers. A major effect of protected cultivation had been observed where productivity was increased around three times as compare to open field. The adoption of protected cultivation technology was found to increase average farm and household income by about 25% and 11%, respectively (Table 15.5). The income variability of protected growers was also lower as compared to open-field growers, indicating lower risk and higher income stability for the protected growers.

TABLE 15.5 Difference of Income Generated by Growers from Protected Cultivation and Open Field (Rs./farm).

Particular	Type of income	Mean income (Rs.)	Coefficient of variation (%)
Polyhouse growers	Farm	2,51,005	23.53
	Household	5,88,819	42.07
Open field growers	Farm	2,24,513.9	49.76
	Household	4,38,166.8	50.99

Source: Adapted from Mehta et al., 2020.

15.4 TOOLS FOR TECHNO-ECONOMIC ASSESSMENT OF PCT

15.4.1 DEPRECIATION

An actual reduction in the reasonable value or usefulness of an asset is termed as depreciation. Any fixed assets have depreciation, that is, the value for its wear and tear or use. Any equipment, farm machinery, implements, building, infrastructure, etc. has depreciation over the period of time which can be calculated using the formula given below (eq 15.1):

$$\text{Annual depreciation} = \frac{(\text{Purchase value} - \text{Junk value})}{\text{Economic life of asset}} \tag{15.1}$$

Assets average as indicated by the experts (engineers) is used for the computation of the depreciation. The depreciation cost should be apportioned to the various crops, based on its percentage use. This depreciation cost component has to be included in fixed cost in the profitability analysis of PCT.

15.4.2 AMORTIZED COST OF ESTABLISHMENT

Amortized cost of establishment is the proportion of the total cost of establishment of protected structure (PC) which can be a polyhouse/ greenhouse etc. made in initial years to be included under fixed cost component. Amortization of costs considering an interest rate which should be a representative of the present prevailing inflation rate. The below mentioned formula (eq 15.2) is employed to estimate the amortized establishment cost of polyhouse/any other protected structure.

$$\text{Amortization cost} = \text{Investment X}\left(\frac{(1+i)^{AL}\,X\,i}{(1+i)^{AL\,-1}}\right) \tag{15.2}$$

Where,

I = interest rate and AL = average life of polyhouse.

This final amortized cost of PC structure establishment should be considered in the economic feasibility analysis of PCT.

15.4.3 PARTIAL BUDGETING TECHNIQUE

This method is used for the impact assessment of technologies which can be used in the studies made for thc cconomic impact analysis of adoption of PCT by the farmers. This technique involves the estimation of initial investment incurred in the development of the selected technology in a given time which includes cost of the structure, labor, etc. Further, the probability of performance of PCT against open field cultivation can be assessed by considering changes in the cost and returns structure. The economic worthiness of the PCT over open field cultivation is determined using the following format of analysis:

Debit (A+B)	**Credit (C+D)**
A. Added cost due to the adoption of PCT	C. Reduced cost due to the adoption of PCT
B. Reduced returns due to the adoption of PCT	D. Added returns due to the adoption of PCT
Total debit = (A+B)	Total credit = (C+D)

Economic worthiness of the new technology developed = Credit – Debit = (C+D) – (A+B)

A difference of higher than zero between credit and debit exhibits that the technology or innovation is economically worth.

15.4.4 DISCOUNTED CASH FLOW TECHNIQUES

Net present value (NPV), benefit cost ratio (BCR), and internal rate of returns (IRR) are used to evaluate the economic feasibility of investing in protected cultivation technologies. The benefit cost analysis will be carried out based on the investment on PC structures and additional returns due to PC structures.

NPV: It denotes the difference between the current value of profits and the present value of costs incurred in PC. NPV of an investment depicts the current value of projects cash inflows less its investment. If NPV is greater than zero, then the investment in protected cultivation is feasible. NPV is calculated using the formula given below (eq 15.3):

$$NPV = \sum_{t=0}^{n} B_t(1+d)^{-t} - I \tag{15.3}$$

BC Ratio: Ratio of the present worth of the benefits to the present worth of costs. The decision criterion is to accept if the value is more than or equal to one.

The benefit cost ratio is stated mathematically as (eq 15.4):

$$BCR = \frac{\sum_{t=o}^{n} \frac{B_t}{(1+d)^t}}{\mathrm{I}} \tag{15.4}$$

IRR: It is the discount rate that NPV of an investment to zero and benefit cost ratio to one. The formal selection criterion is to accept all independent projects having IRR equal to or higher than opportunity cost.

Mathematically, IRR can be represented as given below (eq 15.5):

$$IRR: \sum_{t=1}^{n} B_t(1+r)^{-t} - I = 0 \tag{15.5}$$

Where,

r = Internal rate of return
B_t = Incremental net cash flow from protected cultivation in 't' year
n = Economic life of PC structures
d = Discount rate
I = Initial investment
t = 0,1,2,...n years

Economic feasibility study of flower and vegetable cultivation in polyhouse by using the discounted cash flow measures was conducted by

Bhoopesh et al. (2017) and the results indicated the economic feasibility of carnation cultivation was with BC ratio of 1.43, net present value of Rs. 20,02,285 and 73% IRR. Similarly, the study also indicated the economic feasibility in cultivation of carnation with capsicum and carnation with other flower crops under polyhouse with 31 and 38% of IRR which is much higher than the discount rate. Study also indicated the payback period ranging between 3 and 5 years.

Sharma et al. (2014) studied on the economic feasibility of major flower crops grown in the state of Himachal Pradesh, India under naturally ventilated low cost polyhouses where they found that the BC ratios for chrysanthemum, carnations, roses, gerbera, and lilium, were 2.39, 2.37, 2.89, 2.01, and 1.89, respectively. The economic feasibility of flower and vegetable cultivation in polyhouse constructed @ Rs.1000/m^2 is demonstrated in Table 15.6.

TABLE 15.6 Economic Viability of Vegetables and Flowers Cultivation under a Polyhouse Constructed @ Rs.1000/m^2.

Crop (s)	Total cost (Rs.)	Gross returns (Rs.)	BC ratio	NPV	Pay-back period (year)	IRR (%)
Carnation	59,49,470	89,98,267	1.43	20,02,285	3	73
Carnation and capsicum	59,99,488	73,91,500	1.21	9,68,780	5	31
Carnation and other flowers	63,82,479	82,81,600	1.27	13,79,627	4	38

Source: Adapted from Bhoopesh et al., 2017.

15.5 INDIAN GOVERNMENT SCHEMES FOR THE PROMOTION OF PROTECTED CULTIVATION

National Horticulture Mission (NHM), National Horticulture Board (NHB), Rashtriya Krishi Vikas Yojana (RKVY), and Mission for Integrated Development of Horticulture (MIDH) help the farmers, by providing financial support to them, creating awareness and promoting the protected technology of farming to cultivate high-value vegetable and cut flowers.

Establishment of the NHB by the Indian Government was done in 1984 as an independent organization and registered as a society with the Registrar of Firms and Societies, Chandigarh. It has its headquarters currently in Gurugram and 29 field offices located throughout the country. The board's broad goals and objectives are to establish production hubs for commercial horticulture development, post-harvest infrastructure, cold chain facilities, and growers' associations. The NHB provides credit linked back ended subsidy @ 50% of the whole cost of project which is restricted to Rs. 5.6 million for one project for protected cultivation with more than 2500 m^2. Table 15.7 has the year wise allocation of funds under NHB which exhibited that the fund allocation is increased from Rs. 12,247 lakhs to Rs. 27,725 lakhs in the year 2008–2009 to 2016–2017. The allotted funds under the NHB program increased annually at a compounded growth rate of 11.63% during 2008–2009 to 2016–2017. Likewise, the government's share of funds used decreased at a rate of −1.37% per year. The findings revealed the government's low level of fund utilization, which ranged from 85 to 100%. Efforts must be made by the government to use the maximum amount of available funds.

TABLE 15.7 Allocation of Funds for Polyhouse Construction Under NHB Programme in India (Rupees in. Lakhs).

Particular	2008–2009	2012–2013	2016–2017	*CAGR
Funds allocated	12,247	14,066	27,725	11.63
Funds available	12,435.25	15,008.34	30,216.44	12.46
Amount	12,435.18	13,412.74	25,956.42	10.92
% funds utilized	100	89.37	85.9	−1.37

**CAGR, compound annual growth rate.*

Source: Adapted from Prakash et al., 2019.

15.5.1 MISSION FOR INTEGRATED DEVELOPMENT OF HORTICULTURE (MIDH)

Special emphasis is being placed on the production of hybrid vegetables under MIDH in protected cultivation. Indian Government launched the MIDH in 2014–2015 as a centrally sponsored scheme for promoting holistic progress in horticulture sector of the country through various schemes such as NHM, NHB, National Bamboo Mission (NBM), Horticulture Mission for North East & Himalayan States (HMNEH), Central Institute for Horticulture (CIH), and Coconut Development Board (CDB). Until recently, these schemes functioned independently, with no inter-scheme coordination. However, as of 2014, all these schemes have been consolidated under the Mission for Integrated Development of Horticulture (MIDH) scheme as subsystems.

15.5.2 NATIONAL HORTICULTURAL MISSION (NHM)

It is the main program through which funds were allocated in the form of subsidies for the development of the country's horticulture sector. The NHM provided a 50% subsidy for the construction of protected cultivation structures, as well as a 50% subsidy for the purchase of planting materials and the cultivation of vegetables and flowers in a polyhouse/shade net house. The support of funds for protected cultivation of vegetables and flowers in greenhouses (fan and pad polyhouses) was up to Rs. 2.8 millions, and it was up to 16.88 lakhs in naturally ventilated polyhouses (tubular structures). The fund support for shade net house (tubular structure) and wooden structure protected cultivation was 1.42 and 0.9894 million, respectively. According to Prakash et al. (2019), total area under protected cultivation in India was estimated around 215809.97 ha by NHM during 2005–2006 to 2017–2018. Plastic mulching took the lead with 88.72% of the total area, followed by 3.24% under anti-bird/anti-hail nets, 2.26% under plastic tunnels, 2.24% under shade nets, and 1.92% under naturally ventilated polyhouses (Table 15.8). The use of protected cultivation for the production of quality planting materials accounts for less than 1% of total area, and policy should prioritize the production of quality planting materials using protected cultivation technologies. The NHM funds grew at a compound annual growth rate of 1.14% (Table 15.9). According to NHM, the cost norms for a polyhouse construction with 50% subsidy are presented in Table 15.10.

TABLE 15.8 Area Under Protected Cultivation under NHM Scheme.

Protected structure (s)	Area in hectare (% share)
Plastic mulching	1,91,472.4 (88.72)
Anti-hail/anti-bird nets	6999.99 (3.24)
Plastic tunnels	4868.35 (2.26)
Shade net houses	4827.5 (2.24)
Naturally ventilated polyhouses (NVPH)	4136.81 (1.92)
Shade net/planting material for polyhouse flowers	1674.55 (0.78)
High-value vegetable planting material grown in a polyhouses	1527.96 (0.71)
Greenhouse with fan and pad system	298.62 (0.14)
Walk-in tunnels	3.84 (≈0.00)
Total area in hectare (% share)	2,15,809.97 (100%)

Source: Adapted from Prakash et al., 2019.

TABLE 15.9 Allocation of Funds for Polyhouse Construction Under NHM Program in India.

Year	Funds available (Rs. Lakh)	Amount (Rs. Lakh)	Percent of funds utilized
2008–2009	2338.81	2332	99.71
2012–2013	141.5	134.08	94.76
2016–2017	349.9	297.59	85.05
CAGR (%)	**−12.78**	**−10.2**	2.95

Source: Adapted from Prakash et al., 2019.

TABLE 15.10 Cost Norms by NHM for Polyhouse Construction with 50% Subsidy.

Sr. No.	Item	Polyhouse size (000, m^2)	Investment (000, Rs./m^2)	Investment (Rs. in lakh)	Subsidy (Rs. in lakh)
1	Greenhouse structure				
	a) Greenhouse with natural ventilation (naturally ventilated system)				
	i. Tubular structure	0.500	1.060	5.30	2.65
		0.500–1.008	0.935	4.67–9.42	2.33–4.71
		1.008–2.080	0.890	8.97–18.51	4.48–9.26
		2.080–4.000	0.844	17.55–33.76	8.77–16.88
	ii. Wooden structure (except hilly areas)	4.000	0.540	21.60	10.80
	iii. Bamboo structure (except hilly areas)	4.000	0.450	18.00	9.00
	b) Greenhouse with fan and pad system	0.500	1.650	8.25	4.13
		0.500–1.008	1.465	7.32–14.77	3.66–7.38
		1.008–2.080	1.420	14.31–29.54	7.15–14.77
		2.080–4.000	1.400	29.12–56.00	14.56–28.00
2	Shade net house				
	a) Tube-shaped structures (except hilly areas)	4.000	0.710	28.40	14.20
	b) Wooden structures (except hilly areas)	4.000	0.492	19.68	9.84
	c) Bamboo structure (except hilly areas)	4.000	0.360	14.40	7.20
3	Poly tunnel (excluding hilly areas)	1.000	0.060	0.60	0.30
4	Walk-in tunnels	4.000	0.600	24.00	12.00

TABLE 15.10 *(Continued)*

Sr. No.	Item	Polyhouse size (000, m^2)	Investment (000, Rs./m^2)	Investment (Rs. in lakh)	Subsidy (Rs. in lakh)
5	Anti-hail/anti-bird nets	5.000	0.035	1.75	0.88
6	Investment on planting materials and cultivating high-value vegetable crops in polyhouses	4.000	0.140	5.60	2.80
7	Investment on planting materials and cultivating anthurium and orchid in a polyhouse or shade net house	4.000	0.700	28.00	14.00
8	Investment on planting materials and growing gerbera and carnation in a polyhouse or shade net house	4.000	0.610	24.40	12.20
9	Investment on planting materials and cultivating lilium and rose in a polyhouse or shade net house	4.000	0.426	17.04	8.52
10	Polythene mulching (excluding hilly areas) @ 32,000/ha	–	–	0.32	0.16

Source: Adapted from open source: https://midh.gov.in/PDF/Annexure-V.pdf.

15.6 ASSISTANCE PATTERN AND COST NORMS UNDER THE NATIONAL HORTICULTURE BOARD

The MIDH has proposed assistance pattern and cost norms under NHB (Table 15.11). Credit linked back-ended subsidy at 50% of cost limited to 56 lakhs per project is being assisted for commercial horticulture development in a protected cover. Subsidy of Rs.17.5 lakh to Rs. 20.13 lakh is being provided for fan and pad system of greenhouse structure in hilly areas. The amount of subsidy varies between Rs. 5.63 lakh and Rs. 12.13 lakh for different types for naturally ventilated greenhouse structures. Similarly, the pattern of assistance for various types of shade net house is also presented in Table 15.11.

TABLE 15.11 Cost Norms and Patterns of MIDH Assistance for NHB Activities during the XII Plan

Item	Cost norms/m2	Assistance/subsidy on construction of polyhouse of size >2500 m2	Loan amount per project and subsidy amount (rupees in lakh)
1. Protected cultivation structure			
A. Naturally ventilated system			
a. Tubular structure	Rs. 844 (Rs. 970 for hilly area)		21.01–24.25 and 10.51–12.13
b. Wooden structure	Rs.540 (Rs. 621 for hilly area)	50% (in each case)	30.5–15.68 and 15.25–7.84
c. Bamboo structure	Rs.450 (Rs.518 for hilly area)		11.25–12.95 and 5.63–6.48
B. Fan and pad system	Rs.1400/(Rs.1610/m² for hilly area)	50%	35.00–40.25 and 17.5–20.13
2. Shade net house			
a. Tubular structure	Rs.710 (Rs.816 for hilly area)		17.75–20.40 and 8.88–10.20
b. Wooden structure	Rs.492 (Rs.566 for hilly area)	50% (in each case)	12.30–14.15 and 6.15–7.07
c. Bamboo structure	Rs.360 (Rs.414 for hilly area)		9.00–10.35 and 4.50–5.17
3. Plastic tunnel	Rs.60 (Rs.75 for hilly area)	50%	1.50–1.88 and 0.75–0.94
4. Walk-in tunnel	Rs.600	50%	1.50 (7.50)
5. Anti-bird/anti-hail net	Rs.35	50%	0.88(0.44)
6. Cost of planting materials and cultivating high-value vegetable crops in polyhouses	Rs.140	50%	3.50 (1.75)
7. Cost of planting materials and cultivating anthurium and orchid in a polyhouse or shade net house	Rs.700	50%	17.50 (8.75)
8. Cost of planting materials and growing gerbera and carnation in a polyhouse or shade net house	Rs.610	50%	15.25 (7.63)

TABLE 15.11 *(Continued)*

Item	Cost norms/m2	Assistance/subsidy on construction of polyhouse of size >2500 m2	Loan amount per project and subsidy amount (rupees in lakh)
9. Cost of planting materials and cultivating lilium and rose in a polyhouse or shade net house	Rs.426	50%	10.65 (5.33)
10. Plastic mulching	Rs.3.200 (Rs.3.680 for hilly areas)	50%	
11. Development of horticulture at commercial level under protected covering	Rs. 112 lakh/project (area >2500 m^2)	50% of total budget per project and restricted to 56 lakh	

Source: Adapted from open source: http://nhb.gov.in/schemes/NewGuidelines.pdf

The guidelines for cost of protected cultivation under the NHB Scheme (Rs. lakh per acre) are presented in Table 15.12.

TABLE 15.12 Cost Guidelines of Protected Cultivation Under the NHB Scheme (Rupees in Lakh/Acre).

Crop type	Cost of structure (tabular structure) with a drip and fogging system*	Cost of cultivation	Cost ceiling per acre (plus add on components under project mode)
High-value vegetable crops (in polyhouse)	33.76	5.60	47
High-value vegetable crops (under shade net)	28.40	5.60	40
Orchid and anthurium	33.76	28.00	70
Chrysanthemum, lilium, and rose	33.76	17.04	60
Carnation and gerbera	33.76	24.40	66

**Cost in flat/plain areas; however, it will be 15% higher in hilly areas.*

Source: Adapted from http://nhb.gov.in/schemes/NewGuidelines.pdf.

15.7 LIMITATIONS OF PCT

In spite of assured crop return and enhanced productivity, the high initial investment for constructing polyhouse is one of the utmost concerns to the farmers for adopting the technology (Jadhav and Rosentrater, 2017). The PCT is a developing technology to grow high-value cash crops, vegetables, and cut flowers; however, some constraints (Table 15.13) are also faced for adopting this technology by the farmers.

High initial investment for infrastructure (capital cost) is more due to the establishment cost incurred on fixing the protective structures like shade net/polyhouse/greenhouse. Nonavailability of skilled labor for working in protected structures, lack of scientific and technical knowledge of growing crops under protected structures, requires close supervision and monitoring, infestation by pests and soil-borne pathogens, nonavailability of minimum support price (MSP) for fruits and vegetables, nonavailability of market information, low output price, higher perishability, poor access to subsidy in spite of huge support from the government and lack of insurance facility were the main constraints reported in various studies.

TABLE 15.13 Constraints in Protected Cultivation Technology.

Sl. No.	Constraints	Authors
1	High initial investment for infrastructure	Sanwal et al. (2004); Ghanghas and Rati (2015); Kaur et al. (2017); Prabhakar et al. (2017); Prakash et al. (2019)
2	Nonavailability of skilled labor for working in protected structures	Kaur et al. (2017); Prabhakar et al. (2017); Prakash et al. (2019)
3	Lack of technical know-how of protected cultivation and lack of literature in local language	Ghanghas and Rati (2015); Kaur et al. (2017); Prabhakar et al. (2017); Mehta et al. (2020)
4	Infestation by pests and soil-borne pathogens	Ghanghas and Rati (2015); Prabhakar et al. (2017); Kumar et al. (2018); Prakash et al. (2019)
5	Non-assured marketing/Nonavailability of minimum support price (MSP)/low price	Kumar et al. (2016); Kaur et al. (2017); Kumar et al. (2018); Prakash et al. (2019)
6	Relatively higher perishability of flowers/vegetables	Prabhakar et al. (2017)
7	Poor accessibility to subsidy	Prabhakar et al. (2017)
8	High cost of planting materials/seeds	Prabhakar et al. (2017); Prakash et al. (2019)
9	Nonavailability of insurance	Prakash et al. (2019); Mehta et al. (2020)

15.8 DETERMINANTS FOR THE ADOPTION OF PCT BY THE FARMERS

The selection of a site is critical for profitable and long-term greenhouse production. The type and level of greenhouse technology (structure and internal climate control equipment) and subsequent crop production conditions are influenced by the climate, which in turn influences product cost and quality. Distance to markets, particularly in export-focused production, can be a barrier to profitable greenhouse cultivation. To produce proper quality commodities at a competitive level, an economic compromise between the investment costs of greenhouses and equipment and their agricultural performance is required. Farmers are motivated to adopt protected cultivation due to a lack of land, the possibility of a high income, the availability of subsidies, and the availability of technology (AERC, 2017).

15.9 CONCLUSIONS

Protected cultivation is the hope for meeting the food demands of the increasing population. Increased productivity leads to achieve doubling farmers' income besides achieving the food security. Despite of support from the government, area under protected cultivation is still meager. Hence, large-scale extension activities need to be taken to create awareness and to provide technical guidance to the farmers. Skill development of the labors and use of MGNREGA labors for protected cultivation should be encouraged so that problems related to nonavailability and lack of skilled labors can be curbed. Provision of minimum support price should be extended even to fruits and vegetables. Crops cultivated under protected structures needs to be included under the government crop insurance scheme known as "Pradhan Mantri Fasal Bhima Yojana" (PMFBY).

KEYWORDS

- **protected cultivation**
- **economic viability**
- **benefit cost ratio**
- **internal rate of return**
- **net present worth**

REFERENCES

AERC (Agro-Economic Research Centre). An Economic Analysis of Protected Cultivation Under MIDH in Sikkim; Study No. 186, Agro-Economic Research Centre (For the States of West Bengal, Sikkim and Andaman & Nicobar Islands), Visva-Bharati, Santiniketan, West Bengal, 2017; pp -x+66.

Bhoopesh, P.; Suresh, P.; Girish, K. J.; Pramod, K. Economics and Institutional Aspects of Protected Cultivation of Carnation in Himachal Pradesh. *Agric. Econ. Res. Rev.* **2017,** *30*(1), 73–80.

Chakraborty, H.; Sethi, L. N. Prospects of Protected Cultivation of Vegetable Crops in North Eastern Hilly Region. *Int. J. Basic Appl. Biol.* **2015,** *2*(5), 284–289.

FAO. FAO: Unlocks the Potential of Protected Agriculture in GCC Countries. A Stakeholder Workshop to Discuss Cutting Water Consumption and Improving Nutrition and Food

Security, 2018. http://www.fao.org/neareast/news/view/en/c/1158858/#:~:text=Protected%20 Agriculture%3A%20Greenhouses,diseases%20for%20high%2Dvalue%20crops

Ghanghas, B. S.; Rati, M. Protected Cultivation (Polyhouse) in Haryana: Problems and Prospects. *Indian J. Appl. Res.* **2015,** *5*(8), 684–685.

Jadhav, H. T.; Rosentrater, K. A. Economic and Environmental Analysis of Greenhouse Crop Production with Special Reference to Low Cost Greenhouses: A Review. In *Conference Proceeding, ASABE Annual International Meeting,* 2017; pp 1–6.

Kaur, K.; Kaur, P.; Singh, K. Problems Faced by the Farmers in Adoption of Protected Vegetable Cultivation in Punjab. *Int. J. Farm Sci.* **2017,** *7*(3), 106–109.

Kumar P.; Chauhan, R. S.; Grover, R. K. Economic Analysis of Capsicum Cultivation Under Poly House and Open Field Conditions in Haryana. *Int. J. Farm Sci.* **2016a,** *6*(1), 96–100.

Kumar P.; Chauhan, R. S.; Grover, R. K. Economics Analysis of Tomato Cultivation Under Poly House and Open Field Conditions in Haryana, India. *J. Nat. Appl. Sci.* **2016b,** *8*(2), 846–848.

Kumar, P.; Chauhan, R. S.; Tanwar, N.; Grover, R. K. Status and Constraints in Vegetable Cultivation Under Polyhouse in Haryana. *Adv. Biores.* **2018,** *9*(2), 61–66.

Laxmi, N. S.; Chakraborty, H. Agricultural Economics of Protected Cultivation Structures with Drip Irrigation for Tomato Production in a Hilly Terrain of Assam. Paper Presented at the *International Conference on Integrating Climate, Crop, Ecology–The Emerging Areas of Agriculture, Horticulture, Livestock, Fishery, Forestry, Biodiversity and Policy Issues,* 2017; pp 63–67. ISBN: 978-81-930585-9-6.

Mehta, K.; Thakur, R. K.; Guleria, J. S. Socio-Economic Impact of Protected Cultivation on Tomato Growers of Himachal Pradesh. *Econ. Aff.* **2020**, *65*(1), 01–07.

Mehta, K.; Thakur, R. K.; Guleria, J. S. Constraints in Adoption of Protected Vegetable Cultivation in Mid-Hill Areas of Himachal Pradesh. *Indian J. Econ. Dev.* **2020,** *16*(2s), 372–377.

Nair, R.; Barche, S. Protected Cultivation of Vegetables-Present Status and Future Prospects in India. *Indian J. Appl. Res.* **2014,** *4*(6), 245–247.

Nimbrayan, K. P.; Chauhan, R. S.; Mehta, V. P.; Bhatia, J. A Review on Economic Aspect of Protected Cultivation in India, 7th ed. In *Research Trends in Horticulture Sciences*; Meena, M. L., Ed.; Akinik Publications: New Delhi, 2018; Chapter: 3, Vol. 7, pp 43–59.

Prabhakar, I.; Vijayaragavan, K.; Singh, P.; Singh, B.; Janakiram; Manjunatha, B. L.; Jaggi, S.; Sekar, I. Constraints in Adoption and Strategies to Promote Polyhouse Technology Among Farmers: A Multi-Stakeholder and Multi-Dimensional Study. *Indian J. Agric. Sci.* **2017,** *87*(4), 485–90.

Prakash, P.; Pramod, K.; Amit, K.; Awani, K. S.; Anbukkani P. Progress and performance of protected cultivation in Maharashtra. *Indian. J. Econ. Dev.* **2019,** *15*(4), 555–563.

Sanwal, S. K.; Patel, K. K.; Yadav, D. S. Vegetable Production Under Protected Conditions in NEH Region. *Envis Bull. Himal. Ecol.* **2004,** *12*(2), 1–7.

Sharma, M.; Thakur, R; Mehta, P. Economic Feasibility Analysis of Major Flower Crops in Himachal Pradesh State of India. *Int. J. Adv. Res. Manage. Soc. Sci.* **2014,** *3*(9), 24–40.

Singh, B. Protected Cultivation of Horticultural Crops in India: Challenges and Opportunities. *Agrotechnology* **2014,** *2*(4), 51.

Singh, B.; Tomar, B. S.; Hasan, M. Plug-Tray Nursery Raising Technology for Off-Season Cucurbits Cultivation. *Acta Hort.* **2010,** *871*, 279–282.

Singh, M. C.; Kachwaya, D. C., Kalsi, K. Soilless Cucumber Cultivation Under Protective Structures in Relation to Irrigation Coupled Fertigation Management, Economic Viability and Potential Benefits - a Review. *Int. J. Curr. Microbiol. App. Sci.* **2018b,** *7*(03), 2451–2468.

Singh, M. C.; Singh, A.; Singh, J. P.; Singh, K. G. Economic Viability of Soilless Cucumber Cultivation Under Naturally Ventilated Greenhouse Conditions. *Ind. J. Hort.* **2020b,** *77*(2), 315–321.

Singh, M. C.; Singh, J. P.; Pandey, S. K.; Mahay, D.; Srivastava, V. Factors Affecting the Performance of Greenhouse Cucumber Cultivation-a Review. *Int. J. Curr. Microbiol. App. Sci.* **2017a,** *6*(10), 2304–2323.

Singh, M. C.; Singh, J. P.; Singh, K. G. Optimal Operating Microclimatic Conditions for Drip Fertigated Cucumbers in Soilless Media Under a Naturally Ventilated Greenhouse. *Ind. J. Ecol.* **2017b,** *44*, 821–826.

Singh, M. C.; Singh, K. G.; Singh, J. P. Interactive Effects of Fertigation and Varieties on Plant Growth Attributes and Yield of Soilless Cucumbers. *Int. J. Sci. Technol. Soc.* **2020a,** *8*(3), 60–65.

Singh, M. C.; Singh, K. G.; Singh, J. P. Nutrient and Water Use Efficiency of Cucumbers Grown in Soilless Media Under a Naturally Ventilated Greenhouse. *J. Agric. Sci. Technol.* **2019b,** *21*(3), 193–207.

Singh, M. C.; Singh, K. G.; Singh, J. P. Yield of Soilless Cucumbers Planted Under Partially Controlled Greenhouse Environment in Relation to Deficit Fertigation. *Ind. J. Hort.* **2018b,** *75*(2), 259–264.

Singh, M. C.; Singh, K. G.; Singh, J. P.; Mahal, A. K. Performance of Soilless Cucumbers in Relation to Differential Fertigation Under Naturally Ventilated Greenhouse Conditions. *J. Plant. Nutr.* **2019a,** *42*(11–12), 1316–1332.

Index

L

M

R

S